ENVIRONMENTAL POLICY IN MINING

Corporate Strategy and Planning for Closure

ENVIRONMENTAL POLICY IN MINING

Corporate Strategy and Planning for Closure

Alyson Warhurst

Director
Mining and Environment Research Network
University of Bath
United Kingdom

Ligia Noronha

Research Fellow
Tata Energy Research Institute (TERI)
Panaji, Goa
India

LEWIS PUBLISHERS

Boca Raton   London   New York   Washington, D.C.

Library of Congress Cataloging-in-Publication Data

Environmental Policy in Mining: Corporate Strategy and Planning for Closure/ Alyson Warhurst and Ligia Noronha.
     p. cm.
     Includes bibliographical references and index.
     ISBN 1-56670-365-4 (alk. paper)
     1. Mineral industries - - Environmental aspects Case studies.   2. Environmental management Case studies. 3. Mine
closures. 4. Best management practices (Pollution prevention).
     I. Noronha, Ligia. II. Title.
     TD195.M5W3597 1999
     363.738--dc21   99-29715
                                                             CIP

© 2000 by CRC Press LLC
Lewis Publishers is an imprint of CRC Press LLC

No claim to original U.S. Government works
International Standard Book Number 1-56670-365-4
Library of Congress Card Number 99-29715
Printed in the United States of America 1 2 3 4 5 6 7 8 9 0
Printed on acid-free paper

Preface

This book began life as a research project that set out to describe and compare the different practices of ecological management by mining firms worldwide. As the study progressed, and particularly on account of the interdisciplinary network approach to undertaking this research, a pattern emerged. Those firms that demonstrated improved efforts towards environmental management tended to be newer operations where reclamation and pollution management had been built into the project from the outset. Those firms demonstrating a lack of effective environmental management and poor community relations were more closely associated with older operations; located in regulatory regimes where environmental laws were poorly enforced or absent; and where, on account of a lack of new investment, there were few environmental or social conditions attached to the provision of finance.

The fact that our network comprised researchers with varied disciplinary training meant that not only were we able to adopt an interdisciplinary approach to the complex challenges posed by managing the environmental impacts of mining, but also that we were able to analyse in depth some of the issues that need to be addressed. These ranged from the ecotoxicological effects of metal residues to the land use effects of mining, and from socioeconomic impacts to environmental regulation.

As a result, this book takes neither an exclusively sociological or technical perspective. Rather, it brings together a collection of research studies by research specialists in mineral-producing countries worldwide that, in combination, highlight the wide range of issues which need to be addressed in managing the environmental impacts of mineral projects, and describe how different firms in different countries have approached this challenge.

Contributors

Dr. Peter C. Acquah
Environmental Protection Agency
Accra
Ghana

Kathleen Anderson
Mining and Environment Research Institute
Queens University
Kingston, Ontario
Canada

Prof. A.K. Barbour
Bristol
United Kingdom

A. Boateng
Environmental Protection Agency
Accra
Ghana

J. Box
Wardell Armstrong
Newcastle under Lyme
United Kingdom

Dr. Gavin Bridge
Department of Geography
University of Oklahoma
Norman, Oklahoma

Dr. C.P. Broadbent
Wardell Armstrong
Newcastle under Lyme
United Kingdom

Dr. Ian Clark
Research Centre for Environmental and
  Recreation Management
Faculty of Engineering and the Environment
University of South Australia
Adelaide
Australia

N.J. Coppin
Wardell Armstrong
Newcastle under Lyme
United Kingdom

Luke Danielson
Mining Policy Research Initiative
International Development Research Centre
Montevideo
Uruguay

Prof. Bharat B. Dhar
Department of Mining Engineering
Banaras Hindu University
Varanasi
India

Prof. Ge Feng
Institute of Zoology
Chinese Academy of Sciences
Beijing
People's Republic of China

John Hollaway
John Hollaway & Associates
Harare
Zimbabwe

Dr. Richard Isnor
Science Policy Division
Ecosystem Science Directorate
Environmental Conservation Service
Environment Canada
Hull, Quebec
Canada

Prof. Gao Lin
Department of Systems Ecology
Research Center for Eco-Environmental
  Sciences
Chinese Academy of Sciences
Beijing
People's Republic of China

Dr. Fernando Loayza
Fondo Nacional para el Medio Ambiente
  (FONAMA)
La Paz
Bolivia

Magnus Macfarlane
School of Management
University of Bath
United Kingdom

Dr. Isidro R.V. Manuel
Faculty of Science
Eduardo Mondlane University
Maputo
Mozambique

Dr. Paul Mitchell
Technical Director
KEECO (U.K.) Limited
Penryn, Cornwall
United Kingdom

Dr. David R. Morrey
DMA LLC
Consultants in Natural Resource Development
Littleton, Colorado

Dr. Tomás Muacanhia
Faculty of Science
Eduardo Mondlane University
Maputo
Mozambique

Marily Nixon
Dorsey & Whitney LLP
Denver, Colorado

Dr. Ligia Noronha
Tata Energy Research Institute
St. Inel, Panaj, Goa
India

Edward Nyamekye
Minerals Commission
Accra
Ghana

Jochen Petersen
Department of Chemical Engineering
University of Cape Town
Cape Town, South Africa

Prof. Jim Petrie
Department of Chemical Engineering
University of Sydney
New South Wales
Australia

Meredith Sassoon
Editor – Mining Environmental Management
60 Warship Street
London
EC2A 2HD
United Kingdom

Prof. I.C. Shaw
Centre for Toxicology
University of Central Lancashire
Preston
United Kingdom

Mary Stewart
Department of Chemical Engineering
University of Sydney
New South Wales
Australia

Dr. Solomon Tadesse
Department of Geology and Geophysics
Faculty of Science
Addis Ababa University
Ethiopia

Prof. Alyson Warhurst
Mining and Environment Research Network
School of Management
University of Bath
United Kingdom

Senior Engineer Zhang Wenmin
Beijing General Research Institute of Mining
  and Metallurgy
Beijing
People's Republic of China

Prof. Geof Wood
Centre for Development Studies
Department of Economics and International
  Development
University of Bath
United Kingdom

Senior Engineer Wang Yishui
Department of Science & Technology
  Development
China National Non-Ferrous Metals Industry
  Corp.
Beijing
People's Republic of China

Acknowledgement

The research reported in this book was undertaken by international members of the Mining and Environment Research Network (MERN). Much of the research and the writing of this book was supported by the UK's Department for International Development (DFID), and specifically the Environment Research Programme of DFID.

In addition, grants from DFID, SIDA, IDRI and the MERN Industry Club of Sponsors, specifically Rio Tinto, Anglo American plc, and BHP, enabled many of the authors to meet in Zimbabwe in August 1996 for a research workshop to discuss their findings and to undertake collaborative fieldwork.

The authors acknowledge that kind support. In addition, the authors would like to thank Dr. Paul Mitchell for his painstaking work during the final editing stages of this book.

Alyson Warhurst
Director, Mining and Environment Research Network
University of Bath
United Kingdom
1999

Dedication

This book is dedicated to our families:
from Ligia Noronha with love to Peter, Siddharth, Gayatri, and Roshni;
from Alyson Warhurst with love to William and Edward.

Contents

1 Planning for Closure: Towards Best Practice in Public Policy and Corporate Strategy in Managing the Environmental and Social Effects of Mining

Alyson Warhurst

CONTENTS

1.1 INTRODUCTION

Before overviewing the chapters of the book, it is important to explain in more detail the research process which underlies it. These chapters also represent a process of research capacity building in the area of mining and environment, undertaken within the framework of the Mining and Environment Research Network (MERN).

Seed funding from DFID and the OECD Development Centre enabled the establishment of MERN, which began in 1991 as a collaborative project involving researchers from the following institutions: the University of Sao Paulo, and the Centro de Tecnologia Mineral (CETEM) in Brazil; the Institute for Research on Public Health (INSAP) and the Universidad de la Catolica in Lima, Peru; the Centro de Estudios Mineria y Desarrollo (CEMYD) in La Paz, Bolivia; and the Centro de Estudios del Cobre y de la Mineria (CESCO) in Santiago, Chile. The collaborative research project which was subsequently developed won the prestigious John D. and Catherine T. MacArthur Foundation Collaborative Studies competition in 1991 and, together with complementary funding for the Bolivia project from the International Development Research Centre (IDRC), this launched the first phase of MERN. This is the subject of our first network book, "Environmental Management and Sustainable Development: Case Studies from the Mining Industry," to be published in 1999 by IDRC Books, Canada. MERN rapidly expanded to include a range of different types of interdisciplinary centres of excellence from both mineral-producing developing and industrialised countries. The current list of members is summarised by institution and country in Table 1.1. The Network now involves centres of excellence throughout Argentina, Australia, Bolivia, Brazil, Bulgaria, Canada, Chile, China, Colombia, Ethiopia, France, Germany, Ghana, India, Italy, Malaysia, Mozambique, Netherlands, Norway, Pakistan, Papua New Guinea, Peru, Poland, Republic of Guinea, Russia, South Africa, Sweden, Tanzania, Thailand, Ukraine, U.K., U.S., Zambia, and Zimbabwe.

1-56670-365-4/00/$0.00+$.50
© 2000 by CRC Press LLC

TABLE 1.1
Mining and Environment Research Network: Member Institutions

Institution	Country
International Centre for the Environment, University of Bath; Universities of Sussex, Surrey and Dundee; Camborne School of Mines; and Royal School of Mines	U.K.
Gerencia Ambiental; CIS; University of Buenos Aires	Argentina
Universities of South Australia and Murdoch	Australia
CEMYD	Bolivia
University of Sao Paulo; CETEM	Brazil
Geological Institute of Bulgarian Academy of Sciences	Bulgaria
Centre for Resource Studies	Canada
CESCO; Catholic University of Chile	Chile
Eco-Environmental Research Centre Academia Sinica	China
Universidad Pontificia Bolivariana; Instituto de Estudios Regionales	Colombia
Mineral Resources for Exploration and Development; Addis Ababa University	Ethiopia
Ecole des Mines	France
Projekt Consult; Oeko Institute for Applied Ecology; Universitat Gesamthochscule	Germany
Minerals Commission; Institute of Management & Public Administration; Environmental Protection Agency	Ghana
Tata Energy Research Institute; National Institute of Small Mines; Central Mining Research Institute	India
Trinity College, Dublin	Ireland
Universita degli Studi di Cagliari	Italy
National University of Malaysia	Malaysia
Eduardo Mondlane University	Mozambique
University of Amsterdam	Netherlands
University of Oslo	Norway
University of the Punjab	Pakistan
Department of Environment & Conservation	Papua New Guinea
INSAP/Pontificia Universidad Católica de Peru	Peru
University of Mining and Metallurgy	Poland
Ministry of Mines & Geology	Republic of Guinea
Institute of Economic Forecasting, Russian Academy of Sciences	Russia
MEPC; Mintek; University of Cape Town	South Africa
Raw Materials Group; University of Lund	Sweden
Ministry of Energy and Minerals	Tanzania
Thailand Development Research Institute Foundation; Prince of Songkla University	Thailand
Centre for Scientific and Technological Potential and Science History Studies (STEPS)	Ukraine
Colorado School of Mines; Massachusetts Institute of Technology (MIT); East-West Centre	U.S.
Mining Sector Coordinating Unit; South Africa Development Community	Zambia
University of Zimbabwe; Institute of Mining Research; John Hollaway Associates	Zimbabwe

From the outset, the aim of MERN is to provide research analysis to assist mining companies in achieving environmental compliance and improved competitiveness in the context of growing environmental regulation and technological innovation, and to inform environmental public policy. The international collaborative research programme first set out to examine the relationship between environmental regulation, technical change, and competitiveness in the non-ferrous minerals industry. In particular, it investigates the ways in which the process of technological innovation and organisational change can be harnessed to prevent environmental degradation while enhancing productivity and sustainability. Appendix 1.1 provides summary information about the goals, scope, and research programmes.

The underlying rationale for the international focus of the research effort in the developing and industrialised countries was the perceived need to extract lessons from the experience of more competitive and environmentally proficient firms, while focusing on the challenges for achieving environmental best practice in each participating country. The overall goal of adopting a network approach, as opposed to following independent research objectives, was the building up of an international pool of interdisciplinary research competence. In other words, research capacity building was fundamental to the nature and scope of MERN from the outset.

There were a number of environmental concerns underlying our decision to work collaboratively towards these research objectives. First, we had developed our early research profiles in the area of mineral policy and technological change in the mining sector, and in each of the countries where we worked we increasingly observed environmental damage that was associated with minerals extraction or processing activities. Although in each country and at each mine site environmental damage effects differed — and that in itself raised the question of *why* it differed — some common explanatory factors seemed to be emerging. Some of these, we immediately recognised, "flew in the face" of conventional wisdom.

For example, conventional wisdom is epitomised by the "Pollution Haven" thesis, which suggests that international firms will locate their production activities where they can most easily externalise the environmental damage costs associated with their production. Geographically, these havens are thought to occur in developing countries, where environmental regulation is either limited or poorly enforced. However, an early common observation, which was explored in these studies, was that environmental damage was not evenly distributed within the minerals sector of each developing country studied, but that it seemed to vary according to a number of factors. These included: type of mineral; vintage of technology; stage of investment; stage of operation; level of integration; effectiveness of environmental regulation and its enforcement; socioeconomic context, including poverty within local communities and the educational and training status of the workforce; and most of all, environmental performance varied according to the firm's inherent technological dynamism.

These studies address real environmental problems and solutions, and are therefore unique in that they do not *a priori* accept conventional wisdom, but rather seek to investigate actual environmental performance and its determinants at the mine site, and across a range of very site-specific circumstances.

The advantage of a network approach to this research exercise was that the intellectual benefits of working together and collaborating across a range of research and dissemination activities achieved greater impact than the combined sum of each institution working alone. The type of networking activities which underlie the research process reported here include coordination through an increasingly electronic database and information system. As we grew from an initial six institutions to fifty-six, so did our use of information technology grow in sophistication. The majority of our network activity is now undertaken using electronic communications. The network produces a Bulletin and Newsletter twice annually, which contains progress reports from each group, policy updates, industry news, professional articles, and a conference calendar. We have recently completed Bulletin Numbers 13/14 and, collectively, the Bulletins form a record of the Network's growth over the years.

Once a year, the Network meets for a research workshop on a theme-by-theme basis. Workshop themes have included: mining and sustainable development; pollution prevention; risk and responsibility; planning for closure and best practice in the management of the ecological impacts of mining; and the development of environmental and social performance indicators and sustainability markers. The recommendations that emerged from the final research dissemination workshop for this project, which took place in Harare, Zimbabwe in 1996, are included in the concluding chapter (Chapter 29). Each workshop incorporated specialist sessions for research feedback and dissemination, and each of the chapters in this book has undergone a peer-review process to facilitate quality control.

1.2   ORGANISATION

It remains to briefly review each chapter and describe the perspective each author adopted to investigate the theme of *Environmental Policy in Mining: Corporate Strategy and Planning for Closure.*

Part I lays out the rationale for the book. Chapter 1 provides the introduction to the research process. Chapter 2 (by Prof. Alyson Warhurst, Professor of Environmental Strategy at the University of Bath, and Dr. Ligia Noronha, an environmental economist at the Tata Energy Research Institute in India) explains the planning for closure perspective of the book and the focus on integrated environmental management from the outset. The chapter discusses the benefits of closure plans, highlighting the time factor in that the greater the time lapse between the occurrence of damage and its remediation, the greater will be the resources necessary to address the problem. The chapter also discusses the objectives of a plan, the components of planning for closure, and the constraints and challenges faced by regulators and the industry in operationalising the plan. Chapter 2 draws on the analysis presented in subsequent chapters to put forward a case for planning for closure from the outset based on an Environmental Management System (EMS) approach.

Part II provides a focus on the issues that must be addressed in closure planning. The chapters, written by specialists from different fields, deal with the impacts associated with mining activities, the prediction and prevention of impacts and the management, and the treatment and rehabilitation of polluted waste materials. The issues are highlighted through references to both the technologies being developed to mitigate those impacts and to operations worldwide. In most cases, these issues are analysed through reference to primary data generated during empirical case-study research involving structured interviews and field visits.

Part II begins with Chapter 3, by Prof. Alyson Warhurst, which reviews the environmental impacts of mining and minerals processing. This chapter overviews the mining process, mineral processing, and hydrometallurgical and pyrometallurgical processing, and their impacts on land, water, and air resources. It draws a distinction between different types of mines and minerals in order to introduce the diversity of the mineral production process and the environmental complexity of its impacts, both of which are explored in detail in later chapters.

Prof. Joe Barbour (Honorary Professor at the University of Bath) and Professor Ian Shaw (Centre for Toxicology, University of Central Lancashire) in Chapter 4 address the ecotoxicological impacts of mining. The authors, both eminent practitioners and researchers in the field of environmental management and ecotoxicology, examine the routes of metals contamination in ecosystems and indicate the methods that might be used to monitor such contamination. They assess the environmental management challenges of mining and provide a detailed analysis of the specific physiological and ecological problems and thresholds associated with copper, zinc, lead, arsenic, cadmium, mercury, aluminium, chromium, and other metals and minerals.

Chapter 5 is written by a team of researchers from the International Centre for the Environment at the University of Bath who, collectively and individually, are engaged in research to manage and alleviate the social impacts of mine closure and who draw on the perspectives of sociology, management, and public policy. Prof. Alyson Warhurst, Magnus Macfarlane and Prof. Geof Wood argue that effective planning for closure needs to incorporate corporate strategies and public policies that address the economic and social dislocation which can result from mine closure. They review recent research into the economic, social, and psychological impacts of planned and unplanned closure, and point towards emerging best-practice in incorporating community stability and social welfare objectives into mine closure and regional development plans. The historical roles of government, company, and other stakeholders in approaches such as diversification, selective recruitment, and re-skilling are reviewed. The authors conclude by outlining the possibilities for new partnerships to reduce risk and provide long-term stability in the post-closure environment, based on the concept of a "sustainability contract" agreed upon at the outset between key project partners amongst company, government, and community stakeholders. The criteria for its design, monitoring, and evaluation would be defined by ongoing social impact assessments aimed at predicting and setting mitigation targets for mine developments through closure.

Meredith Sassoon, an environmental consultant with experience in the environmental management of mining projects in many different mineral producing countries, addresses the design of effective Environmental Impact Assessments (EIAs) in Chapter 6. Although EIAs are now required in most countries as a statutory prerequisite for the development of a mining project, the author argues that their role in planning for closure is limited by the fact that in practice they are reactive, static, and short-term in design. Conventional EIAs are therefore unable to take account of the evolutionary nature of mine development and fail to sufficiently integrate environmental considerations into mine planning throughout the project life cycle. She proposes Strategic Environmental Assessment as an alternative and explains how this longer-term, region-wide assessment of environmental implications can provide a more effective means of planning for closure by decreasing companies' compliance costs and improving levels of environmental protection.

Dr. Paul Mitchell reviews the significance of acid rock drainage (ARD) in planning for closure in the metal mining industry in Chapter 7. Dr. Mitchell is Technical Director of KEECO (U.K.) Limited and has undertaken extensive research on ARD and solid waste management over the past decade, including three years as a Clean Technology Fellow at the University of Bath. He reviews research on the prevention, formation, control, migration, and treatment of ARD, paying particular attention to the significance of biologically mediated reactions in the process of ARD generation, and the significance of ore mineralogy and geochemistry in determining potential natural *in situ* neutralisation prior to migration. Prediction of ARD generation and migration by mineralogical and petrographic investigations, static and kinetic tests, and empirical and deterministic modelling is examined with reference to recent advances. The role of prediction in planning for closure and in aiding prevention and treatment regimes is highlighted.

In Chapter 8, Prof. Jim Petrie and Mary Stewart apply the methodological tools of life cycle assessment to the management and disposal of mining wastes to develop an assessment tool with which companies can target the most effective process operations for waste reduction. At the time of writing, the authors were part of the Department of Chemical Engineering at the University of Cape Town, which has established a reputation for innovative applications of Life Cycle Analysis (LCA) to the development of cleaner technologies in the minerals industry. The authors have since relocated to the Department of Chemical Engineering at the University of Sydney, Australia. The authors illustrate how environmental performance in mining and minerals processing operations is dependent on effective waste management and discuss how the generation of waste is a direct consequence of the thermodynamic limitations under which the industry operates and the choice of technology that is implemented. They argue that improvements in environmental performance are realised primarily by changes in technology, but that current environmental management systems offer little in the way of technical guidance for mining and minerals processing companies intent on improving their environmental performance. To resolve this problem, the authors introduce a systemic model of mining and minerals processing in which critical resource consumption and waste flows are linked directly to unit operations. Sectoral case studies of the South African mineral industry are used to examine in detail the relationship between income generation and waste generation potential for a given process requirement. The authors conclude that the model may be able to assist in planning for closure by identifying constraints on the part of operating companies to fund closure management practices.

Dr. Gavin Bridge discusses the challenges associated with the protection of water quality during the operation and closure of leaching facilities from the perspective of regulatory policy in Chapter 9. Dr. Bridge, now an Associate Professor at the University of Oklahoma, completed this work while a Research Officer with the Mining and Environment Research Network in the School of Management at the University of Bath, where he researched corporate strategy and public policy approaches towards the development and diffusion of cleaner technologies and management practices in the minerals industry. Drawing on the experience of copper and gold leaching operations in the United

States, he assesses best-practice approaches to decommissioning and reviews the rapidly changing public policy framework within which closure is taking place. He argues that the high costs of groundwater remediation, long-term monitoring, and the possibility of incurring environmental and health liabilities provide a strong argument for planning closure from the outset through the design and management of leach facilities to prevent contamination. He discusses recent initiatives in the framework of U.S. environmental legislation relating to leaching practices and outlines emerging best-practice management and technological approaches by U.S. mineral producers. His chapter concludes with an assessment of current obstacles to the effective transfer and diffusion of best-practice approaches.

In Chapter 10, Dr. Chris Broadbent and Nick Coppin, environmental consultants with Wardell Armstrong, with extensive research and advisory experience in the area of waste minimisation and pollution prevention, review recent developments in the treatment and utilisation of metallurgical residues. They argue that growing environmental and feed constraints, coupled with changing social perceptions of acceptability, have required waste producers to reconsider the disposal of metallurgical residues as waste products. They point to the need to consider the most appropriate chemical and physical forms for eventual disposal. This leads them to discuss the suitability of silicate slag as a disposal medium for heavy metals and evaluate a number of extractive industry residues with a view to recovering valuable by-products. In particular, they focus on the results of recent research in which mercury is recovered from natural gas production sludges, alumina from colliery spoil, as well as metal and alloy from nickel smelter slags. Based on this evaluation, the authors are optimistic that waste recovery schemes can be realised commercially within a short time frame.

In Chapter 11, Prof. Jim Petrie (Professor, Shell Chair of Environmental Engineering, Department of Chemical Engineering, University of Sydney, Australia) and Jochen Petersen (Department of Chemical Engineering, University of Cape Town, South Africa) introduce their recent research into effective methodologies and modelling tools to guide planning for closure in connection with ferro-alloy waste deposits. Focusing on the question of technology choice, they examine how the rehabilitation of ferro-alloy waste deposits, including slag/dust heaps and tailings impoundments, is motivated by concerns about the long-term leachability of salts and heavy metals (particularly chromium). The authors argue that operator liability with respect to waste management practice can be assessed only with the aid of a comprehensive model of the physical and chemical processes that affect leach resistance within waste deposits. They assess the effectiveness of such a model as a tool for predicting leachate generation and mobility, and present results which suggest that common methods of waste treatment and disposal (such as neutralisation and precipitation of metal hydroxides followed by slimes disposal) are inadequate to ensure long-term waste stability. By characterising the stability of specific waste products, the authors are able to draw attention backwards in the material chain to the waste generation processes themselves, identifying opportunities for process improvements to minimise waste production.

In Chapter 12, Nick Coppin and Dr. John Box, of the consultancy firm Wardell Armstrong, discuss the rehabilitation and revegetation of land utilised for mineral extraction, waste disposal, and processing, with respect to sustainability. They identify how both rehabilitation and revegetation are in fact processes, rather than events, and require careful consideration of the final land use. This, in turn, determines the type and duration of the inputs required for those processes. The factors which determine the ability of a site to achieve a given post-mining land use are identified and linked through a land suitability classification which can be used to suggest final land uses and the amelioration required to achieve them. Ecological principles and their application to the revegetation of abandoned mineral workings are also explored.

Dr. David Morrey, an environmental consultant with extensive experience in mineral projects, assesses the costs and benefits of environmental management and planning for closure in Chapter 13. He describes how environmental and cost management can be integrated as early as the pre-feasibility assessment stage, and advocates a process of reducing costs by incorporating sequential reclamation and partial closure concurrently with mining operations. He discusses the selection of reclamation and closure technologies, closure design and the process of cost-benefit analysis, and addresses uncertainty and risk as factors that must be included in closure design and cost management.

Dr. Paul Mitchell examines the potential role of cleaner mining, milling, processing, and refining technologies in reducing long-term liabilities in Chapter 14. Current waste management practices are summarised and contrasted with possible technology-based preventative measures. Although recent advances in cleaner mineral processing offer potential benefits for both operational performance and the closure process, case studies also support the view that waste management has an equal role to play in specific circumstances and that this approach should not be discounted purely on the basis that a clean technology approach has become increasingly dominant in many other industrial sectors.

Dr. Kathleen Anderson, Director of the Mining and Environment Research Institute at Queens University, Ontario, examines the role of financial mechanisms in managing the environmental risks of mining projects in Chapter 15. She reviews the range of public policies which have historically been adopted to manage the environmental impacts of mining and illustrates how many governments are now considering an alternative approach, that of financial assurances. Financial assurances typically require mining companies to guarantee financial responsibility for reclamation of mine sites and therefore serve to redistribute the risks of environmental damage from the public sector to the private sector. Different types of financial assurance are reviewed and their effectiveness assessed. The author concludes with a series of questions to assist in the design of appropriate public policy mechanisms that will provide for both a high level of environmental protection and a dynamic mining sector.

Part III comprises a collection of case studies that MERN researchers have undertaken in their own countries to examine the factors that explain different levels of environmental degradation associated with minerals production, and the environmental management approaches promoted by companies, the state, or both, depending on the researcher's perspective. Part III includes authors with quite different areas of expertise, which include mining engineering, mineral processing, ecology, economics, sociology, geography, and law. While each author was encouraged to investigate issues in terms of their own disciplinary strengths or research focus (e.g., gold, the small-scale sector, or a geographical region), there is a common methodological approach that underlies each study

In summary, each study was, first, to include a literature review and appraisal of research to date. It should be appreciated, however, that in some developing countries such as Ethiopia, China, Mozambique, and to a lesser extent Ghana, no such "groundwork" research was available. As a result, some of those author's chapters are more general and focus on describing the overall mining and environment context. Second, each study included an appraisal of the current approaches to mine closure and reclamation adopted by companies and national or regional governments. Third, each study investigated the elements that explained success and failure within those approaches. Fourth, the goal of each study was to draw out the elements of corporate strategy and public policy that are more likely to move best practice towards integrated environmental management through a planning for closure approach. Each study concludes with suggestions for further research, the identification of policy implications, or the drawing out of policy recommendations.

Prof. B. B. Dhar is a mining engineer at the Department of Mining Engineering, Institute of Technology, Banaras Hindu University, with a long-established reputation as one of India's leading academics in the area of mining and the environment. In Chapter 16 he provides an overview of the environmental degradation arising from mining, including the effects on biodiversity conservation, which he links to the lack of a rational and scientific planning process. His research covers case studies of good practice regarding effective revegetation and mine water treatment, which he associates with the need for underlying good science in the area of soil quality, mineralisation, and seed screening. He concludes by drawing out lessons for further research in the areas of restoration, biological intervention, biodiversity conservation, agrochemical use, and technology innovation.

Prof. Luke Danielson and Marily Nixon introduce a lawyer's perspective on current regulatory approaches to mine closure in the U.S. In Chapter 17 they review the historical legacy of widespread environmental damage from mining and highlight specific pollution events that have promoted growing public concern. They argue a case for taking a long-term and dynamic approach to mine closure. In doing so, they examine the evolving U.S. regulatory system, drawing examples from a range of different state laws and, in particular, focusing on the negative effects for industry of retroactive liability, as incorporated within U.S. Superfund legislation. They conclude that even the U.S. system leaves legislative loopholes and is full of uncertainty. In particular, they refer to long-term obligations for conserving water quality; retrospective care in the future as technology advances and scientific knowledge of hazards and their effects improve; and the inability of current technology to enable some mine operators in some environmental contexts to deliver a "walk away" site at the end of the closure period. Notwithstanding these challenges, the authors describe the potential for regulatory change to assist the diffusion of clean technologies in terms of practical closure plans, workable financial guarantee mechanisms, and information disclosure and exchange.

Dr. Fernando Loayza is an economist who also has extensive research experience within MERN, beginning with a Ph.D. supervised by Prof. Alyson Warhurst, working on the relationship between technological change and environmental performance. He has developed a new Bolivian mining policy that includes an environmental regulatory framework, the reform of the taxation regime, and a new mining code. His study, presented in Chapter 18, provides an overview of the tin mining industry and its role in the Bolivian economy, and describes the development of the Catavi region. Catavi was developed at a time when there was little environmental concern expressed and no competition for land, water, and other natural resources from either agriculture or industry. Dr. Loayza's chapter tells a story of a large world-class mining operation in a low-income developing country giving rise to temporary economic expansion followed, once the ore is depleted, by severe economic recession, manifested most acutely in the former mine workers' population and surrounding local communities which served the mine. He describes a "technological retrogression" on the part of cooperative workers that took over the closed mine to rework it for marginal ore and explains how this generated negative health and environmental effects. As a policy maker, Dr. Loayza explains the impossibility of imposing a closure plan after the event. He argues that the policy lesson is clear: there is need for prior socioeconomic and environmental impact assessments and it is important to integrate the results of these assessments into the design of closure plans.

John Hollaway, a mining engineer with expertise in small-scale gold mining throughout Africa, analyses lessons for best practice at small- and medium-scale mines in Zimbabwe in Chapter 19. His paper presents a powerful argument for the need to convert the fast-growing artisanal mining sector into a formal, regulated sector if control over environmental effects, such as river degradation and mercury toxicity, is to be achieved. He argues that there is a need for an incentive structure to promote self-regulation, given the weak institutional capacity within Zimbabwe, and to assess and monitor environmental impacts and closure planning.

Dr. Isidro Manuel and Dr. Tomás Muacanhia are, respectively, a geologist and an ecologist at the University Eduardo Mondlane in Maputo, Mozambique. In Chapter 20 they describe alluvial gold mining in Mozambique and identify some of the opportunities and constraints for sustainable development. They focus on the need for environmentally sensitive mineral development that would address the protection of biodiversity and the need for reform of the weak legal and institutional framework. This leads them to argue for the state to develop alliances with local communities and recognise their land title rights so as to avoid potential future conflict and accelerate capacity building at the institutional level. Significantly, the overriding concern of poverty, ecological vulnerability, political instability, and economic crisis constrain the attainment of sustainable development goals and the authors are able only to explain why environmental mismanagement has occurred in this context, rather than identify any good practice.

Dr. Peter Acquah is Director of the Environmental Protection Agency in Ghana and a longtime research member of MERN. In Chapter 21 he overviews the environmental problems associated with both minerals extraction and the cessation of mining activities. He then discusses the need for financial provisions to be made from the outset to cover a closure plan as an integral part of the general mining plan. Dr. Acquah describes the legal framework that is being developed in Ghana to ensure best practice in this area and illustrates his case with the plans of two gold mines in Ghana.

Dr. Edward Nyamekye is Director of Small Scale Mining in the Minerals Commission of Ghana. In Chapter 22 he focuses on the socioeconomic impact of mine closures. He reviews the historical rationale for planning for closure by describing in detail the results of failing to manage environmental impacts adequately. Dr. Nyamekye concludes by recommending that mining projects need to be conceived as opportunities to solve social and economic problems in mining regions, but need to be designed accordingly.

Dr. Solomon Tadesse, an exploration geologist, undertook his case study research in Ethiopia, focusing on gold and rare metals. His work represents a valuable description of the mineral production process and its environmental impacts. In Chapter 23 Dr. Tadesse draws on this analysis to argue for the simultaneous incorporation of environmental management techniques, for example in dust control, water treatment/recycling, and reforestation schemes, and into the mining process so as to avoid potentially costly problems of post-closure cleanup.

Prof. Gao Lin (Department of System Ecology Sciences, Research Centre for Eco-Environmental Sciences, Chinese Academy of Sciences, Bejing), Wang Yishui, Ge Feng, and Zhang Wenmin focus on environmental management practices in the Chinese non-ferrous metals industry in Chapter 24. Following a review of environmental damage effects, Prof. Gao Lin describes the trend in corporate strategy away from end-of-pipe control towards pollution prevention. She also describes a new trend towards metals recovery from waste and ongoing reclamation as part of the implementation of a best-practice plan to achieve integrated environmental management at Chinese mines.

Dr. Ian Clark, Director of the Research Centre for Environmental and Recreational Management at the University of South Australia, argues in Chapter 25 that mining in Australia is generally regarded as an activity that leaves land unfit for future use. He describes, however, how the problem of sustainable development is fast requiring mining firms in Australia and the policy framework within which they operate to address the issue of reuse. Dr. Clark describes the new "close out" criteria in Australia which defines the success of rehabilitation programmes, and reviews both the supporting legislative process and industry codes of practice that are being developed. This includes more flexible performance bonds, environmental auditing, and self-regulation. Dr. Clark also supports issues that will become increasingly important with regard to mine closure. These include long-term liability, heritage value, and land contamination.

Dr. Clark's research project — Smelter in the Park — is presented in Chapter 26 and describes Portland Aluminium's innovative project to recreate plant communities and manage diverse wildlife habitats around the smelter. Dr. Clark also demonstrates that this project was not only good for the environment but also good for business. Notwithstanding the immediate apparent success of the project, Dr. Clark identifies some physical and psychological constraints that still require addressing. The most important is that environmental management is a long-term process, not a "promotional concept" for which a company receives an award. It requires a long-term commitment to continuous improvement at the levels of both company culture and the core business activity.

Dr. Richard Isnor is a technology policy researcher for Environment Canada with specialist expertise in mining and environment. His Ph.D. was also supervised by Prof. Alyson Warhurst. In Chapter 27 he describes a shift in public policies towards mining and environment to emphasise mechanisms that are anticipatory and preventive, replacing or reinforcing those that are reactive and curative. However, while such a shift in turn requires a new policy focus on technological change to promote pollution prevention from the outset, Dr. Isnor argues that traditionally explicit and implicit closure policies in Australia, Canada, and the U.S. have adopted a narrow regulatory approach. Nonetheless, the author has identified an emerging trend for governments to consider closure as an integral part of the mining process, through the development of state-funded research and development networks that are broadening the technological choices open to mining firms to encompass more environmentally proficient techniques.

Dr. Ligia Noronha, an economist with expertise in environmental costing and natural resource economics, undertook a major empirical and pioneering research project on the explanatory factors underlying different levels of environmental performance within the iron ore mining industry in Rajasthan. In Chapter 28 she explains how she found environmental regulation to be a major factor in inducing improved environmental practices. Over time, she documented proactive responses beyond environmental compliance by some companies as a function of either the innovative capacities of firms on their willingness to change. She found that investment in environmental training was an important factor in explaining positive improvement. However, she noted negative effects for the firms she studied as a result of failure to include local participation in either the assessment or mitigation of environmental impacts. Her conclusions indicate that environmental policy has been defined in terms of constraints rather than incentives, and she suggests the need for policy incentives in order to promote proactive and improved environmental practices.

This collection of case studies is particularly instructive since, in combination, they present a strong argument for planning for closure from the outset to achieve integrated environmental management throughout the life cycle of a mining operation. The strength of their argument rests on the diverse disciplinary and country perspectives they bring to bear on the generic problem of how to manage the environmental and social effects of mining. In the final chapter, Chapter 29, Prof. Alyson Warhurst draws some conclusions and presents the recommendations drawn up by the chapter authors following a review and discussion of research results in Zimbabwe in August 1996. These recommendations cover policy mechanisms and institutional support, EIAs, technology, company strategy, and education and training.

Several of our authors are now policymakers in government office in their own countries, while others are increasingly called upon to play an advisory role at either the national level or for multilateral organisations. The task ahead is to ensure that these recommendations are implemented and to continue our collaborative policy research and dissemination activities to ensure the diffusion of best environmental practice in future mineral development.

APPENDIX 1.1

THE MINING AND ENVIRONMENT RESEARCH NETWORK (MERN)

The **Network** is an international collaborative research programme, involving centres of excellence in the major mineral-producing countries of the world. It was established in 1991 with the aim of generating analysis to facilitate the improvement of environmental and social performance and competitiveness of mining companies, in the context of growing environmental regulation, "voice of society" concerns, and technological innovation.

Our **current research** examines the relationship between environmental regulation, technical change, social policy, and competitiveness in the minerals industry. It investigates how the process of technological innovation and organisational change can be harnessed to prevent environmental degradation, while enhancing productivity and sustainability. This focus also includes the effective integration of social impact analysis within environmental management planning. The liberalisation of investment regimes worldwide, combined with growing environmental regulation and the increasing conditionality of loans on prior environmental and social impact analysis, indicates that there is an urgent need for objective and well-documented policy analysis to contribute to decision-making in industry, donor agencies, government and non-governmental organisations. This programme of collaborative research aims to facilitate the global diffusion of such policy analysis and contribute to the building of international research competence in this important area.

Taking this context into account, and building on our diagnostic research, the next phase of Network Research covers **four** inter-related themes:

Comparative Analysis of Environmental/Social Performance and Its Relationship with Production Efficiency

Network research has demonstrated that good environmental management in the firm is more closely related to production efficiency and capacity to innovate than simply to a regulatory regime. Environmental degradation and negative social impacts tend to be greatest in high-cost operations working with obsolete technology, limited capital, and poor human resource management. Since the latter problems are characteristic of much of the mineral production of developing countries, they are a special, but not exclusive, focus of analysis within the Network. A major area of empirical research is an international benchmarking exercise that investigates environmental performance.

Analysis of International Environmental Regulations and the Definition of Improved Policy Options

Building on an international comparative analysis of the effectiveness of current environmental regulation, researchers are investigating a range of policy approaches to achieve sustained and competitive improvements in environmental management, and to achieve pollution prevention rather than pollution treatment. An original contribution of the research will be to evaluate the potential of technology transfer and training, particularly within joint-venture agreements and linked to credit conditionality, to accelerate the development and diffusion of improved environmental management practices. Researchers are also analysing the environmental implications of new trade policies and agreements, such as GATT and NAFTA.

Towards "Best-Practice": Corporate Trends in Environmental Management and Social Responsibility

A preliminary conclusion of the Network research is that technical change, stimulated by the drive for improved competitiveness and the "Environmental Imperative" is reducing both production costs and environmental costs to the advantage of those companies that have the resources and capacity to innovate. Our current phase of research is to evaluate and compare trends in environmental best practice for non-ferrous minerals production in different socioeconomic and policy contexts, drawing out the lessons for both corporate strategy and government policy. This includes empirical research into planning for closure within the minerals industry.

Towards Corporate Social and Environmental "Best Practice"

An increasing proportion of current research on "best practice" is focused on social impact prediction, prevention, analysis, and mitigation.

At the University of Bath's International Centre for the Environment, researchers are working on various aspects of the interface between mining and the social environment:

- A comparative study of partnerships and participation in mining projects (Assheton Carter).
- Development of appropriate methodological techniques for social impact assessments at mining operations (Magnus Macfarlane).
- Ethical considerations for mining companies (Amy Lunt).
- Development of sustainability indicators for corporate reporting and social policy in the minerals industry (David Uglow).

Each of these four projects benefits significantly from links with the Centre for Development Studies at the University of Bath and each project is co-supervised by Dr. Geof Wood, an eminent sociologist with the Centre.

Other researchers are indirectly linked with the programme of work on mining and the social environment:

- Biodiversity conservation indicators and minerals development (Kevin Franklin).
- The development of an environmental management system and environmental/social performance indicators for the international aggregates producer, Aggregate Industries (Miles Watkins).
- Financial drivers of environmental/social performance (Alyson Warhurst and Nia Hughes).

Within the wider Network, other researchers are also examining mining and a number of related key social issues:

- Dr. Ligia Noronha, using extensive results from an in-depth case study in Goa (India), is developing a methodology using environmental economics to analyse, evaluate, and integrate ecological and health impact data and reclamation work.
- In Colombia, Cristina Echavarria is working to define cultural and biophysical indicators for minerals development that relate particularly to indigenous peoples.

The above and additional research form the basis of a DFID-supported project entitled Environmental and Social Performance Indicators (ESPIs) and Sustainability Markers (SMs) in Minerals Development. This is a three-year project that involves the development of a series of indicators to be used throughout the life cycle of a mining operation. The research will identify a framework of indicators of environmental and social performance, in the context of sustainable development goals that is meaningful to stakeholders both internal and external to the industry. Guidelines for the implementation of new business strategies and cost-benefit analysis will also be included.

For further details on membership, sponsorship, and research, please contact: Professor Alyson Warhurst, Director of the Mining and Environment Research Network, International Centre for the Environment, School of Management, University of Bath, Bath, BA2 7AY, U.K., Tel: +44 (0)1225 826156; Fax: +44 (0)1255 826157.

2 Integrated Environmental Management Through Planning for Closure from the Outset: The Challenges

Alyson Warhurst and Ligia Noronha

CONTENTS

1-56670-365-4/00/$0.00+$.50
© 2000 by CRC Press LLC

ABSTRACT

This chapter provides a framework for this collection of papers on integrated environmental management through planning for closure from the outset. It provides an overview of the environmental impact of mining and underlines the imperative of improved environmental management and planning. It argues that pollution prevention, through planning for closure, can lead to cost-effective strategies for operationalising sustainable minerals development. This seems to be most true for greenfield sites since, generally, the earlier closure planning and pollution prevention is built into a project, the more cost-effective and environmentally benign closure will be. Furthermore, for greenfield sites, pollution prevention techniques can be employed from the outset, at the stages of exploration and mine development, and then monitored and improved through the operation stage to closure.

The chapter then discusses how global changes in the industry, following the liberalisation of investment regimes, and mergers and strategic alliances between key firms, has, by virtue of the diffusion of new technology, led to further opportunities to prevent pollution through planning for closure from the outset. The objectives and components of closure plans are also reviewed as the chapter draws on the case studies to highlight some of the possible constraints and challenges to pollution prevention that may be faced in operationalising such plans at the level of both public policy and corporate strategy. Finally, the chapter concludes by suggesting a forward-looking approach to integrated environmental management based on a dynamic model for an environmental management system.

2.1   ENVIRONMENTAL IMPERATIVES

Integrated environmental management through planning for closure involves planning in anticipation of operations and implementing this plan throughout the life of the mine so that potential environmental effects associated with each stage of the mineral production process are avoided or reduced and managed. The waste products generated at each stage from initial exploration and extraction through concentration, smelting, refining, fabrication, and finally decommissioning, and the potential hazards to which they give rise are summarised in Figure 2.1.

The extent of environmental hazards and the costs of mitigating them are site-specific and will be affected by local geology, geography and climate (see Warhurst, Chapter 3). The mineralogy and geochemistry of different mineral deposits and therefore their pollution potential may also vary considerably. For example, copper can be found in association with lead, bismuth, arsenic, and sulphur compounds, any of which may complicate the environmental impacts of emissions or leakage from tailings ponds or rock dumps. The composition of the rocks and soils underlying dumps and tailings ponds will determine the dispersion pathways of natural seepage and the need for expensive plastic membrane linings. Whether the mineral deposit is located on a mountainside or in a basin or whether it intersects the water table or is close to an urban centre will influence the extent of hazard created through its exploitation. Climate also exerts an influence on, for example, the direction of prevailing winds carrying pollutant compounds and particulates, as well as the extent to which rising dusts from dumps, tailings, and mine pits cause problems.

In this book, the environmental imperative encompasses far more than simply the hazards posed by mining. Rather, it includes: stricter environmental regulation, a growing voice of society, increasing pressures driving environmental management through the supply chain, and the environmental conditionality that is increasingly being attached to credit, equity investment and insurance investment, and insurance.

2.1.1   BENEFITS AND COSTS OF PLANNING FOR CLOSURE

Planning for closure can generate benefits at two levels: first, at the level of the local community or region; and second, at the level of the mining company. Closure of mines, in the absence of any planned closure process may have the most serious consequences for local communities, because of their greater dependence on a livelihood provided by the mining activity (see Warhurst et al., Chapter 5). A forward-looking approach for miners, ancillary staff, and related family units can involve developing viable economic alternatives, transforming mined land for productive purposes, such as for growing cash crops, and timing new mining projects in the region to follow consecutively (see Noronha, Chapter 28). Planning can also provide the framework for the psychological adjustment to unemployment (see Warhurst et al., Chapter 5). Planning for closure can enable the region's resources to be mined in a manner that does not result in a net impoverishment at the close of mining operations, and avoids long-term or permanent damage to ecosystems that have a positive value to local communities and business activities. These resources have both "use" values and "option" values. They are use values in the sense that they are inputs in production. In addition, environmental resources have another value in that society's future options are enhanced.

The benefits of planning for closure to the mining firm are that it:

- Ensures that a stock of water, air, and land (components of the natural environment necessary for production) is available for future operations.
- Provides a healthy environment in which miners can work and live, thereby maintaining/increasing the productivity of the workers.
- Reduces the extent and expense of final remediation, as is discussed below.
- Reduces future risk of more strict regulation.
- Improves the company's profile and track record through greener operations.
- Reduces tensions and conflicts within local communities.

The benefits of the first two points to a company's operations are evident; the weight of the remaining points as benefits will depend on the regulatory regime in place, societal preferences and attitudes to the environmental impacts of mining, and company strategy.

The attractiveness of the concept of planning for closure as a whole life cycle approach to managing the environmental impacts from mining from the development stage through operations to closure is rooted in what may be referred to as the time factor. The greater the time lapse between the occurrence of environmental damage and its remediation, the greater (in most cases) will be the resources (both human and financial) needed to address the problem (see Figure 2.2).

In the absence of adequate closure planning, the demand for financial resources to complete the closure process will occur at a time when the firm is experiencing a reduction in cash flows. Integrating environmental management by planning for closure from the outset will enable the firm to provide for the costs of closure during the period of positive cash flows. This is illustrated in Figure 2.3. If the firm operates in a regime in which there are no closure responsibilities, then the curve will follow the trajectory of Yb. If the firm has closure responsibilities, and there has been no planning for closure, the firm may need to incur costs at the time of closure that will burden the cash flows at a time when they are substantially reduced. The cash flow curve will then have the shape of Y. An integration of environmental costs from the outset implies that the burden of costs will be spread throughout the project life and thereby the burden at the time of closure is reduced. Yec represents the cash flow curve that integrates closure costs into operations through the life of the project. As a result of this integration, the burden on the company at the time of closure is substantially reduced. Planning for closure would require that investments in environmental quality be put in place, or allowed to ensure that at the time of closure the site is rehabilitated and in a condition that is commensurate with the expectations of the community. Investments would involve: expenses to reduce pollution at source, i.e., treat pollutants before they are discharged into the environment and/or adopt a cleaner technology that reduces the amount of pollution which is generated; expenses to clean up the pollution/degradation that occurs anyway; and expenses involved in monitoring predicted and unpredicted impacts; and expenses involved in developing human capacity to initiate and sustain environmental management practices.

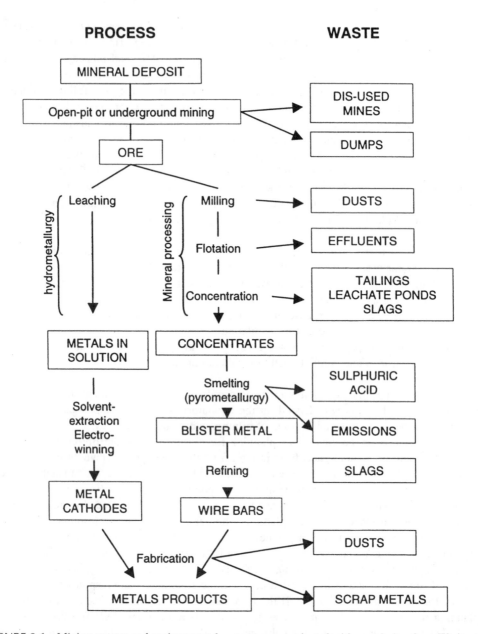

PROCESS   **WASTE**

FIGURE 2.1 Mining, wastes and environmental consequences, adapted with permission from Warhurst and Franklin[18], Biodiversity conservation, minerals extraction and development: a realistic partnership, in *Strange Bedfellows: Extractive Industries, Infrastructure, and Biodiversity Conservation*, Prickett, G.T. and Bowles, I.A., Eds., Conservation International: Oxford University Press, 1999.

The premise that planning for closure can actually have positive benefits to the mining firm must provide the principal rationale for forward planning at the firm level, within the context of evolving frameworks of legislation that increasingly require firms to take responsibility — even retrospectively — for cleaning up their environmental damage from the exploration stage through operations to post-closure. In this emerging situation, it makes sound business sense to prevent pollution from the outset and to be socially responsible in practices towards local communities.

Another driver for integrating environmental management into planning for closure from the outset are the environmental conditions increasingly being attached to the provision of credit and

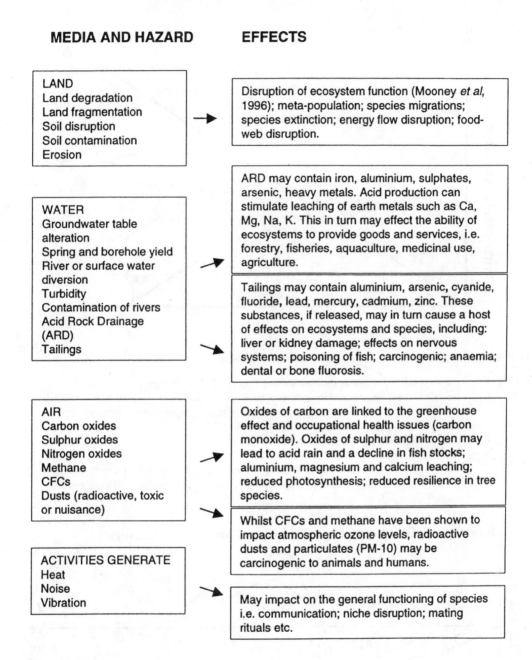

MEDIA AND HAZARD

EFFECTS

LAND
Land degradation
Land fragmentation
Soil disruption
Soil contamination
Erosion

Disruption of ecosystem function (Mooney *et al*, 1996); meta-population; species migrations; species extinction; energy flow disruption; food-web disruption.

WATER
Groundwater table alteration
Spring and borehole yield
River or surface water diversion
Turbidity
Contamination of rivers
Acid Rock Drainage (ARD)
Tailings

ARD may contain iron, aluminium, sulphates, arsenic, heavy metals. Acid production can stimulate leaching of earth metals such as Ca, Mg, Na, K. This in turn may effect the ability of ecosystems to provide goods and services, i.e. forestry, fisheries, aquaculture, medicinal use, agriculture.

Tailings may contain aluminium, arsenic, cyanide, fluoride, lead, mercury, cadmium, zinc. These substances, if released, may in turn cause a host of effects on ecosystems and species, including: liver or kidney damage; effects on nervous systems; poisoning of fish; carcinogenic; anaemia; dental or bone fluorosis.

AIR
Carbon oxides
Sulphur oxides
Nitrogen oxides
Methane
CFCs
Dusts (radioactive, toxic or nuisance)

Oxides of carbon are linked to the greenhouse effect and occupational health issues (carbon monoxide). Oxides of sulphur and nitrogen may lead to acid rain and a decline in fish stocks; aluminium, magnesium and calcium leaching; reduced photosynthesis; reduced resilience in tree species.

Whilst CFCs and methane have been shown to impact atmospheric ozone levels, radioactive dusts and particulates (PM-10) may be carcinogenic to animals and humans.

ACTIVITIES GENERATE
Heat
Noise
Vibration

May impact on the general functioning of species i.e. communication; niche disruption; mating rituals etc.

FIGURE 2.1 Continued.

insurance for mining projects. For example, in November 1995, the withdrawal of political risk insurance cover was announced in part on environmental grounds by the Overseas Private Investment Corporation (OPIC — an insurance firm and U.S. government agency) from the international mining operation of Freeport McMoran Copper and Gold at Grasberg, Irian Jaya in Indonesia. Then in March 1996 came the news of the reinstatement of the political risk insurance on condition that the operation establish a trust fund to finance environmental remediation, beginning immediately and accumulating to a total of U.S. $100 million, the amount of the original risk cover. Therefore, these drivers are clearly all the more significant in the context of the global changes currently affecting the structure of the world mining industry.

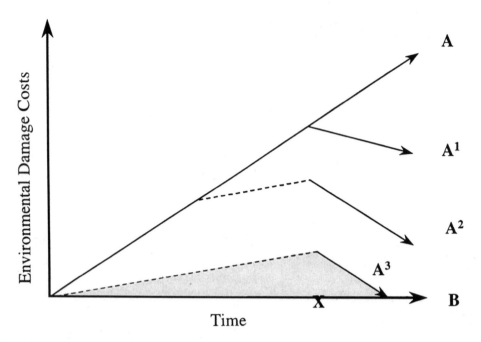

FIGURE 2.2 Environmental damage and the remediation challenge.

KEY **A**   Environmental damage costs from mining (not necessarily linear)
A^{1-3} Describes the trajectories of operations associated with different levels of environmental
damage costs incurred over time depending upon resources invested in remediation and
point of starting remediation (not necessarily linear)
B   Axis commensurate with no environmental damage costs from mining
X   Point of decommissioning and close-down
The shaded area from axis **B** to lines A^{1-3} symbolises the extent of internalisation of
environmental damage costs required under these different scenarios to reduce environmental damage from mining

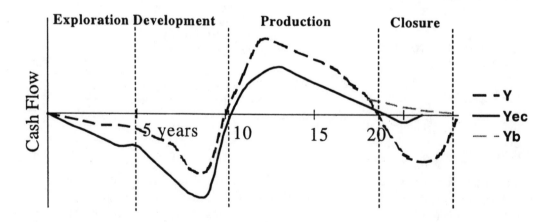

FIGURE 2.3 Annual field cash flow (net).

Y - cash flow curve uncorrected for closure costs
Yec - cash flow curve incorporating closure costs
Yb - cash flow where there are no closure responsibilities

2.1.2   ECONOMIC LIBERALISATION AND NEW MINING INVESTMENT PROJECTS: OPPORTUNITIES TO INTEGRATE ENVIRONMENTAL MANAGEMENT FROM THE OUTSET

(This section is extracted with permission from Warhurst and Bridge).[1]

The international mining industry is undergoing a transformation. Over the past decade, dynamic mining companies have begun to restructure their operations in response to new opportunities arising from the liberalisation of investment regimes for mining in many developing countries, to develop or acquire new production technologies, and to respond to heightened environmental awareness and scrutiny of their operations. This bundle of technological and organisational changes has the potential, if effectively managed, to contribute to economic growth and improved environmental performance in developing countries.

Since 1989, over 75 countries have liberalised their investment regimes for mining. Economic and political reforms have opened up new opportunities to the international mining industry in areas that were formerly closed, either because of *de jure* political restrictions or closed *de facto*, since economic and political risks were sufficiently high to deter prudent investment. Private investment flows to developing countries have increased in response to these opportunities, with mining playing a significant role as a proportion of the total direct foreign investment. For example, in 1995, direct foreign investment in developing countries was valued at $90,000 million, while capital expenditures on mining alone were estimated at $20,000 million for the period 1995–2000.[2] This transformation of investment regimes and patterns of investment flows is occurring at a time of significant technological change within the mining industry as firms respond to increasing market and regulatory pressures. Incremental improvements in process control and optimisation, or the application of existing technologies at increasing scales of operation to capture greater efficiencies, have proved fundamental to maintaining the competitiveness of major mineral producers. Not only have innovations in processing technology improved productivity and efficiency but, by improving process control, increasing recovery rates and reducing waste, several key processing innovations have enabled firms to combine gains in competitiveness with improved environmental performance. Direct foreign investment through joint ventures with state-firms and/or newly privatised entities in developing countries may, under certain conditions, provide an effective vehicle for the transfer of these innovations generating improvements in production efficiency and environmental performance.

In this context, the development and/or acquisition of cleaner production technologies could be especially attractive to governments in developing countries since they hold the promise of reducing environmental damage costs while at the same time maintaining the social and economic benefits of mining (e.g., jobs, taxes, foreign exchange earnings, skill and technology transfer, and royalty payments). The opportunities for technological leapfrogging through the transfer of innovative production techniques are rapidly expanding as many less developed countries encourage large exploration programmes and design new mining laws while establishing codes of environmental practice. As the mining industry enters a new phase of globalisation, facilitated by economic deregulation and the privatisation of formerly state-owned mining concerns, mining companies and equipment suppliers are forming a range of joint-venture agreements with remaining state-owned and newly privatised operations. New investment and strategic partnering provide an opportunity for technology tie-ins in which recipient companies use externally acquired technology to leverage technological and organisational innovation for their competitive advantage. Greenfield investment provides opportunities to select state-of-the-art-processing technologies from the outset and integrate new production methods with pollution prevention techniques and environmental management systems to achieve lower cost and environmentally proficient production.

Since the late 1980s, many countries have begun to pass legislation designed to improve the investment climate and, in particular, to encourage foreign interest in mineral resources. The policy challenge for many developing countries is to find a balance between the investors' objectives of profitability and the government's objectives of revenue generation and positive social and environmental externalities from mineral development. Ecuador, for example, has received a U.S. $24 million World Bank loan to improve the mining sector by rewriting the Mining Law, creating a new framework for mining concession titles, promoting foreign investment and the sale and transfer for government-held mining rights.[3] There is now considerable competition between liberalising countries for foreign mining capital as each country seeks to create a competitive fiscal regime, attractive investment policies, and transparent and expeditious permit processing.[4] In Argentina, for example, the Mining Investments Act (Codigo de Minera) was signed in 1993. This covers prospecting, exploration, mineral production, and other activities and includes a range of measures which guarantee stability of the municipal, provincial, and national tax regime for 30 years, allow 100% income tax deductions for the costs of prospecting and exploration, cap royalties, and provide import duty exemptions for mining equipment. The country also signed a Mining Integration Treaty with Chile in June 1996 to facilitate the free flow of mining materials and equipment in the border area where several major projects and prospects are located.[2,5] In addition to these specific measures aimed at the mining industry, Argentina has passed relatively liberal investment legislation which allows unrestricted transfer of currency overseas and does not preclude the repatriation of capital or the transfer of profits. As a consequence of these measures, Argentina has become one of the world's principal targets for exploration expenditure, with an anticipated annual total of U.S. $135 million by the year 2000.

2.1.3 PRIVATISATION OF STATE MINING FIRMS

Associated with the liberalisation of the economy and increased economic and political stability for investors is the transfer of state assets to the private sector through a process of privatisation. Key national assets that were formerly state-owned and operated, including natural resources, are being opened up to private and foreign investment. Extensive programmes of state disinvestment from mining operations have been initiated in Europe (e.g., Britain, Portugal, France), Latin America (e.g., Bolivia, Nicaragua, Peru), and Africa (e.g., Ghana, Guinea, Tanzania). Privatisation can take a variety of forms: liquidation of state assets through the full divestment of state interests; partial sale of state ownership; and the injection of private capital for the development of new ventures.[6] Privatisation has been identified by many countries as a means of modernising old plants and equipment through the injection of new capital and skills with which to promote technological change, improve efficiency, and enhance international competitiveness. Guinea's Minster of Mines and Geology, for example, stated in a recent interview that the mining sector would be the standard bearer for foreign investment in his country and that his country was "keen to attract international mining companies not only for the economic benefits they bring with them, but also as a means of developing Guinea's extremely limited infrastructure."[7] Following the collapse of the Soviet Union, a principal trading partner for Guinean bauxite, Guinea is seeking to diversify its mineral production by reducing the state's holdings in principal assets such as the Friguia alumina smelter, and encouraging foreign investment. International gold companies have expressed interest in Guinean reserves, including Golden Shamrock Mines of Australia and the Lero joint venture between the French BRGM and Kenor of Norway which began production in 1995.[8]

The privatisation of natural resources, energy, and infrastructure sectors in both industrialised and developing societies reflects a fundamental reappraisal of the state's role in the development of natural resources for national benefit, and is often associated with a shift from an import substitution model of development to one based on the promotion of exports. Rather than being the owner, operator, and regulator of mineral production, the state's direct role is reduced, transformed to that of a facilitator of private mineral production. Although in most cases mineral resources remain state property, the development of those resources into productive assets is increasingly undertaken by the private sector. The state aims to ensure social benefits from mineral extraction, not through controlling the means of production but through the mechanisms of regulation and taxation. The privatisation of state mining companies has been underway for some time, but has accelerated in recent years in response to rising mineral prices and increased interest from potential buyers. Privatisation of mineral assets world-wide (excluding the former COMECON countries) raised over $2.2 billion in 1995, twice that of 1994.[9] Bolivia, for example, began privatisation of state industrial assets in 1985, Venezuela initiated privatisation in 1994, and following an unsuccessful attempt in 1992, Peru renewed its efforts to privatise CENTROMIN in 1995. India opened up 13 minerals industries to private investment in 1993, while Portugal confirmed its privatisation of Empress de Desenvolvimento Mineiro, the state holding company for mineral activities in 1995. The privatisation of Zambia's Consolidated Copper Mines (ZCCM) has also been proposed, although the weak position of ZCCM and the considerable investment required to restore competitive production have deterred foreign interest.

The globalisation of mining activity is reflected in the dramatic increase of the resources dedicated by international mining firms to exploration and new investment in developing countries. Between 1990 and 1995, new mining investment in Latin America totalled $238 billion, while exploration expenditure increased five-fold to $500 million per year, with the devaluation of the Mexican peso in 1994 appearing to have only a short-term effect.[10] Private sector copper projects initiated in Chile over the past five years include Rio Algom's investment at Cerro Colorado, Cominco at Quebrada Blanca (see below), Phelps Dodge Candelaria, Outokumpu at Zaldivar, and Anglo American at Namtoverde.[3] Asia and the Pacific have increased their share of foreign direct investment over the last few years, increasing the competition between mineral producing regions for foreign investment. Although mineralogical potential remains the principal determinant of the location of new investment, the wide range of mineralogical properties now potentially available for investors means that government policies on investment, taxation, and the environment play a significant role in determining where investment is made. A substantial proportion of the foreign direct investment in mining has taken the form of joint ventures with newly privatised firms or with those companies that have remained under state ownership. In May 1992, Chile passed the so-called CODELCO Law, enabling the state-controlled copper mining giant CODELCO to acquire capital, equipment, and expertise by entering into joint ventures with the private sector. Cyprus Amax became the first foreign private sector partner of CODELCO, acquiring a 51% stake in the El Abra copper mine for $330 million in June 1994. In the same year, Cyprus Amax also acquired a 91.5% stake in the Cerro Verde copper leaching operation as part of Peru's privatisation programme. Formerly off-limits to foreign investment, Cuba extended allowances for foreign investment to mining in 1992 and has subsequently seen considerable interest from Australian and Canadian mining investment. The first Joint Enterprise Agreement between foreign capital and the Cuban government in mining is the solvent extraction-electrowinning copper oxide project at Mantua, developed by Miramar Mining Corporation.[3] In December 1994 Sherrit signed a 50/50 joint venture agreement with the Cuban state enterprise La Compania General de Niquel of Cuba to mine, refine, and market nickel and cobalt in Holguin Province, including a $125 million investment at Moa Bay to double production to 24,000 tonnes per year.[10]

2.1.4 TECHNOLOGICAL CHANGE, PRODUCTION EFFICIENCY, AND ENVIRONMENTAL PERFORMANCE

Recent evidence suggests that environmental degradation resulting from mining is more closely related to production efficiency and capacity to innovate than to firm size, ownership, location, or regulatory regime.[11-14] Environmental degradation tends to be greatest in low-productivity operations with obsolete technology, limited capital, inefficient energy use, and poor human resource management. These problems are endemic in, although not exclusive to, many developing countries.

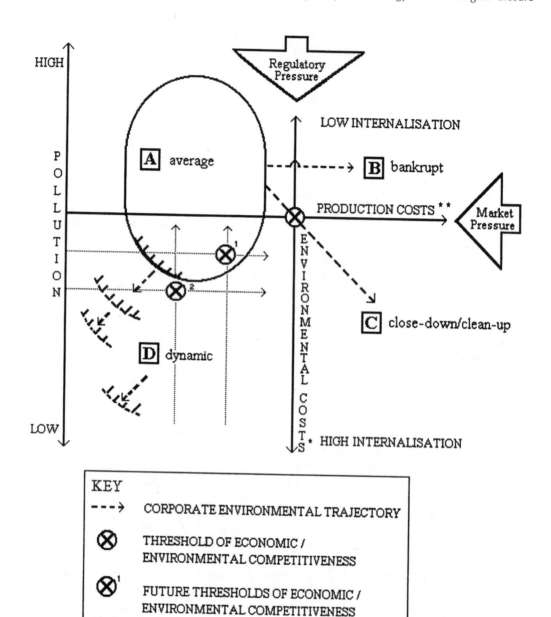

FIGURE 2.4   Corporate environmental trajectories.

*   Environmental costs refer to environmental damages
**   Production costs include abatement costs

Conversely those companies which have the resources and capacity to innovate are able to harness technological and organisational change to reduce both the production and environmental costs of their operations. This suggests that there is much scope for improving environmental performance by adopting public policies and corporate strategies which promote the development and mastery of technological processes and which facilitate organisational structures encouraging process control, continuous improvement, and organisational learning. It is in this respect that technology transfer, involving the building up of managerial innovative capacity and improving human resource development, could have a more significant impact on environmental problems than changes in regulatory regime. Figure 2.4 summarises these relationships by describing the different environmental trajectories that firms follow as they respond to the environmental imperative by either internalising their environmental damage costs and/or incurring pollution abatement costs.

Hence, we have the challenge to explore and define concepts of the development and diffusion of best practice in integrating environmental management within the whole life cycle of mining projects. Figure 2.4 categorises the "environmental trajectories" that different mining enterprises might take in response to environmental and market pressures. The average enterprise is competitive (i.e., with trajectories to the left of the threshold of economic competitiveness, $\otimes$), although to a greater or lesser extent produces environmental pollution and, in response to regulation, has internalised the cost of the environmental degradation associated with its production. (The threshold of "environmental competitiveness" for a given regulatory context is also $\otimes$, and operations in compliance have environmental trajectories in the quadrants below).

However, market pressures — a real decline in metal prices, combined with economic inefficiencies — mean some companies are going bankrupt (a move to quadrant B). They will leave a legacy of environmental pollution, and the burden of cleanup will fall on the state and society. Other companies will respond by innovating; moving into quadrant D; building into the new generation of technology both improved economic and environmental efficiencies (protecting themselves in the process from relatively more costly add-on, incremental technical change, and rehabilitation at later stages in their operation). Freed from the costs of retrofitting sunken investments, greenfield plants in particular display new levels of dynamism — the latest best-practice technology incorporates improved economic and environmental efficiencies. Nonetheless, a growing group of companies, if obliged to "add-on" environmental controls in line with new regulations would have to close down since the costs required would render their operations uneconomic. The environmental trajectory of these companies is "step-like" towards quadrant C. Currently, such examples are few. However, that group will grow in number, since combined market and regulatory pressures will lower the threshold of economic and environmental competitiveness such that the average company will only survive in the new regime if it innovates. Even previously dynamic companies will need to keep their environmental trajectories moving ahead of the encroaching threshold of economic/environmental competitiveness ($\otimes^1$ and $\otimes^2$).

2.2 OBJECTIVES OF A PLAN TO MANAGE THE ECOLOGICAL IMPACTS OF MINING THROUGH MINE DEVELOPMENTS, OPERATIONS, AND CLOSURE

The main objective of planning for closure is to reduce the extent of environmental damage and waste generation by reducing the time lapse between damage occurrence and remediation. The other objectives of a plan for closure are quite straightforward and are often site-specific. The challenges relate more to operationalising the plan and integrating the different elements. This is where we would expect to draw generic lessons from best practice (see Bridge, Chapter 9; Clark, Chapter 26; Danielson and Nixon, Chapter 17; Dhar, Chapter 16; Mitchell, Chapter 14; and Hollaway, Chapter 19). In summary, the main objectives of planning for closure and whole life cycle integrated environmental management of a mine include:

- The reduced generation of waste and encouragement of efficient recycling.
- The efficient use of energy.
- The efficient use of chemicals and minimisation of any hazards related to their supply, use, and disposal.
- The stabilisation of residues and reduction for acid mine drainage or water contamination from the outset.
- Disposal and containment of waste to achieve zero discharge over time.
- Progressive remediation and revegetation.

Objectives should also:

- Ensure viable post-mining land use for the region.
- Ensure that there are no health impacts detrimental to local community.
- Ensure that local communities are not impoverished as a result of this activity or at the end of the mine life.
- Spread costs over life of mine.
- Enhance capacity utilisation over life of mine.

These objectives, by necessity, should be goals and expected performance outputs, not specified technologies or approaches. Mine projects often last decades, and such a framework which is defined by performance goals rather than detailed procedures provides flexibility and allows technological and organisational innovations to be promoted and incorporated along the way. This is an important finding based on past MERN research.

2.3 COMPONENTS OF PLANNING FOR CLOSURE

2.3.1 INTRODUCTION

Mineral extraction can have a number of impacts, both socioeconomic and ecotoxicological. These impacts may continue after closure as, for example, acid rock drainage (see Warhurst, Chapter 3; Barbour and Shaw, Chapter 4; Broadbent and Coppin, Chapter 10; Mitchell, Chapter 7), or arise as a result of the closure, e.g., local people displaced from work at the mine site or from services that supported the extraction activity (see Warhurst et al., Chapter 5).

A mining company planning for closure has to address both these types of impacts. As a result of such a plan, environmental impacts that would continue beyond the life of the mine, for example, acid rock drainage or surface erosion, may be avoided or reduced through timely and appropriate action during the life of the mine (see Mitchell, Chapter 7). Those that would arise because of closure can be planned in anticipation of their occurrence and therefore reduce the extent of negative impacts and, consequently, the costs of remediation (see Warhurst et al., Chapter 5; Broadbent and Coppin, Chapter 10; Bridge, Chapter 9).

From the outset, the plan needs to identify the potential effects arising as a result of closure or those following from operations. Having identified possible effects, the company needs to put in place mitigation measures. Progressive rehabilitation and the management and monitoring of impacts during the operational phase of the mine will reduce the extent and expense of the final remediation required (see Morrey, Chapter 13).

A plan for closure has to address the following issues:

- What are the kinds of environmental and social problems that may continue or arise at the end of the life of the mine? (see Bridge, Chapter 9; Warhurst et al., Chapter 5, Barbour and Shaw, Chapter 4; Nyamekye, Chapter 22)

- How can these be mitigated or avoided? (see Broadbent and Coppin, Chapter 10; Mitchell, Chapter 7; Petersen et al., Chapter 11)
- When should these problems be addressed in order to be most effective from an environmental, social, and economic perspective? (see Bridge, Chapter 9; Stewart and Petrie, Chapter 8)
- Who should be involved in the process of planning for closure? (see Warhurst et al., Chapter 5)

These issues generate the following components of planning for closure, each of which are discussed in turn below:

- An extended environmental impact analysis (EIA) incorporating a social impact analysis (SIA) to identify continuing and new impacts at the time of closure.
- A monitoring plan for predicted and unpredicted impacts.
- An environment management system in place to ensure the identification and mitigation of impacts.
- A progressive land rehabilitation plan.
- A consultative procedure in place that would allow interaction among the stakeholders.

2.3.2   EXTENDED EIAs

An EIA is a process by which the likely impacts of a project on the biogeophysical environment and the local community's health and welfare are identified. This information is conveyed to the decision-makers at a stage when it can materially affect the decision to undertake the project. At present, EIAs tend to be limited to a prediction of impacts at the development and the operational phase of a mining project, but do not extend beyond the proposed life of the mine (see Sassoon, Chapter 6). Planning for closure would require that the time frame for an EIA be extended to the time of closure and the period after closure when impacts are expected.

A careful EIA would be able to identify those impacts that could continue beyond the life of the mine, for example:

- Soil erosion from exposed surfaces.
- Subsidence from underground mining.
- Deterioration in surface and ground water quality because of acid rock drainage, heavy metal contamination, erosion, and sedimentation.
- Solid waste and tailings-related impacts.
- Site safety issues.

It would also be able to identify new impacts: physical and socioeconomic impacts that may arise upon mine closure, for example: unemployment amongst the local community members who depend on the mining activity directly or indirectly for their livelihood; disruption of the supply and services industry whose main market was the mining operation; and the discontinuation of the social services established and supported by the mining company, e.g., hospitals, schools, and sports facilities (see Warhurst et al., Chapter 5).

2.3.3   A MONITORING PLAN FOR PREDICTED AND UNPREDICTED IMPACTS

Monitoring of the impacts predicted in an EIA, and attentiveness to impacts that were not predicted by an EIA but emerge during the life cycle of a project are crucial to the success of a plan for closure (see Sassoon, Chapter 6). Since the main objective of planning for closure is to reduce the occurrence of damage/waste generation and the time lapse between damage occurrence and remediation, a continuous monitoring of impacts (predicted and unpredicted) is a necessity both for action and information feedback to the management system in place. This feedback will enable the mining company to adjust and modify its plan to address the new problem, as an iterative process of learning and adjustment. One implication of this approach is that appropriately skilled personnel must be in position to undertake the monitoring and manage the feedback process.

2.3.4   ENVIRONMENTAL MANAGEMENT SYSTEMS

Environmental management systems in place allow a company to identify opportunities for more effective use of resources and to address constraints and risks through continuously addressing the problem (see Sassoon, Chapter 6). An environmental management system in place would mean that action plans drawn up to mitigate or avoid the impacts predicted by an environmental impact assessment are implemented by the site staff, and there is continuous communication between the site staff and the management to enable the latter to track actions. It would involve listing of practices and operational procedures and the establishment of organisational structures that would be able to identify any ill effect at each stage of its occurrence. One important environmental management tool that can be used by industry is life cycle analysis, which in the context of a multi-objective optimisation regime can be used to link environmental and economic performance (see Stewart and Petrie, Chapter 8).

2.3.5   PROGRESSIVE LAND REHABILITATION AND WASTE MINIMISATION

As discussed earlier, the longer waste deposits and degraded land are left unattended the greater will be the resources and the attention required to deal with the problem. It is more cost-effective to have a progressive rehabilitation plan for the mine site in order to reduce the costs of final remediation, both physical and financial. A progressive rehabilitation can involve a land-use plan for the site, before mining operations start, which either returns the land to its pre-mine use or to any other purpose that the local community considers beneficial (see Morrey, Chapter 13; Noronha, Chapter 28). However, the net result is a land use that is more beneficial to the local community than the pre-mining land use. This is necessary not only to ensure that the local community is not impoverished because of the loss of the mining activity and the services that it supported, but also to compensate the community for the loss of the resources that in a way was part of its heritage.

2.3.6   CONSULTATIVE PROCEDURE

The main stakeholders in mining operations are the mining firm, the government, the creditors and insurers, and the local community. To ensure that mining operations evolve in a manner that is of benefit to all, it is necessary to involve all stakeholders in the planning process so that views, interests and constraints encountered by each agent, can be considered jointly and through a process of consultation (see Noronha, Chapter 28; Warhurst et al., Chapter 5).

There is evidence to suggest that there are benefits to all if the local community is involved with the rehabilitation planning and process from the start. If the local community is involved in the scoping process during the EIA phase of the project, whereby their input is sought to identify impacts and compensate appropriately, and to develop rehabilitation plans, then the company ensures a reduction in tensions and conflicts from the outset. The company also ensures that the plan that they arrive at for the post-mining land is one that has the approval of the local community.

Research suggests that the main advantages of involving local communities are (see Warhurst et al., Chapter 5)

- The valuable input about impact details and possible remedial actions or avoidance measures that the local population may be able to suggest and implement. Moreover, local communities are a possible source of ideas and innovative ways to handle a problem that may arise as a result of the disturbance caused to the ecosystem by mining, and to participate in the monitoring process.

- The reduction in costs to the mining companies due to local participation in mitigating and monitoring of environmental impacts.
- The avoidance or minimisation of local protest leading to delays and temporary closures after work on the project has begun.

The integration of all mining staff from the outset—not just the management—into the process, through ongoing environmental awareness programmes and training, ensures continuous attention to problem identification, site-specific solutions, and fine tuning of the closure plan. After all, the mining staff encounters the problems and, in a way, is best able to identify potential solutions. The Indian country case study for studying ecological management in mining revealed that better practices were observed in those cases where the mining staff had been involved in environmental programmes (see Noronha, Chapter 28). In companies where there had been a conscious attempt to inculcate environment awareness in the staff and provide them with training, a culture has been oriented towards better ecological management of operations and often-spontaneous innovative solutions to environmental problems.[15]

2.4   OPERATIONALISING THE PLAN: CONSTRAINTS AND CHALLENGES

In operationalising the plan for closure a mining company, a number of constraints may be faced. Some may relate to the natural systems that it has to contend with, some to the environmental baseline information, and others to its own internal financial, technical, and managerial capacity. These constraints pose a number of challenges to both the mining company and the regulators, which need to be addressed. These challenges arise from the existence of a number of innovative financing arrangements that are available, or can be introduced or devised to suit the specific needs of the company and the regulators (see Danielson and Nixon, Chapter 17); the need to formulate policy mechanisms that create incentives for improved environmental performance; the challenge of identifying the training and capacity building requirements for the various parties to the plan; and the need to put in place policy incentives by which technological and organisational innovations are commercialised, disseminated, and adopted (see Isnor, Chapter 27).

2.4.1   UNCERTAINTY IN PREDICTION

The discussion in Jongh[16] (pp. 67–70), relating to how uncertainty can be incorporated at each stage of the predictive process in environmental impact assessments, is of relevance here.

- In the scoping exercise, a decision has to be made on how to describe an effect and on what to predict. Considerable uncertainty may arise about the required quantity of information and the criteria for judging the relative importance of impacts.
- In developing the baseline study, there is a need to decide on what data are to be collected relating to the activity and the environment. Uncertainty arises because it may be difficult to judge whether a given resource use pattern — mineral and environmental alike — is sustainable or not because the natural variation is already so great.

Hirsch[17] defines a baseline study as a "description of conditions existing at a point in time against which subsequent changes can be detected through monitoring." This links the baseline study to monitoring.

Lack of baseline information about relevant ecological parameters, or uncertainty about the baseline data, makes it difficult to profile those zones that may be vulnerable to the activity being cited. Therefore, impact prediction becomes difficult and most often not attempted. When no impacts are predicted there is naturally no meaningful plan to mitigate the adverse effects.

- A method to obtain the prediction must be selected and developed. Uncertainties arise because the models developed for prediction do not exactly reproduce reality.

Uncertainties create a tendency to be overcautious and forego development opportunities that may have impacts that are actually reversible and may create an over-investment in environmental protection measures unwarranted by the realised impacts.

The best way to reduce uncertainty is to invest in research directed at the problem and to establish systems by which there is a continuous information flow between researchers and the users of the research.

2.4.2   Lack of Institutions and Capability for the Generation and Analysis of Scientific Information

An extended EIA that tries to incorporate environmental and socioeconomic concerns beyond the operational life of the mine will need considerable technical input from the scientific community. While far more facilities and skilled personnel are available in developing countries than there were some years ago, the infrastructure still remains weak in many cases and incapable of making full and effective use of the research capabilities available, or even looking elsewhere for assistance in strengthening this capability (see Dhar, Chapter 16). In some cases, good research institutions are caught up in bureaucratic red tape or inter-institutional rivalry, making it difficult to have access to the scientific information necessary to carry out the decision analysis. In part, this is due to a number of these institutions being state funded and monopolistic and not having to face any competitive pressure to deliver. Also, the problem arises due to a lack of vision on the part of the institutions themselves that have not, as yet, been able to appreciate the value of accessing cross-institutional capabilities.

2.4.3   Lack of Technical and Managerial Resources Required for the Implementation of Closure Plans

Effective transfer and adoption of best practice for environmental management requires the acquisition of capacity to effect and sustain the managerial process that promotes high environmental quality through the project life cycle (see Bridge, Chapter 9). For small mining companies especially, all of the available resources may be tied up in carrying out the routine operations. There may be no excess resources available to take up issues of impact monitoring, progressive rehabilitation, and continuous improvement of the environmental management system in place (see Hollaway, Chapter 19).

2.4.4   Costs

In planning for closure, a company needs to have an estimate of the costs that would be involved. In general, closure plans would involve objectives such as protection of water quality to desired levels, reduction of surface erosion, removal of equipment, and other mining-related structures that could pose threats to the safety of the local community and its habitation of the land, in terms of quality and economic uses. For this, a company would need to have some estimate of the costs involved. However, given the long time frame of a closure plan, the company would need to allow for uncertainty in its impact predictions and arrive at cost estimates based on scenario development: base, high, and low scenarios. Along with this, the company would need to allow for periodic reviews of its cost estimates if circumstances change. Such circumstances include:

- Successful exploration which expands the life of the mine.
- Significant changes in product price that affect profitability and decisions about the timing and rate of resource extraction.
- Introduction of cleaner technologies or management approaches, and innovations in modelling and prediction which indicate potential environmental problems different from the initial plan (see Mitchell, Chapters 7 and 14).

2.4.5 POLICY UNCERTAINTY

An uncertainty that may inhibit the successful implementation of closure plans relates to policy. This may arise because of new regulation requirements becoming more stringent, requiring changed approaches to closure, or because of the lack of clear terms for the release of the financial mechanism used by the company to cover closure costs. Hitherto, regulators have attached a low importance to closure issues relative to other aspects of mining activity. This is especially so in developing countries where a number of problems have a higher place than the environment in a decision-maker's agenda, simply because there is more information available about these other concerns or because decision-makers have a more immediate interest in addressing those concerns (see Dhar, Chapter 16; Acquah, Chapter 21). The weight of importance of foreign exchange to be earned from the export of minerals, the creation of jobs, and meeting domestic requirements of the mineral tends to make regulators less concerned with issues of closure as the time horizon extends beyond the immediate benefits of mining. This translates into less pressure for closure plans as of now, but as the environmental imperative grows, regulators may swing to the other end, requiring closure practices that could well push some companies out of business.

2.4.6 CHALLENGES

2.4.6.1 Policy instruments to deliver improved environmental performance

Unclear, inadequate legislative provisions that do not clearly specify tenure or ownership rights on operators tend to reduce the motivation for investing in environmental management practices. This was a finding in the Indian case study.[15] Whilst this is not making a case for vesting ownership rights in the mining company everywhere it mines, there may be a case for such legislative reform in regions which are particularly difficult to operate, are already degraded for other reasons, or bear the scars of earlier mining activity. Another policy challenge lies in determining long-term liability for environmental issues at closure of mine sites. Clark (Chapter 25) provides an interesting account of how liability in Australia varies from state to state. This case study illustrates the need for national guidelines for some aspects of closure, and site-specific criteria for others.

The focus of policy instruments to encourage the adoption of environmentally sound technologies at the level of the firm has to be on the depth and breadth of technical knowledge acquired (see Isnor, Chapter 27). Breadth refers to the skills and know-how for operating and maintenance; depth refers to the knowledge, expertise, and experience to generate and manage technical change. There is a need:

- To ensure that companies employ qualified and experienced engineers and technologists who are specialists in specific technical fields.
- For close and extended relationships between supply firms and importers, to allow for the transfer of experience-related knowledge.
- For the exposure of mining company staff to technical advances, through sponsorships of technology conferences, technology fairs, workshops, and so on.
- For greater interaction between firms and research and development (R&D) institutions, both local and foreign.

Policy instruments should also focus on rewarding creativity and innovative practices. In general, schemes that are devised should focus on promoting and rewarding innovation, rather than on standards of compliance to rules and regulations. Apart from stimulating technical change, this approach helps reduce the tendency of firms to divert resources from conventional business R&D towards compliance-related R&D. The importance of this lies in the fact that conventional R&D can result in innovations which go beyond regulatory requirements, whereas the compliance-related R&D used the regulation as a benchmark against which to measure its R&D performance. Expenditure programmes for R&D in selected areas of pollution prevention technologies would also need to be supported.

2.4.6.2   Commercialisation and dissemination of technological and organisational innovations

In the course of operationalising the closure plan, the mining company may come up with organisation or technological innovations that could prove of immense value to others in the same line of business. There has to be some way by which these innovations can be commercialised and transferred to other interested parties. For a regulator, there is a need to create incentives where it makes economic sense for the innovating company to commercialise and transfer.

In a developing country context, the most significant constraint is the absence of linkages to bring together the producers of innovative ideas for protecting and rehabilitating affected resources, often local people and users of innovative ideas. Encouraging contractual arrangements between companies and local communities, for tapping local knowledge about resource uses, is one way of commercialising local solutions to the problem and providing a way of rewarding the local innovator. Joint venture arrangements between mining innovators in other countries and local mining companies in the developing country provide another mechanism for dissemination of innovations.

2.4.6.3   Training, information, and environmental awareness requirements

Training and information increases environmental awareness requirements. Effective management would require this to be at the level of all stakeholders.

For the regulator, training is required:

- To enable them to evaluate EIAs effectively.
- To be able to monitor EMPs and closure plans.
- To be able to create interfaces between mining companies, local communities, and service companies; between mining companies and R&D institutions; and between environmental innovators and users of such innovations.

For the mining companies, training and information is necessary:

- On technologies and approaches available to deal with environmental problems and rehabilitation.
- To evaluate environmentally sound technologies in mining for their cost effectiveness and applicability to the region.
- On methods to deal with acid mine drainage.
- On the importance of scientifically sound mining.

For local communities, there is a need:

- To raise awareness about the environmental problems that mining can create.
- For realisation that local initiatives may be an important source of sustainable solutions to some of these problems.
- For the realisation of the benefits, monetary and non-monetary, of participatory rehabilitation.

2.5 CONCLUSION

Integrated environmental management and the parallel concept of planning for closure aims to promote competitive and sustainable development in industrial production. This chapter has argued that to be successful, this approach will require that regulatory objectives are underpinned by technology policy mechanisms designed both to stimulate technological innovation and best practice in environmental management within firms, and to encourage the commercialisation and diffusion of these innovations across the boundaries of firms and nations. Previous MERN research and the case studies reported here indicate that the firms that pollute most mismanage the environment precisely because of their inability to innovate. Environmental degradation post-closure is greatest in operations with low levels of productivity, obsolete technology, limited capital, and poor human resource management. Secondly, in the context of poorly formulated policy regimes, the most efficient firms are generally better environmental managers because they are innovators. They are able to harness both technological and organisational change to plan for closure from the outset, and reduce the production and environmental costs of their operations. Furthermore, where environmental compliance represents a cost that could reduce competitiveness, the dynamic firm is able to offset abatement costs with improved production efficiency. The requirement that these firms reduce pollution at source — which necessarily involves planning, as well as changing the technology and organisation of their production processes — assumes that they would not already be searching for new ways to improve metal recovery, reagent use, energy efficiency, water conservation, and remediation as part of a corporate strategy to increase competitiveness.

This chapter also makes a case for the training of regulators, since technological advancement is an important indicator of the effectiveness of environmental regulation and the subsequent "ratcheting" of regulation would further enhance the competitive advantage of firms. Regulators, as well as corporate analysts, might also be assisted in their strategies to achieve competitive environmental best practice by the definition of "corporate environmental trajectories" in different economic and regulation contexts. This would evaluate the evolutionary development of a firm's competitiveness in response to changing market conditions and regulatory requirements. Broadening the range of regulatory mechanisms, and the technology policy mechanisms and economic instruments which would need to be in place to support them, as proposed here, would in totality comprise a more integrated and effective policy approach towards both regulating and promoting closure planning.

An increasingly relevant and integrating concept to consider the many facets of closure planning is the EMS. Indeed, a best-practice EMS for a mine site ought to incorporate the capacity of the operation to deliver optimal environmental performance within a given geographical, geological, and socioeconomic context.

The key components of a functional EMS for a mining operation are summarised in Figure 2.5. The items are not a list, but rather form key components of a system that requires investments in capital, technology, and human resource development in order that the system delivers the required environmental performance. This chapter argues that, increasingly, an important measure of environmental performance will be the extent to which an operation put in place a plan for closure from the outset. That is, a plan — or an EMS — aimed at and capable of managing in a forward-looking and integrated way the environmental impact of mining activities on land, air, water, and local communities throughout the life cycle of its operations.

REFERENCES

1. Warhurst, A. and Bridge, G., Economic liberalisation, innovation and technology transfer: opportunities for cleaner production in the minerals industry, *Natural Resources Forum,* 21,1, 1997.
2. Anonymous, Argentina: exploring the final frontier, *Mining Journal,* 327 (26 April), 1996.
3. Suttill, K., Cuba turns to the pragmatists, *Environment and Management Journal,* May 29, 1994.
4. Andrews, C. B., Mineral sector technologies: policy implication for developing countries, *Natural Resources Forum,* August, 16, 212, 1992.

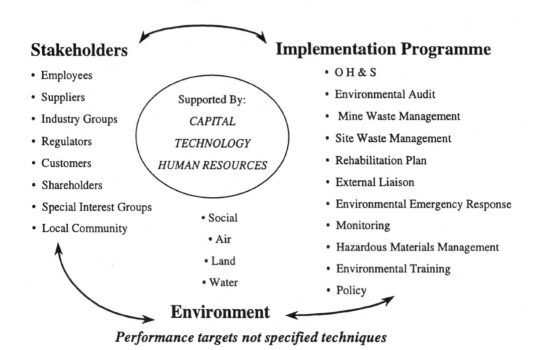

FIGURE 2.5   Environmental Management System: key components for a mining operation.

5. Anonymous, Argentina/Chile mining treaty, *Mining Journal*, 327 (3 May), 1996.
6. Wasserflauf, M. and Condy, A., Techno-economic Evaluations of Thiosalt Treatment Processes, Proc. CIM 15th Annual Hydrometallurgical Meeting, Vancouver, Canada, no date, 31.
7. Gooding, K., Rich seam for foreign investors, *Financial Times*, 12th October, 1995.
8. Prast, B. and Thomas, A., Mineral investment trends and issues: specific regions, in *Proceedings International Mining Investment and Regulation Conference*, Centre for Petroleum and Mineral Law and Policy, Dundee, 1995.
9. Anonymous, Latin America, still emerging, *Mining Journal*, 327 (19 April), 1996.
10. Anonymous, Investing in the Americas, *Mining Journal*, 326 (24 March), 1995.
11. Warhurst, A., Environmental management in mining and mineral processing in developing countries, *Natural Resources Forum*, 16, 39, 1992.
12. Lagos, G., Mining and the Environment in Chile, Mining and Environment Research Network Working Paper Series, Number 23, University of Bath, U.K., March, 1992.
13. Loayza, F., Environmental management of mining companies in Bolivia: implications for environmental and industrial policies aiming at sustainable growth in low-income developing countries, unpublished report, Science Policy Research Unit, University of Sussex, U.K., 1993.
14. Acero, L., The case of bauxite, alumina and aluminium industry in Brazil, Mining and Environment Research Network Working Paper Series, Number 1, University of Bath, U.K., February, 1992.
15. Noronha, L., Ecological management in mining: the Indian case study, Mining and Environment Research Network unpublished research report submitted to the U.K.'s Overseas Development Administration (O.D.A., now Department for International Development), 1996.
16. Jongh, P. D., Uncertainty in EIA, in *Environmental Impact Assessment: Theory and Practice*, Wathern, P., Ed., Routledge, London and New York, 1992.
17. Hirsch, A., The baseline study as a tool in environmental impact assessment, Council Of Environmental Quality: Proceedings of the Symposium on Biological Evaluation of Environmental Impacts [Fws/Obs - 80/26, P87-93], Department of the Interior, Washington D.C., 1980.
18. Warhurst, A. and Franklin, K.C., Biodiversity conservation, minerals extraction and development: a realistic partnership, in *Strange Bedfellows: Extractive Industries, Infrastructure, and Biodiversity Conservation*, Prickett, G.T. and Bowles, I.A, Eds., Conservation International: Oxford University Press, 1999.

3 Mining, Mineral Processing, and Extractive Metallurgy: An Overview of the Technologies and Their Impact on the Physical Environment

Alyson Warhurst

CONTENTS

1-56670-365-4/00/$0.00+$.50
© 2000 by CRC Press LLC

3.1 INTRODUCTION

The availability of rich geological resources, in tandem with market conditions largely outside the control of metal-producing enterprises, have resulted in a sector characterised by a low level of technological innovation. Corporate strategies have traditionally focused on the discovery and acquisition of high-grade and easily accessible mineral deposits, and on increased scale of production to offset declining ore grades. In the past, health and safety regulation has been a major driver behind technological change, alongside the need to improve process energy efficiency. However, the emphasis has now altered: after a period of limited technological change, a spur to technology development in the minerals industry has been applied by public concern over adverse environmental effects and the design of environmental regulation that obliges firms to mitigate or prevent such effects.

Although permanent, temporary, or transient environmental impacts are common, their association with mining and milling operations is not inevitable. For example, of 33 mining and milling operations surveyed in 1989–90 by Environment Canada, approximately half had no adverse environmental impacts, while a further 15% had only minor effects.[1] However, as is the case for many heavy, process-orientated sectors, mining has had limited success in altering the widely held perception that it must be an intrinsically "dirty" and polluting activity. Despite proactive initiatives in the fields of waste management, pollution prevention (see Mitchell, Chapter 14) and multi-party dialogue with stakeholders (see Warhurst et al., Chapter 5), societal perception of the general industry continues to revolve, to a large degree, around consideration of "sins of the past" rather than current and state-of-the-art operations.

Dynamic companies have responded to societal perceptions and other drivers by improving environmental performance as part of the renewed drive for competitive edge. In many cases, environmental performance and standards continue to improve. Companies are now beginning to operate within the regulatory standards and guidelines of their home countries at overseas sites where less stringent legislation or enforcement would otherwise apply. This shows that a commitment to minimising environmental impacts is growing and that competitiveness and enhanced environmental performance are not mutually exclusive. Similarly, it is becoming increasingly clear that technological incompetence is not compatible with sustainable development. Consequently, mining, mineral processing, and extractive metallurgical technologies all play a significant role in determining the environmental performance of an operation and its contribution to regional and national development in a sustainable manner. However, limitations arising from the choice of technology can potentially undermine even the best efforts in environmental and human resource management and stakeholder consultation.

This chapter, which should be read in conjunction with Chapters 9, 10, and 14 to explore recent innovations in mineral processing and extractive metallurgy in greater detail, examines the environmental consequences of mining (ore extraction), mineral processing (physical separation of minerals), and extractive metallurgical technologies (recovery of metal products), particularly in the context of base and precious metal operations, where some of the most complex and intransigent impacts occur. Although health and safety and energy efficiency play a significant part in determining overall environmental performance, it is the interaction between *technology* and the *external physical environment* that is the focus here.

3.2   MINING: A WASTE DISPOSAL INDUSTRY?

Metals and other mineral resources are rarely found in a sufficiently pure state to be sold in an "as-mined" form. Metals are often found in chemical combination with oxygen (as oxides), sulphur (as sulphides), or other elements (e.g., chlorides, carbonates, arsenates, phosphates, etc). Non-metal mineral resources (e.g., coal, industrial minerals) also normally contain physically or chemically entrained impurities in their undisturbed state.

The major source of solid waste from mining and subsequent processing is *gangue* (valueless or subeconomic minerals associated with the target or economic mineral(s)). Depending on the point at which they are rejected from the process, gangue may be disposed of in an as-mined state (e.g., waste rock), as tailings (e.g., following mineral processing), as slags (e.g., after smelting), or as other waste products (e.g., dusts, sludges from water treatment, spent ore from leaching, etc). These various wastes may also contain significant quantities of the target mineral or metal due to inefficient processing, technological limitations, or mineralogical factors.

In non-ferrous metal mining, gangue is normally the major component of an orebody. Nowhere is this more apparent that in the case of gold, where the concentration of valuable material is so low (e.g., normally 5 g t^{-1} or less) that, in effect, all of the mined ore is disposed of as waste (unless other valuable components such as base metals are also present)[1]. Other mineral resources may have less gangue relative to the target mineral, but disposal of gangue-related wastes normally remains a significant issue. "Average" figures (based upon a survey of Canadian metal mines) indicate that 42% of the total mined material is rejected as waste rock; a further 52% from the mill as tailings; and an additional 4% from the smelter as slag, leaving a valuable component of just 2% of the originally mined tonnage.[2,3] In effect, such sites are as much about waste disposal as they are about resource extraction (see also Mitchell, Chapter 14).

Table 3.1 shows the ore and waste production figures (not including overburden) for a range of mineral resources exploited in the U.S. during 1991.

TABLE 3.1

Estimated Ore Production, Average Grade and Waste Generation (1991)[4,5]

Mineral	Ore (million tons)	Average Grade (%)	Waste (million tons)
Copper	1000	0.91	990
Gold	620	0.00033	620
Iron	906	40.00	540
Phosphate	160	9.30	140
Potash	160	17.00	130
Lead	135	2.50	130
Aluminium/bauxite	109	23.00	84
Nickel	38	2.50	37
Tin	21	1.00	21
Manganese	22	30.00	16
Tungsten	15	0.25	15
Chromium/chromite	13	30.00	9
Total	**3200**	**—**	**2700**

TABLE 3.2

Anthropogenic Inputs to Aquatic Ecosystems from Base Metal Mining and Dressing (million-kg year^{-1})

As	Cd	Cu	Mn	Ni	Pb	Zn
0–0.75	0–0.3	0.1–9	0.8–12	0.01–0.5	0.25–2.5	0.02–6

TABLE 3.3

Metal Content of Mine Tailings (million-kg year^{-1})

As	Cd	Cu	Ni	Pb	Zn
7.2–11	2.7–4.1	262–787	22–64	130–390	194–620

More recently, Thornton[6] adapted figures from Nriagu and Pacyna[7] to estimate the anthropogenic input of a range of metals to aquatic ecosystems from base metal mining and processing (Table 3.2) and disposal of mine tailings on land (Table 3.3).

It is apparent that the process of waste disposal related to mining activity is a significant source of potentially harmful elements in the natural environment. However, inputs do not necessarily result in damage to the environment: many mitigating factors may exist, some of which relate to the process (e.g., the chemical and physical characteristics of the wastes) and others which relate to the external environment (e.g., climate, topography, and ecosystem characteristics). It is therefore essential to understand the distinction between a *release* and an *impact*.

3.3 THE ENVIRONMENTAL IMPLICATIONS OF MINING-RELATED ACTIVITY: AN OVERVIEW

3.3.1 ENVIRONMENTAL RELEASES, EFFECTS, AND IMPACTS

The terms *environmental releases* (also known as *process releases*), *environmental effects*, *environmental impacts* (or *environmental damage*) are often used interchangeably. However, this abuse of terminology promotes the erroneous belief that a release will inevitable cause an impact, and that cause-and-effect factors can always be clearly identified. The concepts should be seen instead as a logical sequence, which begins with the environmental release, and *may* end with environmental damage, depending on other mitigating factors such as dilution, biodegradation, or attenuating mechanisms.

Environmental (or *process*) *release* can be defined as a transfer of material or energy to the external environment (i.e., everything outside the process). However, the absence of an environmental release should not be taken to imply that a process has no detrimental impact. Occupational health can be severely impacted by releases within the process boundary (e.g., fibrotic pneumoconiosis caused by exposure to biologically-active silica; carcinogenic and toxic effects of chronic exposure to dusts containing mercury, copper, zinc, lead, and manganese).[8,9]

Examples of process releases include:

- The transfer of a solid waste or liquid effluent outside of the mill building.
- The discharge of gaseous emissions from a stack.
- Transfer of a metal from the solid mineral phase to water in underground and open pit workings.
- Dust from an open pit.

In most cases, potential *environmental effects* are linked to the material being transferred. However, some may be related to the consequences of making that transfer (e.g., subsidence in underground mining or slope stability in open pits relate to the *removal* of ore and overburden).

Environmental effects can be defined as a measurable change in the external physical environment (i.e., a measurable disturbance in the existing system) that results from an environmental release. Therefore, certain environmental releases may cause no measurable effect due to mitigating processes in the external environment. However, there is also a strong argument for employing the precautionary principle: the absence of effect should not be seen as a *carte blanche* for continuing releases. The recent controversy surrounding oestrogen-mimicking chemicals in water can be used as a case in point. Others argue that as it is not possible to prove the absence of an effect, there are practical limits to the precautionary principle. Ultimately, societal concerns will dictate priorities and the level of acceptable (and perceived) risk, and industry and regulators will be required to respond accordingly. Effects, while measurable, may have no discernible consequence for ecosystem and/or human health.

Environmental impact (or *damage*) implies that the effect of an environmental release is such that mitigating processes are not sufficient to protect the external environment. In these cases, either the external environment cannot return to its previous state without human intervention or the release has a direct and measurable consequence for ecosystem and/or human health.

To ensure a high standard of environmental performance, an understanding of potential releases, effects, and impacts is essential and should underpin and inform decisions throughout the life cycle of an operation (see Stewart and Petrie, Chapter 8).

3.3.2 POTENTIAL SOURCES OF ENVIRONMENTAL RELEASES, EFFECTS, AND IMPACTS

Any unit operation within the life cycle of a mining operation has the potential to produce an environmental effect or impact. Typically, the potential arises from the deliberate (regulated) and accidental (non-regulated) discharge of solid, liquid, and gaseous waste products. The characteristics of the discharges, the nature of the receiving environment and the distance over which the discharges are transported are major factors in determining the magnitude of the effect or impact. Societal values and preferences also play a significant role in determining how certain discharges are viewed by various stakeholder groups: this more subjective adjunct to the quantifiable and measurable discharge and receiving environment characteristics therefore sets, in part, the site-specific environmental "footprint" of an operation.

It should, however, be remembered that environmental releases could be localised and contained within the site boundary. Impacts may be *transient* (often acute, associated with spills or accidental discharges), *temporary* (related to operational discharges, ceasing when operations are terminated) and *chronic* (long-term, often those arising from ore extraction and waste disposal). Typical transfer mechanisms are wind and water erosion of contaminated solids or dissolution of contaminants into the aqueous phase. The movement of intermediate products off-site for further processing (e.g., the transport of concentrates to a smelter that may be hundreds or thousands of miles away) is another mechanism by which the spatial and temporal environmental impact of an operation can be extended. Therefore, any analysis of the environmental burden arising from a particular operation must address the entire life cycle, from run-of-mine ore through to the final product, including any and all off-site operations. This may lead to the choice of a technology and its site that would not otherwise have been considered appropriate or necessary.

3.4 MINING

3.4.1 INTRODUCTION

Mining is the first operation in the commercial exploitation of a mineral resource. It can be defined as the act of tunnelling and digging out of the ground in order to recover one or more component parts of the mined material.[10]

In broad terms, there are three types of mining: surface, underground, and *in situ* (solution mining). The latter is somewhat limited in its application, although it is sometimes used to exploit residual mineralisation as grades drop at surface or underground mines. Surface mining is dominated by open pit (e.g., base and precious metal ore extraction) or open cast (e.g., coal operations). Surface and underground mining usually occur independently of one another, although open pit mining does occasionally occur in areas already partly worked by underground methods. Similarly, underground methods are sometimes used to extract ore from beneath or in the vicinity of pits, where further extension of the pit itself is not economic or technically feasible.

Irrespective of the method employed, mining is always accompanied by processing of some description. For relatively pure or homogeneous materials, processing may be limited to crushing and sizing (e.g., some natural zeolite extraction, quarried rock) or washing (e.g., some coal operations). Simple processing such as this is only possible where the target mineral forms the majority of the material mined. In these cases, the main environmental releases, effects, and impacts are associated primarily with the mining itself rather than subsequent processing.

Factors that influence the choice of mining method include the size, shape, dip, continuity, depth, and grade of the orebody; topography; tonnage; ore reserves; and geographic location.

3.4.2 SURFACE MINING

Surface mining can be divided into three subgroups:

- Open pit or bench mining — deposits that are deep but have restricted width.
- Open cast or strip mining of relatively shallow deposits.
- Alluvial mining of unconsolidated near-surface deposits on land or beneath water.

Compared to underground mining, surface mining methods are generally considered to allow for a higher degree of worker safety, greater flexibility in extraction (the capacity to practice selective mining and grade control), and lower development and maintenance costs (due to the requirement for fewer specialised systems). However, the major benefit lies in economy of scale, as large capacity earth-moving equipment can be used to generate high productivity.

3.4.2.1 Open pit mining

Open pit mining is normally used for steeply dipping beds or veins, or for massive irregular deposits. In open pit mining, overburden (non-mineralised soil/rock that covers an orebody) and waste rock (poorly mineralised or very low-grade soil and rock that are within the orebody or surrounding it) are first removed from the entire area of the final planned pit, often in very large quantities. Some crushing of overburden or waste rock may be required for efficient handling and/or use in bunding. Data for the U.S. in 1989 showed that surface mines produced eight times as much waste per ton of ore as underground mines. Sub-economic ore may also have to be removed to expose the economic reserve. The ore itself is then removed for further processing.

The mine shape is formed by a series of benches or terraces arranged in a spiral or in levels with interconnecting ramps. Open pit mines may reach several thousand feet below the surface. The pit is deepened in a sequential manner using benches that also serve as haulage roads for the removal of ore from the pit. Restoration at the end of the operation can be very expensive particularly if it involves backfilling, as the wastes are often dumped at some distance to avoid obstructing the removal of economic ore.

Overburden and waste rock are often used during the operation and closure of a mine (e.g., inert waste rock for bunding, soils for reclamation at closure). However, if they are contaminated with significant concentrations of potentially harmful minerals (albeit at sub-economic concentrations), they must be disposed of as wastes.

3.4.2.2 Open cast mining

In open cast mining, extraction proceeds laterally rather than vertically and this method is therefore most suitable for shallow deposits. A "strip" of overburden is first removed to expose the underlying deposit. The deposit is then worked out, following which another strip is prepared, with the overburden being placed preceding (worked out) strip. Thereafter, restoration is done in a rolling fashion, with only two "active" strips at any one time (one being stripped, one being backfilled).

3.4.2.3 Alluvial mining

Alluvial mining is the cheapest method, and when applied on a large-scale can be used to mine ores of very low contained value (i.e., low grade, low metal value, or both). For example, in Southeast Asia, tin ores containing as little as 0.01% tin are mined by alluvial methods although the ores have a contained value of less than £1 per tonne.[11] Alluvial mining does not involve drilling, stripping or blasting, but rather the recovery of the deposit through physical means such as scraping, high-pressure water jets, or suction (for underwater deposits).

As alluvial deposits often contain liberated or partly liberated target mineral(s) (i.e., they occur in particles physically separated from the gangue minerals), crushing and/or grinding is often unnecessary, dramatically reducing the cost of processing (see Mineral Processing, Section 3.5).

3.4.3 UNDERGROUND MINING

Underground mining methods are usually employed to mine richer, deeper, and smaller ore bodies where open pit methods would be impractical. Underground mining operations are a complex combination of tunnelling, rock support, ventilation, electrical systems, water control, and hoists for the transportation of people, ore, and materials. Consequently, underground mining is less flexible than surface working, and deviations in the production rate are more difficult to accommodate. Because mining may be proceeding in several different underground locations to ensure an appropriate level of production, forward planning and control of mine development are extremely important.

Mines in shallow dipping ore deposits take the form of flat or shallow dipping parallel-sided cavities. In *room and pillar* mining, pillars of unworked ore are used to support the roof. The size of the pillars is dependent on the strength of the ore being exploited. Areas that have been worked out can be backfilled with waste rock, or tailings, often with the addition of cement to strengthen the backfill pumped as slurry from the surface. Temporary supports may also be used to allow the pillars to removed ("pillar robbing"): once work in an area is completed, the temporary supports are removed and the roof is allowed to collapse.

For "weak" ores, other forms of support (in addition to, or as alternatives to the pillars) may be required to ensure a safe working environment. This supplementary support is then removed as the working face proceeds, allowing the roof to collapse in the worked out area. This is known as *longwall* mining.

Ore deposits that dip steeply or which extend over a great vertical distance are mined in such a way as to make use of gravity to transport ore to one or more central haulage points within the mine and in some cases to break the ore under its own weight. A mining method called *stoping* is used to extract the ore in sequential blocks. The cavity formed is called a stope, with subsequent ore being blasted from the roof to the stope floor and transported to an ore haulage area via ore passes (vertical or near-vertical channels).

3.4.4   IN SITU

Also sometimes known as solution mining, this method involves the synthesis of expertise from a number of disciplines including hydrometallurgy, hydrology, geology, geochemistry, rock mechanics, chemistry, and environmental management.[12] *In situ* mining is accomplished by the injection of a suitable leaching agent into a porous and permeable mineral deposit. During its passage, the leaching agent (normally water-based) dissolves the metal(s) of interest. The metal-laden leachate is then pumped to the surface, where it is purified and further processed to recover a saleable product. At present, its use is relatively limited, particularly outside the recovery of uranium from permeable sandstone deposits. It has, however, been used to recover copper from oxide mineral remnants in underground and open pit workings and from caved-in stopes. *In situ* leaching has certain advantages over conventional mining and milling, including lower capital investment, lower operating costs, and faster start-up times.

3.4.5   POTENTIAL RELEASES, EFFECT, AND IMPACTS

There are two major releases from surface and underground mining: first, those associated with mineral wastes (i.e., overburden and waste rock) (examined above in some detail — see Sections 3.2 and 3.3.1 and Tables 3.1 to 3.3); and second, the generation of contaminated waters. Other releases may be locally significant depending on the characteristics of the site and external environment: for example, noise, vibration and dust (from blasting and road haulage), surface subsidence, diesel-related fumes (from mechanical plant), water contamination by nitrates (arising from ammonium nitrate used in blasting), and visual disturbance from operational activities, waste disposal, and ore extraction.

Mine water is generated when water collects in mine workings as a result of inflow from rain or surface water and from groundwater seepage. During the active life of the mine, water is pumped out to keep the mine relatively dry and to allow access to the orebody for extraction, while surface water is controlled using engineering techniques to prevent water from flowing into the mine. Pumped water may be used in extraction and beneficiation activities (including dust control), pumped to tailings impoundments, or discharged as a waste. The quantity of mine water generated varies from site to site, and its chemistry is dependent on the geochemistry of the orebody and the surrounding area. Water exposed to sulphur-bearing minerals in an oxidizing environment, such as an open pit or underground workings, may become acidified and contaminated with metals (e.g., acid rock drainage). Sulphide-rich waste rock piles that are permeable to both air and water may also generate acid drainage.

Acid rock drainage is widely considered the most serious environmental problem caused by the mining of sulphide ore deposits. It results from the reaction of iron sulphides with oxygen and water (see Mitchell, Chapter 7). The sulphides oxidise to generate acidity and other chemical species that are capable of oxidising other metal sulphides into water-soluble sulphate form. The resulting water can be both acidic and highly laden with iron and other metals. The oxidation of the iron sulphides is catalysed by naturally occurring bacteria which thrive in the acidic conditions. When mining ceases and the pumps are removed, water levels can rebound to natural levels, often producing a large volume of contaminated water, which is continually replenished as fresh (oxygenated) water enters the workings (and contaminated water leaves).

If acid drainage is left untreated, it can contaminate groundwater and local watercourses, restricting water use and damaging ecosystem and human health. The drainage can be treated, but this is expensive and preventative approaches are more cost effective (for more information, see Mitchell, Chapter 7).

The main release from *in situ* mining is contaminated water, which may contain a wide range of dissolved species from the ore (similar to acidic drainage) *and* chemical species added to the leaching solution. Although there are other technical limitations to the use of *in situ* mining, the major unresolved issue is the question of losses of metal-laden leachate and subsequent contamination of groundwater resources. Experience with heap and dump leaching (surface-based variants of *in situ* mining) indicates that there are good reasons to be cautious about extending the use of *in situ* mining until this issue is fully resolved (see Bridge, Chapter 9).

3.4.6 INNOVATION IN MINING

The major thrust of innovation in mining has been the development of open pit techniques and exploitation of the economies of scale created by developing larger mining equipment and larger-scale extraction techniques. For example, major developments have included: improved ventilation equipment and improved visibility; larger and more efficient hoisting machinery, compressors, pumps, etc.; improved mine development equipment and procedures for mechanised shaft sinking, shaft freezing, raising, and drifting; larger and heavier equipment with better drill bit and blasthole drilling techniques; trackless mining equipment, initially to mechanise ore loading and transport; improved blasting agents and techniques for greater efficiency and safety; electric mining shovels and draglines with significantly greater bucket capacities; very large (200 ton) diesel electric haul trucks and front loading equipment; portable and mobile in-pit ore crushing and overland conveyor systems, allowing more continuous mining operations and the elimination of truck haulage where hauls are long and adverse; self-propelled scalpers and scrapers to integrate extraction and transport operations; on-line/automated production monitoring and control to optimise extraction and truck and shovel capacity utilisation through scheduling/dispatching and route selection; and computerised and remote control systems for underground train haulage and mine pumps. Some of these, of course, also apply to underground mining.

In underground mining, advances have also been made in the use of "right-in-space" mining methods,[13] which seek to ensure that drill holes, stopes, and other underground workings are placed more accurately in relation to the orebody during both development and production. This allows high ore recovery and reduces dilution by unwanted gangue, reducing waste that requires disposal and management.

3.5 MINERAL PROCESSING

3.5.1 INTRODUCTION

Mineral processing is defined as the physical processing of minerals. It does not result in any chemical changes to the mineral components of the ore, but is a means of achieving the physical separation (and concentration) of different mineral phases (e.g., target minerals from gangue minerals or of one valuable mineral from another).[14] For non-metal mineral resources, mineral processing can produce a final product, but in the case of metals it is an intermediate stage as it does not affect the chemical combination of the metal with oxygen, sulphur, and so on. Mineral processing is normally an intermediate stage between mining and extractive metallurgy, although there are exceptions: for example, heap and dump leaching of "as-mined" ore. The outputs from mineral processing (concentrates) form the inputs to extractive metallurgy (either hydrometallurgical or pyrometallurgical processes; see Extractive Metallurgy, Section 3.6).

Mineral processing methods can be divided into two groups: size reduction and separation of mineral phases.

3.5.2 Size Reduction

Size reduction is undertaken using *crushers* (e.g., jaw, gyratory, and cone crushers) and *grinding mills* (rod, ball, hammer, and impact mills). Different types are employed depending on the planned throughput (tonnes per hour), ore characteristics (e.g., friable, "sticky," "elastic") and the degree of size reduction required. Crushing is used to reduce incoming ore from boulder-size down to 25mm or less. This material may then pass to grinding mills or may be processed directly if the valuable mineral is sufficiently liberated. Grinding reduces particle size, down to a lower limit of about 10µm, below which particles become increasingly difficult to handle and/or separate. In contrast to crushing, which is carried out on run-of-mine ore, grinding is almost always done wet and the output from grinding is slurry (a suspension of fine particles in water).

Crushing and grinding (in tandem with sizing) are used to liberate economic and non-economic minerals from one another and thereby produce suitable feeds for subsequent processes in which separate mineral phases can be generated. However, grinding does not normally produce a completely liberated mineral product and some particles may be mixture of two or more mineral species, with physical or chemical characteristics representative of that mixture. This can result in target minerals being disposed of with the gangue minerals or the dilution of the mineral concentrate by gangue minerals. Separation and regrinding of "mixed" particles to a finer size is often used to improve liberation, while avoiding overgrinding of particles that are already liberated. However, grinding (and crushing) are energy intensive and there is a trade-off between the cost of additional grinding and the value of the extra mineral likely to be recovered. Grinding accounts for nearly half of the total energy used in producing a mineral concentrate by mineral processing.[15] Energy consumption in mineral processing can therefore be reduced by optimising the process flowsheet to minimise regrinding. Of particular importance to optimisation is the capacity to run the plant at steady-state, irrespective of changes in feed mineralogy and grade.[15] Recycled streams (also known as *circulating loads*) at all stages of the mineral processing flowsheet plant are, however, used to smooth out changes in feed characteristics that occur periodically, and therefore the need to minimise recirculating material through the grinding mills must also be balanced against the stability of the overall circuit.

3.5.3 Separation of Mineral Phases

Mineral separation can be achieved by employing differences in mineral characteristics, based on:

- Particle size.
- Particle density.
- Magnetic properties.
- Electrical properties.
- Surface chemistry characteristics (flotation)..

The separation technique of choice is based upon a number of site-specific factors, including the size of the operation, particle size at which minerals are sufficiently liberated, mineralogy, the overall processing circuit (from mining to extractive metallurgy), and so on. Ultimately, each process produces a *concentrate* (rich in the target mineral or minerals), *tailings* (containing mainly sub-economic or non-economic minerals) and *middlings* (particles in which target and gangue minerals have not been fully liberated). Middlings are normally reground or sent to an alternative process to liberate or otherwise recover the valuable mineral, while tailings are disposed of as waste. Of the various separation processes, flotation is now the dominant method for the production of mineral concentrates, particularly from sulphide ores.

3.5.3.1   Particle size

Different minerals have a variable resistance to crushing, abrasion, and deformation during crushing and grinding, which can therefore result in "soft" minerals being reduced to smaller sizes than "harder" minerals. Consequently, by the use of sizing it is theoretically possible to begin the process of separation and concentration of the valuable mineral(s). However, sizing is more normally used to split streams with a broad size range into two or more products with a narrower size range. These more narrowly sized products are then fed to subsequent processes.

Standard separation by size involves sieves, screens, spiral/rake classifiers, and thickeners. Classifiers and thickeners work on the principle that in water a larger particle will drop faster; therefore, the sediment will contain coarse particles, while the water above it will contain smaller particles (still in suspension). Subsequent separation of the sediment and water (as a continuous process) therefore allows particles to be separated by size. However, differences in particle density (e.g., base metal minerals are typically denser than the major associated gangue minerals) results in small dense particles having similar settling characteristics to large light particles. Therefore, a certain amount of material is misplaced during the separation process, and this can be a source of inefficiency in subsequent processes and increased environmental releases.

Separation by size is increasingly conducted using *hydrocyclones*, which apply centrifugal force to accelerate the settling process. Hydrocyclones require less area than other size separation techniques; by placing them in parallel, very high volumes of slurry can be sized. However, as for the other size separation techniques, different particle densities result in misplaced material (i.e., the coarse fraction may contain small, but dense, particles).

3.5.3.2   Particle density

The accelerated settling of dense particles relative to lighter particles is used in the process of separation by density. Typical methods employed include *dense medium separation, jigs, shaking tables* and *spiral concentrators*. Jigs and shaking tables employ mechanical vibration to accelerate the separation process. As is the case for separation based on size, the similar behaviour of light large particles and small dense particles results in misplaced mineral. However, this can be minimised by using density separation on narrowly sized feeds.

3.5.3.3   Magnetic properties

Selection of different mineral phases can also be achieved by employing the differing magnetic properties of minerals. Broadly speaking, minerals fall into two groups: diamagnetic (those minerals repelled by magnetic fields) and paramagnetic (those minerals attracted by magnetic fields). Separation can be undertaken using high and low intensity magnetic fields, but is limited to ores in which there is a demonstrable difference in the magnetic properties of the target and gangue minerals (i.e., it is not applicable where all the minerals are diamagnetic).

3.5.3.4   Electrical properties

Utilising the forces acting on charged or polarised particles, conducting and non-conducting minerals can be separated. As in the case of magnetic separation, this approach is limited to a relatively small number of mineral species.

3.5.3.5   Flotation

Flotation involves the passing of air bubbles through dilute slurry of fine mineral particles held in large *cells*. Minerals that are hydrophobic ("water-hating") become preferentially attached to the air bubble and float to the surface of the slurry to form froth, where there can be removed. Hydrophilic ("water-loving") minerals remain in the bulk slurry. The key activity in flotation is the control, alteration, or accentuation of the hydrophobic or hydrophilic characteristics of target and gangue minerals. In broad terms, two options are possible with flotation: separation of valuable minerals from gangue minerals and the separation of one valuable mineral from another. During the flotation process, minerals may be altered from hydrophobic to hydrophilic by the use of chemicals termed *depressants*, which alter the surface chemistry of the mineral in question. Hydrophilic minerals may be made hydrophobic by the addition of other reagents known as *activators*. Other chemicals added during the processing include those that physically stabilise the froth (to ensure that the bubbles do not burst before the froth is removed); those that adjust the pH of the slurry (to aid in the control of selectivity for one mineral over another); and those that ensure that mineral particles do not aggregate (*dispersants*).

Flotation technologies have been used since the 1920s, but it was not until the early 1980s that automated control systems became available, which enabled a fine-tuning of the flotation process to optimise its performance.

3.5.4   POTENTIAL RELEASES, EFFECT, AND IMPACTS

Many of the unit operations within mineral processing are of limited interest in terms of their potential to impact the external environment as they operate within what are, effectively, closed circuits (i.e., many unit operations pass 100% of their input to the next process, and therefore they involve only very limited transfers of materials from the process to the external environment, such as dust, noise, and vibrational energy). However, this may not be true for small-scale mining operations or operations in a developing country context, where low capital investment means that relatively simple processing is used to "cherry pick" the most easily extracted and valuable component, leaving a high level of residual minerals in the associated wastes which do not undergo further processing. Neither is it true for flotation, which is the biggest single source of tailings in mineral processing.

Tailings generated during flotation are mainly composed of finely sized gangue minerals and varying amounts of the target mineral(s). The concentration of target minerals in the tailings is mainly related to process economics, process efficiency, and mineralogical constraints. The tailings are commonly discharged in the form of slurry to tailings impoundments, which may cover tens of hectares and present a serious problem in terms of dusts in drier climates. If the tailings are rich in sulphides (particularly pyrite), impoundments can also generate significant volumes of acidified water, which may require collection and treatment during the operation of the impoundment and after its closure. Other associated effects include those arising from the chronic discharge of suspended solids, the leaching of associated flotation chemicals, and failure of impoundment walls. The release of cyanide is also a major issue, but this relates to tailings generated by hydrometallurgical treatment of ores and concentrates for gold recovery, and is addressed separately below.

Tailings, by virtue of their semi-liquid composition, are difficult to secure, particularly over long periods of time extending beyond the active life of the mine. Two aspects of tailings disposal are crucial:

- Dam stability, which requires close control of the separation of coarse and fine tailings, installation of adequate drainage, control and monitoring of seepage, geotechnical control of the dam slope, and filtering mechanisms.
- Risk analysis, leading to the building of extra capacity and strength into the reservoir to withstand freak earthquakes and weather events. It was the failure to undertake proper risk analysis — a direct reflection of cost cutting and lack of regulation — which accounted for many disasters associated with tailings in, for example, Papua New Guinea and Peru. Indeed, it was the relatively liquid composition of the tailings from the Ok Tedi and Porges Project in Papua New Guinea, combined with the lack of storage sites, which led to the tailings from those mines being disposed of (after treatment) in the local river, resulting in serious environmental degradation.

3.6   EXTRACTIVE METALLURGY

Extractive metallurgy can be subdivided into two major disciplines, namely hydrometallurgy and pyrometallurgy. A third discipline, *electrometallurgy*, is not considered here as its use is relatively limited in the mining sector (e.g., mainly the production of aluminium and some zinc). Hydrometallurgy involves the use of *solvents* (normally water-based) to dissolve the metal(s) of interest, producing a dilute, metal-laden solution which is then further processed to recover the metal. Typically, hydrometallurgical processing involves temperatures below 50°C (although some pressurised leaching systems do operate up to 250°C.[15] In contrast, pyrometallurgical processes are normally operated at temperatures in excess of 800°C.

3.6.1   HYDROMETALLURGY

Hydrometallurgical methods of ore treatment are most commonly used for gold, uranium, and aluminium and to a lesser extent copper, zinc, and nickel. In particular, ores containing oxide material (about 10% of non-ferrous ores) are treated by leaching. Ore is first crushed and processed according to the requirements of the subsequent processes (e.g., limited crushing for heap and dump leaching, processing to concentrate stage for vat leaching). A leaching agent (*lixiviant*) is then used to extract the valuable metal(s) in the form of a dilute, metal-laden solution. This solution then passes to the metal recovery stage, which may involve precipitation, solvent extraction, or electrowinning.

3.6.1.1   Leaching reagents and solvents

3.6.1.1.1   *Acids and oxidants*
Acid leaching of ores and concentrates is the most common method of hydrometallurgical extraction, particularly for the recovery of copper. Typical acidic leaching agents include hydrochloric acid (HCl), sulphuric acid (H_2SO_4) and ferric sulphate ($Fe_2(SO_4)_3$). Oxidised copper minerals such as azurite, malachite, tenorite, and chrysocolla are completely soluble in sulphuric acid at room temperature. Other, less oxidised minerals such as chalcocite, bornite, covellite, and chalcopyrite require the addition of ferric sulphate and oxygen (as oxidants) to accomplish leaching.

3.6.1.1.2   *Alkalis and ammonia-based reagents*
For certain copper minerals, alkaline (or basic) leaching is more effective. Alkaline leaching is more selective than acid leaching and particularly appropriate for ores with large amounts of acid-consuming carbonate rocks. However, this selectivity often results in lower recovery if the metals are not fully liberated during crushing and grinding. Silica- and silicate-rich ores can be treated using alkaline leaching agents at raised temperatures. The principal reagents used in alkaline leaching are the hydroxides and carbonates of sodium and ammonia, but potassium hydroxide and calcium hydroxide are also used. Those metals, which can form amines (e.g., copper, cobalt, and

nickel) can be dissolved in ammoniacal ammonium carbonate or ammoniacal ammonium sulphate solutions at atmospheric pressure. The Sherritt Gordon ammonia process for copper is the only process that has been used commercially for processing the entire output of a mine.

3.6.1.1.3   Bacterially-mediated leaching

This is applied to low-grade sulphide ores in dump and heap leaching operations. Leaching is much slower than typical acid or basic leaching and relies upon the capacity of bacteria such as *Thiobacillus ferrooxidans* and *Thiobacillus thiooxidans* to oxidise ferrous iron to ferric iron (which in turn oxidises other metal sulphides, producing water-soluble sulphates). Sulphuric acid is also a product of the bacterial activity. The main requirements for bacterial activity are oxygen, ammonia, nitrogen, phosphate, a suitable temperature (approximately 30°C), and acidity (approximate pH of 2.0). Lower or higher temperatures (5°C or 50°C) or pH (0.5 or 4.5) do not tend to kill the organisms, but instead dramatically reduce their activity.

Bacterially mediated leaching is also used to process refractory gold, hitherto unrecoverable due to its crystalline association with pyrite (which the bacteria can readily dissolve). Advances in biotechnology combined with the environmental and economic advantages which bacterial leaching technologies appear to have over other larger scale, more capital intensive, and more polluting traditional processes — like reverberatory furnace smelting — may herald substantial changes in the structure of the minerals industry,[16] although recent improvements in SO_2 capture technology in smelters are undermining the potential competitive advantages of hydrometallurgy.

Although, to date, bioleaching has only been applied commercially to the recovery of gold, uranium, copper, and nickel, it has also been suggested as a route to effective heap leaching of low-grade zinc ores. However, to follow this route further advances are required in the solvent extraction of zinc from iron-contaminated liquors. Electrowinning of zinc, while more complex than is the case for copper is technically feasible.[17] The capacity to leach low-grade zinc ores would improve overall resource recovery.

3.6.1.1.4   Cyanidation

Cyanide (as a sodium or potassium cyanide solution) is used to dissolve gold. Alternatives to cyanide do exist, but few have been used commercially (see Mitchell, Chapter 14); for example, Newmont Gold's innovative approach to bioleaching combines bio-oxidation with a patented ammonium thiosulphate treatment as an alternative to cyanidation for refractory ores.[18]

3.6.1.1.5   Amalgamation

Mercury is used in the amalgamation of gold. This practice is widely applied at small-scale mining operations, principally in developing countries, and causes considerable damage to ecosystem and human health (see below).

3.6.1.2   Leaching methods

3.6.1.2.1   Dump leaching

Dump leaching refers to leaching that takes place on an unlined surface. The term "dump leaching" derives from the practice of leaching materials that were initially deposited as waste rock; however, now it also is applied to run-of-mine, low-grade sulphide or mixed grade sulphide and oxide rock placed on unprepared ground specifically for leaching. Copper dump leaches are typically massive, with waste rock piled into large masses ranging in size from 20 to over 100 feet in height. These may cover hundreds of acres and contain millions of tonnes of waste rock and low-grade ore.[19]

3.6.1.2.2   Heap leaching

In contrast to dump leaching, heap leaching refers to the leaching of low-grade ore that has been deposited on a specially prepared, lined pad constructed of synthetic material, asphalt, or compacted clay. In heap leaching, the ore is frequently pre-treated using size reduction (e.g., crushing) prior to placement on the pad. Site-specific characteristics determine the nature and extent of the crushing and the leaching operations used.[20] Cyanide heap leaching is a relatively inexpensive method of treating low-grade gold ores while vat leaching is used for higher grade ore. Heap leaching is generally used to treat ores containing less than 0.04 oz gold t^{-1}. Tank or vat methods are generally used to treat ores containing more than 0.04 oz/t^{-1}. The cutoff value for the different treatment routes is dependent on many factors, including the price of gold and an operation's ability to recover the precious metal.[21]

3.6.1.2.3 Vat leaching

For copper, the vat leaching process works on the same principles as the dump and heap leaching operations, except that it is a high production-rate method conducted in a system of vats or tanks using concentrated lixiviant solutions. It is typically used to extract copper from oxide ores by exposing the crushed ore to concentrated sulphuric acid. The vats are usually run sequentially to maximise the contact time between the ore and the lixiviant.[20] Vat leach units may be large drums, barrels, tanks, or vats. The design capacity of the leaching units is dependent on the amount of ore to be leached. For example, a 25-meter-long, 15-meter-wide, and 6-meter-deep vat unit is capable of leaching between 3000 and 5000 tons of ore per cycle. Vat leaching of concentrates is also used to extract metals (e.g., the cyanidation of gold-rich sulphide concentrates).

3.6.1.3 Metal recovery

For copper, there are two main methods of recovering the metal from dilute aqueous solutions: cementation onto scrap iron and solvent extraction/electrowinning.

3.6.1.3.1 Cementation

Typically, precipitators are shallow-round or stair-stepped wooden or concrete basins.[22] The simplest and most common precipitation systems used in the copper mining industry are open-launder-type cementation systems where copper-rich solution flows through a trough filled with scrap iron. Metallic copper precipitates on the iron (with a concomitant dissolution of the iron). The copper precipitate is invariably contaminated with iron and requires further refining before it can be sold.

3.6.1.3.2 Solvent extraction and electrowinning (SX/EW)

For copper, the solvent extraction operation is conducted in two stages. In the first, dilute and impure leach solution containing dissolved copper, iron, and other base metals (from the leaching stage) is passed to a mixer for extraction of the copper. In the mixer, the aqueous solution is contacted with an active organic extractant (chelating agent) in an organic diluent (usually kerosene), forming a copper-organic complex. The extractant is designed to selectively extract only the desired metal (in this case copper), while impurities such as iron are left behind in the aqueous phase. The organic phase (containing the organic-copper complex) is then physically separated from the barren aqueous phase.[23] The latter is recirculated back to the leaching units while the copper-loaded organic phase is transferred to the stripping section where the copper is removed by mixing the organic phase with concentrated sulphuric acid solution (spent electrolyte from the electrowinning stage) to produce a clean, high-grade solution of copper for electrowinning. Copper is then plated out of solution onto inert (non-dissolving) cathodes made of lead alloyed with tin and calcium or of stainless steel. The obvious advantage of solvent extraction is that cathode copper of saleable quality can be produced directly from leach solutions; therefore, further purification is not required. However, there are signs that traditional SX/EW for copper may be under threat from systems that remove the need for SX by allowing direct recovery from low concentration copper solutions.[24]

3.6.1.4   Potential releases, effects, and impacts

3.6.1.4.1   *Releases from heap and dump leach piles during and after closure*

When heap leach operations are concluded, a variety of constituents remain in the wastes. These include cyanide not removed during rinsing or neutralization (for gold leach operations), acid or alkalis (for base metal operations), as well as heavy metals and sulphides. After the operation has been closed or reclaimed, in the absence of proper design and control measures, runoff from the spent ore may occur. This runoff may contain constituents associated with the ore, such as heavy metals, and suspended solids. Depending on the method and completeness of detoxification, leachate from spent ore may also maintain a high pH (gold leaching) or low pH (copper leaching) over an extended time period.

3.6.1.4.2   *Releases from active heap and dump leach units*

The release of leaching agents (e.g., cyanide in gold operations, sulphuric acid in copper operations) from active leach piles or leachate collection ponds may occur during snowmelt, heavy storms, or failures in the pile or pond liners and associated solution transfer equipment, with severe implications for surface and groundwater quality.

3.6.1.4.3   *SX/EW sludge*

Sludge is the semi-solid gelatinous material that can accumulate in solvent extraction/electrowinning tanks. These sludges are colloids of suspended material that cannot be easily settled or filtered. The solvent extraction process generates sludge consisting of a solid stabilized emulsion of organic and aqueous solutions. It is located at the organic/aqueous interface in the settlers, periodically removed from the system, and centrifuged or otherwise treated to remove the organics. The aqueous solutions and the solids are disposed of and the organics are returned to the solvent extraction circuit for reuse. Depending on the characteristics of the orebody, the sludges may contain base or precious metals in quantities sufficient for recovery.

3.6.1.4.4   *Spent electrolyte*

Spent electrolyte is generated during electrowinning activities. Historically, electrolyte went through a stripping step and was subsequently discharged to a tailings pond. Today, due to economics, this effluent is recycled to reduce capital costs associated with the electrolytic acids used in these operations. Over time, electrolyte in the electrowinning cells becomes laden with soluble impurities and copper. When this occurs, the solution is removed and replaced with pure electrolyte (to maintain the efficiency of the solution and prevent co-precipitation of the impurities at the cathode). Purification of the spent electrolyte is done by electrowinning in "liberator cells." Liberator cells are similar to normal electrolytic cells, but they have lead anodes in place of copper anodes. The electrolyte is cascaded through the liberator cells, and an electric current is applied. Copper in the solution is deposited on copper starting sheets. As the copper in the solution is depleted, the quality of the copper deposit is degraded. Liberator cathodes containing impurities (such as antimony) are returned to the smelter to be melted and cast into anodes. Purified electrolyte is recycled to the electrolytic cells. Any bleed electrolyte is neutralized, usually with mill tailings and disposed of in a tailings pond.[25]

3.6.1.4.5   *Mercury releases*

A growing area of concern relates to mercury toxicity resulting from its amalgamation with gold metal as a recovery method in small-scale gold mining in developing countries. The small-scale gold miner often works for himself, putting in long hours under hazardous conditions (e.g., often standing in insect/disease-ridden waters). The miner frequently moves from plot to plot in nomadic fashion without a long-term perspective on the environmental impacts of his activities. These conditions induce excessive use of mercury reagents to amalgamate the mercury in the erroneous belief that the greater the amount of mercury used, the greater the rate of gold extraction;[26] rather, it is a precise ratio between mercury and gold that should be sought to maximise gold recovery. A number of countries have sizeable small-scale alluvial gold mining sectors, including Brazil, Colombia, Ecuador, Ghana, Papua New Guinea, Peru, Philippines, and Zimbabwe. In each case, the volume of gold produced by such means is estimated to exceed 15 tonnes per year, of which a combined total of 100 tonnes are produced using mercury. With an estimated usage rate of 4 kg of mercury per kg of gold produced, the total environmental loading of mercury from such sources is roughly 400 to 500 tonnes per year. It is the release of this excess mercury into both water systems and the atmosphere that accounts for the fast-spreading incidents of mercury poisoning which are being reported in many mining regions of gold-producing countries. In tropical climates, mercury also tends to have a higher methylisation rate (involving the reaction of elemental mercury to form easily absorbed and highly toxic organic compounds). Furthermore, the detection and control of the problem is made more difficult by the complexities of the food chain and the fact that affected fish can be eaten hundreds of miles downstream from the mercury's point of entry to the ecosystem.

Unlike cyanide, mercury does not degrade in the environment, but is transformed into organo-mercury species that are both toxic and bioaccumulators (i.e., they tend to accumulate in the food chain). There are a number of safer, cleaner alternatives to mercury use; for example, gravity separation and cyanidation and simple procedures to reduce losses of mercury to the environment during gold recovery (e.g., closed retorting). However, there are many reasons for the dominance of mercury in the small-scale sector: mercury is relatively cheap, it is simple to use, and requires limited capital outlay on equipment. There is often a lack of education on the part of the users as to the health and environmental implications, and the sector is largely unregulated; or where regulation has been enacted, enforcement is not possible due to lack of resources, civil unrest, or other sociopolitical factors. The question of how to reduce the use of mercury in the small-scale mining sector is one that requires input from social scientists, politicians, economists, and engineers if a viable and lasting solution is to be found.

3.6.2 PYROMETALLURGY

Pyrometallurgical processes are currently the backbone of the recovery of copper, zinc, nickel, and lead from sulphide deposits. The process route includes mineral processing (normally flotation) to produce a concentrate, followed by smelting, which breaks down the crystalline structure of the minerals by heat-fuelled oxidation. This produces a *matte* (containing up to 40% metal content) which in its molten form is converted and separated into *blister* (approximately 97 to 99% pure) and an iron-silicate slag — which may have sufficient economic value to be worth processing. This waste is deposited in "tailings ponds" which, as noted above, are a potential source of toxic leakage and dust pollution if not carefully contained and managed. Since blister metal is too impure for most industrial applications, refining is necessary. This is usually undertaken by a fire process (using a reverberatory furnace) if the feed has a low by-product content, or by electrolysis if additional metals are to be recovered. The resulting cathodes which are usually around 99.8 to 99.9% pure are marketed directly to the semi-fabricators (at the semi-fabricating stage, refined metal is worked into forms like sheets, strips, tubes, and wire for subsequent manufacture into metal-bearing goods) or cast into shapes (e.g., wire bar).

Widespread concerns over the effects of smelter emissions during the last two decades, particularly with increasing research data on the acid rain phenomenon, combined with economic concerns relating to energy availability and cost, and low efficiency levels, have brought smelter design to the forefront of recent innovation efforts in the mining industry. Technical change resulting in a cleaner continuous smelting process, often with electric furnaces replacing reverberatory furnaces, is at the heart of much of the innovation in lead, copper, and tin production. Closed systems avoid SO_2 emissions while high SO_2 concentrations in the effluent gases enable its recovery in the form of sulphuric acid. At the same time, such systems are often more efficient in their use of energy and enable higher recovery of metals and by-products.

Considerable progress towards pollution prevention in the smelting industry has been made over the last few years through the redesign of the production process for sulphide ores to facilitate sulphur dioxide capture and its efficient conversion to sulphuric acid. Together with a steep rise in energy prices during the 1970s, the demonstration of the linkage between sulphur dioxide emissions and acid precipitation challenged the smelting industry to find ways of reducing sulphur dioxide emissions, while remaining viable in a very competitive world market. In seeking to meet the challenge posed by competitive market pressures and regulatory and societal demands for better environmental performance, new technologies have improved process efficiency and cut emissions by reducing the number of stages in the smelting process, increasing the concentration of sulphur in the off-gas, and enclosing the process so as to make the capture of off-gases as efficient as possible. Noranda Minerals, Inc., for example, reduced SO_2 emissions at its seven metallurgical facilities from 800,000 tonnes per year in 1970 to 270,000 tonnes per year in 1990 by adopting smelter technologies that reduced SO_2 production, and by increasing the conversion to sulphuric acid, sold as a by-product.[27]

3.6.2.1   Techniques and technologies

3.6.2.1.1   Copper

For copper, the traditional three-phase roasting, converting, and smelting process is currently being replaced by a combined step direct matte smelting and continuous smelting using fluid bed roasters instead of multiple hearth roasters. The concentrate is suspended in a stream of hot air and flue dusts are recovered for re-processing. Closed systems avoid SO_2 emissions and electric or electronic furnaces instead of reverberatory furnaces are also used. The advantages over the latter include: high SO_2 concentrations in the effluent gas enabling efficient recovery as sulphuric acid; greater energy efficiency and reduced fuel consumption; reduced blowing time in the converter; and higher throughputs. The most efficient "flash" smelter currently in use is the Outokumpu process where the concentrate is dispersed in an oxygen stream. Of the continuous copper-making processes, the Noranda and Mitsubishi processes are considered state-of-the-art, although recently the Isasmelt process (Australia) and Cyclomelt process (Holland) put into production in the late 1980s demonstrate further improvements in fuel efficiencies, environmental control, waste gas, dust, and by-product recovery, as well as in the economics of the smelting process itself.

However, these have been superseded by the development of a new generation of flash smelting/flash converting by Kennecott and Outokumpu Oy at Garfield, Utah. The smelter at this site has been heralded as the "cleanest smelter in the world" and as such, one of the most significant innovations in extractive metallurgy in recent times.[28] The new smelter and converter complex replaces an existing facility which was able to handle only 60% of the concentrates produced at the Bingham Canyon mine. To meet increasingly tough air quality regulations, the company was faced with a choice of investing $150 million in pollution control technology and being constrained by the existing smelter capacity, or investing $880 million on a new process. The new process increased the capacity of the smelter to handle 100% of the concentrates, thereby eliminating transportation and processing costs associated with the shipment of concentrates to Pacific Rim smelters, and enabling the plant to meet or exceed all existing and anticipated air quality regulations. It is anticipated that the new plant will reduce operating costs by 53%.[29] The principal features of the new complex are the replacement of traditional Pierce-Smith converters with a patented flash converter, the total enclosure of the converter, and the replacement of open-air ladle transfer of molten matte with a solid-state transfer. Molten matte is cooled with water into a granulated form prior to transfer to the converters, significantly reducing the release of sulphur dioxide and other gases in the transfer process. Although the cooling of the matte involves a loss of heat energy, "waste" heat is captured as steam and fed to a co-generation unit. The selection of flash converting enables a continuous high throughput of material and a much-increased concentration of sulphur in the off-gas, greatly improving the efficiency of sulphur capture. In combination with the world's largest double contact acid plant, annual average emissions of sulphur dioxide will be reduced from 3600 pounds per hour to 200 per hour.[29-31]

3.6.2.1.2 Lead

The traditional primary lead production route is based on sintering, reduction of the sinter in a blast furnace, and refining of the bullion (which can be undertaken following pyrometallurgical or hydrometallurgical routes). Environmental controls to existing plants of this type require expensive add-on controls, such as dust recycling systems; automatic pressure controls inside the sinter machine's gas collecting system; electrostatic dust precipitation; de-drossing; higher shaft furnaces with new cooling systems; filter bags; and hoods to collect waste gas.

Relatively recently, several new, more environmentally efficient smelting processes have been developed which apparently demonstrate energy, operating, and investment cost savings over traditional smelting technology. These, however, are still at the demonstration stage. They all follow the principle, first developed in new copper production technology, of autogenous smelting where the natural heat generated by the oxidation of sulphide raw materials during roasting is used to smelt the charge in one single feed. This reduces overall toxic emissions through built-in dust and gas collection systems and by eliminating a heating stage (which produces off-gases) and indirect SO_2 emissions from the carbon fuel used in reverberatory smelters:

- The Kivcet Process uses an electric shaft furnace and collects vaporised metals in the off-gases in oxidic form. It contains built-in cleaning, ventilating, and dust recovery systems at several points within the process.
- The Boliden (Top Blown Rotary Converter) Process involves dry-feeding ore to a furnace fuelled by injecting compressed air and oxygen through a lance. The furnace itself is located within a ventilated hood in which furnace tilting and pouring of lead bullion and slag takes place. Slag granulation is undertaken, ventilated by bag filters if dry fumes are generated and by wet cleaning systems if moist gas is generated.
- The Outokumpu Process involves placing dry feed into a flash furnace via a special burner with oxygen injection under a ventilated conveying system. Waste gas from the flash furnace is neutralised by a special boiler and a hot electrostatic precipitator. After final cleaning in a wet electrostatic precipitator, gas is sent to an acid conversion plant. Dust is recycled in a closed conveying system to the flash furnace. Built-in filters, dust collectors, and vacuum floor cleaning systems are located at different points in the process train.
- The QSL Process—currently scheduled for industrial operation at several European plants—works with moist feed. The charge is sent along ventilated conveyors to the furnace, which is equipped with one or two gas uptakes leaving the furnace at the oxidising zone or at the oxidising and reduction zone. Again, it contains built-in gas cleaning and dust collecting systems.

3.6.2.1.3 Zinc

One potential leap forward in the pyrometallurgical processing of zinc has been the Warner process, developed at the University of Birmingham (U.K.). Here, zinc sulphide is converted directly to zinc metal using molten copper (which is thereby converted to copper sulphide). The zinc is vaporised and subsequently condensed under reduced pressure on a barrel condenser.[32] Energy given out during the conversion of the copper sulphide back to copper is transferred back to the process, and therefore large amounts of purchased energy are not required.[33] However, the final zinc product would require refining to produce high grades.

3.6.2.1.4 Tin

Tin poses special environmental problems on account of its relatively high value. As a consequence of this, lower grades are processed involving many by-products such as copper and lead and impurities such as selenium, tellurium, antimony, bismuth, and, of course, sulphur. Distinct smelting processes for different grades have been traditionally employed. Innovation has proceeded in two directions. First, it has aimed to remove and recover by-products more effectively. Second, it has aimed to use short rotary or electric furnaces depending on comparative fuel prices and series of filters and precipitators to collect toxic dust, particularly lead. Two new processes are currently under development: the Siromelt and TBRC smelting processes, which offer the advantage of consecutive treatments of metals and slag in the same vessel, avoiding the transfer of hot phases and reducing fugitive gas emissions.

3.6.2.1.5   Nickel

INCO's development of oxygen flash smelting technology is an example of radical technical change necessitated by the exhaustion of possibilities for further efficiency improvements in conventional technologies. One of the world's highest cost nickel producers until recently, INCO was the greatest single source of environmental pollution in North America as a result of an aged and inefficient reverberatory furnace smelter at Sudbury, Ontario which emitted excessive volumes of SO_2. Having reached the limits of efficiency improvements and unable to meet increasingly stringent regulations as part of an intensive acid rain abatement programme by the Ontario Environment Ministry, INCO invested over C\$3000 million in research and development. The INCO oxygen flash smelter produces a concentrated SO_2 off-gas stream that can be efficiently captured and fixed as sulphuric acid. In addition, the flash smelting process utilises the exothermic properties of sulphide ores and requires very little additional fuel. The process efficiencies stemming from the application of the technology have not only reduced SO_2 emissions at Sudbury by over 100,000 tonnes per annum, but have helped transform the company into one of the world's lowest cost producers.[34]

3.6.2.2   Potential releases, impacts, and damage

Smelting gives rise to four potentially pollutant products: waste gas, fugitive gas, smelter dust, and effluents. The extent of damage can generally be related to four variables: first, the overall efficiency and type of the smelter process (smelters may last for decades and many particularly polluting ones date back to the 1940s and 1950s); second, the environmental impact controls imposed upon the smelter and implemented by its operators; third, the levels of naturally occurring impurities in the copper or metal concentrates (e.g., arsenic and lead, typical of Peru; bismuth, typical of Bolivia; and molybdenum, typical of Chile); and fourth, the effect of smelter location and geographical features on emission controls.

The pyrometallurgical industry has within its power the opportunity to, using existing technologies, design and produce wastes with specific characteristics.[35] This could entail minimising the concentration of potentially hazardous components, or adjusting the physical and chemical characteristics of the wastes to endow them with greater resistance to detrimental changes once disposed of to the external environment, thereby reducing disposal costs. The wastes could also be designed as a resource (e.g., use as an aggregate or for potential recovery of contained values at some future time).

3.6.2.2.1   Sulphur dioxide emissions and acid rain

Worldwide, the smelting of copper and other non-ferrous metals releases an estimated 6 million tonnes of sulphur dioxide into the atmosphere each year — which constitutes 8% of the total emissions of the sulphur compounds that cause acid rain.[36] Such emissions produce "dead zones" where little or no vegetation survives. Such an area around the Ontario nickel smelters (of INCO and Noranda) measures 10,400 hectares, and acid fallout from the smelters has destroyed fish populations in lakes 63 kilometres away. In the U.S., the dead zone surrounding the Copper Hills smelter in Tennessee covers 7000 hectares.

The first difficulty in determining emission factors from smelting relates to the complexity of the relationship between sulphur dioxide emissions and local sulphate concentrations, the latter being largely responsible for the associated environmental hazard. The Community Health and Environment Surveillance Study made by the Environmental Protection Agency (EPA) in 1974 systematically documented the health effects of sulphate. For example, when sulphur dioxide is emitted into the atmosphere it reacts with water to form a sulphuric acid aerosol and with various metallic ions to form metallic sulphates, which are of respirable size range. The rate of oxidation of sulphur dioxide to various sulphate-related species varies between 0.1 and 30% per hour (i.e., by a factor of 300). Moreover, the conversion is catalysed by photochemical smog and varies with sunlight intensity, ambient temperatures, humidity, and the presence of particulate matter.[37] The suggestion that health problems are the result of sulphates rather than sulphur dioxides has dramatic implications for sulphur dioxide discharge policy, since there is little relationship between sulphur dioxide discharge and local sulphate concentrations, which have more to do with other ambient pollutants, prevailing winds, and humidity levels; the related subject of acid rain therefore becomes important.

The most comprehensive documentation for pollution from smelting contributing to acid rain is for North America, where in 1980 SO_2 emissions were 28.9 million tons of which 3.5 million tons (or 12%) originated from non-ferrous smelters. The nickel and copper smelting facilities in Canada of INCO at that time were the continent's largest single source of air pollution, and facilities such as the Falconbridge and Kidd Creek smelters followed close behind, accounting alone for more than 19% of North American acid rain. Ontario, Canada is unique in that it is a significant source and recipient of acid rain, due in the main to the processing of local nickel, copper, and iron ores. In addition, the rock formation that provides this mineral wealth is the granitic Pre-Cambrian Shield, which is characterised by thin acid soils that provide little chemical buffering to counteract the incoming acid from the atmosphere. Consequently, acid rain damage to the natural environment in Ontario is severe and widespread. Similar damage is found in the granitic Andes regions that host the mineral wealth and smelting facilities of Chile, Peru, and Bolivia. The problems in the latter regions are, however, less well documented.

3.6.2.2.2 Arsenic

Arsenic is mainly produced as a by-product during the production of other more important metals such as copper, lead, zinc, gold, silver, and tin. Commercial-grade arsenic trioxide is recovered from the smelting or roasting of non-ferrous metal ores or concentrates in at least 18 countries.[38] Other countries also collect arsenic-rich residues such as smelter or roaster dusts, but these are not processed further to produce the commercial grade. For example, Ghana produces estimated 9000 short tons of crude arsenic trioxide from gold ore roasting operations, some of which is exported to France for further refining (to a purer trioxide). Future market opportunities do not look good for arsenic, given the replacement of lead-arsenic alloys in batteries with lead-calcium and pressure to reduce its use in low-grade applications, such as the treatment of timber. Therefore, it is certain that in the future, potential supply will far outweigh demand (see Mitchell, Chapter 14 for further data on the recovery, disposal, and environmental impact of arsenic).

3.7 CONCLUSIONS

As a means of summarising the environmental releases, effects, and impacts from mining, mineral processing, and extractive metallurgy, it is useful to take the perspective of the receiving (external) environment (Table 3.4).

Increasingly, based on a more thorough understanding of these potential impacts, the pressure for environmental protection may outweigh the justification for exploiting particular mineral reserves,[39] and the industry as a whole, in partnership with the broadest possible spectrum of stakeholders, must strive towards ever higher levels of performance in control and management of the mining process throughout the life cycle of an operation.

TABLE 3.4
Physical, Chemical and Biological Environmental Effects of Mining and Milling Operations (after Eaton et al.)[1]

Environmental effect/impact	Cause
Water	
Turbidity; smothering of benthic ecosystems	Suspended solids from creation of infrastructure, mine dewatering, and surface run-off
Groundwater contamination (e.g., degradation of potable water resources); surface water contamination (e.g., rivers, streams and springs); acute and chronic bio-toxicity (e.g., fish kills, growth, and reproduction defects)	Acid rock drainage (ARD) (e.g., from surface wastes, ore stockpiles, open pits, and underground workings); deliberate or accidental release of process chemicals (e.g., cyanide); spills or leakage of other materials (e.g., mineral oil lubricants, petroleum and derivatives, cleaning agents)
Eutrophication	Nitrates derived from wash-down of explosive residues
Oxygen consumption	Presence of inorganic and organic chemicals that consume oxygen during changes in chemical speciation or during degradation
Depression of water table	Dewatering of mine workings; hydrological and hydrogeological disruption of surface and at-depth aquifers
Contamination of riverine and estuarine sediments	Erosion and dispersion of solid wastes
Soil (land)	
On- and off-site contamination of top and subsoil horizons	Wind and water erosion and dispersion of metal-bearing solids (e.g., wastes, run-of-mine ore); transfer from contaminated waters to soil components (e.g., clay minerals, organic matter)
Land sterilisation and/or destruction of vegetative cover (including rare and endangered species)	Disposal of contaminated and/or inert wastes; open pit mining; "footprint" of the processing plant and associated infrastructure
Inhibition of vegetative regeneration	Surface and sub-soil contamination; soil acidification by acid waters
Aesthetic impact	Lack of vegetation; high profile waste disposal sites; open pits; severe topographical disruption (e.g., re-routing of rivers); transient and regular noise
Air	
Dust	Creation of infrastructure; wind erosion and dispersion of fine solids (e.g., tailings); crushing; movement of site vehicles; surface blasting
Contaminant emissions	Minerals processing operations (particularly SO_2) from smelting, but also localised occurrences of other process chemicals or degradation products (e.g., hydrogen cyanide from cyanide-contaminated gold ore tailings)
Noise	Creation of infrastructure; blasting; operation of heavy plant

ACKNOWLEDGEMENTS

The author would like to acknowledge the assistance of Liz Smith and Yvette Haine in the preparation of early drafts of the manuscript. Sincere thanks go to Dr. Gavin Bridge and Dr. Paul Mitchell, who provided useful inputs and insights throughout the development of this chapter.

REFERENCES

1. Eaton, P. B., Gray, A. G., Johnson, P. W., and Hundert, E., *State of the Environment in the Atlantic Region*, Environment Canada, 1994.
2. Boldt, J. R., *The Winning of Nickel*, Longmans, Toronto, 1967.
3. Godin, E., Ed., *1990 Canadian minerals yearbook — review and outlook*, Energy, Mines and Resources Canada, Ottawa, Ontario, 1991.

4. Worldwatch Institute, Mineral Commodity Summaries 1992, Figures based on production estimates in U.S., U.S. Bureau of Mines, Washington, D.C., 1992.

5. Rogich, D. G., Trends in material use: implications for sustainable development, Unpublished paper, Division of Mineral Commodities, U.S. Bureau of Mines, April 1992.

6. Thornton, I., *Metals in the Global Environment: Facts and Misconceptions*, The International Council on Metals and the Environment, Ottawa, Canada, 1995.

7. Nriagu, J. O. and Pacyna, J. M., Quantitative assessment of worldwide contamination of air, water and soils by trace metals, *Nature*, 333, 134, 1988.

8. World Health Organisation, Occupational health of miners in the non-ferrous metal industry. *UNEP Industry and Environment*, 8, January–March, 1985.

9. Axelson, O. and Sundell, L., Mining, lung cancer and smoking, *Scandinavian Journal of Work Environment and Health*, 4, 42, 1978.

10. Dunster, J. and Dunster, K., *Dictionary of Natural Resource Management*, CAB International: Wallingford, U.K., 1996.

11. Wills, B. A., *Mineral Processing Technology*, 6th edition. Butterworth-Heinemann: London, 1997.

12. Hiskey, J. B., *In situ* leaching recovery of copper: what's next?, in *Hydrometallurgy '94*, Chapman & Hall, 1994, 43.

13. Almgren, G., Almgren, T., and Kumar, U., Just-in-time and right-in-space, *Minerals Industry International*, No. 1033, September, 26, 1996.

14. Hayes, P. C., *Process Principles in Minerals and Materials Production*, Hayes Publishing Company: Australia, 1993.

15. Rankin, W. J. and Wright, J. K., Greenhouse strategies for the metallurgical industry, in *Minerals, Metals and the Environment*, Elsevier Applied Science, London, 1992, 172.

16. Warhurst, A., Environmental management in mining and mineral processing in developing countries, *Natural Resources Forum*, 16, 39, February 1992.

17. Grant R. M., Emerging developments in zinc extraction metallurgy, in *Metallurgical Processes for Early Twenty-First Century*, Sohn, H.Y., Ed., The Minerals, Metals & Materials Society, 1994, 125.

18. Warhurst, A. C. and Bridge, G., Improving environmental performance through innovation: recent trends in the mining industry, *Minerals Engineering*, 9, 1996.

19. Biswas, A.K. and Davenport, W. G., *Extractive Metallurgy of Copper*, Pergamon International Library, International Series on Materials Science and Technology, Volume 20, 1976, Chap. 2.

20. USEPA, Final Report: Copper Dump Leaching and Management Practices that Minimize the Potential for Environmental Releases, Prepared by PEI Associates, Inc. (Hearn, R. and Hoye, R.) under USEPA Contract No. 68-02-3995, 1989.

21. Van Zyl, D. J. A., Hutchison, I. P. G., and Kiel, J. E., Eds., *Introduction to Evaluation, Design and Operation of Precious Metal Heap Leaching Projects*, Society for Mining, Metallurgy, and Exploration, Inc.: Littleton, CO., 1988.

22. U.S. Congress, Office of Technology Assessment, Copper: Technology and Competitiveness, OTA-E-367. U.S. Government Printing Office, Washington D.C., September 1988.

23. Anonymous, Phelps Dodge Corp. has announced the discovery of a substantial open-pittable copper mineralization, *Engineering and Mining Journal*, 192, 71, March 1991.

24. Clifford, D., Mineral processing roundup, *Mining Magazine*, 177, 176, September 1997.

25. USEPA, Overview of Solid Waste Generation, Management, and Chemical Characteristics, Prepared for USEPA under Contract Nos. 68-03-3197, PN 3617-3 by PEI Associates, Inc., 1984.

26. CETEM, Analysis of Garimpeiro in Brazil, unpublished research reports, Centro de Tecnologia Mineral, Rio de Janeiro, Brazil, 1990.

27. Noranda Minerals Inc., Environmental Report, Noranda, 1990.

28. Emmons, C. S. and Gabb, P. J., Factors influencing a modernised copper smelter impurity control policy, in *Pyrometallurgy '95*, Institution of Mining and Metallurgy, London, 1995, 125.

29. Dimock, R., Kennecott has modern mettle for mining, *The Salt Lake Tribune*, April 16, 1995.

30. Chiaro, P., Waste minimisation and pollution prevention at Kennecott, in *Proc. International Conference on Pollution Prevention in Mining and Mineral Processing*, Anderson, K. and Purcell, S., Eds., Colorado School of Mines, 1994, 100.

31. Kosich, D. Y., Kennecott's vision may revolutionize smelter technology. *Mining World News*, 1, 2, 1995.

32. Turner, J. and Warner, N. A., Assessment of reduced pressure condensation of zinc on a barrel condenser, in *Pyrometallurgy '95*, Institution of Mining and Metallurgy, London, 1995, 219.

33. Warner, N. A., Davies, M. W., Holdsworth, M. L., and Turner, J., Direct zinc smelting with virtually zero gas emission, in *Metallurgical Processes for Early Twenty-First Century*, Sohn, H.Y., Ed., The Minerals, Metals & Materials Society, 1994, 233.

34. Warhurst, A., The limitations of environmental regulation in mining, in *Mining and the Environment: International Perspectives on Public Policy*, Eggert, R., Ed.,. Resources for the Future: Washington D.C., 133, 1994.

35. Coppin, N. J., Bryson, F. E., and Brown, C. W., Pyrometallurgy and the environment: at what cost?, in *Pyrometallurgy '95*, Institution of Mining and Metallurgy, London, 1995, 1.

36. Young, J. E., Mining the Earth, Worldwatch Institute, Paper 109, July, Washington, D.C., 1992.

37. Sawyer, J. W., Environmental quality and the extractive industries: the sulphate issue, in *Proceedings of the Economic Council*, American Institute of Mining Engineers, 1977.

38. Broad, A., Arsenic down but not out, *Metal Bulletin Monthly*, No. 315, 54, March 1997.

39. Hodges, C. A., Mineral resources, environmental issues, and land use, *Science*, 268, 1305, 1995.

4 Ecotoxicological Impacts of the Extractive Industries

A. K. Barbour and I. C. Shaw

CONTENTS

ABSTRACT

The major stages in the extraction of metalliferous minerals are outlined as extraction, crushing, grinding/milling, and concentration. The major environmental challenges from each stage, and the disposal of waste rock and tailings, are described briefly. Indicative management schemes are provided to ameliorate the environmental effects derived from:

  (i)   Erosion of waste-rock dumps.
  (ii) Liquid effluents from milling/concentration.
  (iii) Tailings deposition, dam safety, and liquid effluent control.
  (iv) Minimisation of dust-blow.
  (v)   Visual impact, remediation, and restoration.

The conversion of sulphidic concentrates to refined metal requires effective management of:

  (i)   Particulate emissions to atmosphere.
  (ii) Sulphur dioxide fixation, usually by conversion to sulphuric acid.
  (iii) Control of liquid effluents.

This general scheme is expanded in detail for copper and in outline for zinc, lead, nickel, mercury, and toxic by-products such as arsenic and cadmium.

In addition to being the major tonnage non-ferrous metal, primary aluminium is unusual in being derived exclusively from an oxidic mineral, bauxite, and its conversion to metal is by an electrothermal reduction process using a fluoride flux. Thus, the current embodiment of this process (the prebake process) requires the management of:

  (i)   Gaseous and particulate fluorides.
  (ii) Pitch fume and polycyclic aromatics (PAHs).
  (iii) Minor concentrations of sulphur dioxide and perfluorinated alkanes in emissions.
  (iv) Safe disposal of spent pot linings and anode stubs.

Following each account of extractive aspects, toxicological and ecotoxicological details are provided for each metal. Particular emphasis is given to those metals whose adverse impact is perceived to be greatest, either at the extraction stage or in end-uses. Examples include lead, mercury, cadmium, arsenic, nickel, and chromium.

Lead is of particular concern because very young children may suffer damage to the developing nervous system and older children are much more susceptible than adults to long-term exposure to low levels of environmental lead.

The toxicity of cadmium arises from its accumulation in liver and kidney and its potential as a human carcinogen, a property that may also be significant for arsenic (oxide), chromium, and nickel. The role of mercury in "Minamata" disease is outlined; this incident was of particular interest in concentrating attention on environmental speciation since alkyl mercury compounds were unexpectedly found to be the active species.

The chapter article concludes with details of toxicology and ecotoxicology of the major minerals fluorspar, borax, silica, asbestos, and coal.

4.1 INTRODUCTION

The extractive industries form the basis of many materials essential to our present way of living. When non-ferrous ores are mined, concentrated, smelted to metals, sometimes alloyed with other metals, and then fabricated by the engineering industry, they form vital components in automobiles; consumer durables such as washing machines, refrigerators, freezers, and dishwashers, and the equipment for cooking and housing; as well as vital components in energy and information generation and transmission. In this sense, metals are largely competitive with materials based on oil (plastics) and on renewable cellulosics (paper and timber). Minerals are also essential for concrete, paints and pigments, plasterboard, bricks, detergents, and many medicinal products.

Minerals, metals, and other valuable commodities have been sequestered in the earth since time began. Most of them are so strongly bound to rock or so physically isolated by being surrounded by impervious rock that they are very unlikely ever to find their way into terrestrial or aquatic ecosystems in concentrations that might cause harm to these systems. They do, of course, leach into water as it percolates through the ground as part of the water cycle. Generally, the very low concentrations of metals and other ions from the bed rocks which are dissolved in ground water are not harmful, but rather are essential nutrients and components of water systems. Indeed, some waters are revered for their metals content. Spa towns such as Bath and Budapest have developed around natural water sources which have apparent healing properties that have been in part, attributed to their metal content. A swim in the Roman Kings' Bath in the city of Bath results in one's skin turning brown due to the deposition of oxidised iron species.

Man's mining activity and the use of extracted metals has upset this balance and released numerous metals in forms and amounts, which if not carefully controlled, could seriously contaminate the environment. An example of environmental contamination with copper can be seen at the Victorian copper mines high in the Cumbrian copper mines of the English Lake District. Although the mines have been defunct for many years, the lakes near to Coniston are bright blue and devoid of plant life. The blueness is due to copper (II) and there are few aquatic plants in the lakes because copper inhibits the formation of chlorophyll, which is vital for photosynthesis.

Before exploring the routes of metals contamination of ecosystems and methods that might be used to monitor such contamination, it is important to have a basic understanding of the mechanisms and signs of toxicity of some of the more commonly encountered metals. It is important to note that many metals are essential for cellular function, but that they are usually required at very low concentrations. It is only when these concentrations are exceeded that toxicity might be seen. Paracelsus, a 16th Century German scientist (arguably the first toxicologist), first noted this concept:

All things are poisons
There is nothing which is not a poison
It is the dose
Which makes a thing safe

A measure of the toxicity of a substance in an aquatic environment is the EC_{50} (the concentration of a chemical required to kill 50% of a population of a particular animal or plant). The EC_{50} is often determined using a marine phosphorescent bacterium (*Photobacterium phosphoreum*) in a commercially available test called Microtox®. Copper's EC_{50} in the Microtox® test is approximately 20 mg L^{-1} in sea water and 0.2 mg L^{-1} in fresh water. This aquatic toxicity and other considerations led the International Maritime Organisation (IMO) to categorise copper as dangerous to the marine environment.

When considering toxicity of metals in aquatic environments it is crucial to determine which form the metal is in, because different ionic forms (species) have different levels of toxicity. Copper can occur in three main forms in solution:

- Cu^o (copper (0))
- Cu^+ (copper (I))
- Cu^{2+} (copper (II))

The metal (Cu^o) is ostensibly non-toxic. Copper (I) is difficult to study because it is rapidly oxidised to copper (II) in water. The copper (II) form of copper is definitely toxic and is generally the form found in plants and animals which have succumbed to copper poisoning.

Environmentally important metals, which could have serious human health effects if excessive absorption occurs, arise mainly as by-products from the smelting/refining of other metals. Therefore, close control of the primary smelting/refining process is necessary. Examples include: arsenic, a common by-product in the smelting of copper concentrates to metal which also arises from the smelting of secondary tin hardened with arsenic; and cadmium and mercury obtained as by-products from zinc and lead smelting. Zinc smelting is the sole primary source of cadmium, which occurs only as a minor constituent of other ores. Some mercury continues to be produced by the "social" mines of Brazil and in other parts of South America. However, most markets for arsenic, cadmium, and mercury have been severely reduced by regulations designed to limit the use of these metals in end-products and the likely consequent uncontrolled dispersion of small quantities throughout the environment. Releases from smelting continue to require careful management.

Thus, essentially all of these substances have toxicological characteristics that impact upon the environment and human health to a greater or lesser extent. These impacts require management and control/amelioration at each stage of metal extraction, fabrication, use, and disposal. The following discussion presents in general terms the technologies by which the major non-ferrous metals and some minerals are extracted and converted into products (metals and minerals) (see also Warhurst, Chapter 3) which can then be converted into saleable end-products, and outlines their toxicological and ecotoxicological characteristics.

Whilst the extractive industries operate in many countries worldwide, their environmental impact is quite localised, even for major mines. It is through their conversion into products utilised by the general public and their disposal at the end of their useful life that metals are dispersed widely throughout the environment. This is why it is so important to monitor concentrations (and preferably identify species) so that the risk and degree of harm can be estimated at each location throughout the cycles involving these materials.

Most non-ferrous metals occur in sulphidic deposits, usually in a mixture with relatively large amounts of iron and variable amounts of undesired metals. The average content of the desired metal can also vary greatly; economic extraction of copper can be undertaken at concentrations as low as 0.5%, whereas zinc concentrations may need to approach 20%, and bauxite, the source of primary aluminium, 58% alumina.

Extraction technology is site-specific, differing according to whether extraction is carried out underground or by open-pit methods. Thereafter, the extracted ore is invariably crushed, graded for size, and usually milled before concentrated, frequently by flotation.

Iron is often present in ores being worked for copper, zinc, lead, tin, and nickel. Trace impurities of other metals are also present, some of which have value (e.g., silver, bismuth, manganese, and cadmium), whereas others accumulate and must be permanently disposed of at a cost (e.g., arsenic and mercury). Thus, the processing of non-ferrous concentrates has to be designed to separate the desired metal from impurities (the refining process) and to manage the disposal of those impurities which cannot be sold on the market. This also applies to the non-metallic constituents: sulphur and silica. The former is usually converted to sulphuric acid for sale, or is occasionally disposed of by costly alternative methods, whilst silica must be managed as a waste, since abundant supplies of pure material are available naturally in the form of sand. Since all of these metals are more or less toxic, the extraction and processing of metalliferous ores present significant environmental management problems.

As in many other respects, aluminium differs from other non-ferrous metals because it occurs as the oxide (bauxite) admixed with iron, and usually occurs in deposits that are near to the surface. It is usually processed to alumina by the Bayer Process, which uses alkali to separate it from iron. This process, *inter alia*, creates something of an environmental problem in the disposal of the iron-rich "red mud." Few other impurities are usually present and most of the other environmental management challenges come from the use of fluoride fluxes in the subsequent electrothermal reduction process, which converts alumina to aluminium metal.

Two other exceptions to the generalisation about non-ferrous metals are uranium and gold because they occur in low concentrations and have highly restricted markets, but are very valuable. After extraction of the ore and crushing to size, both are concentrated by solvent extraction methods, but only gold is processed through to pure metal in the commercial sense.

4.2 NON-FERROUS AND INDUSTRIAL MINERALS: ENVIRONMENTAL MANAGEMENT CHALLENGES

Since the concentration of desired metals is frequently small, undesired (waste) materials are correspondingly large, generating a considerable tonnage of waste rock and tailings for permanent disposal. Historical operations have in most cases caused severed landscape dereliction, particularly where extraction has been by open-pit methods, albeit over a small, localised area. Dust-blow has been inevitable and, frequently, stream contamination as well, particularly where oxidation and hydrolysis have generated metal-dissolving sulphuric acid from the sulphuric materials. In more recent times, developing technology has permitted amelioration of these problems, both during the operational lifetime of the mine and in the increasingly common phase of managed remediation after closure. However, it is doubtful whether extraction operations can be environmentally neutral.

The extraction of minerals such as kaolin, borax, talc, and silica sands is generally by open-pit techniques and is greatly simplified by the much higher concentrations of desired materials present. Thus the flowsheet is correspondingly simpler and, in particular, the need for concentration by flotation is generally absent. Processing is usually restricted to crushing/grading, classification, and perhaps washing with waste-dumps comprising only a relatively small tonnage of overburden. However, dust-control from unconfined areas remains a difficult management problem, particularly if the prevailing wind is in the direction of populated areas. For the same reason, bulk transportation of materials needs careful control during loading and movement.

Extraction operations for all non-ferrous metals have to manage the following main environmental impacts to varying degrees depending on the specific site and extraction technology used:

- Adverse visual impact on the local landscape with probable adverse effect on the local ecology. The large excavations inseparable from open-pit mining makes visual impact from these mines greater than that from underground mines.
- Pollution and possible volume effects on local streams and the water table, particularly where sulphidic ores are being extracted. Significant issues requiring positive action are likely to run-off from waste rock and tailings and from processing chemicals.
- Mine safety and dam stability.
- Uncontrolled dust-blow whose effects are usually visual/environmental but, in severe cases, could cause adverse health effects for near-neighbours.
- Noise nuisance from blasting, operational equipment, and transportation.

Mine managers, stimulated in some countries by permitting requirements, are beginning to appreciate that mine remediation and rehabilitation, including the dismantling and disposal of buildings, machinery, and other equipment should be planned continuously from the start to the closure of the operation, including the minimisation of local environmental impacts. Such planning must, of course, be soundly based on financial as well as social/technological factors. Detailed analysis of these issues and proposals for their amelioration form the content of the Environmental Impact Assessments now required in most countries before operating licenses are granted.

4.2.1 EROSION OF WATE-ROCK DUMPS

Unlike chemical and metallurgical processing operations, mines have to be located where economic mineralisation naturally occurs. Since large tonnages of extracted low-value materials have to be transported for upgrading, a concentration plant associated with the mine also has to be located nearby. Extraction operations naturally break up the terrain and hence increase greatly the surface area of material exposed to rainfall which, in many parts of the world, falls as intense storms of relatively short duration, giving a high risk of flash flooding.

In these circumstances, "wash-out" from waste rock piles is inevitable. Fortunately, by defini- tion, waste rock contains only low concentrations of the desired elements, which are often relatively toxic, but the clays and silts eroded can cause local streams to become opalescent due to the high burden of suspended solids. No ameliorative measures of general applicability are available, though sometimes it is possible to channel run-off streams via the tailings impoundment. Fortunately, another feature of tropical weather is the occurrence of relatively long periods of dry weather when erosion by water is small. However, during such periods dust-blow, particularly from dried-out tailings areas, can become a significant problem.

4.2.2 GENERATION OF ACID ROCK DRAINAGE

The most intractable environmental problem in the extraction of sulphidic concentrates is acid mine drainage where sulphuric acid generated by the oxidation reactions can result in stream contami- nation, either directly by the sulphuric acid generated or by the toxic metals which it has dissolved (see Mitchell, Chapter 7 for further information).

Many methods have been proposed for dealing with acidic run-off including deep injection, neutralisation with lime, and dilution. None are of general applicability; such treatments can only be applied where run-off follows well-defined channels. Recycling to process will usually improve treatment economics; recycling is a common practice in extraction operations, particularly when located in arid areas. Often, recycled water is used to ameliorate dust-blow from haul roads, waste- rock dumps, and blasting operations.

4.2.3 LIQUID EFFLUENTS FROM MILLING

Milling is the comminution of the extracted ore into particles that can be subjected to a recovery process that separates the valuable materials (concentrate) from the valueless (gangue). The term is now usually used to cover the flotation process which is now an essential stage of all non-ferrous mining operations, other than the production of aluminium from bauxite, and the extraction pro- cesses for gold and uranium which are based largely on leaching followed by solution concentration.

After primary and secondary crushing and screening, milling operations start with grinding in a multiplicity of ball and rod mills. After classification, the ground material passes to the flotation units where a variety of reagents may be used depending on the chemical composition, density, and other characteristics of the mineral being concentrated (see Mitchell, Chapter 14).

Froth flotation, by far the most widely used concentration method, is based on conferring hydrophobicity to the individual particles and hence assisting them to become attached to air bubbles. Particles with higher mineral content then rise to the surface in a froth that is skimmed. The remaining barren particles become tailings. The flotation reagents used tend to be specific for particular processes; general examples are shown in Table 4.1.

Separation by multi-stage cyclones and thickeners provides a concentrate of the desired metal- bearing minerals and slurry of process water, unwanted gangue, and the reagents used during the

TABLE 4.1
Common Flotation Reagents

Class	Use	Compound
Collectors	To selectively coat particles with a water-repellent surface attractive to air bubbles	Water-soluble polar hydrocarbons, such as fatty acids and xanthates
Modifiers pH regulators	To change pH to promote flotation, either acid or basic	NaOH CaO Na_2CO_3 H_2SO_4 H_2SO_3
Activation and depressants	To selectively modify flotation response of minerals present in combination	Metallic ions Lime Sodium silicate Starch Tannin Phosphates
Frothers	To act as flotation medium	Pine oil Propylene glycol Aliphatic alcohols Cresylic acid
Oils	To modify froth and act as collectors	Kerosene Fuel oils Coal-tar oils

flotation stage. The latter passes to the tailings impoundment area, the location of which is of great importance from ecological and environmental standpoints, including the safety and security of near-neighbours.

4.2.4 TAILINGS MANAGEMENT

Unlike waste rock, which has to be transported expensively in haultrucks or extended conveyor-belt systems, tailings can be pumped economically as slurries over quite considerable distances. The extreme choices for location can be summarised as deep valleys or plain land with additional factors such as the potential for stream or groundwater contamination (and the use of such waters) and safety/security as the major additional considerations.

Narrow, deep valleys are usually easier to dam and do not disturb agricultural land, although they may, for example, obliterate small ecologically important areas of tropical jungle. They are usually visually unobtrusive, partly because they are exposed to the vision of few people. On the other hand, such valley locations are often relatively elevated, thus increasing pumping costs and often increasing hazard in seismic areas if tailings liquefaction ever caused break out; land at lower elevations is usually more valuable agriculturally.

For streams and surface waters, the choice of the best practicable environment option involves weighting and balancing factors such as the degree of treatment (and its cost) required for tailings water disposal into a particular stream versus, perhaps, discharging untreated into a more distant but environmentally and commercially unimportant stream.

Schemes for the purification of tailings effluent have to focus on parameters such as heavy metals including arsenic; chloride; sulphate; occasionally fluoride and, increasingly, nitrate; suspended solids; and pH, together with the flow characteristics and uses of the receiving bodies of water. Dependence is usually placed on utilising the dilution and absorptive powers of the receiving bodies of water. Conventional treatment, e.g., liming to precipitate heavy metals and pH adjustment is used where it is necessary to preserve existing uses of the receiving body. If cyanide is used in an extraction circuit, as in most gold concentration processes, it is desirable to convert it to relatively innocuous cyanate by oxidation.

However, in view of the very large volumes of water involved in most tailings operations, particularly where recycling of supernatant is not practiced, sludges and other wastes from treatment processes usually have to be disposed separately in small impervious impoundments; this is often not a preferred environmental option compared with dilution into streams as it creates a toxic "hot-spot" which may be difficult to manage after general operations have ceased. Whatever disposal option is selected, adequate monitoring should be practiced to ensure that any significant changes in the quality of the receiving body are quickly detected and assessed.

Contamination of groundwater, though less apparent visually, can be very important, particularly if it is used as a potable supply. Consequently, the analysis of environmental impacts must include detailed hydrological assessment of seepage flows from the tailings impoundment area as well as the extraction site.

Tailings dam failure can be disastrous for both human beings and any other activities located nearby. Whilst the design parameters for tailings dams are now well-developed and incorporate safety factors relating to predicted severity and frequency of earthquake and storm, it remains prudent to locate tailings impoundments well away from human activities. Regular inspection of such dams by specialists is a normal part of virtually all mine safety regulations. Discharge through professionally designed outfalls to deep sea avoids most of these problems, but it is not an option favoured by most regulatory authorities because of localised damage to benthic organisms and, sometimes, local fisheries.

4.2.5 DUST-BLOW

Fallout of dust onto areas adjacent to mining and processing operations not only spreads ecological effects but can also affect adversely the quality of life of near-neighbours, in rare cases constituting a significant effect on human health. Total elimination of dust-blow is impracticable but considerable amelioration will result from the adoption of common-sense practices such as dampening all areas liable to generate dust and the paving of haul roads and the provision of effective coverings for conveyors, trucks, dumps, and storage facilities. To minimise the effects of residual dust on operators and other personnel, efficient respiratory protein and/or filtered air supplies should be provided and atmospheric monitoring conducted on a planned basis in all areas where significant exposure is likely.

4.2.6 IMPACT, REMEDIATION, AND RESTORATION

Many regulatory authorities worldwide now require the submission of detailed Environmental Impact Assessments (EIAs) for consideration and approval before a license to operate is granted.

Progressive managers now appreciate the value of extending the mine planning process in this way so that the optimum balance is achieved for extraction, concentration, and waste disposal. Because the tonnages of waste for disposal are so large, it is very important to explore all options for location, visual impact and, ultimately, revegetation before operations commence.

Disposal back into the worked-out pit or underground is generally not a practicable option, though some regulatory authorities have considered it. The remediation and restoration of tailings and waste-rock deposits thus has to be managed as a surface disposal issue and that is why location from the inception of operations is so important.

During the period of extraction operations tailings are usually submerged beneath mill effluent leaving only the beaches exposed; this is an important factor in minimising wind erosion that can become serious during prolonged dry spells.

Regulations in several countries now require systems for the management of tailings and waste-rock deposits on a long-term basis following mine closure. The general objectives are to ensure that the areas present no health hazard to persons or animals; windage loss is minimised; continued contamination of watercourses does not occur, and aesthetics are improved wherever possible.

Particularly where the tailings contain high proportions of very fine slimes, drying out can be accelerated by transpiration from suitable tree plantations. Even with such barren and hostile substrates vegetation can usually be established by regimes involving liming to establish a suitable pH followed by the application of nitrogen and phosphorus-containing plant nutrients. Indigenous grasses are often successful in the initial stages followed by legumes once a limited humus cover has been established. Such systems are usually much less expensive than proceeding by importing topsoil which is often sparse in mined-out areas.

Options for the pit itself are restricted to making it secure from trespass with the second option of encouraging organised visits through tourism, depending on the ultimate use of the closed-down operations.

Depending on the weather and hydrology of the area, it may be possible to allow the pit to fill with water, provided it is acceptable for recreational or fishing purposes and does not contaminate local surface or groundwaters. The minerals extraction industries have now developed many leisure complexes, thus providing community value from completed operations. Clearly, however, location and the ethos of the local population can be a major factor in restricting such developments. In general, the processing buildings, foundations, and contained equipment will have to be dismantled carefully and either sold or disposed in an environmentally acceptable manner. The inevitable contamination of the plant areas with heavy metals and/or chemical reagents will have to be assessed by specialists and remediated according to their recommendations. A much larger issue both physically and in terms of ultimate responsibility concerns the disposal of the "mining towns," some quite substantial, which have developed, with more or less company participation, near to most significant mining operations. It is outside the scope of this review to do other than note these restoration issues, but they are major in scope and not always the subject of clear regulations, particularly as most mines predate the requirements of modern mining regulations including EIAs.

4.3 COPPER

4.3.1 EXTRACTION

Much of the greatest proportion of the world's supply of primary copper is obtained from sulphidic minerals, particularly chalcopyrite ($CuFeS_2$), chalcocite (Cu_2S) and bornite (Cu_5FeS_4). Copper concentrations in ore bodies are generally low with large-scale extraction from open pits being economic down to 0.5% copper in ore. Underground mines may be in the range from 1 to 2% copper with occasional higher occurrences generally in deeper mines.

Large-scale open pit operations occur in the U.S.; South America, particularly Peru; South Africa and the "Copper Belt"; and in Europe, particularly Spain. Such excavations can become enormously large; the Kennecott Copper open-pit mine at Bingham Canyon near Salt Lake City, Utah, U.S. is now $2\frac{1}{2}$ miles wide and a maximum $\frac{1}{2}$–mile deep. The concentrator processes more than 70,000 tonnes per day. Several other copper mines approach this scale of operation. In these circumstances, it is not surprising that the waste rock dumps amount to many millions of tons.

Since the concentration of desired metal in the extracted ore is so low, concentration by flotation is mandatory for the smelting of copper, otherwise energy usage and plant size would be prohibitive economically. Consequently, the safe disposal of the waste from the milling process, or "tailings," is also a major management problem, particularly because of their finely divided nature, residual water and metal content, and toxic process chemicals.

4.3.2 SMELTING

In the processing of sulphidic concentrates, environmental control requires management of the following issues:

- Control of particulate emissions to atmosphere.
- Systems for fixing sulphur dioxide, usually by conversion to sulphuric acid.
- Maintenance of acceptably safe in-plant working atmospheres.
- Limiting liquid effluents to environmentally acceptable concentrations of heavy metals.

As will be apparent in a later part of this chapter, similar issues arise in the production of primary aluminium metal except that sulphidic feed materials are not involved. However, fluoride emissions require equally stringent control at primary aluminium smelters.

Current generally used technology for the conversion of copper concentrates into metal can be summarised in the following three stages after charge preparation:

- Concentrate smelting to matte (a mixture of Cu_2S and FeS).
- Converting of matte to blister copper.
- Electrorefining of blister to cathode copper.

Technologies that are widely practiced and meet likely medium-term environmental criteria are outlined below.

4.3.2.1 Concentrate smelting to matte

The most widely used technology is flash smelting, usually the Outokumpu process, in which the dried concentrates are blown into the furnace using preheated air and are rapidly reacted, mainly in the gaseous phase, before falling into the molten matte at the base of the U-shaped furnace. The matte thus formed contains 55 to 60% copper and is discharged to the converting section; the slags are discharged at the opposite end of the furnace for cleaning and/or disposal. From an environmental standpoint, the main factor in the success of this technology is that it delivers a sulphur dioxide gas well into the range required for conversion into sulphuric acid (5 to 7%); with oxygen enrichment of the preheated air, the off-gas strength can be in the 30 to 40% range. The process is significantly (15%) more energy efficient than conventional reverberatory smelting.

Where cheap electric power is available, electric furnace smelting can meet modern environmental requirements economically. In a few locations, smelting and converting have been combined in continuous smelting processes. Where these processes are provided with adequate technology for sulphur dioxide capture, they will also meet likely medium-term environmental criteria.

4.3.2.2 Conversion of matte to blister copper

With the widespread acceptance of modern smelting technology, the converting stage is generally recognised as the stage requiring closest attention and control to comply with modern environmental standards both in-plant and externally.

The Pierce-Smith converter is the most widely used system. After charging in a cylindrical, tiltable vessel with molten matte, converting takes place in two separate stages by blowing large volumes of air into the matte. In the first stage, the ferrous sulphide in the matte is converted selectively to iron oxides and sulphur dioxide; this stage continues until the matte contains 1% FeS or less. In the second stage, the remaining sulphur associated with copper is oxidised and blister copper formed. The slags formed are immiscible with the matte and are removed by tilting the converter.

The converter is provided with hooding through which sulphur dioxide can be collected and conveyed to the acid plant. Since its strength can vary through the cycle from below 1% to around 10%, a multi-converter system provided with closely programmed control is necessary to provide the acid plant with the reasonably constant-strength has it needs for efficient operation.

Several impurities, such as lead, arsenic, antimony, cadmium, and mercury are evolved during the converting stage and are collected, usually in electrostatic precipitators or the flues leading to them, for further processing or disposal.

4.3.2.3 Electrorefining of blister to cathode copper

Blister copper is melted and cast into anodes after deoxygenation by "poling" with timber or by reduction with methane, the system being provided with appropriate capture/arrestment equipment. The anodes are then electrorefined using a sulphuric acid electrolyte and, usually, pure copper cathodes. The remaining impurities in the anode copper accumulate as tankhouse slime. This may contain the precious metals silver, gold, and platinum together with selenium, tellurium, and lead; it is removed periodically for further processing. Other impurities in the anode copper, particularly nickel, cobalt, arsenic, and bismuth, accumulate in the electrolyte from which they are bled and the bleed stream separately purified.

From both the external environmental and occupational hygiene standpoints, the main issues requiring continual attention are:

- Sulphur capture and control.
- Impurity processing, particularly arsenic.

Flash smelting, coupled to well-controlled converting, provides well-proven technology for controlling sulphur since it generates a feed gas suitable for sulphuric acid manufacture in a double contact plant. This system should meet most envisaged environmental criteria and could be provided with tail-gas scrubbers should emission criteria tighten further. Developing technology necessarily has to meet these environmental criteria to be worth pursuing.

Of the other impurities, disposal of arsenic is probably the issue of greatest environmental concern (see section on arsenic below, and Mitchell, Chapter 14).

As noted earlier, arsenic is present as an impurity in most copper concentrates and requires very careful management from the standpoints of both "in-plant" occupational health and the external environment. Following the categorisation by IARC of "white arsenic" (arsenic trioxide) as a likely lung carcinogen, other regulatory agencies have implemented this hazard assessment as a risk evaluation to impose very tight in-plant criteria on all arsenic compounds. Consequently, copper smelters operate correctly to extremely low in-plant levels for arsenic compounds. In addition, most external uses for arsenic and its compounds have now disappeared with the exception of copper-chrome-arsenates for timber preservation. A parallel situation for zinc producers may be developing for markets for co-product cadmium where several applications are being regulated out of existence for perceived environmental effects. Thus, it is becoming increasingly difficult to dispose of by-product arsenic as a raw material for marketable end products. Increasingly, disposal to permanent, environmentally acceptable waste areas will be necessary. Clearly, such disposal areas should be in impervious basins, preferably clay-lined, remote from potable or recreational water courses. Compounds of arsenic, which are inert and relatively easy to make, are iron and calcium arsenate and these are some of the products of choice for disposal (see also Mitchell, Chapter 14). Recent work has indicated that even these compounds solubilise slowly if exposed to moist air and/or carbon dioxide, thus necessitating clay capping and revegetation of such disposal areas when full. Since they will remain permanently toxic, long-term surveillance and monitoring is essential and is generally required by regulation. Whether disposal to highly engineered landfills of this type is preferable environmentally to dispersion through regulated products or, at extreme dilutions, to controlled water-courses remains a moot point. From a resource standpoint, high-arsenic copper deposits such as those in Mexico are comparatively uneconomic to work and will probably remain as reserves into the distant future.

Most copper is refined electrochemically and it is essential to maintain operating conditions which do not allow the formation of the extremely toxic gas arsine whose effects are rapid, difficult to treat, and frequently fatal. Fortunately, these operating conditions are well defined and problems are extremely rare.

4.3.3 TOXICOLOGY AND ECOTOXICOLOGY

Copper is an important and essential element for both plant and animal life. It is an integral part of the molecular structures of a number of important cellular enzymes (e.g., cytochrome oxidases which are important in the cellular energy liberating process; without these, the cell would very quickly die). Copper is present in normal human serum at 120 to 140 μg L^{-1}; if its concentration rises significantly above this, signs of toxicity will be seen. Acute copper poisoning results in death following vomiting, hypertension, and coma. At postmortem examination, the liver shows centri-lobular necrosis (i.e., death of the cells around the vessels bringing oxygen and nutrients to the liver cells; clearly in copper poisoning, these vessels also bring copper, which is why the liver cells in the immediate vicinity are the first to die). Chronic exposure to copper often results in cirrhosis (a chronic liver disease caused by liver cell death and replacement by fibrous material; eventually, so much of the liver is lost that it can no longer function and death from liver failure results).

Copper poisoning in man is rare because high doses are necessary to cause toxicity. The body is adept at excreting copper in the bile and therefore toxic levels rarely build up.

In mammals, the target organ for copper is the liver. However, there are many creatures who do not have a liver but who are still profoundly susceptible to copper poisoning. Many marine invertebrates (e.g., coelenterates) are very susceptible to copper poisoning; this might be due to their having a symbiotic relationship with algae, by which copper kills the algae so indirectly adversely affecting the marine invertebrate. These fatal effects occur at low copper concentrations (in the mg L^{-1} range). The effects of relatively low environmental concentrations of copper might have profound effects upon, for example, marine ecosystems because they kill, debilitate, or reduce the reproductive potential of the inhabitants of the ecosystem.

The speciation pattern for copper in the marine environment is more complex because of the potential for complex ion formation (e.g., [CuOCl]). Little is known of the toxicity of these species, but it is possible that the greater toxicity of copper to marine organisms may reflect the greater toxicity of these complex ionic species.

Interestingly, copper is less toxic in the marine environment than in freshwater ecosystems (see above). This is very likely to be due to its being completed by (for example) amino acids present and nM concentrations in sea water. The copper complexes formed (e.g., glycine copper complex, CH_3 (COOH) NH-CuHN (COOH) CH_3) are less toxic than free copper ions.

4.4 ZINC

4.4.1 EXTRACTION

Zinc is becoming of increasing importance because of its more widespread use in the galvanic protection of steel, particularly in automobiles and truck bodies. This follows a period of depression in its uses for other forms of galvanising; in zinc-based diecasting alloys of the Mazak and Zamac types, and in fabricated forms such as roofing and rainwater goods.

Since production is predominantly by the electrolytic process (see below) requiring a high-purity concentrate, rather little progress has been made in the production of mixed zinc-lead concentrates which are appropriate feeds for the zinc-lead blast-furnace process. From an environmental standpoint, the most significant of non-ferrous minor constituents are cadmium and mercury where regulation has now severely restricted the uses of end products containing these metals.

The technology of extraction is generally as outlined earlier, but most of the non-ferrous impurities are not removed in the concentration processes and are dealt with by the smelter circuits.

4.4.2   SMELTING

The electrolytic process dominates primary zinc production but the blast-furnace (pyrometallurgical) ISF process that is usually operated to produce primary lead simultaneously also makes 10 to 15%. The characteristics of these two processes differ in almost every respect, but both achieve the major environmental objective of capturing essentially all input sulphur, usually as sulphuric acid, necessarily removing cadmium and mercury in the process. Since the electrolytic process cells require very pure zinc sulphate for their efficient operation, complex circuits are required to remove concentrate impurities such as iron, lead, and manganese as chemical precipitates, most of which are processed further at external specialist plants. The blast-furnace process can accept a wide range of metals in its feed concentrates, including lead and copper; but with the exception of cadmium and mercury, impurities are not recoverable and report, along with the input iron, to a fused slag of low leachability.

The main steps in the electrolytic process are

- Concentrate roasting, usually in fluid-bed roasters, to give a zinc oxide calcine together with sulphur dioxide which, after cleaning, is converted conventionally to sulphuric acid.
- Calcine leaching with sulphuric acid to give an impure zinc sulphate.
- Solution purification.
- Electrolysis of the solution to give zinc metal together with sulphuric acid, which is recirculated to the leaching stage.

The basic stages of the blast furnace process are:

- Sintering by the up-draught method, in which the sulphidic concentrate feed is de-sulphurised and de-cadmiumised to give hard sinter together with sulphur dioxide, which after purification is converted conventionally to sulphuric acid.
- Blast furnace reduction in which sinter is reduced by carbon monoxide to zinc vapour, which passes out of the upper section of the furnace to the condenser, and molten slag with lead bullion which collect for tapping at the hearth.
- Condensation, in which the zinc vapour issuing from the shaft is absorbed in circulating lead, thrown up as a shower of droplets, for subsequent zinc separation by cooling in launders. This shock chilling of zinc is essential to freeze the reversion reaction by which zinc vapour undergoes rapid oxidation.

4.4.3   TOXICITY AND ECOTOXICITY

Zinc is an essential trace metal. Deficiencies are noted in cases of skin disorders, skeletal defects, and stunted growth. Recommended daily dietary intake for zinc is around 0.3 mg kg^{-1}. Zinc is involved in several fundamental mechanisms including protein synthesis, the efficient functioning of many enzymes, and insulin storage.

Zinc is relatively non-toxic and there is a wide margin between the toxic level of zinc and normal dietary intake. Probably the widest-known toxic effect is "metal-fume fever," formerly a not uncommon occupational exposure hazard at poorly operated zinc smelters and galvanizing plants. Common symptoms included fever, nausea, depression, headaches, and dryness of the mouth and throat; fortunately, this is now a very rare condition. There is some evidence that zinc may be involved in potentiating carcinogenicity and teratogenicity of other molecules.

In the broader environment, many plants have been shown to develop tolerance to the uptake of zinc, though this is sometimes accompanied by chlorosis.

4.5 LEAD

4.5.1 EXTRACTION

Pure lead is derived approximately equally from primary sources (concentrates) and recycled metal, mainly from time-expired car batteries. Apart from the 5 to 10% which is made by the combined zinc lead blast furnace, most of the remaining metal is produced in lead blast furnaces or by one the several new processes (e.g., Kivcet, Isasmelt, QSL; see Warhurst, Chapter 3) which have been developed over the past 10 to 15 years.

Many of its well-known uses in the sheathing of heavy-duty underground power cables, pigments, shot, and collapsible tubes have been heavily depressed in recent years. This is largely because of its human- and ecotoxicity-generated particularly by its widespread dispersion arising from the use of alkyl-lead compounds as octane-enhancing petrol additives. The only buoyant uses for lead are in automotive batteries that are, to a very large extent, reprocessed after the end of their useful lives, and in roofing applications.

4.5.2 TOXICITY AND ECOTOXICITY

Lead is toxic to the nervous system, its severity depending upon its chemical form. Lead in organic compounds is very much more toxic than inorganic lead because organic forms are absorbed much more readily by the body and can cross biological membranes into cells, thus exerting its toxicity more readily. Inorganic lead is emphasised in this section since it is more pertinent to the mining industry but is must be remembered that bacteria can convert inorganic to organic lead compounds and therefore influence significantly its toxicity to higher animals.

Lead can be ingested with food and water or absorbed through the lungs. The latter route is important for occupational exposure and, for the general public, through exposure to atmospheric lead from car exhaust emissions. However, generalised environmental pollution leads to residues in food and water. For example, most potable water contains <0.05 μg L^{-1} lead which leads to a total daily intake (in the U.K.) of about 10 μg. Children are far more susceptible to lead toxicity than adults; this is in part due to the fact that children absorb a greater proportion of ingested lead. In very young children (i.e., 0 to 1 years old) the nervous system is still developing and therefore is more susceptible to neurotoxic chemicals. Lead is present in the blood of most people (in the U.S., men aged 30 to 40 have blood lead concentrations ranging from 160 to 180 μg L^{-1} whole blood). It is transported in the erythrocytes partly in association with haemoglobin (Hb) and it eventually resides in one of two body pools. These pools represent long-term (bone, half-life of excretion = 20 years) and short-term (soft tissue [e.g., liver] half-life of excretion = months). Excretion is mainly in the urine; indeed, the kidney is an important soft tissue pool of lead which accumulates with age. The accumulation of lead in the kidney is important to consumers of offal, because kidneys from old animals might contain levels of lead above those generally accepted as being suitable for consumption in large quantities. Following serious toxicity related to occupational exposure, controls were introduced in the workplace in 1911 and these have been maintained ever since.

A very important facet of lead's toxicity is its passage across the placenta into the developing embryo/foetus; this is particularly important because its presence in the embryo/foetus might inhibit the development of the nervous system, resulting in a neurologically defective offspring.

There are many theories relating to the mechanism of neurotoxicity of lead. Perhaps the most likely is that lead mimics calcium in the body and interferes with the transport of calcium across the neuronal membrane, inhibiting neurotransmission which is dependent upon ionic exchange across the axonal membrane. Lead is also known to inhibit the function of the inhibitory neurotransmitter, γ-aminobutanoic acid (GABA) which is extremely important in initiating nervous impulses in the central nervous system (CNS).

Lead is also an inhibitor of Haem (the central oxygen-carrying component of Hb) synthesis because it inhibits the enzyme aminolevulinic acid synthetase (ALA-S), resulting in increased levels of aminolevulinic acid (ALA) in the blood. This is important for two reasons: lead poisoning results in anaemia following exposure to high lead levels, and measuring serum ALA is an important marker of exposure to lead.

Lead has similar effects on other inhabitants of ecosystems. Lead will affect vertebrates, with nervous systems and a Hb-based oxygen-carrying circulatory system in much the same way as man. Other creatures and plants are quite different. The most important moderator of lead toxicity is its uptake by primitive animals and plants. If it is absorbed, it might behave like calcium and have wide-ranging toxicological properties because calcium is an important element in cell signaling, nervous transmission, cellular growth control, and so on. In plants, lead replaces magnesium and so inhibits the synthesis of chlorophyll, without which green plants cannot survive.

In ecosystems, lead occurs as several species which are readily interconvertible. As discussed above, alkylation of lead by bacteria is an important reaction because this generates a very well-absorbed, profoundly neurotoxic organolead species.

In conclusion, lead is the most worrying of our metallic environmental contaminants because it is the most abundant toxic metal in the environment and its toxicity might only become manifest (i.e., reduced IQ) after many years of exposure. It is probably the metal with the smallest margin of safety from the human absorption standpoint.

4.6 ARSENIC

Arsenic compounds are intensely toxic and have been associated with fatalities for many hundreds of years. They were poisons of choice for people with murderous intent in the times of Queen Victoria; they have even been the subjects of classic plays and films (for example, *Arsenic and Old Lace*). Arsenic has very limited applications in medicine (it was formerly used in the treatment of syphilis), industry, and as a fungicide (very rarely used); its main use is as a component of copper-chrome-arsenate timber preservatives.

The speciation of arsenic is extremely complex; it occurs in oxidation states 0, +3 and +5, but some of its cationic forms are more toxic than others, irrespective of the oxidation state of the arsenic within the compound. Arsenic acid (H_3AsO_4 – arsenic's oxidation state = +5) is extremely toxic (LD_{50} [Rabbit] = 6 mg kg^{-1}), probably because it is well absorbed by animals, whereas arsenic trioxide (As_2O_3 – arsenic's oxidation state = +3) is less toxic (LD_{50} [Mouse] = 39 mg kg^{-1}) because it is insoluble and less well absorbed. Arsenic is metabolised by micro-organisms in ecosystems to form organoarsenic compounds (e.g., dimethylarsenic – $As(CH_3)_2$) which are generally less toxic than their inorganic counterparts. Note that this is in opposition to lead and mercury where the organometals are very much more toxic than the inorganic forms). Indeed an important route of detoxification of ingested arsenic in mammals is methylation.

Generally, trivalent arsenic is the most toxic form of arsenic (despite the example given above which illustrates the importance of absorption in toxicity); its mechanism of toxicity is akin to that of mercury. Arsenic forms complexes with sulphydryl groups and so inhibits enzymes and interferes with important structural proteins. Its main toxic target is the mitochondrion where it inhibits biochemical respiration and results in cell death.

Arsenic accumulates readily in animals and prolonged exposure leads to neurotoxicity (as is the case with other metals whose mechanisms of action involve sulphydryl binding) and liver toxicity (because it inhibits important hepatic enzymes). In addition, it is teratogenic (causes birth defects) and IARC have categorised arsenic trioxide fume as a suspect lung carcinogen.

Relatively low levels of arsenic have devastating effects upon ecosystems because of its multifaceted mechanisms of toxicity. At the high dose end of the spectrum it simply kills animals and plants, whereas at the lower dose end of its toxicity spectrum it reduces the production of viable offspring which significantly upsets the balance of the ecosystem.

4.7   CADMIUM

Compared with other metals dealt with in this chapter, cadmium is a relatively recent "introduction" into the environment. Although it was discovered in 1817, it was not utilised industrially until several decades after that date. Although its uses are now subject to severe regulatory control, it is still used in small quantities in electroplating, as a thermally-stable yellow-orange-red pigment, as a stabiliser for some plastics, and in nicke-cadmium rechargeable batteries. Since it arises inevitably from the production of zinc, its disposal in a environmentally acceptable manner is a significant problem.

The gastrointestinal absorption of cadmium is less efficient than for many other toxic metals (indeed, occupational poisoning in industry is usually due to breathing cadmium fumes from poorly controlled industrial operations) and so its uptake from dietary sources is not very efficient. On the other hand, cadmium is well absorbed by plants (unlike lead) and so levels in plants grown on contaminated land can be high. Another important facet of cadmium's disposition is that it accumulates in the kidney and liver. For this reason, animals grazing on cadmium-contaminated land will accumulate high kidney and liver cadmium concentrations which could imply that it would be inadvisable to eat such offal in large quantities.

It requires surprisingly high dietary intakes of cadmium to cause acute toxicity (e.g., beverages containing 16 mg L^{-1} were at the lower end of a concentration range associated with toxicity). The signs of toxicity are severe gastrointestinal disturbances and, after chronic exposure, pulmonary dysfunction which might lead eventually to emphysema. In addition, renal tubular dysfunction might lead to kidney failure after long-term exposure (as would be expected, because cadmium accumulates in the kidney). It is important to remember that cadmium is accumulated and long-term low doses can result in significant chronic toxicity.

Cadmium mimics calcium (as does lead) in biological systems and therefore it might be expected to interfere with biological processes dependent upon calcium. For example, osteomalacia is associated with chronic exposure to cadmium, as is myocardial disease.

A very interesting toxicological property of cadmium is its association with tumour development. Cadmium is one of the few metals (including nickel) which is regarded as carcinogenic. However, the doses necessary to induce tumours are far in excess of those which would be expected following environmental exposure (cadmium's carcinogenicity might be a worry for men working in factories where airborne cadmium concentrations might be elevated).

4.8   MERCURY

Mercury is important toxicologically because of its continued use in gold recovery. Small-scale gold panners use mercury to extract gold from river silt by utilising mercury's important property of amalgamation (this involves gold "dissolving" in the mercury, thus facilitating its isolation from the silt). The amalgam (a liquid) is removed from the silt mixture and the mercury is evaporated off to leave the gold. The problem is that some of the mercury enters the river being panned and the rest enters the atmosphere when it is evaporated off. This is a potentially severe problem; panners are beginning to show the neurological signs of mercury toxicity and it is only a matter of time before ecotoxicity is seen.

Mercury is a classic poison. The Greek and Romans were well aware of its deleterious effects. Lewis Carroll's depiction of the Mad Hatter was based upon the fact that hat felters in Victorian times used mercuric chloride to treat wool as part of the felting process. Many of them became mad because of the neurotoxicity of mercury. In more recent times, Japan has had a major case of mercury environmental pollution which led to human deaths. Minimata Bay is a fishing area in Japan. In the late 1950s, many fishermen and their families (and later, members of the general public) began to show neurological degeneration. After very considerable study, it was shown that their fish diets were heavily contaminated with mercury because local industry had decided to discharge its waste (containing mercury) directly into the bay. The discharged mercury was inorganic, but the bacteria in the silt of Minimata Bay methylated the mercury (in the same way as described above for lead) to methyl mercury. The fish (which have an oily flesh) absorbed the methyl mercury and the population around Minimata Bay relied upon fish as a dietary staple. Methyl mercury is extremely toxic because its hydrophobic molecular structure facilitates its uptake into the brain. This was the first large-scale indication of the importance of bio-methylation of heavy metals in promoting their absorption and toxicity to animals and man.

The mechanism of toxicity of mercury is not fully understood but it almost certainly involves binding to the sulphydryl group (-SH) contained in the host target. Many important enzymes and neurotransmission processes rely upon intact -SH groups and therefore mercury disrupts these processes by forming sulphydryl complexes.

In man, signs of mercury poisoning involve generalised neurological degeneration associated with neuropathy. Other mammals have very similar symptoms. In theory, any animal with a nervous system is susceptible to mercury poisoning and, more broadly, mercury profoundly affects all living creatures because of its thiol-complexing property. Thiol compounds are important in maintaining protein structures and so are crucial for enzyme activity.

In conclusion, mercury is a poison which manifests its toxic effects at relatively low concentration (doses).

4.9 ALUMINIUM

Aluminium is the third most common element in the earth's crust and is mined extensively, generally from shallow deposits. For many years, it has had the most rapid growth rate of the non-ferrous metals and has many important industrial and commercial applications because of its ease of fabrication, low density, and high heat capacity.

Primary aluminium is universally manufactured by the electrothermal reduction of alumina dissolved in a cryolite (Na_3AlF_6) melt, both contained in carbon-lined steel cells (or "pots") using carbon electrodes. The American Hall and the Frenchman Herault developed this process independently in the 1880s: it operates at about 950°C. The process chemistry has remained essentially unchanged every since, though many efficiency improvements have been made by close attention to important detail.

Two different embodiments of the process have been used on the industrial scale:

- The prebake process, in which the carbon anodes, manufactured in a separate plant, are fitted to each cell and replaced as they are eroded by the electrothermal reduction process.
- The Soderberg process, in which the anodes are continuously made and renewed at each pot as erosion takes place.

The former process predominates, a tendency that has accelerated in recent years as pressures on both efficiency and in-plant hygiene have increased. Modern prebake cells operate at around 300 kA and might produce in the order of 2000 kg metal per pot per day. Power utilisation is obviously a key factor in the economics of the process and is now running typically around 13 kWh kg^{-1} aluminium.

In contrast to the feed and product complexities of copper, zinc, and lead smelting, primary aluminium production is based on a single high purity feed (alumina) to manufacture a single high-purity product. In comparison with other non-ferrous operations, neither feed nor product is seriously toxic.

The major environmental issues requiring management are

- Gaseous and particulate fluorides.
- Pitch fume and polycyclic aromatic hydrocarbons (PAHs).
- Minor emissions of sulphur dioxide.
- Minor emissions of perfluorinated alkanes because of their potential as high impact "greenhouse" gases.
- Spent pot linings and anode stubs.

Since most modern processes utilise dry scrubbing of cell gases using feed alumina for fluoride removal, such plants no longer face the problem of wet scrubbing of these gases and the disposal of the resultant aqueous effluent. Waste water for disposal at "dry" plants is restricted to that used for cooling mechanical equipment.

Dry scrubbing of gaseous and particulate fluorides has now reached a very high state of efficiency after development by many companies over the past 30 years. In its most recent embodiment, the total incoming alumina is passed into a vertical reactor, through which the cell gases pass. Efficient adsorption of the fluorides occurs after which the alumina/fluoride solids are recovered in a bag-filter plant and returned to the cell. This forms an efficient type of internal recycling as well as removing emissions; a total capture efficiency of 97 to 98% for fluorides is now claimed. Ejection to atmosphere is usually through a tall stack, though some countries now require a final chemical treatment. Such treatment, sometimes in the same plant, is used to neutralise the small volumes of sulphur dioxide, which arise from the pitch utilised in anode making and in miscellaneous heating processes.

Great care also has to be exercised to minimise fugitive emissions from the cells by the provision and effective utilisation/maintenance of cell hooding. This is also a major factor in the control of in-plant hygiene.

Volatile emissions derived from the pitch have significant carcinogenic potential and therefore, require careful management, starting with efficient collection. In recent technology, collected fumes is adsorbed on finely divided coke that is then utilised in the manufacture of anodes. Coal-tar pitches are now available with much lower volatile contents than hitherto, but attempts have been unsuccessful to replace coal-tar pitch with alternative resin binders that might not evolve potential carcinogens during the anode-making process. Control of pitch volatiles at the anode manufacturing plant is generally simpler than at Soderberg plants.

The recent realisation of the very high "greenhouse" or global warming potential of fully fluorinated alkanes has uncovered another previously unappreciated environmental concern in the production of primary aluminium. Since these compounds are so inert chemically and present in relatively low concentrations in the cell emissions, they have to be minimised at source rather than by chemical treatment or physical absorption. This has been done by minimising the frequency of so-called "anode effects" resulting when the intermittent feeding of alumina to the cell results in below-optimum quantities of alumina entering the cell. The management of anode effects is greatly facilitated by the close control of cell conditions made possible by modern computer control technology.

The main solid wastes arising from primary aluminium production are eroded anodes ("anode stubs") and the cathode cell linings. The latter absorbs cryolite over its operational life (which in modern cells might be 5 or 6 years) eventually precluding stable operation. Such time-expired cell linings might contain as much as 50% cryolite as well as cyanides and aluminium carbide, and require permanent disposal in secure landfills. Some countries continue to allow disposal to designated areas of the deep sea; whichever system is used, no adverse environmental effects have been found. In spite of this, disposal to secure landfill is not permitted in the U.S. and storage in secure aboveground facilities is the only option at present. Many technologies for purification and recycling have been reported, but none are apparently in commercial use, though in some circumstances it is practicable to utilise the materials in the steel and cement industries.

4.9.1 TOXICOLOGY AND ECOTOXICOLOGY

The ecotoxicological impact of aluminium was, for many years, thought to be minimal. Recently, however, some evidence has accumulated to indicate that a decrease in the pH of rain (acid rain) driven by the emission of sulphur dioxide from coal burning and sulphide mineral smelting has raised levels of aluminium in rivers and lakes. These high concentrations of aluminium in water have resulted in fish dying because aluminium hydroxides precipitate on their gills where the pH is alkaline and the hydroxides are insoluble. This toxicological effect of aluminium originates from natural (soil) sources rather than metal introduced into the environment from mining and smelting but its release is probably due ultimately to the effects of man-made sulphur dioxide in the atmosphere.

The apparent non-toxicity of aluminium may be challenged by the views of some scientists that it is implicated in the development of pre-senile dementia (Alzheimer's disease). Its presence has been demonstrated in the plaques in the brains of Alzheimer's disease patients. It is possible that long-term, low dietary intake of aluminium might be important in the development of this disease that is becoming increasingly prevalent amongst the elderly, with severe consequences for their quality of life and that of their careers.

The mechanism of toxicity of aluminium is not well understood. Concentrations of aluminium in the human brain are normally in the region of 2 mg kg^{-1}. In aluminium-sensitive species (e.g., cats), brain concentrations of only twice the "normal" value appear to result in behavioural changes and defective motor function. It may be, therefore, that safety margins for human beings are rather narrow.

An unusual case of accidental mass exposure to aluminium in drinking water occurred in July 1988, when 20 tonnes of concentrated aluminium sulphate was discharged into the Lowermoor-treated water reservoir which supplied the inhabitants of Camelford, Cornwall, and nearby Port Isaac, Tintagel, and Boscastle (aluminium sulphate is in widespread use as a coagulant in the purification of water to potable standards). In spite of the astringent taste of this highly contaminated water at the taps of householders, a few persons received high doses of aluminium and there were reports of nausea, vomiting, diarrhoea, headaches, fatigue, itchiness, sore eyes, mouth ulcers, and even changing of hair colour, the last effect probably resulting from aluminium interfering with artificial hair dyes. However, no substantiated cases of acute toxicity have been found, and it remains to be seen whether there will be any long-term effects on those persons who drank the contaminated water. This incident suggests that the acute toxicity of aluminium to man is quite low, otherwise many more acute toxicity cases would have been reported.

4.10 CHROMIUM

Chromium is an important metal in many industrial and manufacturing processes. For example, it is a constituent of stainless steel and is an important component of many pigments. It is mined as chromite ore.

Chromium's speciation is complex; it occurs in oxidation states 0 to +6, although only trivalent chromium (the most common form) and hexavalent chromium are of biological importance. Trivalent chromium is an essential ion, involved with glucose metabolism and particularly in the action of insulin. Hexavalent chromium appears to be the "toxic" form of the metal, but his may merely reflect the fact that if the hexavalent form is elevated, then the trivalent form must be reduced.

At high inhalation levels, chromium is carcinogenic and causes lung cancer. This facet of its toxicity is very unlikely to have any environmental relevance but it is a serious concern to workers in the smelting and related industries. In general, chromium is of low environmental toxicity unless very large amounts are deposited on land or discharged into watercourses.

4.11   NICKEL

Nickel is the 10th most abundant element in the earth's core, making up some 10% of the core's total mass. It is far less abundant in the earth's crust. It is mined as pentlandite, which is a nickel iron sulphide $(Ni,Fe)_9S_8)$ or garnierite, a magnesium nickel silicate $((Mg,Ni)_6[Si_4O_{10}](OH)_8)$.

Nickel is an essential element being present in the metallothionein enzyme, urease. Urease catalyses the decomposition of urea to ammonia and carbon dioxide and is particularly important to plants and bacteria. Nickel has not been found to be involved in mammalian enzyme systems, but if rats are deprived of the metal in their diets they fail to grow at the normal rate. Clearly nickel has a yet-to-be-discovered role in mammalian metabolism.

Nickel is present in the environment only at very low concentrations (e.g., in freshwater, $1 \mu g\ L^{-1}$). If environmental nickel concentrations rise marginally above their "normal" values plant growth is severely inhibited. For example, nickel in soils around nickel mines is often very significantly elevated, in such regions the flora is often limited to one or two species that are known to tolerate high soil nickel concentrations.

In mammals, nickel is carcinogenic via the inhalation route. Workers exposed to nickel dust are at increased risk of developing lung cancer. In addition, nickel is associated with contact dermatitis, which sometimes develops in people who wear jewellery containing nickel.

4.12   THE TOXICITY AND ENVIRONMENTAL IMPACT OF MINERALS

There are many commercially important minerals that are mined for use in industry, agriculture, and as fuels. A selection of the most commonly extracted minerals will be dealt with here in order to illustrate their impact upon the environment.

4.12.1   LIME

There are two forms of lime: quicklime (calcium oxide—CaO) and slaked lime (calcium hydroxide—$Ca(OH)_2$). Slaked lime is formed by reacting quicklime with water.

Quicklime is the basis of cement, which when mixed with sand and water undergoes a chemical reaction involving the formation of calcium hydroxide with the liberation of much heat and results in the formation of concrete (or mortar).

Slaked lime is used to neutralise acid soils and improve clay soils in agriculture and gardening. Calcium hydroxide is a fairly strong base that explains its use in neutralising acid soils. Its improving effect upon clay soils is due to calcium binding to negatively charged clay particles, thus binding the particles together and forming a better soil crumb structure.

Quicklime is also used in the iron smelting industry where it acts as a reducing agent at the extremely high temperatures of the blast furnace and in the presence of a carbon source.

The environmental impact of lime is very difficult to assess because it is ostensibly non-toxic. Its application to acid soils changes their pH so preventing growth of acid-loving species (e.g., *Ericacea*). This impact, however, is minor because such species would be unlikely to grow on cultivated land that is regularly ploughed and sown with crops. The quarrying of lime has a major local environmental impact visually because it results in the destruction of the immediate environs of the quarry. This issue is currently a matter of great controversy, as the number of limestone habitats in the British Isles are gradually diminishing and with them goes a very characteristic and unique flora.

4.12.2 BARITE

Barite is barium sulphate ($BaSO_4$). It is extracted for the oil and gas industry where it is used as a mud slurry lubricant in the drilling process. Soluble barium salts are exceptionally toxic. However, $BaSO_4$ is of very low toxicity because it is almost water insoluble and not absorbed via the gastrointestinal tract (indeed, barium sulphate is radio-opaque and is pumped through the intestine in barium meal X-ray investigations in clinical diagnosis).

Barium does not appear to be an essential element in either animals or plants, but some primitive marine animals accumulate crystals of insoluble barium salts that are thought to help the creatures to sink on account of the high density of barium salts. This is a unique approach to the problem of buoyancy!

The environmental impact of barite is likely to be very low in purely toxicological terms. However, if discharged to watercourses it might have deleterious physical effects upon aquatic ecosystems because the fine powder has a high density and sinks to the bottom of water systems, forming a smothering blanket on the bed. This would kill bottom-living and feeding organisms.

4.12.3 FLUORSPAR

Fluorspar is calcium fluoride (CaF_2). In comparison with the other minerals discussed in this chapter, its scale of production is relatively small. It is used as a flux in iron blast furnaces with smaller quantities being used in the ceramics and glass industries. A major chemical use is as the raw material for the production of hydrogen fluoride, the basis for fluorine-containing refrigerants, solvents, heat transfer agents, and air-conditioning fluids. However, since the banning of the "hard" chlorofluorocarbons, this use has dropped sharply. Very small amounts are used to produce the highly reactive fluorine gas.

Surprisingly, fluorspar is almost insoluble in water (150 mg.dm^{-1} at 18°C) and consequently is of very low toxicity. From the point of view of direct toxicity to either man or the environment fluorspar is insignificant, but some deposits contain relatively high concentrations of lead and/or cadmium.

When assessing the environmental toxicity of a mineral, it is important to consider the environmental impact of the processes for which the mineral might be used and the products that might be obtained from the mineral. All soluble fluorides are intensely toxic (e.g., sodium fluoride, NaF; oral LD_{50} [rat] = 0.18 gkg^{-1}, 5 g has caused death in man). Sodium monofluoroacetate is one of the most toxic substances known and was banned from use as a rat poison. Even though these fluorides are so toxic, their environmental impact is small because of their limited industrial and medicinal use, e.g., the prevention of tooth decay by the fluoridation of drinking water in some areas and as an additive to some toothpastes. However, emissions of both particulate and gaseous fluorides from primary aluminium plants have to be controlled rigorously to prevent damage to local vegetation and to the bone structures of any neighbouring cattle ("fluorosis").

The widespread use of chlorofluorocarbons (CFCs), formerly in hundreds and thousands of tonnes annually, was banned internationally by the Montreal Protocols of 1987 and 1990. These compounds were developed in the early 1930s as non-inflammable, non-toxic fluids for ultimate use in refrigeration, air conditioning, solvents, aerosol propellants, and blowing agents for increasing the effectiveness of plastic heat insulation systems. In 1975, Molina and Rowland postulated that these highly stable substances could remain in the upper atmosphere for many decades and could slowly initiate radical-catalysed reactions promoted by high intensity ultraviolet radiation. This reaction results in reduction in the concentration of ozone (O_3) and has caused a "hole" to develop in the ozone layer above the earth's poles, the existence of which was shown in 1985 by dramatic photographs taken in the Antarctic. If this is allowed to continue it could result in environmental devastation and severe effects upon human health. The hole in the ozone layer allows low wavelength UV radiation to reach the earth's surface; this will result in warming and will increase the incidence of skin cancer (particularly melanomas). Fortunately, these effects were discovered in the early 1990s and have led to the Montreal Protocol that requires that CFCs be withdrawn by 2000. So far, no proven effects on the earth's environment or population have been documented but, clearly, the stratospheric effect had to be halted. The CFCs provide another example of the complexity of ecotoxicological effects and the difficulty of defining them accurately, particularly over a prolonged time-span.

4.12.4 BORAX

Borax ($Na_2B_4O_7.10H_2O$) is the best known of a large group of borax-containing compounds whose annual production worldwide is about 2 million tonnes. The commonest minerals are $Na_2B_4O_7.10H_2O$, known mineralogically as tincal, kernite ($Na_2B_4O_7.4H_2O$), and colemanite ($Ca_2B_6O_{11}.5H_2O$). Extraction is both by conventional open-pit methods and by the processing of lake brines. The major application of borax and its related compounds is in the manufacture of borosilicate glasses, e.g., Pyrex, which contain 10 to 13% B_2O_3 in ceramic glazes and enamels; as a metallurgical flux; in detergents (usually as sodium perborate $NaBO_3$); and in timber preservation.

Borax is of relatively low toxicity to mammals (oral LD_{50}[rat] = 6 g kg^{-1} although 10 g has caused death in a child). Its environmental toxicity is obscure; however, the borate ion is known to be essential for plant life. Borax's high water solubility (60 g dm^{-3}) facilitates its rapid transport in the terrestrial environment following rain and into watercourses. Its insecticidal properties will result in deleterious effects upon crustacea in these environments, and so cause imbalance in many ecosystems.

4.12.5 SILICA

Silica (silicon dioxide, SiO_2) is almost inert and forms about 27% of the mass of the earth's crust (e.g., quartz). Sand is primarily silica, which illustrates its ubiquitous occurrence and low chemical reactivity. Silica melts to form glass, hence its most important industrial use. In addition to the manufacture of glass, silica is used to form the moulds in metal castings, as an anticaking agent, and in the form of silicone polymers to prevent liquids from foaming.

Whereas silica is essentially non-toxic, inhalation of fine particle silica (particle size <5 μm diameter) is associated with the development of silicosis which may lead to lung cancer. This is an example of non-genotoxic carcinogenesis and is likely to be due to the irritant nature of silica particles in the lung which results in cell proliferation and carcinogenesis.

The environmental impact of silica is very low indeed; the only adverse effect is the destructive effects of its extraction (for example, as quartz). It is interesting to note that some 2×10^{11} kg of silicon (mostly as silica) is cycled through the hydrosphere per year; this is an incredible amount, bearing in mind the exceptionally low water solubility of silica. It is important for some simple plants (e.g., diatoms) whose cell walls are based upon SiO_2. A recent finding suggests that silica might have toxicity reducing capacity in the biosphere; it is thought to reduce the toxicity of aluminium (see aluminium, above), probably by sequestering the aluminium in a form not bio-available to fish.

On balance, silica is beneficial to the environment. The relatively small amount extracted for industrial purposes, put in the context of the massive quantity that forms the sea and river beds, will have a negligible environmental impact.

4.12.6 ASBESTOS

Asbestos is a group of complex silicates of sodium, calcium, iron, or magnesium. The group is broadly divided into two groups, namely blue asbestos (crocidolite — $Na_2Fe^{2+}_2(Si_8O_{22})(OH)_2$) and white asbestos (e.g., chrysotile — $Mg_6(Si_4O_{10})(OH)_8$). They are used as heat-resistant insulators (e.g., in pipe lagging, roofing, brake linings, and fireproof gloves) but substitutes of lower toxicity are now widely used. Asbestos usage is now strictly regulated.

Blue asbestos is a potent non-genotoxic carcinogen (see silica, above) to the lung, pleura, and chest wall when the asbestos fibres are smaller than 5 μm long (i.e., it is respirable). It causes a very specific, highly malignant tumour (mesothelioma) which is untreatable and invariably causes death, usually within a year of diagnosis. As a result of this toxicity, blue asbestos is no longer used and there is a major programme aimed at removing all blue asbestos from all buildings.

On the other hand, white asbestos is very considerably less toxic (this is probably because its fibres are less sharp, and therefore if inhaled into the lung cannot penetrate the pleura and trigger the inflammatory response, which is thought to initiate non-genotoxic carcinogenesis). Despite this, its use has diminished very considerably and much has been removed from previous applications.

The removal programmes for asbestos have resulted in a major problem of disposal. Asbestos is inert; being non-inflammable and resistant to all acids, it is therefore not possible to incinerate or chemically destroy it. The only way in which it can be disposed of is to landfill. There are strictly monitored regulations governing how it may be disposed of; it must be contained in approved packaging and its burial position must be recorded. This is all that can be done, but asbestos is resistant to decay, and therefore will remain in the ground for many hundreds of years. Practicable methods for the destruction of asbestos should be an environmental priority, otherwise a risk exists that buried material will inadvertently be exposed and again become a hazard to health. Asbestos is only a problem to man and animals possessing lungs, but it is of major significance whose adverse effects to health are difficult to exaggerate.

4.12.7 COAL

Coal is formed by the compression and fossilisation of the remains of plants from swampy environments, such as river deltas. The process of coal formation proceeds through several stages over thousands of years. It commences with peat and progresses via lignite and bituminous coal to anthracite. The final coal is mainly carbon and represents a very important energy source. As discussed, coal is derived from plants, and therefore contains other elements present in its precursors; for example, sulphur (which originated in part from sulphur amino acids in proteins), nitrogen (which originated from the amino groups of protein amino acids), and small concentrations of several heavy metals. These "contaminants" are important when assessing the environmental impact associated with the burning of coal.

In terms of environmental impact, the burning of coal can be considered at two levels Primarily, burning coal releases a considerable amount of energy mainly as heat. The plants that form the coal laid down this energy resource by the process of photosynthesis. While the formation of the energy source occurred over many years, its release happens over a very short time period (i.e., minutes to hours). The energy had been sequestered deep in the earth until the coal was mined. This release of energy adds to the problem of overheating the earth.

In addition to the energy release, which is of course why coal is burned, the combustion process consumes atmospheric oxygen and results in the release of oxides of carbon (carbon monoxide — CO; carbon dioxide — CO_2), sulphur (sulphur dioxide — SO_2) and nitrogen (nitrous oxide — N_2O); nitric oxide — NO; nitrogen dioxide — NO_2). These oxides have profoundly deleterious effects upon the environment.

Carbon dioxide is a greenhouse gas; it increases the internal reflection of heat from the earth, so causing global warming. The burning of coal therefore contributes to global warming in two ways: by producing heat on burning and by releasing CO_2, which reduces radiant loss of heat from the earth. Informed scientific opinion suggests that the earth's atmosphere is currently warming at the rate of 2 to 3°C per century. If this rate of increase is sustained, thawing of icecaps will result and low-lying areas (e.g., Holland) will be inundated. It has even been speculated that changes in the earths weight distribution might occur, resulting in changes in the earth's axis of tilt with consequential major climatic changes. Of course, the burning of coal is not the only source of "greenhouse" gases; the burning of oil and gas also releases such gases from both stationary and mobile sources, such as motorcars and aircraft.

A more immediate environmental problem associated with the burning of coal is the liberation of nitrogen oxides (commonly referred to as NO_x or NOX) and SO_2. These gases are forced high up into the atmosphere from chimneystacks and are transported on the prevailing winds. During their journey, they dissolve in water in the air and become oxidised further by atmospheric oxygen to form acids. These acids precipitate with rain many miles away from the chimney of their origin. For example, the Scandinavians and Germans have blamed Britain for forest deaths and acidification of lakes and river due to this acid rain. This has now become a political "hot potato" and much diplomacy is needed to secure international agreement for minimising emissions which might result in acid rain (see also aluminium, above), particularly because such agreements will seriously affect modern lifestyles.

4.13 CONCLUSIONS

This chapter provides a brief account to the technologies for extracting and refining the major non-ferrous minerals. Outlines are also given of the toxicology and ecotoxicology of each, including the major minerals that are used in a form similar to that extracted, rather than for any metalliferous content; for example, fluorspar, borax, silica, asbestos, and coal. Real and/or perceived adverse environmental impacts are described for each, together with outlines for their amelioration by good management and ultimate remediation/restoration.

No attempt is made to quantify the ecotoxicological impacts of these industries. Extractive operations are necessarily located at sites where economically workable deposits exist. Accordingly, their impacts are highly site-specific since deposits exist in all parts of the world, reflecting the complete range of weather and terrain. Some exist in developed countries where they may be uncomfortable neighbours for relatively large numbers of sophisticated populations; many others will be in relatively virgin and undisturbed areas where conservation interests may or may not become activated.

Whilst the generally small "footprint" of even the largest extraction operation is usually overlooked because of its potentially significant local impact, the distribution of the products of the extractive industries is worldwide. It is the use and disposal of these products, and their real or perceived adverse toxicological or ecotoxicological impact, which has provided the main driving force for concern, and hence resultant restrictive legislation applied to the industry as a whole.

Although non-ferrous metals have been substituted heavily by plastics, particularly in decorative or unstressed applications, it should be noted that they remain essential for many aspects of modern life in the developed world. Our judgment is that their extraction and use can never be entirely without adverse environmental impact. Nevertheless, significant avenues exists for amelioration of such impacts. It is judged that the modern industry is fully aware of the need to apply both efficient technology and progressive management techniques. In this way, it is possible to achieve the continual improvement necessary for the industry to prosper in an environmentally acceptable manner.

5 Issues in the Management of the Socioeconomic Impacts of Mine Closure: A Review of Challenges and Constraints

Alyson Warhurst, Magnus Macfarlane, and Geof Wood

CONTENTS

ABSTRACT

There is a dearth of empirical literature on the subject of the socioeconomic effects of mine closure. This chapter draws on a limited literature and from fieldwork observations and analysis on the part of the authors to define some of the issues and possible policy solutions. The chapter analyses the range of impacts relevant for both individuals and communities across the issues of income, mobility, skills, health, well-being, and alternative work options. It considers implications of closure for both the formal mine workforce and also the broader network of subcontracted suppliers of goods and services. Examples of "best practice" are highlighted, and some constraints to their "take up" are analysed. The principal conclusions and policy implications include:

- Social Impact Assessment (SIA) needs to be ongoing throughout the life phases of the mine and planning for decommissioning, downsizing, and closure needs to begin at the outset based on criteria developed through the SIA.
- Closure planning should address effects and solutions for the remote community involved in supplying the industry, as well as the formal and informal workforce of the mine.

- Profiles of recruits, recruitment strategies, and human resource development through the mine life need to be included in closure planning to facilitate transition for redundant workers and their families, and broaden the possibilities of future work options.
- A post-closure environmental management plan needs to be in place where there is a threat of ongoing contamination/environmental damage, such as acid rock drainage or tailings leaks/slippages, so as to improve possibilities of alternative land uses, particularly farming.
- Closure planning could include alternate uses for housing, facilities, and equipment, as well as policies to protect the continuation of social networks and community activities.
- Continuous consultation with the community is paramount, assisted by participative approaches to forward planning, so as to involve the community from the outset in addressing eventual closure and future options.
- Financial mechanisms need to be in place to ensure sufficient resources exist at the end of the mine's life to implement closure plans and fund appropriate compensation and redundancy schemes. Bonding regulatory systems could cover social as well as environmental issues.
- More research is required on the socioeconomic effects of mine closure and their mitigation; and case study analysis needs to inform the drawing of lessons to design best practice corporate strategy and improved public policy.
- What capabilities do companies need to develop, and how might different areas of expertise be integrated, to ensure improved planning for closure from the outset and its subsequent implementation?
- How might research contribute to the development of indicators that might define the quality of closure plans with regard to the predictions made and the effectiveness of mitigation efforts and responses?

5.1 ANALYSIS OF THE PROBLEMS AND CHALLENGES

The socioeconomic impacts resulting from the closure of a mine site are numerous and often dramatic. Best-practice in coping with these impacts, and constraints to the adoption of such practice, will be evaluated in the course of this chapter. What immediately follows, however, is an overview and analysis of the kind of socioeconomic impacts that typically result from the downsizing and closure of mine operations, and the variables affecting their breadth and scale. They can be broadly identified and analysed as occurring both at the level of the community and at the level of the individual. The community in this respect can include both the mine workers and their families and also the remote, but nonetheless affected, community members whose livelihoods depended indirectly on services and supplies to the mining project and its employees. These impacts are the types of problems and challenges that industry, society, and government need to tackle and overcome in the future planning of mine decommissioning.

5.1.1 IMPACTS ON THE INDIVIDUAL

The socioeconomic impacts of worker displacement caused by the closure of any major productive unit are serious, and have been well documented by behavioural scientists. Research indicates that displaced workers subsequently tend to obtain jobs with less occupational status and income, have increased blood pressure and cholesterol levels, experience increased alcohol and drug consumption, and are more likely to get divorced and abuse their spouse or child. A 1% rise in unemployment has been linked to increases of 4.1% in suicides, 5.7% in homicides, 1.9% in stress-related illness, and 4.3% in admissions to mental hospitals.[1]

The problems of coping with closure on an individual level among mine workers are made particularly problematic because of what has been described as the "lonely mountain man culture," a prevailing masculine trait amid mine workers against self-disclosure and an unwillingness to seek help.[2] This is arguably compounded because of "the increasing destruction and dissolution of a worker's culture, which eased unemployment through collective methods of coping,"[3] resulting in unemployment today becoming a problem of individual concern.

While coping with unemployment has become less of a community matter, a study by McKee and Bell[4] suggests that it is no less a family matter. They found that male unemployment has a profound effect on the status of the spouse, with women becoming more likely to postpone a return to work or resign from their present employment activities. This was seen to be the result of a reinforced attitude toward traditional men's and women's bread-winning roles as a result of male redundancy. Given that a strongly male-dominated orientation has been observed in mine communities, such polarisation is likely to be particularly acute within mining areas, highlighting the importance of establishing support services that are able to address the direct effects of closure on women.

Individual reactions to unemployment have been identified by Nygren and Karlsson[2] to correspond to three idealised typologies. These reactions will in turn determine the coping mechanisms the individual faced with imminent closure adopts, and the amount of psychological stress experienced. Some will react passively, which will be manifest in behavioural traits like denial, inactivity, displacement, or intellectualisation of the problem. Some will react in what is called an "an active and adequate" way. Here, the crisis-affected person is successful and adaptive in their utilisation of remaining resources, effectively changing consumption patterns and/or engaging in alternative income-raising strategies. The final type of reaction is "active but inadequate," in which the individual experiences a transformation of needs of a different or negative quality. This can manifest itself in the types of destructive behaviour and actions described in the first paragraph. The passive reaction will require one kind of support, while the active, but inadequate reaction needs another. The former is in need of a strategy that motivates the individual — through job-searching groups, for example, while the latter will need more directional support for an already-active strategy. Because these typologies are idealised, however, in reality policy makers will need to establish support systems and labour adjustment mechanisms that can deal with individuals who have a mixture of the above reactions in the site-specific socioeconomic context in which they live.[2] Ultimately, any account of the effects of mine closure on individuals has to be understood within the context of "the social meanings which actors attach to their actions, their definitions of their situation at particular points, and the ends which they are pursuing."[5]

With significant changes in the employment policies of mining companies, there has been a corresponding shift in the "social meaning" individual employees attach to their work. One major change has been the recruitment of non-miners to the industry. For example, in northern Scandinavia, mining companies have provided jobs in areas that have undergone restructuring problems as a result of the mechanisation of forestry and the rationalisation of farming. In addition, there has been a growing trend of mining companies subcontracting work to smaller firms associated with transport or quarrying. Many mines now operate on a fly-in, fly-out basis. In Australia, open cast mining largely utilises the labour of workers from the city, who are temporarily based at the site in order to save enough money to leave and establish themselves in an alternative occupation.[28]

As well as transitions in the nature of the employee base, there have also been significant changes in the nature of the work itself. The sheer scale and ever-increasing mechanisation of mine operations has depreciated the ability of workers to have autonomy over how their tasks are performed, and has further alienated the worker from interaction with his colleagues. Overall, this process has been accentuated, in a number of countries, by a decline in the number of unionised operations. Collectively, such changes can have a massive impact on the identity and solidarity of the labour force, and in turn on the reaction of mineworkers to mine closure.[28] Following automation and industry rationalisation, mining work forces (direct and indirect employees) worldwide have declined drastically since 1945, sometimes by as great a factor as a ten-fold decline. The British coal industry previously supported more than 700,000 workers and now employs an estimated 1500.

For Neil[6], it is this decline in labour force solidarity that helps to explain why, given the potentially devastating effects of mine closure on mine communities, there is not more effort on the part of mine employees to expand the life of mines through the control of production levels. Such changes in unionisation also help to explain an increased lack of occupational commitment to mining, arguably making the process of reabsorbing individuals back into the general labour force easier.

5.1.2 IMPACTS ON THE COMMUNITY

In addition to the impacts on the affected individual, unemployment has been shown to have a very significant impact on community cohesion, separating the employed and their families from the unemployed and their families.[4] Moreover, as the mine downsizes in the closure phase, groups of workers can progressively lose their jobs, exacerbating divisions amongst them and leading to insecurity and potentially harmful rivalries between workers and between their families. In rural mining communities, where research suggests that divisions between mine workers and the original community existed during boom periods, the problem is particularly exacerbated following redundancy. It is essential, therefore, that mine communities both during the operation of mines, and particularly during the inception of closure, are able to reinforce the social infrastructure of their environment as recipients and participants of comprehensive and ongoing community development programmes.

The potential socioeconomic impacts of a mine closure on the community are likely to vary considerably, depending on the settlement type and its social composition and cohesiveness. The effects of closure or downsizing on people in spin-off employment will be significantly affected by the degree to which the local economy has become dependent on the mine and have been given opportunities for developing capabilities relevant to alternative employment. In their analysis of the social impacts of closure in Australia, Maude and Hugo[8] distinguished between nine different types of mining community settlements ranging from "communities very highly dependent upon mining" to "service centres with a small mining component."

Where mining activities are isolated but closely integrated into the local economy, either through taxes to the local government, the provision of facilities, or the economic spin-offs from mine worker demand for services, mine closure and downsizing can have a potentially devastating effect, particularly if the community has lost a commitment to self-help. It is in these "capital intensive enclaves, lacking any significant economic or social links to their surrounding regions," that the most serious local socioeconomic impacts are felt. However, while the closure of the new company towns would have major consequences for their own populations, it might have little economic or social impact on their surrounding regions. In contrast, mine closures or downsizing in towns possessing stronger economic links with their regions would have a much greater regional impact.[8]

The lifespan of the mine will also influence the degree of social and economic problems within the impacted community. Many of the new Canadian and Australian mining towns have been constructed with a known limited lifespan. On the whole, these towns are populated by people who were aware on arrival of the limited life expectancy, and will settle regardless of whether the industry is in boom or bust cycle.[9] Occupants of these communities tend to be relatively young and well educated and as a result are in a relatively better position to find alternative employment post-mine closure.[7]

Mine communities that have evolved over a number of generations, where diversification opportunities are limited; the lifestyle that has evolved is very distinctive; the population, in relative terms, is older; and where there is a high proportion of home ownership, are far more likely to be impacted with severe social problems by the closure of a mine than communities in which there has been a relatively recent boom in mining. These are some of the conditions that existed in Bolivia, where a formal work force of over 35,000 was reduced to an estimated 5000 over the span of a few years.[24]

The case of Bolivia following the drastic mine closures subsequent to the 1985 tin price collapse underlines the fact that closure has implications that extend far beyond the immediate workforce and their families[9]. Particularly in remote regions, as in the Andes, a whole range of indirect services and supply activities have been built up around a burgeoning mining industry. Many of these activities are in the so-called "informal sector." They are rarely unionised and workers have no eligibility for social welfare or compensation/redundancy benefits from either the mining company or government. The vulnerability of this group of affected families is often overlooked in analyses of industry closure. Given current estimates suggest that the indirect, including informal, workforce that supports a mining project may be as significant as a ratio of 6:1,[19] then this raises important questions of responsibility and policy at the regional level, which are beyond the usual remit of the mining company.

Another key issue relevant to closure, particularly the large state-owned mining operations such as COMIBOL (Bolivia), MINEROPERU (Peru), etc., is the withdrawal of a range of social welfare benefits provided as part of the "social wage" of being a state employee. In COMOBOL during the 1960s–80s, the workforce and their families, in spite of low wages and exposure to health and safety hazards, had access to four staple products at subsidised prices, and were provided with schools, health clinics, medicines, and medics. The collapse of this subsidised social infrastructure also contributes to increasing the vulnerability of the workforce and their families post-closure. The reduction of company-maintained transport systems reduces workers' mobility in searching for new employment. It also affects family mobility in purchasing from and supplying to local markets, with further implications for nutrition in remote mining areas. High mountains and deserts, where minerals are commonly found, do not necessarily support a varied agriculture.

Figure 5.1 notes some of the social, economic, and biogeophysical issues, and short- and long-term effects of mine closure on local communities.

Note that the arrows in Figure 5.1 go two ways to demonstrate a) socioeconomic effects of mine closure and b) the elements of a closure plan required for each specific mine project.

5.2   ANALYSIS OF BEST PRACTICE IN COPING WITH CLOSURE

The issues discussed above have important implications for the design of policies aimed at assisting local communities post-closure. Those affected by mine closures cannot be treated as a generic case or a homogenous group. Every community has its own distinct characteristics and history. Mining settlements vary dramatically in the proportion of their labour force employed in mining; the extent of employment and capacity to generate alternate employment in agriculture, manufacturing, and servicing; and the extent of public sector employment and support. Some mine settlements are small and isolated, while others are major urban commercial centres. This economic diversity of mining communities is matched by wide variations in their social and demographic characteristics, as well as a labour force that is more varied and dynamic than ever. This diversity must be the starting point for the evolution of community, state; union, and mining company policies for handling mine closures and downsizing.

Another pertinent point is raised by Davis,[12] who argues that there is a need for consultation with local communities to occur at both earlier and later stages of the mineral project than has traditionally been the case. Consultation needs to move from an almost exclusive focus on the operational period to be given a similar emphasis during the exploration and closure phases (see Figure 5.2).

Keyes[10] comments that in Canada the stereotyped image of mine closure is one of short notice, ghost towns, displaced workers and businesses, and crumbling infrastructure: "many past closures have seen workers dismissed on two weeks or less notice, usually without severance pay, little help with relocation or a job search, minimum unemployment insurance, and no pension benefits." While he admits that this scenario has changed, owing to the increased sense of responsibility on the part of all players, in general, company, community, and government responses to dealing with the

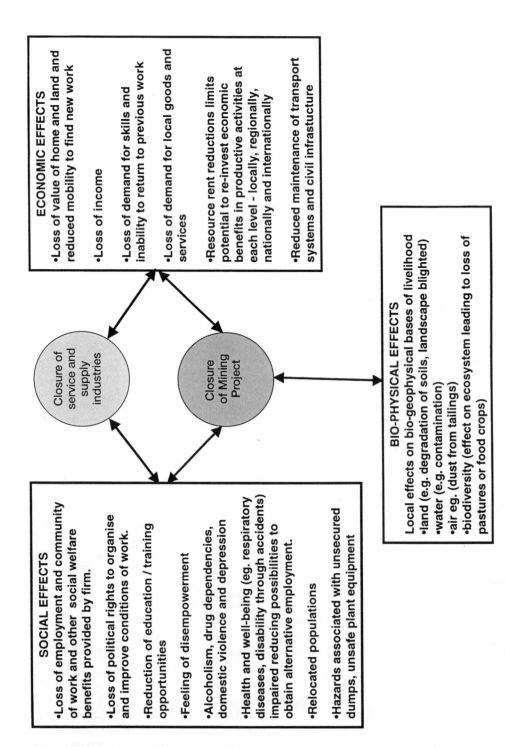

FIGURE 5.1   Scheme of some of the socioeconomic effects of mine closure.

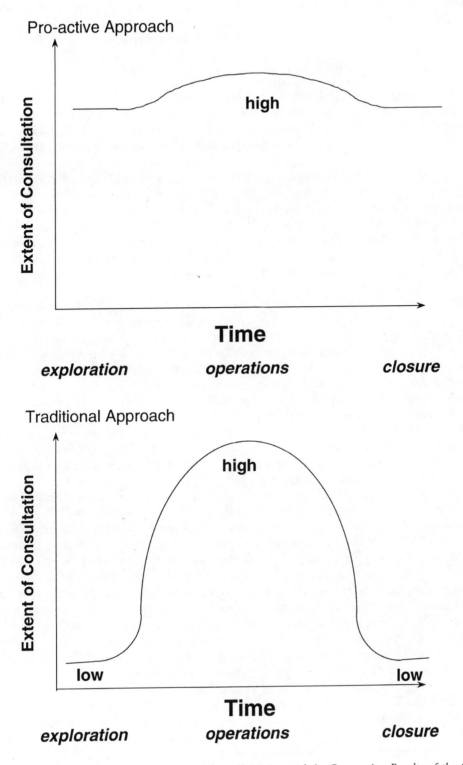

FIGURE 5.2 Source: Warhurst[34] adapted from Davis[12] *Mining and the Community: Results of the Quito Conference*, McMahon, G., Ed., World Bank EMT Occasional Paper No. 11, April 1998.

problems of closure are still very much reactive rather than proactive. In the following section, some of the more proactive responses to closure, at pre-production, production, and closure stages will be assessed, highlighting the elements relevant to the planning process.

5.2.1 BEST PRACTICE AT PRE-PRODUCTION

With the exception of personal and family trauma, many of the problems associated with the decommissioning of mines and mining communities could be prevented, or at the very least effectively managed, with advance planning. A plan for mine closure could be built into the mine development process at the outset to minimise the eventual impact. (see Chapter 2, Warhurst and Noronha).

A tool that is specifically developed to facilitate this process is Social Impact Assessment (SIA). Becker[36] defines SIA as *the process of identifying the future consequences of a current or proposed action which are related to individuals, organisations, and social macro-systems*. The SIA process can be regarded as providing direction in: (1) the identification and understanding of the likely impacts stemming from societal change; (2) the prediction and mitigation of likely impacts from change strategies or development projects that are to be implemented; and (3) the development of appropriate monitoring programs to identify and manage unanticipated social impacts that may develop as a result of social change.[31]

The general practice of Environmental Impact Assessment (EIA) is already a well-established activity within the mining industry and is a requirement of many legislative environments and multilateral and bilateral agencies. However, both within and outside the mining industry SIA was, until recently, an inconspicuous component of EIA process. In the last few years, recognition of the potential contribution of SIA to industrial project developments like mining and to the philosophy of corporate social responsibility has grown considerably.[32] This has been reinforced by mounting social upheaval within mining communities and pressure from government, lending agencies, and non-governmental organisations (NGOs) globally.

The cost of integrating SIA into all aspects of the environmental management systems of mining companies are estimated by the industry to be less than 2% of both the capital and operating costs for new mining projects. Increasingly, it is being realised that these costs can be offset against the many financial benefits this best practice can bring. These can include direct and immediate cost savings by avoiding unacceptable social impacts caused by mining operations. Longer-term gains include reduced financial liability at decommissioning, improved acceptability and financing of future mining proposals, and reduced outlays in security deposits and political insurance premiums.[33]

A useful matrix relating the decommissioning/closure project stage to social impact assessment variables is provided by the Interorganisational Committee on Guidelines and Principles for Social Impact Assessment. The committee lists a set of social impact variables against the four project stages of planning implementation, operations, and decommissioning. This is reproduced in Figure 5.3. The introduction of these criteria explicitly to help predict the effects of closure and implicitly to inform the design of an appropriate management plan is just one example of the way in which SIA can benefit planning for closure at the outset of mine development. But a purely ex ante approach may be inadequate. Most experienced social scientists feel that if social analysis is to succeed in human terms, it must occur very early in the project cycle and recur in subsequent phases.[29] A mining project can have some consequences which cannot be predicted in advance and there is a need for SIA to operate continuously, assessing and re-assessing up to and even beyond closure.[30]

According to Wolfe,[11] the first aim of advanced planning is to avoid the establishment of communities based solely on non-renewable resources in the first place. This might be resolved through the consideration of the following options:

- Integrated resource centres serving more than one industry.
- Fly-in/fly out operations.

Social Impact Assessment Variable	Planning/Policy Developments	Implementation/ Construction	Operation/ Maintenance	Decommissioning/ Abandonment
Population characteristics				
Population change				
Ethnic and racial distribution				
Relocated populations				
Influx or outflow of temporary workers				
Seasonal residents				
Community and institutional structures				
Voluntary associations				
Interest group activity				
Size and structure of local government				
Historical experience with change				
Employment / income characteristics				
Employment equity of minority groups				
Local / regional / national linkages				
Industrial / commercial diversity				
Presence of planning and zoning activity				
Political and social resources				
Distribution of power and authority				
Identification of stakeholders				
Interested and affected parties				
Leadership capability and characteristics				
Individual and family changes				
Perceptions of risk, health and safety				
Displacement/ relocation concerns				
Trust in political and social institutions				
Residential stability				
Density of acquaintanceship				
Attitudes towards policy / project				
Family and friendship networks				
Concerns about social well being				
Community resources				
Change in community infrastructure				
Native American tribes				
Land use patterns				
Effects on cultural, historical and archaeological resources				

FIGURE 5.3 Matrix relating project stage to social impact assessment variables.

Even when a long lifespan is anticipated for a mine, the concert of an isolated community as a model for settlement is becoming increasingly unpopular. According to a recent industrial professional,[15] "there is a move away from the creation of townships that are solely dependent on the mining operation."

Recent developments have, where possible, included the establishment of integrated resource centres. These are settlements designed to serve a number of economically productive sources. Tumbler Ridge in Canada is one such community, which was constructed to house workers in two mines and potential hydro/forestry developments. Originally serving just one mine, Marathon in Canada was expanded into an integrated resource centre servicing three new gold mines in the area.

Even more preferable, from the perspective of public utilities, is a centre with a fully diversified industrial base. Any expansion of the economic base in this way will make settlements less vulnerable to economic swings and cycles, making social disintegration and worker displacement less likely prospects if a mine closes.[6]

Fly-in/fly-out mine operations, in which workers commute, are also increasingly being seen as a solution to the ultimate social problems of mine closure. Because workers on such operations generally come from widely dispersed areas, closure usually presents fewer problems than those associated with a single mine community. However, as Gagnon[22] suggests, the problems are merely dissipated away from a collective centre but not actually eradicated; instead, there is the danger that they are concealed. Such operations do, however, help to resolve a number of problems such as resettlement and displacement, which can be served by a more far-reaching labour force and diversified economic base, as well as avoiding large economic losses resulting from a depleted tax base and wasted infrastructure.

However, if the establishment of a single resource-dependent community is the only viable option for the exploitation of the mineral resource, then planning for closure prior to the commencement of town construction is all the more imperative. This emphasises the point above regarding the integration of closure and decommissioning, with any initial social impact assessment carried out on the proposed mine site. Without conducting such an assessment from the outset, the potential impacts of closure prior to operation will not be understood by relevant stakeholders. This could dramatically affect whether construction proceeds, how it proceeds, and the mitigation and compensation mechanisms to be adopted when production is achieved. Once this has been conducted, inexpensive ongoing participatory monitoring on behalf of the community and the company will facilitate the iterative identification of deviations from the proposed actions; evaluation of changing needs; unanticipated impacts; and potential impacts through to when closure or downsizing are eventually realised. This is an essential precursor to the next stage of advanced planning which, suggests Wolfe,[11] would involve the drawing up of a contract between the employees, community, and the mine company:

- Establishing a specific fund, served perhaps by royalty payments, to tackle the future problems of decommissioning (Manitoba in Canada already operates such a fund).
- Stating the company policy toward closure with regard to notice, moving costs, severance payments, retraining, and related issues.
- Including an undertaking to develop with the community at the earliest stage (given that there is typically up to a seven-year lead in the establishment of new ventures) alternative sources of economic productivity.
- Developing an appropriate recruitment policy.

Kidston gold mine in northern Queensland, Australia is apparently an excellent example of planning for closure prior to the inception of the project. This is particularly true with regard to its recruitment policy. It is a fly-in/fly-out mine, given a life expectancy of 20 years. It was first decided that workers would be employed from a number of small rural settlements. This was intended to distribute opportunities over a wide area, as well as dissipating the effects of closure.

Secondly, it was decided that the age of recruitment should average 35 years so that by the time that work had ceased workers would be at retirement age. This marks a trend in recruitment policy, towards employing older workers who can retire and remain in the community. An additional policy shift has been to attract target workers who come to the community on an agreed short-term basis, whose objectives are to work to save rather than work to subsist. This maintains a high turnover in the workforce, but avoids many of the social problems normally associated with retrenchment following closure.[6]

5.2.2 BEST PRACTICE AT PRODUCTION

The company and local government policies in place during operations will also have important implications for the adjustment process in the event of mine closure. The first concerns, for example, the company's policy on the sale of housing. The consequences of home ownership with little market value for displaced miners are very serious issues. It reduces mobility — an essential prerequisite of new job acquisition. To help deal with this problem, companies may be able to employ a buyback policy, or at least secure a nominal minimum for which the value of the property can be protected. Secondly, job-skilling by the company can provide additional support to other economic initiatives of the community after the mine has been decommissioned, as well as increasing the individuals' prospects for employment in other areas. Thirdly, company investment in the social infrastructure of the community, through the provision of key facilities such as sports centres or business centres, means that the likelihood of social disintegration following closure is reduced, and community and industrial relations are enhanced. The company could exclusively fund such projects, jointly fund with the community, or provide credit on reasonable terms, if it is felt this would cultivate a more positively empowered residency.[6]

Through self-help schemes like downstreaming, multi-skilling, and the provision of social infrastructure, mentioned previously, the company can help foster the conditions necessary for community development initiatives, encouraging local people to sustain themselves beyond the closure of the mine. Again, this could be complemented by government initiatives to make appropriate credit readily available. Further, through the participatory appraisal of community needs, integrated within an ongoing social impact assessment, a reciprocal environment can be established with the community, in which residents have a positive feedback into company policy. Participatory appraisal has proved particularly useful in situations where there is competition or conflict concerning resource management, serving a mediatory function by providing previously hostile groups a recognition and understanding of each others' competing pressures. From here, the proposal of regulatory measures or resource negotiation becomes easier and future conflicts are often avoided. For this situation to be realised, the company must adopt a progressive attitude based on the underlying philosophy that:

"...the potential for public acceptance of your project is directly proportional to the amount of control you are willing to relinquish, and inversely proportional to the walls behind which you try to hide."[12]

Sierra Rutile's titanium mine in Sierra Leone has been considered by some observers to provide a very good example of company and community relations in this respect. What makes its operations exceptional is an open development strategy established in partnership with a non-government organisation, reciprocally promoting participation of the community in company policy and of the company in community affairs. As well as successfully fostering beneficial ongoing relations, they have formed an apparently sustainable platform for negotiations through to closure. A review by the World Bank affiliate, the International Finance Corporation (IFC) in Washington D.C., regarded Sierra Rutile in that context as a "good corporate citizen," while Martyn Riddle, manager of the IFC's Environmental Division believes that in this respect the company has done what the World Bank tries to encourage, by "making a conscious effort to sit down with the people, and to talk to the community about what they want."[13]

Advance notice of closure is accepted as another essential element in successful community adjustment. This provides families and workers the lead-time necessary to plan their futures, start new local economic initiatives, begin a job search, and negotiate a severance package. At Pine Point, in the Northwest Territories of Canada, for example, employees were given six months' formal notice of the closure, even though legislation at that time in the region required none. In a questionnaire on the issue of advance notice at this plant, 87% of respondents reported finding it "'adequate.'"[23] Companies argue that if the advance notice is too long, however, worker morale will generally decline and the company will lose their best workers early, with detrimental effects on productivity. But, by providing concessions to key workers, these problems can be avoided. At the closure of the Selco copper and zinc mine in Ontario in the early 1980s, for example, key employees who agreed to stay to the end date were paid 150% of their normal salaries between notification of closure and stoppage, plus three months' supplementary wages, which maintained the required labour force up to the termination of operations.[11]

But, although advance notice will help to reduce the transitional problems for individuals, it nevertheless remains a socially and psychologically wrenching experience to be forced to move into a post-closure situation. Through successful economic diversification efforts, it can be possible for the local population to side-step altogether the dramatic social, psychological, and economic costs associated with forced relocation. Furthermore, it avoids the under-utilisation or wastage of public infrastructure; it displaces the need to fund relocation assistance, unemployment benefits, and retraining programmes; and diminishes the need to fund the establishment of infrastructure elsewhere to accommodate those made redundant.[14]

For many mining areas, facilities left behind could be renovated or transformed into useful workplace or community buildings and equipment could be used for other purposes. This was not the case at a site called Kabwe, that was owned by ZCCM in Zambia, where, post-closure, facilities were vandalised.[37] This also raises the important issue of decommissioning safely. Rio Tinto Zimbabwe has invested considerably at its closed Empress Mine, in ensuring all waste is safely disposed of, with plant and equipment either broken up and taken off site and made safe.

A mining community may diversify its economic activities either "vertically" or "horizontally." When a mining community diversifies vertically, it attempts to increase its economic base by expanding the activities of the mining sector. This has the advantage of utilising existent skills, but does not evolve the community away from dependence on a single sector. However, it means that if a mine is closed, there may be subsequent implications for smelters or refineries, which although they depend on feed associated from several sources, may nonetheless have difficulty sustaining capacity levels and may also be obliged to downsize or ultimately close. For this reason, economic buffers to the potential socioeconomic impacts of mine closure can be better achieved through horizontal diversification. But it would be misleading to neglect the limited successes of the vertical approach. At Lynn Lake in Manitoba, diversification into gold extended the life of the operation for three years after copper/zinc operations had stopped; in Rouyn/Noranda, a copper smelter prolonged employment after other copper mines in the region had closed; and in Wawa, Ontario, a new gold mine will act as an economic backup if the present iron ore mine should close.[10]

Horizontal diversification, as an alternative to vertical diversification, involves the broadening of the economic base into entirely new sectors, such as manufacturing or services. While this has the advantage of decreasing the vulnerability of mining communities, it has often proved difficult given the small size and remoteness of many mine settlements. Nevertheless, there are well-documented examples, where the planning and commitment of residents and government have made horizontal diversification possible. Atikokan, a mining town in northern Ontario, is a small and remote settlement that was subject to the closure of its two iron ore mines. Through complementary small business initiatives such as tourism, light manufacturing, and forestry, and the expansion and upgrading of regional infrastructure, it was able to find alternative economic activities prior to actual closure and so sustain the survival of its community. Thus, although closure of the iron ore mines rendered over half the community's labour force unemployed, it subsequently lost less than a fifth of its total population.[14]

Between 1981 and 1985, six of the seven mines of Tennant Creek, in Australia's Northern Territory closed. However, while employment in mining fell by 64%, total employment fell by only just over 30%, allowing Tennant Creek to survive as a functional community. The reason for this was a concurrent expansion in the tourist and food processing industry and a continued commitment to maintain the town as a regional service centre. In this way, government programmes helped support the public sector which in turn supported employment in the private sector service industries.[14] However, although there have been successful diversification efforts in mining areas, there have also been failures. Many governments are reluctant to subsidise former mining communities to engage in what is often a futile struggle of transition. But many representatives of mining communities feel that if their governments and companies provided more support to diversification efforts through greater planning and preparation prior to closure, there would be more success stories.

In 1982, a Canadian government-industry-labour task force was assigned to engage with the problems of mining communities and report on the primary determinants of diversification,[16] which identified the following:

- Opportunity potential, primarily in terms of resource endowment.
- Lead-time within which to plan and act.
- Designation of a party or agency responsible for diversification initiatives.
- Attitude of senior governments with respect to the community.
- Attitudes and resources of the community and its leadership, and community access to financial and other resources.

Most importantly, the task force recommended that diversification assessments should be conducted at the earliest possible stage of mine development or operation, and that diversification planning should be included as an integral element of the planning of a new mining town. Evaluation of opportunities at the local level, either by members of the community themselves or consultants employed by the community, is a critical preliminary step before any economic initiatives are embarked upon. Given that local people often do not possess the resources to undertake this kind of assessment, a government or company fund could be considered for such purposes. The role of the state as facilitators of formal services and financial backers must not be underestimated in this process. In Canada, for example, government programmes such as the federal Community Futures Programme now offer technical and financial support to communities in the assessment of their futures, as well as providing capital backing for new businesses.

One of the main principles espoused by the Canadian task force report was that communities and individuals within those communities take a degree of responsibility for their own futures. As Keyes[10] writes:

"To a large degree workers and residents are often portrayed as the innocent victims of corporate decisions. At times this is the case. On the other hand, people make rational decisions to go to mining communities, fully aware of the vagaries of the mining industry. It would seem appropriate that they assume some personal responsibility for their own situation."

It was with this in mind that, in 1986, the Canadian Association of Single Industry Towns (CASIT) was established. This organisation was specifically commissioned to provide financial, moral, and analytical support for communities to help themselves. For successful economic development to take place, it is important that the affected communities be actively interested in the future of their own towns, with a base commitment to the town and its viability by a significant number of residents. In addition, there is a need for community solidarity which can provide the consensus for strategic planning, as well as representative individuals within those communities whose organisational skills and awareness of the local social structure are capable of mobilising and facilitating collective action.[17] Where this social cohesion, commitment, and vision exists and the necessary external support networks from government bodies are in place, a number of mining communities have successfully contributed to the regeneration of their environments.

During the closure of the Stekenjokk mine in Sweden, community groups collectively planned the development of the area, with creative results. It was suggested that the mine might be converted into Sweden's only underground gymnasium and swimming pool or turned into a tourist attraction; alternatively, that health rehabilitation centres, fisheries, and deer farms could be established in the area.[2]

Kiruna, a mine town in Northern Sweden, provides another good example of a community which, in conjunction with local government, initiated education and training programmes; created a positive climate of investment for industry; established a business centre; organised two international symposiums; and searched for new products and technologies. As such, it managed to successfully diversify industry and mobilise a new labour market into a creative environment of high-technology computer industries, transforming the local economy from mine to "mind," and so securing its economic and social fabric into the future.[18]

5.2.3 BEST PRACTICE AT CLOSURE

In addition to the numerous policies and strategies that can be adopted to ameliorate the social and economic impacts of closure prior to and during operations, there will be further requirements when closure becomes imminent. These will include the need for an adequate company severance package to assist workers in their relocation and compensate for future lost income; counselling to help overcome the psychological stresses that accompany unemployment; retraining to help miners successfully gain employment in another industry; and job search facilities to increase the chances of acquiring a new position.

At Pine Point, severance packages played a significant role in the worker adjustment process. The company provided for severance payments equal to one week's wages for each year of service, in addition to C$500 cash bonus. Medical plans and group life insurance were continued for three months beyond the severance date, and a skills inventory of the laid-off workers was also made available to prospective employers. A special Canada Employment centre was established to provide "customised" assistance to the mining labour force, providing help with resume writing and job search skills. The federal government and the company also jointly funded travel for job search purposes, and provided up to C$10,000 per worker as a relocation package.[23] According to questionnaire results, the worker readjustment programmes at Pine Point were well received, with 93 out of 125 respondents finding the programmes "helpful in coping with the closure and finding a new job." Kendall[23] found that 75% of miners displaced by closure gave continuing in the mining industry as their first priority. Of all the programmes, the relocation programme was regarded as being the most beneficial, with the Canada employment job search service being cited as the next most beneficial.

The worker adjustment package and accompanying services were negotiated between the company and the government at Pine Point. Alternative models for the provision of these services are possible, and may be generated through cooperation between the company and community, company and the union, or government and community. However, company cooperation has become regarded as critical to the successful provision of these services because, as well as allowing for their on-site delivery, the company is familiar with individual worker's needs. Company assistance in the establishment of retraining and job search facilities will largely be attributable to the company's ideology, although the size and geographic location of the company will influence the availability of personnel to assist with the programmes, and the ability of the company to mobilise external support.[6] Whatever model is used, Neil et al.[6] specify a number of components to be included in a comprehensive adjustment scheme:

- Financial support to cover job search and relocation.
- Skills inventory for prospective employers.

- Job vacancy listings.
- Development of job search skills.
- Information concerning costs and standards of living in other areas.
- Retraining in specific job skills.

In 1978, when the Anaconda Uranium mine on the Laguna Pueblo Native American Reservation in New Mexico was faced with closure, four years of worker retraining formed the major element of a closure package for the largely Native American workforce, because of the limited alternative employment opportunities in the area. Consequently, when faced with closure, the Laguna Pueblo workers were able to utilise their closure package payment to form their own company, Laguna Construction. Securing a contract for $43 million from the company, Laguna Construction was able to take full responsibility for the reclamation of the area, finishing under budget and two years ahead of schedule. Laguna Construction is now one of the largest Native American enterprises of its kind, and has won contracts for the reclamation of projects across the United States.[19] The experience of retraining workers at the Anaconda mine benefited both the displaced manpower and also the company through the transference of risk, proving that such policies, while demanding investment and vision, can be rewarding to all stakeholders.

However, in addition to practical and financial assistance in the readjustment process, it is essential that displaced workers and their families receive some psychological support to cope with the stresses of unemployment and the types of negative reactions that can result. In the Nordic countries, such provisions are effectively made by the public sector. Kieselbach[3] suggests that where the public administrative support is not in place for such services, these might be provided through the trade unions, with the advantage that individual problems could be perceived in their social context. Alternatively, where there is no union or governmental provision, the progressive company may wish to utilise the resources of a non-government organisation specifically trained in this capacity. It is further suggested that outreach services are available to those, numerous within mining, who are reluctant to approach agencies themselves for assistance.[6]

5.3   ANALYSIS OF CONSTRAINTS TO TRANSFER OF BEST PRACTICE

The main focus of this chapter has been in the identification of best practice within the mining industry of planning for closure in regard to the socioeconomic environment. However, each and every case has, to a varying extent, been specific to its environment, and is not necessarily transferable to other contexts where they may be rendered inoperable. Many of the practices are contingent on the early warning or prior identification of closure. However, although planning for mine closure before development is imperative, because of the vast arena of variables which can affect anticipated mine life, "neither companies nor government can really hope to anticipate or mitigate all of the effects and impacts".[16] Indeed, even when prediction of closure is made possible, companies are often very wary of revealing this publicly. There is firstly a fear that this will lead to apathy amongst the workforce and the loss of key workers, and secondly that it will result in an abandonment of financial support from shareholders and more central company divisions. Such negative effects could be rectified to the mutual benefit of companies, employees, and local communities, if appropriate management plans are in place to inspire confidence, promise support, and reduce fear.

In addition, communities are often ill-prepared for the effects of closure because employees, unions, and management are unwilling to accept the inevitable. Hegadoren and Day's study[20] showed that "predictions of additional large-scale industrial development in the optimistic climate which often accompanies mine openings is a major factor in limiting adoption of measures from the beginning of mining to reduce the stresses of closure." Moreover, where these predictions are made, there are no assurances that action will ensue. At the Marmoraton mine in Ontario, a four-year warning of closure was given, but there were few formal arrangements made until the final termination of operations. There is also reluctance on the part of government to actively seek predictions of and the preparations for the closure of mines well in advance through local economic initiatives, because of fears that it would be seen to undermine any popular support they accrued for its opening.[6]

Furthermore, as O'Faircheallaigh[14] rightly observes, government faces the continued dilemma of opportunity cost when considering the attempted diversification of single-resource settlements during their productive stage. It is often difficult to predict in an open market system whether investment in diversification now will not at a later stage prove futile. Although the experiences of Atikokan and Tennant Creek were positive in attempts to diversify their social and economic environments, conditions for such a transformation were predisposed. Both possessed a very significant potential for tourism, as well as being located in regions that contained sources of alternative non-mineral resources. Moreover, the expansion of economic initiatives in new sectors was concurrent with (or, in the case of Atikokan, in advance of) mine closure. Finally, the alternative employment opportunities (tourism, forestry, and meat processing) generated for both communities were in industrial areas that did not demand specialised skills or experience.

Diversification is an attractive option, but the ability of a mining community to successfully diversify economically will depend upon a vast range of factors, such as the nature and size of the present economic base, infrastructure of the area, community solidarity, and government support. Because of the range of necessary conditions, diversification efforts have usually proved very difficult to achieve in reality, as the literature will bear witness.[21,26,14] Mining communities are generally very isolated, making the cost of transport high and, therefore, the profitability of manufacturing projects low. Additionally, mining communities are rarely proximal to other exploitable natural resources like fishing or forestry, or based in centres of outstanding natural beauty where tourism could be exploited. To compound these difficulties, miners tend to possess very limited and specialised skills, which are not easily transferred to alternative forms of productive activity. Moreover, there is also the problem of timing. New economic ventures must be realised at the same time as closure if they are successfully to offset the problems of worker displacement. This may be possible if notice of closure is reliably confirmed well in advance, but as discussed earlier, this too can be problematic.

An important issue that affects potential for diversification is the extent to which agricultural land and pastures have been contaminated or access to good quality water supplies has been reduced post-closure. As noted in Chapter 3, acid rock drainage, once initiated, can continue indefinitely if not treated and mine closure may not oblige the continued operation of a water treatment plant.

Fostering development initiatives among local communities will demand not just the prerequisites of alternative resources and timing, but will also depend on the level of social cohesion and individual and collective goals of those involved. There is a temptation to view communities as small, well bounded, and homogenous. As Nelson and Wright [34] observe, "Community is a concept often used by state and other organisations, rather than the people themselves, and it carries connotations of consensus determined within the parameters of the outsiders." In reality, people associate through multiple, overlapping, and competing networks with diverse linkages based around different interests and opinions.[35] It has been shown for example that:

"The social planning and community development approach relies very heavily on the concept of 'process' and 'consensus' through community involvement and participation.... [however] Communities are not homogenous entities, and it is often unreasonable to expect that all citizens will work for a common ideal." (Office of Local Government, 1989, cited in Neil et al.)[6]

There have been cases, including Colquiri and COMIBOL in Bolivia, and Wheal Jane, formerly of Rio Tinto in the U.K., where a company has decided a mine should close on profitability grounds but workers or managers decide to buy out the operation or seek approval to keep operations going. This may aid the transition phase by prolonging closure but it may also mean less support for workers in the final stages of the operation, increased safety hazards, and enhanced risk of environmental damage. In Bolivia, there are cases where closed mines of COMIBOL during the late 1980s/ early 1990s were taken over by workers' cooperatives. They mine old adits and tailings/waste dumps on a small scale but involving a large workforce over a large surface area seeking missed pieces of ore of economic value, which are then sold back to other mines still operating in the region. Again, this avoids the obligation for mining companies to pay "social wages," widows' pensions, and other social benefits, whilst bringing in extra income for the company through recovering metals from otherwise dumped resources. Whilst this meets the prime objective of providing income/earning opportunities for displaced workers, it also takes the pressure off the government, company, and community to seek more systematically to develop new, productive, safer, and more secure work opportunities.

In the developing country context, the large numbers of dependent workers and the lack of alternative sources of income, due to dependence on mining and a failure to reinvest resource rents, makes planning for closure all the more complex, insecure, and hazardous, as well as difficult to implement.

Finally, a continuing debate holding back the transfer of best practices focuses on the notion of responsibility. Despite four years' warning of closure at the Mamoratum mine in Ontario, no preparations were made for eventual termination because of confusion concerning responsibility. The community felt that federal and provincial governments should provide assistance through the securing of alternative industry. However, they felt it was the responsibility of the municipal government, who in turn looked to the company and back to the senior government to provide such assistance.[20] The company normally has the role of a guest in a host community, and as such they may try to build for the future of the community. In some cases, there may be resistance to such moves at the governmental level, as it is seen as undermining the government's role. However, in a non-regulatory environment, there is a strong driving force behind helping to sustain communities even after closure. Ultimately, problems associated with the closure of other operations can be documented and used in evidence to prevent new mine development. In strictly business terms, the creation of disenchanted individuals or communities through inadequate or ineffective planning for closure is not in the industry's long-term interest.

5.4 CONCLUSIONS AND POLICY AND RESEARCH IMPLICATIONS

- Social Impact Assessment (SIA) needs to be ongoing throughout the life phases of the mine and planning for decommissioning, downsizing, and closure needs to begin at the outset based on criteria developed through the SIA.
- Closure planning should address effects and solutions for the remote community involved in supplying the industry, as well as the formal and informal workforce of the mine.
- Profiles of recruits, recruitment strategies and human resource development through the mine life needs to be included in closure planning to facilitate transition for redundant workers and their families, and broaden the possibilities of future work options.
- An environmental management plan, post-closure, needs to be in place where there is a threat of ongoing contamination/environmental damage such as acid rock drainage or tailings leaks/slippages, so as to improve possibilities of alternative land uses, particularly farming.
- Closure planning could include alternate uses for housing, facilities, and equipment, and policies to protect the continuation of social networks and community activities.
- Consultation throughout is paramount, assisted by participative approaches to forward planning so as to involve the community from the outset in addressing eventual closure and future options.
- Financial mechanisms need to be in place to ensure sufficient resources exist at the end of the mine's life to implement closure plans and fund appropriate compensation and redundancy schemes. Bonding regulatory systems could cover social as well as environmental issues.

- More research is required on the socioeconomic effects of mine closure and their miti-gation, and case study analysis needs to inform the drawing of lessons to design best-practice corporate strategy and improved public policy.
- What capabilities do companies need to develop, and how might different areas of expertise be integrated to ensure improved planning for closure from the outset and its subsequent implementation?
- How might research contribute to the development of indicators that might define the quality of closure plans with regard to the predictions made and the effectiveness of mitigation efforts and responses?

ACKNOWLEDGEMENT

In addition to drawing on the authors' own worldwide research, this chapter draws extensively on the excellent, albeit unique, review of issues in mine closure found in Neil, C., Tykkylainen, M., and Bradbury, J., Eds., *Coping with Closure: An International Comparison of Mine Town Experiences*. The authors also acknowledge the kind assistance of Assheton Stewart Carter (University of Bath) and Matthew Shelley (formerly of University of Bath) in the writing of this chapter. Thank you also to Jill Siddall for painstaking final editing.

REFERENCES

1. Kinicki, A., Bracker, J., Kreitner, R., Lockwood, C., and Lemak, D., Socially responsible plant closing, *Personnel Administrator*, 32, 116, 1987.
2. Nygren, L. and Karlsson, U., Closure of the Stekenjokk Mine in North-West Sweden, in *Coping With Closure: An International Comparison of Mine Town Experiences*, Neil, C., Tykkylainen, M., and Bradbury, J., Eds., Routledge, London, 1992, 99.
3. Kieselbach, T., Self-disclosure and help-seeking as determinants of vulnerability, in *Unemployment, Social Vulnerability and Health in Europe*, Schwefel, D. et al., Eds., Springer Verlag, Berlin, 1987.
4. McKee, L. and Bell, C., His unemployment, her problem: the domestic and marital consequences of male unemployment, in *The Experience of Unemployment*, Allen, S. et al., Eds., Macmillan, Basingstoke, 1986, 134.
5. Bulmer, M. I. A., Sociological models of the mining community, *Sociological Review*, 23, 61, 1975.
6. Neil, C., Introduction in *Coping with Closure: An International Comparison of Mine Town Experiences*, Neil, C., Tykkylainen, M., and Bradbury, J. Eds. Routledge. London. 1992.
7. Neil, C., Tykkylainen, M., and O'Faircheallaigh, C., Planning for closure, dealing with crisis, in *Coping With Closure: An International Comparison of Mine Town Experiences*, Neil, C., Tykkylainen, M., and Bradbury, J., Eds., Routledge, London, 1992, 369.
8. Maude, A. and Hugo, G., Mine Settlements in Australia, in *Coping With Closure: An International Comparison of Mine Town Experiences*, Neil, C., Tykkylainen, M., and Bradbury, J., Eds., Routledge, London, 1992, 141.
9. Warhurst, A. and Jordan, R., The Bolivian Mining Crisis, *Resources Policy*, March 1992
10. Neil, C. and Brealey, T. B., Home ownership in new resource towns, *Human Resource Management Australia*, 20, 38, 1982.
11. Eikeland, S., National policy of economic redevelopment, regional development, and how the working class deal with uncertainty, in *Coping With Closure: An International Comparison of Mine Town Experiences*, Neil, C., Tykkylainen, M., and Bradbury, J., Eds., Routledge, London, 1992, 119.
12. Davis, S., Consultation and the East Asian Perspective, *Mining and the Community: Results of the Quito Conference*, McMahon, G., Ed., World Bank EMT Occasional Paper No. 11, April 1998.
13. Keyes, R., Mine closures in Canada: problems prospects and policies, in *Coping With Closure: An International Comparison of Mine Town Experiences*, Neil, C., Tykkylainen, M., and Bradbury, J., Eds., Routledge, London, 1992, 27.

14. Becker, H., *Social Impact Assessment,* UCL Press, London, 1997.
15. Burdge, R. J. and Vanclay, F., Social Impact Assessment, in *Environmental and Social Impact Assessment*, Vanclay, F. and Bronstein, D. A., Eds., Wiley and Sons, Chichester, 1995.
16. McPhail, K. and Davy, A., *Integrating social concerns into private sector decision making*, World Bank, Washington, D. C., 1998.
17. Australian Environmental Protection Agency, *Best Practice Environmental Management in Mining*, Ward, S., Ed., Australian Federal Environmental Department, Canberra, 1995.
18. Cernea, M., *Putting people first: sociological variables in rural development*, OUP, New York, 1985.
19. Macfarlane, M., *Evaluating SIA Methodologies in the Mining Industry*, forthcoming Ph.D. thesis, University of Bath, U.K.
20. Guidelines and Principles for Social Impact Assessment, by the Interorganisational Committee on Guidelines and Principles for Social Impact Assessment, Department of Commerce, Washington D.C., 1994.
21. Wolfe, J. M., Mine closure in single-industry towns and the problem of residual activity, in *Coping With Closure: An International Comparison of Mine Town Experiences*, Neil, C., Tykkylainen, M., and Bradbury, J., Eds., Routledge, London, 1992, 192.
22. Unpublished minutes of "Planning for Closure" workshop, University of Bath, Bath, U.K., 3 April, 1996.
23. Gagnon, J., Native labour commuting to uranium mines in Northern Saskatchewan, in *Coping With Closure: An International Comparison of Mine Town Experiences*, Neil, C., Tykkylainen, M., and Bradbury, J., Eds., Routledge, London, 1992, 291.
24. Taylor, C. N., Goodrich, C., and Bryan, C. H., Social assessment: issues oriented approach to social assessment and project appraisal, *Project Appraisal*, 10, 142, 1995.
25. Bomford, L. and Smith, K., Can mining companies act responsibly? *The Guardian*, 24 February, 1996.
26. Kendall, G., Mine closures and worker adjustment: the case of Pine Point, in *Coping With Closure: An International Comparison of Mine Town Experiences*, Neil, C., Tykkylainen, M., and Bradbury, J., Eds., Routledge, London, 1992, 131.
27. O' Faircheallaigh, C., Mine closures in remote regions: policy options and implications, in *Coping With Closure: An International Comparison of Mine Town Experiences*, Neil, C., Tykkylainen, M., and Bradbury, J., Eds., Routledge, London, 1992, 347.
28. Energy, Mines and Resources, Report of the Task Force on Mining Communities, EMR, Ottawa, 1982.
29. Koch, A. and Gartrell, J., Keep jobs in Kootenays: coping with closure in British Columbia, in *Coping With Closure: An International Comparison of Mine Town Experiences*, Neil, C., Tykkylainen, M., and Bradbury, J., Eds., Routledge, London, 1992, 208.
30. Liljenas, I., From mine to outer space: the case of Kiruna, a town in Northern Sweden, in *Coping With Closure: An International Comparison of Mine Town Experiences*, Neil, C., Tykkylainen, M., and Bradbury, J., Eds., Routledge, London, 1992, 247.
31. Bridge, G., unpublished data from site visit, October 17, 1996.
32. Hegadoren, D. B. and Day, J. C., Socioeconomic mine termination policies: a case study of mine closure in Ontario, *Resources Policy*, 7, 265, 1981.
33. Wolfe, J. M., Last gasp or second wind? Mine closures at Uranium City and Chefferville, in Recession, Planning and Socio-Economic Change in the Quebec-Labrador Iron Mining Region, Bradbury, J. H. and Wolfe, J. M., Eds., McGill Subarctic Research Paper No.38, Centre for Northern Studies and Research, McGill University, 1983.
34. Warhurst, A., Consultative Processes: Workshop Report *Mining and the Community: Results of the Quito Conference*, Gary McMahon, Ed., The World Bank EMT Occasional Paper No. 11, April 1998.
35. Nelson, N. and Wright, S., *Power and Participatory Development: Theory and Practice*, Intermediate Technology, London, 1995.
36. Cornwall, A. and Jewkes, R., What is participatory research?, *Social Science Journal*, 41, 12, 1667, 1995.
37. Kapelus, P. and Pinter-Burger, A., Building Partnership in Managing Resources; a Socially Responsible Approach from Exploration to Post Closure, unpublished paper 1998.

6 Effective Environmental Impact Assessment

Meredith Sassoon

CONTENTS

6.1 INTRODUCTION

Environmental Impact Assessments (EIAs) are now required in most countries as a statutory prerequisite for the development of a mining project. The aim of the EIA is to determine a baseline data set, predict the environmental impacts of the project, and outline a monitoring programme. The environmental predictions are based on the specific design criteria, and other information established for and contained in the Feasibility Study.

In the life cycle of a mining project, the EIA is carried out as a reactive study, once an economically viable mineral deposit has been proved and the mine project concept outlined. This is due to the nature of mining projects and their development, rather than a conscious obstruction on the part of the developer, but does result in the environment being seen as an "afterthought." In addition, the predictions contained in the EIA can alter completely, if the original project design criteria on which they are based, change. Due to the evolutionary nature of mining projects, this is a fairly common occurrence and can happen during construction and/or operation, at an advanced stage of financial commitment.

Both these issues would appear to detract from the role of an EIA; to minimise the overall negative environmental effects of the project both during operations and following decommissioning. Are EIAs, in the form that we know them at present, the best way to ensure that mining is carried out in an environmentally acceptable manner, or are they only paying lip service to legislation?

At present, the majority of environmental aspects are either handled retrospectively or viewed as a hurdle that has to be crossed before the real negotiations for project approval can begin. At a recent conference, a top-level manager from an international company described an EIA as, "an operational tool that is used to assist projects to meet the statutory requirements," This lack of a real commitment to care for the environment is reflected by the dearth of political, government, or company inclination to take a more integrated approach to project development and environmental considerations. This situation is exacerbated by the restrictive nature of existing government structures, environmental legislation, and environmental planning requirements. As such, EIAs often fail to provide an environmentally sound basis for project approval.

This chapter explores the present method of integrating an EIA into the "project cycle," for the development of a mine, and, if pitfalls exist, where they may lie. The case studies are used as examples to consider whether, for an EIA to be successful as a management tool, it should occupy a more central or strategic position in policy and planning systems. The case studies are not used with the intention of criticising the actions of the companies or regulatory authorities involved. In addition, this chapter will briefly look at Environmental Management Systems (EMS) and planning for closure and their potential relationships with EIAs. All these features of environmental control are intrinsically related yet tend to be treated as separate issues. Another aspect of an EIA, that of community involvement and public consultation, is mentioned in connection with the Tara case study. However, the full intricacies of this delicate area will not be included in this chapter as they are not relevant to the central discussion.

There seems to be some confusion in the literature, and in the minds of practitioners, over the use of the terms "Environmental Assessment," "Environmental Impact Assessment," "Environmental Impact Statement," and "Environmental Plan," and definitions abound. This could be due to the uncertainty of the role of the environment in project development or possibly just reflect variations in the use of the English language across the globe. For the purpose of this chapter, the following definitions extracted from Glasson[1] will be used:

- Environmental Impact Assessment (EIA) is a systematic process that examines the environmental consequences of development actions.
- Environmental Impact Statement (EIS) provides documentation of the information (gathered by the Assessment) and estimates of impacts derived from the various steps in the process.

In the U.K. there is now a tendency to leave out the word "impact" because it automatically implies a negative result, but this trend will not be followed here.

6.2   THE PROCESSES IN THEORY

6.2.1   ENVIRONMENTAL IMPACT ASSESSMENT (EIA)

Until 40 years ago there were very few controls in relation to the location, development, and operating practises of the minerals and energy industry.[2] The only exception to this was legislation to protect the health and safety of miners, particularly in underground mines. The 1950s saw the introduction of environmental planning and control for mining projects in developed countries, though it was not until between 1960 and 1975 that a more stringent position began to be taken. Much of the initial concern was related to pollution, though there was still a prevailing attitude that the economic importance of the minerals industry should take precedence over other considerations. Projects were assessed by means of a cost/benefit analysis that had no generally accepted mechanisms to take account of environmental or social issues.

The emergence of the Environmental Impact Assessment (EIA) in its present form was initiated in the United States, with the enactment of the National Environmental Policy Act in January 1970. This legislation aimed to place environmental quality on an equal footing with economic growth, by insisting that "every recommendation or report or proposal for legislation and other major Federal actions significantly affecting the quality of the human environment" must be accompanied by a detailed environmental statement.[3] In August 1973, the Council on Environmental Quality produced guidelines that outlined procedures for the preparation of these Environmental Impact Statements (EIS), to support the more general wording of the Act. However, none of the legislation left any doubt that these environmental goals were only to be pursued when consistent with economic growth and the maintenance of high living standards.[2]

The first EIA legislation was very general and it was left to the relevant agency to identify the range and scope of the study. It has been said that a universally accepted definition of EIA does not and cannot exist,[4] though many have made attempts. In 1975, Munn reported on the Workshop held by the United Nations Scientific Committee on Problems of the Environment,[3] which outlined that an EIA should contain:

- A description of proposed action and alternatives.
- Predictions of the nature and magnitude of environmental effects.
- Identification of relevant human concerns.
- A list of the impact indicators to be used.
- A determination of the values of the impact indicators and the weighting of the total environmental impact.
- Recommendations for acceptance or rejection of the project and remedial action.
- Recommendations for inspection procedures.

The impacts mentioned here include human health and well being, and the preservation of the ecosystems on which human survival depends. Wathern[5] summarises this as "a process having the ultimate objective of providing decision-makers with an indication of the likely consequences of their actions," whether these actions are projects, plans, or policies.

More recently, Clark[4] has identified the purpose and nature of EIA as being:

- To ensure that development options are environmentally sound and sustainable.
- To ensure that environmental consequences are identified early and taken into account in project design and implementation.
- A management tool providing information upon which decisions may be taken.

To these definitions, Heape[6] has added:

- An information source which can contribute towards the construction and operation of projects, and the definition of monitoring and auditing programmes.

Because of their nature, project personnel rarely carry out EIAs. The work is more usually contracted out to a specialist company, who uses a team of experts coordinated by an interdisciplinary manager. Once the work is completed, and the EIS has been approved, it is rare for the company to retain the services of these consultants.

As can be seen by these definitions, the EIA has centred on project development and its approval is usually required prior to the go-ahead being given for a project. As stated in the Berlin Guidelines,[7] "An Environmental Impact Assessment process is utilised to inform governmental decision-making of the environmental consequences of mining activities requiring government approval." In this way, EIA has become part of the planning and decision-making process for the initial design, acceptance and implementation of a mining project. However, the evidence indicates that it is not being used as a management tool throughout the life of the project.

TABLE 6.1

Environmental Impact Assessment (EIA) and the Project Life Cycle

EIA process	EIA action	Project phase
		Exploration
		Identification
Screening	Baseline Studies	Concept
Scoping	Define Issues	Pre-Feasibility
EIA Preparation	Identify/Predict Impacts	Feasibility
EIA Approval		Project Approval
Monitoring		Construction
Monitoring/Auditing		Operation
Restoration	Rehabilitation	Decommissioning

Adapted from [5], Wathern, P., et al., Assessing the environmental impacts of policy, in *The Role of EIA in the Planning Process*, Clark, M. and Herington, J., Eds., Mansell, London, 1988.

6.2.2 PROJECT LIFE CYCLE

The life cycle of a mining project traditionally involves three sequential phases: design, implementation, and closure, which are preceded by a period of exploration. These phases can be subdivided into stages which may overlap and need not be continuous, but do follow an established sequence[8] and are intrinsically reliant on each other.

The first stage in the design phase is the identification and location of the project and the initial concepts. The second stage is the Pre-Feasibility Study, which involves preliminary engineering design and a cost/benefit assessment. If a positive decision on the project is made, then the investigations are stepped up in preparation for the third and final stage in the design phase, the Feasibility Study. This will include geotechnical surveys, engineering design, and a more detailed cost/benefit analysis of the project, including market predictions where relevant. This Feasibility Study will then be appraised by either the financial backers of the project and/or the relevant government authority or agency. If there continues to be approval in principle, then the project will enter the implementation phase and negotiations for its development will begin as the first stage.

When the negotiations are complete and project approval is granted, the project then moves into the second stage of this phase, which is construction. At this time, more elaborate engineering designs are produced and, if relevant, a more detailed geotechnical survey will be carried out. From construction, the third stage is the commissioning and operation of the project. The final phase is closure, with the only stage in this phase being decommissioning.

Table 6.1 shows how the development of an EIA currently fits into the project cycle. Because of the potential importance of environmental issues to the success or otherwise of a new project, they should be incorporated into project planning from the outset.[8] It is essential to initiate the baseline studies as early as possible to ensure that there is adequate information as a foundation for the predictions. Proponents are more frequently being encouraged to start these baseline studies while the project is still at the conception stage, to enable sufficient data to be collected. The identification of impacts, predictions, and mitigation measures will then be established during the feasibility stage, using the data gathered by the baseline studies and during the geotechnical and engineering design stages.

This information is then compiled, along with the outline monitoring programme and auditing procedures, in the EIS that is submitted before or at the same time as the Feasibility Study. Project approval includes the "acceptance" of the environmental impacts and any mitigation measures, though negotiations often result in conditions being attached to this approval. Monitoring would then begin at the start of construction and normally continue in some form for the life of the project. An environmental audit and review of the monitoring and management programme may then be carried out after the project has been operating for some time and can be repeated at set intervals if necessary. When the mine is decommissioned, it is now accepted by most mining companies, and is a requirement of most countries that some form of restoration and rehabilitation is carried out before the mine operator surrenders the lease.

6.2.3 TOWARDS ENVIRONMENTAL MANAGEMENT SYSTEMS

In Europe, there has recently been a growth in use of management systems, both for quality control and improvement in environmental performance. The concept of these systems is to guarantee a certain standard of operations as a starting point and continuous improvement as a goal. Unlike a project cycle, which is designed as a once-only process, a management system is set up as an iterative cycle with each stage being continuously revisited. Environmental Management Systems (EMSs) have grown from a concept known as Total Quality Management. This is a philosophy of management that was developed as a response to the difficulties of reconciling engineering quality with a rapidly expanding, unskilled, or semi-skilled workforce. Its central core was to emphasise the critical importance of customers and the necessity for an empowered, integrated, and well-trained workforce in order to meet their needs. The concept stressed the importance of systematised procedures. Total Quality Management is seen to fit in well with growing environmental concerns and to support environmental management because it encompasses the following aspects:

- It provides a framework for considering and responding to the demands of environmental stakeholders.
- The concept of continuous improvement encourages organisations to move beyond mere compliance with environmental regulations.
- The focus on root causes of problems rather than the symptoms is a better approach to environmental problems than "bolting on" pollution control equipment.
- The belief that quality is everyone's responsibility supports the concept that all employees have to make a contribution to environmental performance.
- Concern with calculating the cost of quality provides a useful framework for considering the total costs and benefits of environmental action or inaction.

From this background, EMSs have been developed in response to increasing concerns about the environmental impacts of industrial and commercial operations. Their concept is to help businesses assess and develop environmental management procedures. This then establishes a recognised system which goes beyond "green" public relations, providing a structure to keep abreast of increasing environmental legislative pressures and to respond to queries about environmental performance from employees, shareholders, investors, or regulators.

The basis of the EMS is the familiar management cycle of establishing a policy, implementing it, assessing and reviewing progress via an audit, and making modifications as required. It is a repetitive cycle with each stage being continuously revisited and improvements made on each visit. These are the stages in an EMS, based on an outline specified by the Australian Environmental Protection Agency:[9]

- Organisational commitment.
- Environmental policy.
- Environmental Impact Assessment.
- Consultation with regulatory authorities and community.
- Objectives and targets.
- Environmental Management Plan.

- Documentation and environmental manual.
- Operational and emergency procedures.
- Training.
- Emission and performance monitoring.
- Environmental compliance audits.
- Reviews.

The objectives and targets must comply with the legislative and regulatory requirements applicable to the environmental aspects of the company's activities, products, and services. Environmental Management Systems are therefore designed to produce continual improvement in environmental performance throughout the life of a project.

6.3   THE PROCESSES IN PRACTICE

The presumption underlying an Environmental Impact Assessment (EIA) is that prevention is better than the cure,[10] and that the consequences of a development can be foreseen during the feasibility and design stages. According to Ricks, from a planning and regulatory point of view, an EIA offers the most effective means of minimising detrimental effects from the outset. However, mining by its very nature is a dynamic process and the majority of projects are evolving from pre-feasibility until decommissioning. To grant a mining project environmental approval, ostensibly for the whole of its life, on the basis of the EIA tacitly implies a static situation.

Environmental Impact Assessment procedures are designed as a "once-only" cycle to aid in the decision-making for the general concept of a project. This means that when the relevant agency grants approval, it is accepting certain stated parameters that include the predicted environmental impacts. On the whole, the regulatory framework does not exist to deal with a project that is constantly evolving, especially where the changes have environmental implications. In some instances, if the developers wish to make any major changes to the original project design, they are required to produce an amendment or addendum to the original Environmental Impact Statement (EIS) or an additional, free-standing EIS. However, there are not always the procedures in place for assessing these impacts in conjunction with those contained in the original EIS.

6.3.1   Western Australian Legislation

The regulatory requirements and procedures in Western Australia provide an example of how the environmental control of mining projects may be handled. Here, the responsibility is shared between the Department of Environmental Protection, on behalf of the Environmental Protection Agency (EPA), and the Department of Minerals and Energy. If a project has the potential to create a "significant effect on the environment" it is forwarded to the EPA for assessment. However, the Department of Minerals and Energy handles most of the day-to-day control of the mining industry with the Department of Environmental Protection having an auditing and referral role.[11]

One of the prerequisites for the granting of a mining or general purpose lease is that the proponent must produce a plan of operations and a programme to safeguard the environment, called a Notice of Intent. The approval of this Notice of Intent is accompanied by the following conditions:

- The construction and operation of the project and measures to protect the environment being carried out generally in accordance with the document titled 'insert document title'. Where a difference exists between the above document(s) and the following conditions, then the following conditions shall prevail.
- Any alteration or expansion of operations within the lease boundaries beyond that outlined in the above document(s) must not commence until a plan of operations and a

programme to safeguard the environment are submitted to the State Mining Engineer for his assessment and until his written approval to proceed has been obtained.

- The lessee submitting to the State Mining Engineer in [*insert month*] of each year, a brief annual report outlining the project operations, mine site environmental management, and rehabilitation work undertaken in the previous twelve months and the proposed operations, environmental management plans, and rehabilitation programmes for the next twelve months.

These conditions allow for the regulatory authority to request an additional Notice of Intent for any change made to the original project design criteria. Each Notice of Intent is assessed on its own merits, but is also considered in relation to previous Notices of Intent and the overall project.[12] Part of this assessment includes a review of the financial surety, known as an "unconditional performance bond," which may be increased or decreased as required. In addition, these bonds are reevaluated at the annual environmental review and, if the performance is unsatisfactory, they can be increased to cover the additional risk.

In this way, the Western Australian regulatory authority is made aware of the environmental implications of any project alterations or expansions. However, this does not change the fact that the project was accepted with one set of environmental predictions that have been changed. Alterations or expansions made to the original design of a project may be based on economic considerations, such as to improve or increase metal production, or may be unintentional, due to geological, geotechnical, or engineering details that are only established during the construction phase or later. It is possible that these changes will substantially alter the environmental impacts that were predicted in the original assessment and, in the worst case, may result in a catastrophic event.

6.3.2 OMAI MINE

In August 1991, the Guyanan regulatory authorities gave approval for the development of the Omai Gold Mine. The Environmental Impact Statement, which was submitted along with the Feasibility Study, was also granted approval. The documents included the design for a tailings retention facility in the valley of the Captain Mann Creek, a tributary of the Omai River that flowed into the Essequibo River. Following approval, the geotechnical engineers who designed the tailings dam were awarded the contract to supervise the construction of Stage I of the embankment to an elevation of 505 m, scheduled for completion at the end of 1992.

The Omai Project developer submitted an addendum to the 1991 Environmental Impact Statement in January 1992. This document contained changes which included the design of the tailings embankment, management of the tailings pond, and a revision of the overall water balance.[13] Included in these changes was zero discharge from the tailings, to the Essequibo River, for the first three years of operation. In December 1992, the original geotechnical engineers were released when Stages IA and IB of the embankment had been completed, to an elevation of 500 m. From this point on, another firm of geotechnical engineers were retained, in an advisory role, for the remaining construction of the tailings retention facility.

In March 1995, the project developers submitted a further addendum which sought official approval for further modifications of the 1991 Environmental Impact Assessment and the 1992 Addendum. The request was for the authorisation of the immediate release of effluent from the tailings pond, with a cyanide content of 8 mg L^{-1}, instead of the 2 mg L^{-1} contemplated in 1991. The motivation behind this addendum was that by this time the tailings dam embankment had reached an elevation of 534 m and the mine operators were reluctant to raise the level any higher. However, the water in the tailings pond had been rising at such a rate that it was anticipated that by the end of June 1995 it would exceed an elevation of 529 m, above the maximum permissible water level when the dam crest was at 534 m. In addition, the mine operators were planning to expand mill capacity by 50% with completion expected in the first quarter of 1996. The regulatory authorities rejected this request on the basis of the elevated cyanide levels.

Two and a half years after commercial production started, in August 1995, Omai had a major piping failure in the tailings embankment which involved the release of approximately 4.2 million cubic meters of tailings, containing cyanide and heavy metals, of which an estimated 2.9 million cubic meters flowed into the Omai River and then into the Essequibo River.

The official report on the failure of the dam wall concludes that it is clear that piping occurred in the dam core as a result of improper construction methods used for the rock fill.[14] The water and slurry that remain in the pond make it impossible to determine when in the construction process the faulty work took place, so it would be pure speculation to think that the changes included in the addendum to the Environmental Impact Statement were responsible for the failure. However, the implication could be that the project was given approval on the basis of the original Environmental Impact Statement, but that changes made during construction and operation may have resulted in a potentially catastrophic environmental disaster. It is interesting to note that on the first page of the 1992 addendum is written the following: "Consistent with the nature of a large scale resource development project, the original development plan has changed as detailed engineering has progressed. Some of the changes are significant with respect to environmental management."[13]

6.3.3 Ok Tedi Mine

Economic considerations often mean that any changes made to a project once it has received initial approval are very unlikely to be refused on purely environmental grounds. This tends to be the case even though the changes may have substantial environmental implications. The Ok Tedi gold/copper mine in Papua New Guinea provides a good example of this concept.

In summary, the regulatory authorities gave approval for the mine on the basis of the tailings being retained in a dam and the waste stored in stable dumps. Construction work on the tailings dam was halted by a landslide and the government was forced to accept an interim tailings scheme, in order to allow the project to start gold production on schedule. This interim scheme involved the treatment of the tailings by oxidation of the cyanide with hydrogen peroxide and the storage of only the sand fraction, which amounted to approximately 30%, in a small impoundment. The remainder of the tailings was then discharged into the river system.

Following a period of negotiation, the government accepted a three-year deferral of the construction of the tailings dam, during which time the mine operators were to carry out an intensive environmental study of the entire Ok Tedi and Fly river systems. This study would establish what waste retention facilities were required, taking economic considerations into account. During this period, the reserves in the "gold cap" were exhausted and the mine started processing copper ore in a conventional flotation circuit, with the remaining gold reporting to the copper concentrate.

At the end of the three years, the scientific data collected failed to prove conclusively that the release of the mine wastes into the river system would have an adverse effect on the biota. In addition, the mine operator proved conclusively that the cost of constructing a tailings dam could result in the closure of the mine.

The government, therefore, decided to allow the project to continue operating without a tailings retention facility, releasing all the tailings into the river system. In addition, the overburden that had been identified in the Feasibility Study as adequate for the construction of the stable waste dumps (competent waste) proved to be unsuitable. This meant that the government also had to accept failing waste dumps, which released additional material into the same river system.

At Ok Tedi, the original concepts for waste disposal were superseded by circumstances, thereby forcing the decisions away from environmental considerations and into the financial and political arena. The mine operator based the location and design of the waste retention facilities on the geotechnical work carried out for the Feasibility Study. The landslide at the tailings dam site, and the lack of sufficient competent rock with which to construct the stable waste dumps, implies that the geotechnical data was insufficient to allow for accurate environmental predictions to be made. However, by the time this had been established, both the private shareholders and the Papua New Guinean Government had already made a substantial financial commitment towards the development of the project. The net result was that the government was forced, by economic and political concerns, to make a decision to allow the mine to continue operating with minimal environmental constraints.

The conclusion can be drawn that the Environmental Impact Assessment totally failed in its aim; that is, to predict the environmental impacts of the mine on the Ok Tedi and Fly River systems. This cannot be seen as the fault of the Environmental Impact Assessment, but more on the quality of the geotechnical work that was carried out for the Feasibility Study, on which the environmental assumptions contained in the Environmental Impact Assessment were based. In addition, in Papua New Guinea the geotechnical aspects of the waste disposal facilities, such as the tailings dam or stable waste dumps, are the ultimate responsibility of the Mines Safety Division. At the feasibility stage of a project, the geotechnical designs are only checked in concept. The mine's inspector would not then be required to make a judgement on any of these facilities until they had been constructed. If they are then found to be unsatisfactory, he can instruct the mine operator not to use them. This, of course, is far too late for environmental considerations to be taken into account.

Even though a project developer is often required to produce some form of Environmental Impact Assessment for any major changes, these later additions rarely undergo the rigorous analysis that the original assessment would have been subjected to. By the time any changes are put forward, the acceptance of the mining project has already been established and financial commitments made. Even though the environment may have been placed on an equal footing with economics during the initial approvals, subsequent decisions are liable to be weighted heavily in favour of financial considerations. The result could easily be a project that was initially only to have a minimal and acceptable environmental impact, resulting in a potentially major impact.

6.3.4 TARA MINE

There are some projects where environmental considerations have been successfully assimilated into the day-to-day management of the mine. One of these is the Tara lead/zinc mine in Ireland. From the outset, the proximity of the town of Navan and the existing prosperity of the region meant that the mine operator had a difficult job in convincing the public that the mine should go ahead. Even though the project was conceived prior to any planning requirement for an Environmental Impact Assessment, extensive biophysical baseline studies were carried out during the feasibility stage.

The community was strongly against any form of mining in the area and so the mine operator initiated an "open-door policy," which included public participation in all stages of planning and design. Rather than holding the usual large public meetings, this was done on an individual basis or with small groups, and an information centre was established which was open to anyone. This meant that many issues that may have been overlooked by a more formal Environmental Impact Assessment were included in the discussions from the start. The result of these open negotiations was that two major design changes were made for environmental and social reasons, before construction had started.

In addition to this innovative approach to public participation, the mine operator also instigated a number of remarkable procedures for the time. One of these was that the Environmental Manager was directly responsible to the Board of Directors and not to the mine manager, which is the normal practice. A second was that when operations started, an environmental training programme was initiated for the entire workforce at all levels. A third was the establishment of the information centre, which was later moved into the town of Navan both for reasons of safety and ease of public access. The combination of these strategies has meant that the environmental issues have been handled as a central point to mine planning and design and included in any subsequent modifications.

6.4   STRATEGIC ENVIRONMENTAL ASSESSMENT

In recent years, there has been some movement towards the idea of Strategic Environmental Assessment. This may be the best starting place for the inclusion of the Environmental Impact Assessment into an evolving Environmental Management System. Strategic Environmental Assessment is defined as the formalised, systematic, and comprehensive process of evaluating the environmental impacts of a policy, plan, or programme and its alternatives. These findings can then be used for publicly accountable decision making.[15] The application of Environmental Assessment at the level of policies, plans, and programmes makes it easier to look at cumulative impacts, alternatives, and mitigation measures. This allows for the environmental implications to be anticipated and developments steered in the right direction from project inception, rather than the more reactive nature of Environmental Impact Assessment.[1]

As Glasson et al.[1] point out, most of the arguments for a Strategic Environmental Assessment are related to problems with the current system of project Environmental Impact Assessment. These are that project Environmental Impact Assessments:

- React to development proposals rather than anticipating them and therefore cannot "steer" development.
- Do not adequately consider the cumulative impacts.
- Cannot fully address alternative developments or mitigation measures (in many cases, a project will already be planned quite specifically, with irreversible decisions taken by the time the Environmental Impact Assessment is prepared).
- Cannot address the impacts of potentially damaging actions that are not regulated through the approval of specific projects.
- Are often carried out in a short period of time, thereby limiting baseline data and the quality of analysis.

A Strategic Environmental Assessment is a tool that the regulatory authority can use to establish that a certain area or district is amenable to mining, prior to issuing an exploration licence. It would also outline the acceptable limits for pollutants and social impacts in that area. This would mean that a company which had been granted an exploration licence would know that any subsequent mining project would be allowed to proceed as long as they were able to achieve the established environmental guidelines.

Although this concept is still in its infancy, and needs considerable development and fine-tuning, it is not envisaged that a Strategic Environmental Assessment would negate the need for a project-specific Environmental Impact Assessment. However, project Environmental Impact Assessments would be carried out within the framework of the Strategic Environmental Assessment and would only contain project-specific information. Project approval would be granted on the basis of this information, with conditions attached tying the company to the parameters set out in the Strategic Environmental Assessment.

At this stage, it could also be established that the environmental parameters would not necessarily remain static and more stringent requirements may be imposed as a greater understanding of the issues became known and technology was developed for their handling. The information contained in the Strategic Environmental Assessment and the project Environmental Impact Assessment would then be the basis for an Environmental Management System. This system would hinge on compliance monitoring and the auditing of environmental performance, but at the same time allow any changes in design specifications to be absorbed without changing the accepted parameters.

The implementation of Strategic Environmental Assessment is seen to be fraught with technical and procedural problems,[1] and a proposed European Commission Directive on the subject has been met with a very chilly reception. However, Strategic Environmental Assessment systems already exist in some form in the U.S., Netherlands, France, Germany, Japan, and New Zealand.

For example, since 1987 the Dutch Government has formally required Strategic Environmental Assessments for sectoral plans on waste management, drinking water supply, energy and electricity supply, and for some land use plans.[15] A number of provincial waste management plans and provincial mineral extraction plans have been subject to a Strategic Environmental Assessment, though it is felt that these early attempts have had little influence on decision making. The main reason for this appeared to be because the Strategic Environmental Assessments were only drawn up after a decision on the plan had already been taken.

Strategic Environmental Assessments are seen as one of the most direct and effective ways of ensuring that human activities are carried out at a level that is environmentally sustainable.

6.5   ENVIRONMENTAL IMPACT ASSESSMENT AND PLANNING FOR CLOSURE

Until recently, an Environmental Impact Statement would include the predictions, amelioration techniques, and monitoring programmes for the life of the project. Closure and rehabilitation were dealt with as a separate issue. The condition in which a decommissioned mine site should be left is still an issue of extensive debate, though there is now general agreement that the area should be made safe with some degree of revegetation.

To ensure that this is the case, mine operators have been persuaded to accept rehabilitation as an integral part of the project concept and design. This is both from an environmental and financial perspective, even though it is not yet included in the legislation of most countries. Both the project proposal and the Environmental Impact Statement are often now expected to include a programme for mine decommissioning and proposed methods of site restoration, otherwise known as planning for closure. Implementation of this concept at the start of a project will minimise the impacts of mining on the environment and may also reduce the financial burden of mine closure and rehabilitation (see Warhurst and Noronha, Chapter 2).

In the past, especially where post-closure planning requirements have been minimal or non-existent, any moral obligations for site restoration and rehabilitation have sometimes been avoided through manipulation of ownership and transfer of physical assets and liabilities. A financial surety indemnifies the authorities against closure and rehabilitation costs, and will come into force should an operator fail to meet his full obligations at the planned time of closure or in the event of a premature, unplanned closure. Operations are not restored to satisfactory standards or not restored at all because of:[10]

- The financial failure of the operator.
- The technical failure of the operator.
- The poor implementation of the planning system.
- A dispute over ownership.

Financial surety can take the form of:

- Irrevocable Letter of Credit — an agreement between the company and the bank whereby the bank will provide cash funds to the authorities if the company defaults.
- Performance Bond — a surety bond issued by an insurance company in which the insurer is responsible for all claims up to an agreed limit.
- Trust Fund — a fund that operates in a similar fashion to a pension fund, with regular contributions being invested by a fund manager.
- Insurance Policy — a special form of performance bond.
- Parent-Company Guarantee — the parent company guarantees to indemnify the government in the event of company default.
- Pledging of Assets — the company assets are pledged to the government.

In his paper "Closure Considerations in Environmental Impact Statements," Ricks[10] states that, "It is because of the limited life of most mines and the poor track record of the industry in the past that the environmental effects associated with mine closure must be included in the Environmental Impact Statement." Mining is a relatively short-term land use and an important aspect of mine planning should be the rehabilitation of disturbed lands to a stable and productive post-mining landform. However, in the past, many mine operators have exhausted the economic reserves, removed any valuable equipment, and turned their back on the site, leaving the clean up to be carried out by others. It is therefore essential that when this does occur there are the funds to pay for the work.

Proponents for a mining project would need to investigate three aspects of planning for closure as part of the Feasibility Study:

- A mine closure plan.
- The predicted impacts.
- The implementation costs.

The first two of these should be included in the Environmental Impact Statement and from these a financial surety could be established.

Although most regulatory authorities that impose a financial surety have some method of adjusting the amount during the life of the project, especially if any substantive changes are made to the original design, there is no guarantee that the sum will be sufficient to ensure adequate rehabilitation. This is not due to any intent of malice on the part of the mine operator or incompetence of the regulatory authority. It is due to the fact that it is almost impossible to satisfactorily plan for closure at the start of a project that is likely to grow and evolve during its period of construction and operation. Again, we have a situation of making decisions at the beginning with no formal method of redress.

6.6   DISCUSSION

Environmental Impact Assessment can, in its current form, probably best be described as a management tool that enables the identification of potential environmental impacts for project approval. In this role, it provides information on the likely consequences of resource allocation and contributes to the overall evaluation of a particular action. As a part of the decision-making process, it has the opportunity of placing the environment on an equal footing with economics, thereby attempting to provide an integrated approach. When used at the project level, it should also ensure that environmental implications are recognised at an early stage and that mitigation measures are incorporated into project design and implementation.

Eggert[16] states that, "Companies also often have to carry out detailed assessments of the environmental impacts of proposed mineral development, which in turn are used by governments in deciding whether to permit mine development at all." This implies that the environmental issues play a leading role in project approval. However, by the time a mineral deposit has been delineated and established as being economically feasible, the decision to go ahead has already ostensibly been made and very few projects are rejected on environmental grounds alone. Confining Environmental Impact Assessment to the project level, as is generally the case, may only serve to undermine its importance in the decision-making process.

The Ok Tedi and Omai case studies show that Environmental Impact Assessments can fail to achieve what is expected of them at a project level; namely, to provide an environmentally sound basis for project approval. In the case of Ok Tedi, the main reason for this was the scarcity of the geotechnical information on which the environmental impacts were based. At Omai, the evolutionary nature of the mining project changed the environmental parameters included in the original Environmental Impact Statement. It is very difficult to argue convincingly that the environmental implications are being given a fair hearing, when they are based on inadequate data and changing parameters. Add into this equation the quality of the predictions and their error margin, and the fears of many are compounded — that Environmental Impact Assessments are being used to protect government and company images, and little else.

This already shaky perception is further exacerbated by the quality of environmental staff employed by a mining company. As mentioned previously, the work for an Environmental Impact Assessment is usually carried out by experienced consultants who are specialists in their field and hired for a particular task. Their work is both coordinated and compiled by an Environmental Impact Assessment expert who has an interdisciplinary understanding of the individual and combined implications of the findings. In contrast, even though some mine operators do now employ specialists to carry out the environmental monitoring, a mining engineer turned environmental engineer almost always heads the department or section. Although this background is very useful in understanding the design and nature of tailing dams and waste dumps, it rarely enables the person to encompass the broader view of the environment. The result is often a lack of continuity between the Environmental Impact Assessment and the environmental management of the project.

The case study of Tara would appear to disprove all these theories, though the reasons for its success are due to a very different set of circumstances. At this project, the proximity of the town of Navan and the economic stability of the area meant that the proponent had a difficult task to convince the regulatory authority and the public that the mine should be allowed to go ahead. At an early stage of negotiations, the mine operator employed a person to specifically handle environmental and social issues. The inspiration and apparent environmental success of this project were the responsibility and work of this one person and his relationship with the mine operator and the public. The result was some of the earliest environmental baseline work ever undertaken and public participation on a scale that is, even now, rarely seen. The continuation of the public participation resulted in ongoing environmental management as a central point to mine development and operation.

There is a need for some guidance at the policy level in the development of the mining industry and the associated levels of impacts that are acceptable. It is a generally established fact that it is not feasible to have a mine without some adverse environmental implications, but, at the same time, a mine "at any cost" is now an unacceptable concept. This policy could be written along the lines that a certain percentage of mine wastes must be retained, subject to a given set of criteria that are established at the planning level. If such a policy had been in place at the time that Ok Tedi were requesting the indefinite delay of any permanent waste retention facilities, it would have given the regulatory authority a much stronger negotiating position and might have resulted in a very different conclusion. However, in reality, the Papua New Guinean Government having a financial interest in the project as a shareholder further complicated this situation.

These policies, translated down to the planning level, could then look at the environmental implications both on a sectoral and area basis. The sectoral (Strategic) Environmental Assessment would cover all the implications and impacts of mining projects and the preferred methods of management. Using the sectoral approach, it would be possible to establish acceptable levels of environmental contamination, in the form of quality standards for such releases as process chemicals, heavy metals, and suspended sediment. Guidelines containing these standards and the preferred methods of achieving them, such as tailings dams and stable waste dumps, could then be drawn up, though the location of the compliance point could remain as negotiable until a more accurate environmental understanding had been established.

At the same time, it would be necessary to look at the environmental implications on an area-by-area basis, to ensure that the cumulative effects of projects were taken into account. This level of planning could be used to refine the more general environmental standards set at the sectoral stage and to take into account existing developments in the area. At the project level, it would then only be necessary for the company to establish how they intended to meet the environmental standards and the monitoring programme that they propose to prove that they are being met.

The major criticism of using Environmental Impact Assessment at the project level is that the decision that the project is desirable has already been taken at a policy and/or planning level,

TABLE 6.2
Implementation of Environmental Assessment at the Policy, Planning, and Project Level

Level	Environmental tool
Policy (SEA)	General environmental assessment
Planning (SEA)	Detailed environmental assessment Criteria and standards Guidelines
Project (EIA)	Mitigation measures Monitoring Auditing

SEA = Strategic Environmental Assessment

without any environmental considerations. In most countries, the existing systems allow for the granting of an exploration permit, with no reference to the relevant environmental regulatory authority, and the location of the mine is then established by the presence of an economically viable deposit. This means that the Environmental Impact Assessment is condemned to a reactive role and is not a true part of the decision-making process. In addition, it denies the environment an equal footing with economics and often the impacts are only viewed in the short-term. The logical conclusion is that for Environmental Assessments to have any meaning they must have a central place in planning and policy, and must be carried out proactively. It is only from this position that sound environmental management strategies can be established.

The tendency to confine Environmental Impact Assessments to projects has come about because, at this level, they show obvious immediate benefits and can be added to existing development processes with minimum disruption, and are therefore easier to handle. To impose an assessment of environmental implications into policy and planning procedures is far more complicated and involves changing long-established bureaucratic conceptions and procedures, as well as transferring the responsibility from the project proponent to the regulatory authority. Table 6.2 gives an example of how Environmental Assessments could be incorporated into the various levels of decision-making as a generic policy or planning tool.

Whichever way it is used, the Strategic Environmental Assessment would not be the short-term, specific, reactive document that has come to be accepted at the project level, but a more general view of the overall environmental implications, a statement of acceptable impacts, and guidelines for implementation and closure. If the type and/or scale of development were approved in principle at the planning stage, then the planning application process for individual projects could be accelerated considerably, saving valuable time and money. The proponent would be responsible for ensuring that the design proposals complied with the planning guidelines, and that all waste disposal methods would meet the criteria. This would mean that the project-specific Environmental Impact Assessment could be reduced to include the mitigation measures, monitoring programmes, and environmental management, making it a much easier document to review.

To take the concept one step further, some sort of cyclical system is required to ensure continuous environmental improvement during the operating life of a project, and to prevent technological stagnation. However, as discussed previously, the project life cycle is not an iterative process but a one-off, with the phases and stages rarely revisited. Once project approval has been granted, the government has committed itself to accepting and supporting the project and, for developing countries, there are often sizeable financial considerations. However, changing technology and further exploration may enable the mine operator to make substantial changes to the original concept such as to:

- Extend life of mine.
- Increase throughput.
- Refine processing methods.
- Change processing methods.
- Re-design the waste retention facilities.

At present, the project-linked Environmental Impact Assessment is tied into the project life cycle and is not designed for change. It can only assess what is known at the time of its production. It does not ensure an active and efficient Environmental Management Programme, nor does it put in place any methods for handling changes in the project. The only environmental aspect of a project that is "ongoing" (apart from monitoring and audits) is the mine operator's environmental policy, which often has little bearing on the actual environmental impacts. If a mine has a ten to twenty-year lifespan, changes in technology may mean that the operation can substantially improve its environmental performance. However, there is little or no incentive for a mine operator to take any notice of these changes, even though many innovations may be cost effective (see, for example, Warhurst, Chapter 3).

One possible alternative is to link the Environmental Impact Assessment into an Environmental Management System, but the question is, would it be possible to integrate them and develop a mutually symbiotic relationship? This may involve a fairly drastic rethink of the structure and nature of the assessment and the overall attitude to the methods of mining in an environmentally acceptable manner. The concept behind an Environmental Management System is to provide a structured method for management to have an awareness and control of the project's environmental performance, at all stages of a project life cycle, from identification to closure. An Environmental Management System is a repetitive cycle with each stage being continuously revisited and improvements made on each visit. A competently prepared Environmental Management System is a useful tool that may assist mine management to meet both current and future environmental requirements and challenges.

At present, the utilisation of the Environmental Impact Assessment by most mine operators denotes an environmental performance that is static. It outlines the expected levels of pollution and predicts the impacts, allowing the regulatory authority to accept those predictions and give approval accordingly. Monitoring, and sometimes auditing, is then carried out for the life of the project, to ensure that those predictions are not exceeded. Not only does this stifle any potential innovative improvements but it also makes no provision for any changes in the development of the mining project or legislation. In Western Australia, changes to the existing conditions or new conditions can be imposed on a mining project at any time and these can take into consideration changes in legislation. The regulatory authority is currently encouraging mining companies to put in Environmental Management Systems, which should improve their ability to handle these and any other changes.

The evidence is that it is rare for mining projects to be rejected on purely environmental grounds. Once a project has been implemented, it would take a very strong regulatory authority to close what is a potentially lucrative investment — a degree of luxury which most developing countries simply cannot afford and many developed countries do not practice, because of the economic factors.

Until such time as some changes can be achieved, it will be necessary to continue to place importance on the Environmental Impact Assessment at the project level, but with some critical changes. The main modifications that need to be implemented are an improvement in the quality of the geotechnical data used as the basis for the Environmental Impact Assessment and predictions, and the inclusion of a "worst-case scenario." For example, if there is a possibility that the proposed waste retention facilities will not be viable, then this should be spelt out early on and not left until approval has been granted. The easiest way to achieve this change would be to upgrade the level of information contained in the Feasibility Study and then make sure that it was adequately referenced in the Environmental Impact Assessment. In addition, procedural amendments to the environmental legislation should be introduced to address the environmental implications of any project plan alterations, following initial approval. This environmental update should cover the impacts of the proposed changes, as well as the overall environmental implications, so that they can be considered prior to any subsequent approvals.

There can be no doubt that Environmental Impact Assessment, even at the project level, has been a major step in the right direction and that considerable progress has been made over the last 20 years in understanding its value and position in development decisions. However, the learning curve still has a long way to go, to achieve a full commitment to care for the environment and place it on an equal footing with economic considerations. One criticism that has been levelled at Environmental Impact Assessments, in their present form, is that environmental experts are increasingly being put in a position of ruling on issues that are essentially political, economic, and legal. This is no bad thing in itself, as long as the environmental implications are honestly given an equal footing. However, reducing the environmental issues to a set of politically neutral issues which can be scientifically categorised and measured may well devalue the non-scientific, non-measurable aspects of the environment and make it difficult to include them in the decision-making process. As we have seen at the project level, the economic and political aspects can and do overrule the environmental considerations. While this continues to be the case, Environmental Impact Assessments will not achieve their proper status and projects will continue to be subjected to evaluations biased towards economic and political factors.

REFERENCES

1. Glasson, J., Therivel, R., and Chadwick, A., *Introduction to Environmental Impact Assessment,* UCL Press, London, 1994.
2. Rees, J., *Natural Resources: Allocation, Economics and Policy,* Routledge, London, 1990.
3. O'Riordan, T., *Environmentalism,* Pion Ltd., London, 1981.
4. Clark, B. D., Environmental impact assessment: origins, evolution, scope and objectives, presented at 11th International Seminar on EIA and Management, CEMP, Aberdeen, 1990.
5. Wathern, P. et al., Assessing the environmental impacts of policy, in *The Role of EIA in the Planning Process,* Clark, M. and Herington, J., Eds., Mansell, London, 1988.
6. Heape, M., Visual impact assessment, presented at 12th International Seminar on EIA and Management, CEMP, Aberdeen, 1991.
7. U.N., *Mining and the Environment: The Berlin Guidelines,* Mining Journal Books, London, 1992.
8. Higgins, R., Environmental management of new mining operations, in *Environmental Management of Solid Wastes,* Salomons, W. and Forstner, U., Eds., Springer-Verlag, USA, 1988, 372.
9. Australian Environmental Protection Agency, Best Practice Environmental Management in Mining, series of booklets produced by the Environment Protection Agency, Australia, 1995-1997.
10. Ricks, G., Closure considerations in environmental impact statements, presented at Institution of Mining and Metallurgy meeting "Planning for Closure," London, November 1994.
11. Lindbeck, K. and Murray, C., Legislating mining operations in Western Australia, *Mining Environmental Management,* 3, 22, 1995.
12. Biggs, J. W., personal communication, 1995.
13. Rescan Consultants, Omai Gold Project: Environmental Impact Statement Addendum, 1992.
14. UNEP, Tailings dam incidents 1980-1996, report prepared for United Nations Environment Programme by the Mining Journal Research Services, London, 1996.
15. Therivel, R., *Strategic Environmental Assessment,* Earthscan Publications, London, 1992.
16. Eggert, R. G., Mining and the environment: an introduction and overview, in *Mining and the Environment: International Perspectives on Public Policy,* Eggert, R. G., Ed., Resources for the Future, Washington, D.C., 1994, 1.

7 Prediction, Prevention, Control, and Treatment of Acid Rock Drainage

Paul Mitchell

CONTENTS

ABSTRACT

Acid rock drainage (ARD)* represents one of the most intractable problems in the non-ferrous metal and coal mining sectors, a fact reflected by the ever-increasing volume of research published in conference proceedings, journals, and books. There is a wide acceptance that ARD is both the most serious environmental impact caused by mining[1] and the industry's greatest environmentally related technical challenge.[2] Based on the substantial published literature and additional discussions with operators, regulators, and technology vendors, this chapter has been developed as an up-to-date review of the prediction, prevention, control, and treatment of ARD. Although much of the research discussed here can be applied at any point within a site's life cycle, the emphasis here is on the *planning* stage. However, there are many lessons that can be learnt from operational sites and these must also be used to inform the development of best practice at future sites.

Following a brief overview of the major sources of ARD, a summary of research on the formation and migration of ARD is presented. Particular attention is paid to the significance of biologically mediated reactions in the process of ARD generation, and the significance of ore mineralogy and geochemistry in determining *in situ* neutralisation prior to migration. Other site-specific factors are also discussed. The critical role of predictive techniques in optimising the closure process is examined. Established, innovative, and pre-commercial approaches to prevention, control, and treatment methods are reviewed with reference to technical, environmental, and economic performance. The implications of site-specific factors for technology transfer and the development of a generic best practice are discussed. Finally, conclusions are drawn regarding best practice and gaps in the current knowledge base. From the perceived gaps, recommendations for the direction and nature of future research are presented.

7.1   INTRODUCTION

ARD results from the abiotic and biotic oxidation and subsequent leaching of pyrite and other metallic sulphides. Generation and migration of ARD can persist for centuries and the volumes of contaminated water can be extremely high, particularly from sub-surface workings that drain large catchment areas (e.g., adits, tunnels, and shafts). An excellent example of this is the County Adit in Cornwall (U.K.), which drains more than 50 abandoned, interlinked underground metal mines, and discharges up to 70 million litres of ARD each day.

ARD is by no means a recent phenomenon; it has been associated with mining activity throughout the industry's long history. The earliest documented case is the Río Tinto region in Spain (probably the oldest major mining area in the world)[3] where activity dates back to the Phoenician period (pre-10th century B.C.). Río Tinto is Spanish for "Red River," and reflects the

---

* Although a number of other terms are also used to describe this phenomenon (e.g., acid mine drainage, acid mine waters or effluents), the author considers ARD to be the most appropriate generic term that represents all mining-related sources of acidic waters, and has used the term throughout this chapter.

discharge of iron-rich, mining-related acidic waters into the local river. In the more recent past, there have been hundreds or thousands of documented cases. One such high-profile example is Wheal Jane mine in Cornwall, U.K. (another, Summitville in Colorado, U.S. is examined in Chapter 17 by Danielson and Nixon).

Wheal Jane was just one of many shallow mines in West Cornwall. Producing tin, copper, zinc, and lead, it was drained by Jane's Adit. A neighbouring mine, Wellington, was also dewatered through Wheal Jane following its closure, due to the underground connections linking the two sites.[4] Wheal Jane operated intermittently until its final closure in March 1991, when dewatering was terminated and water levels began to rise. The first discharge occurred in November 1991, but under extant U.K. law, the operator was not required to meet the costs of treatment; this instead fell to the National Rivers Authority (NRA). Treatment consisted of liming and subsequent disposal of sludge in the tailings dam. However, on 4 January 1992, pumping to the dam was stopped as high winds and rain had prevented settlement of flocculated sludge, taking the dam near to its operational limit. While alternative treatment methods were being investigated, an old plug (in the Nangiles adit) failed, allowing the discharge of millions of gallons of acidic, metal-contaminated water into the Carnon River. Precipitation of ferric hydroxide ("ochre") not far from the discharge point had a highly visual impact, and it was probably this rather than the environmental impacts in an already polluted river system that mobilised a major response from the NRA and the U.K.'s Department of the Environment. The latter authorised in December 1992 the expenditure of £8 million for a pilot scheme to develop a long-term solution to the problem of acid mine drainage at Wheal Jane and for the continued short-term treatment of the water and disposal of the associated sludge. The long-term strategy has involved a detailed study of constructed wetlands, although the latest studies appear to indicate that in isolation this is not a viable response at this particular site.[5] Research is still ongoing at the site in an effort to develop the most cost-effective remedial treatment. It is clear that there will need to be a substantial financial investment at the site for the foreseeable future. Less clear is the role that planning for closure could have played at the site, given the historical nature of mining in the region and constraints to controlling water flow into and through the warren of interconnected underground workings. The question of whether a more stringent regulatory framework (e.g., as in force at Summitville, see Chapter 17) would have changed the outcome is also an interesting, albeit academic, one.

The global depletion of non-acid generating oxide ore deposits and the increasing dominance of sulphide ores means that the *potential* for ARD incidents will continue to increase. Fortunately, consistent improvements in the capacity to predict, prevent, control, and treat ARD have resulted in significant differences between "old" operations where remedial treatments are commonplace and "new" operations where preventative and control approaches are being attempted. However, incidents such as those that occurred at Wheal Jane and Summitville show that there is little room for complacency: further advances are crucial if the industry is to respond properly to public and regulatory concerns in the 21st century. There are signs that some operators, working in partnership with regulatory authorities and other stakeholders, are developing proactive methodologies based in part upon improvements in predictive techniques. It is to these partnerships that industry should look, perhaps, for its definition of best practice.

7.2 SOURCES OF ARD

There are five major sources of ARD, namely drainage from underground workings; runoff and discharges from open pits; waste rock dumps; tailings; and ore stockpiles (which have characteristics analogous to waste rock). Other sources may be locally significant, for example, spent heap-leach piles, stockpiles of segregated sulphides (see below), and natural seeps and springs in areas of sulphidic mineralisation.

7.2.1 UNDERGROUND WORKINGS

Abandonment of underground workings can result in the release of high concentrations of metal salts into the aqueous environment as the water table rebounds and the workings flood. These metal salts accumulate when the mine is pumped "dry" and in-place sulphides are exposed to oxygen and moisture. ARD generation may continue after flooding if there is a persistent source of oxygen (e.g., as a dissolved component of incoming waters).

7.2.2 OPEN PITS

Open pit mining can expose very large areas of sulphide-bearing rock to air and water. Failure to control water flow into open pits can result in large volumes of ARD. As oxidation of the sulphide proceeds, fresh sulphides may be exposed by spalling of the rock face, resulting in the constant renewal of the ARD source.[6]

7.2.3 WASTE ROCK

Waste rock has become a more significant threat as open pit mining has replaced underground mining, particularly in developed countries, and the volumes produced have increased.[6] Waste rock is normally coarse (10 to 20+ cm) with variable sulphide concentrations. Waste rock piles are often highly permeable and oxidation of sulphides can occur rapidly after disposal. During extended dry periods, dumps may build-up "stored" acid products and salts through evaporation and supersaturation processes, which are then released in the form of highly contaminated ARD during the next significant precipitation event.[7]

7.2.4 TAILINGS

Tailings often have a high sulphide content (mainly in the form of rejected pyrite, marcasite, and pyrrhotite) and are much finer than waste rock (e.g. 100% −0.2mm). Although tailings have a much higher specific surface area (measured in $m^2 g^{-1}$) than waste rock, the uniform and fine particle size leads to a much lower permeability than that seen in waste rock piles. Therefore, the increased surface area available for oxidation and leaching reactions is balanced by reduced contact with oxygen due to a slower moving waterfront and slower replenishment by oxygen-rich water. Consequently, tailings often generate ARD more slowly than coarser, but more permeable, waste rock.[8]

7.3 CHEMICAL ASPECTS OF ARD GENERATION AND MIGRATION

ARD generation and migration are complex subjects and intimately linked with the nature of the orebody, host rock mineralogy, and local and regional hydrology and hydrogeology. Generation of ARD does not necessarily lead to its migration, as there are a number of chemical processes that prevent the movement of ARD away from its source. Detailed information on these and other aspects of ARD can be found elsewhere.[1, 9–13] The main points are summarised briefly below.

7.3.1 THE CHEMISTRY OF ARD GENERATION

Pyrite (FeS_2), its dimorph marcasite (FeS_2), and pyrrhotite (FeS_{1-x}) are the principal sulphides responsible for ARD. The oxidation and the subsequent release of acidity (H^+) into the aqueous phase are summarised for pyrite and marcasite in Equations 1 to 4.

$$2\,FeS_2 + 7\,O_2 + 2\,H_2O \quad \rightarrow \quad 2\,Fe^{2+} + 4\,SO_4^{2-} + 4\,H^+ \qquad \text{Eq. 1}$$

$$2 \, Fe^{2+} + \tfrac{1}{2} \, O_2 + 2 \, H^+ \quad \rightarrow \quad 2 \, Fe^{3+} + H_2O \qquad \qquad \text{Eq. 2}$$

$$FeS_2 + 14 \, Fe^{3+} + 8 \, H_2O \quad \rightarrow \quad 15 \, Fe^{2+} + 2 \, SO_4^{2-} + 16 \, H^+ \qquad \qquad \text{Eq. 3}$$

$$Fe^{3+} + 3 \, H_2O \quad \leftrightarrow \quad Fe(OH)_{3 \, (solid)} + 3 \, H^+ \qquad \qquad \text{Eq. 4}$$

The dissociation or oxidation of FeS_2 (Eq. 1) initially releases ferrous iron (Fe^{2+}). After the sequence has been initiated, a cycle is established in which Fe^{2+} is oxidised by oxygen to ferric iron (Fe^{3+}) (Eq. 2). FeS_2 is subsequently oxidised by Fe^{3+}, generating additional Fe^{2+} and acid (Eq. 3). Certain conditions can cause the oxidation and hydrolysis of Fe^{2+} (Eq. 4), with an associated significant generation of acidity. Waters of pH <1 have been recorded (with associated underground seeps registering at pH –3).[14]

FeS_2 oxidation in the absence of Fe^{3+} as shown in Eq. 1 is relatively slow. The rate of abiotic (i.e., non-biological) oxidation of Fe^{2+} to Fe^{3+} shown in Eq. 2 is also slow in acidic environments. However, the presence of certain catalysing bacteria (e.g., *Thiobacillus ferrooxidans*) can increase the rate of Fe^{2+} oxidation by as much as a factor of a million,[15] pushing Eq. 2 to the right. The Fe^{3+} generated is a very strong oxidant, and will readily attack pyrite (generating acidity and Fe^{2+}) and oxidise other metal sulphides (releasing the metals into the acid aqueous phase), while itself being regenerated by biotic and abiotic reactions. Sulphate concentrations can often also rise to high levels. Because of these reinforcing "feedback" loops, ARD generation is considered autocatalytic, and once the process has started, it can be very difficult to halt,[1] indicating the importance of taking *preventative* measures.

The dissolution of non-pyritic minerals can be accelerated also by galvanic interactions causing preferential dissolution in acid solutions,[1] possibly enhanced by the presence of certain bacteria which continuously oxidise the layer of elemental sulphur that would otherwise prevent galvanic action.[11]

The degree of acidity, and concentration and speciation of dissolved contaminants, varies according to a number of site-specific factors, but typical contaminants include iron, manganese, aluminium, copper, lead, zinc, cadmium, arsenic, sulphate, and chloride. Less common dissolved components may also be present, depending on localised and regional mineralogy. Dissolved concentrations can range from below the limits of detection up to thousands or tens-of-thousands milligrams per litre, while pH can vary from near neutrality down to 1 and below.

7.3.2 CHEMICAL FACTORS INFLUENCING THE MIGRATION OF ARD

The aqueous transport and fate of many of the non-ferrous metals and metalloids present in ARD is significantly affected by adsorption by, or co-precipitation with, the various iron compounds that may form under certain conditions. In general, the iron species that precipitate are determined by the iron concentration; redox potential; pH; concentration of complexing anions such as carbonate, sulphate, and sulphide; and partial pressure of CO_2 and O_2.[16] For example, Fe^{3+} is rapidly hydrolysed even at relatively low pH to form iron oxyhydroxides (ochres). These ochres can contain significant concentrations of metals through co-precipitation and adsorption.[9] Similarly, aluminium hydroxide can play an important role in the adsorption and precipitation of other metals. In relatively dry environments (e.g., non-flooded mine workings), various iron sulphates may crystallise; if Fe^{2+} is oxidised to Fe^{3+}, basic sulphates and oxyhydroxides may form. In low pH, high sulphate environments, jarosite ($KFe_3(SO_4)_2(OH)_6$) may form, while at higher pH and lower sulphate concentrations, species such as goethite, iron oxides, and hydroxides may precipitate.[17] These have a large capacity to adsorb non-ferrous metals and metalloids, and will also remove metals from solution by coprecipitation processes. Therefore, it is clear that these types of reactions are capable of preventing the dispersion of dissolved contaminants.[18]

The other main chemical control on ARD migration is the neutralising effect of minerals such as carbonates and silicates. However, even if the host rock contains significant quantities of acid-neutralising minerals, the rate of acid generation may exceed that of neutralisation leading to ARD migration. The rate of neutralisation is influenced by pH, the partial pressure of CO_2, temperature, mineral composition and structure, redox conditions, and the presence of impurity ions in the neutralising mineral's lattice.[10] The significance of reaction kinetics is examined further below.

7.4 PREDICTION OF ARD GENERATION POTENTIAL

A comprehensive and detailed report was written by the United State's Environmental Protection Agency (USEPA) on techniques for predicting ARD generation in December 1994.[7] Within this report, a number of static, kinetic, and modelling techniques were reviewed and this has been a major source of the information set out below. Although specific tests are outlined briefly, it is the *concept* of predictive capacity that is of interest here, as effective control and treatment of ARD are closely linked to this capacity.[10]

7.4.1 THE BENEFITS OF PREDICTION

The capacity to predict ARD generation accurately does not imply that control and treatment will not be required, but it *does* increase the options for planning economically, technically, and environmentally-sound approaches to the problem. During exploration, feasibility, and permitting phases, predictive techniques are critical in (a) proving to the regulatory authorities that suitable plans are in place to deal adequately with the disposal of potentially acid-generating wastes, and (b) ensuring that the economic costs of prevention, control, or treatment have been fully incorporated into the economic evaluation of a project.[19] The natural extension of these two points is the application of predictive techniques in adapting the processing route to minimise or mitigate the generation of ARD.[19]

In carrying out predictive tests, the samples must be statistically representative of the different materials likely to be generated and directly related to the mine development plan. Within the mine plan area, samples should be taken from geological units defined by their physical and chemical homogeneity.[6] In tandem with a thorough understanding of the processing route, predictive techniques allow an estimation of the volume of likely problematic material, and the optimal response (e.g., revision of the processing route to avoid the generation of certain types of waste, early inclusion of preventative or control regimes, or the development of site-specific, cost-effective remedial measures).

7.4.2 STATIC AND KINETIC TESTS

Current static and kinetic predictive methods are based on short-term tests and use simple assumptions to predict the potential for ARD.[10] Each has its own benefits and drawbacks. Typically, data from static tests can be obtained in a short period of time, but are often less consistent and reliable than data from kinetic tests, which are relatively lengthy and therefore more expensive. Kinetic tests give information on the rate and amount of acid that might be generated,[20] but unlike static tests, there are no standard methods for interpreting test results.[7]

Table 7.1 details the main static and kinetic tests.

Of these tests, acid-base accounting remains the most widely used. Using this method, samples are analysed with respect to their Neutralisation Potential (NP) and Acid Generating Potential (AP, expressed as kg $CaCO_3$ t^{-1}). The difference between these yields the Net Neutralisation Potential (NNP). In general, a positive NNP is taken to indicate adequate neutralisation capacity, while a negative NNP shows that there is insufficient buffering capacity to prevent the generation of ARD. However, the ratio NP:AP is more often used now to determine the potential for ARD generation,

TABLE 7.1
Mainstream Static and Kinetic Tests for Acid Generation Potential

Test name or system

Static tests[7,19,21]

Neutralisation potential USEPA standard and modified methods	Well known and popular, low cost, and a quantitative measure of buffering capacity. However, no account is taken of mineralogy and little indication of the magnitude or rate of ARD generation is given.
Alkaline production potential:sulphur ratio (APP/S)	A simple and rapid indicator of ARD generation potential, but as it is based on theory, data needs to be confirmed by experimentation.
Net acid production (NAP) Acid-base accounting (ABA) Modified ABA	These methods can differentiate between sulphide minerals and the significance of carbonates and non-carbonates. However, the procedures are sometimes long and complex, requiring great care in data interpretation. Neither do they give an indication of the rate of ARD generation.
Net acid generation (NAG) Hydrogen peroxide	These methods give good reproducibility and are simple to carry out. However, they may result in an overestimation of net acid production and require pulverised samples which are not representative of the true size distribution

Kinetic tests[7,19,21]

Humidity cell	Useful if many tests are anticipated, procedurally simple, but may overestimate the rate of acid generation. This test also assumes a moist disposal or *in situ* environment
Soxhlet extraction	Rapid, simple and easily interpreted, but as it is an aggressive high temperature test, sample geochemistry may be changed. The effects of bacterial action, temperature and pH cannot be evaluated
Shake flask tests	These allow a wide range of factors affecting ARD generation to be examined, as well as an analysis of various control options. However, the tests tend to overestimate reaction rates (e.g., by enhancing bacterial activity), ignore the significance of secondary salt formation, and require complex data interpretation and modelling
Column tests	These are intended to simulate natural conditions and are simple to construct, operate, and monitor. Various environmental factors can be assessed, as can the influence of various control measures such as cover systems. However, the kinetics of reaction may not be distinguishable from rate limiting transport phenomena, and bacterial populations may differ from those found under field conditions. To be truly representative of field conditions, a period of at least one year is required (based on column experiments on the leaching of metals from pyritic contaminated mine wastes conducted by the author)[22,23]

with incorporation of a factor-of-safety into the ratio to reduce risks arising from unknown parameters. For example, Placer Dome, Inc. uses the NP:AP ratios shown in Table 7.2 to screen samples.[24]

However, still higher NP:AP ratios may be warranted as the dissolution rate of the acid neutralising minerals is generally low compared to the rate of pyrite oxidation. If an apparently "safe" NP:AP ratio overestimates the neutralisation capacity or rate, then there can be severe environmental and, ultimately, cost implications in terms of both site operation and closure.[10]

TABLE 7.2
NP:AP Ratios Used by Placer Dome Inc. for Sample Screening

NP:AP ratio	ARD generation potential
<1	Likely to generate ARD unless sulphides are unreactive
1 to 2	Possible ARD generation if neutralising minerals are preferentially depleted, coated, or otherwise unreactive
>2	ARD generation not expected

7.4.3 FACTORS LIMITING THE USE OF STATIC AND KINETIC PREDICTIVE DATA

Sherlock et al.[10] considered that predictive techniques must be applied on a site-specific basis and take into account the mineralogy of the waste material (e.g., the minerals present; percentage of sulphides, their distribution within the rockmass, along joints, and other discontinuities; and the likely physical durability of the waste rock and likelihood of fresh mineral exposure).[19] This point was also made by the Mine Environment Neutral Drainage (MEND) programme in Canada as long ago as 1991.[25] Supporting research for these views is seen clearly in the following points:

- Analysis has shown a correlation between the surface oxidation of sulphide minerals and the results of static and dynamic tests.[24]
- Kinetic tests generally only attempt to predict what will happen in the early stages (due to the cost limitations of longer-term tests).[10]
- Predictive techniques do not incorporate an assessment of the natural coating of sulphide phases with unreactive layers.[26]
- Predictive techniques alone do not account for the buildup of metal salts that may occur after disposal but prior to final reclamation, which may be periodically "washed out" with severe consequences for receiving waters.
- According to work by Pratt et al.,[26] the pressure and stress associated with the milling process can increase the reactivity of pyrrhotite. This has significant implications for ARD generation if certain types of pyrrhotite are "activated" by milling: alternative milling methods may reduce the potential for ARD generation by reducing the activation. Identification of "high-risk" types of pyrrhotite might have implications for waste management and disposal practices in those cases where ARD from tailings is an issue.
- Accurate data interpretation and modelling are complex, irrespective of the nature of the test (e.g., static or kinetic).

Although the USEPA accepted that mineralogy is site-specific and that, consequently, predictive methods that incorporate mineralogical analysis may be even more costly and complex, it also saw the lack of reliable predictive techniques as a major obstacle to future metal mining in the USA.[7]

7.4.4 EMPIRICAL AND GEOCHEMICAL MODELS

Mathematical models are also available to aid in the prediction of ARD. These can be divided into two groups: empirical and geochemical.[7] Empirical models are based on observed behaviour, while geochemical models rely on an understanding of the physical and chemical processes that control the generation of ARD. Both these types of models are simplifications of reality, and are only as valid as the data and assumptions on which they are based.

As empirical models are based on observation, they do not require a complete understanding of the causal reactions occurring at the waste disposal site and in the surrounding environment. However, extrapolation of conditions into the future (by, for example, the use of "best-fit lines") assumes that the samples used to generate data are truly representative of the situation (which is unlikely, given that spatial variations in terms of sulphide concentration, particle size, mineralogy, and so on are common). This point shows the necessity of developing geostatistics as a means of ensuring sampling is as representative as possible (see for example, Smith,[27] who examined the role of geostatistical characterisation in the remediation of mine wastes). Also, empirical models are necessarily site specific and cannot be transferred to other sites,[7] and the problem of data point outliers must also be addressed to avoid unjustified bias in the model.[28]

Geochemical models are based on a system of equations that represent the factors controlling the generation of ARD.[7] Simulations can be extended decades or centuries into the future. Factors considered by various recent models include diffusion of oxygen, convection driven by exothermic oxidation of pyrite, reaction product transport, water flow pathways, and waste dump geometry.[7]

Even with recent advances in both empirical and geochemical models, the levels of uncertainty are often such that wide application of these models has been blocked. The way forward may be through the validation of geochemical models using comprehensive and detailed sampling data,[7] or comparison of models with kinetic test data (see, for example, White et al.[29]). In addition to assessing the performance of preventative, control, or treatment measures and compliance with discharge consents, monitoring also has a role to play in model validation.

7.5 PREVENTION AND CONTROL OF ARD GENERATION AND MIGRATION

There are two types of prevention and control. The first relates to the generation of ARD, while the second relates to its subsequent migration. Both are interrelated inasmuch as certain approaches to preventing and controlling generation can reduce migration and vice versa, and therefore in some respects the division is an artificial one.

Methods proposed for the prevention and control of ARD generation include treatment of sulphide surfaces via the formation of inert surface "coatings," the use of anti-bacterial agents (bactericides), the segregation of the principal ARD-generating waste fraction, and control of oxygen and/or water infiltration of the sulphide-bearing material. Currently, waste segregation and prevention or control of water/oxygen infiltration dominate, with the other methods having only limited application at full scale. Certain approaches such as inert surface coatings are considered unproven at present, but worthy of further research.

In addition to the use of engineered covers, the most common approaches to preventing or controlling the migration of ARD are the rerouting of water away from the source or the use of subsurface seals and barriers to impede the movement of contaminated groundwater.[6,8] The greater the control achieved, the smaller the volume of ARD that is likely to require treatment. Just as waste minimisation is typically more cost-effective than waste management, prevention or mini-misation of ARD is generally considered a cheaper option than long-term treatment.[8]

7.5.1 PREVENTION AND CONTROL OF ARD GENERATION

7.5.1.1 Hydrophobic coatings

A recent advance has seen research on the use of fatty acid amines,[30] which suppress bacterial activity and chemical oxidation. Treatment with the amine makes the pyrite highly hydrophobic and the pyrite surface consequently repels oxidising ions (i.e., Fe^{3+}). It is, however, unclear whether the effect of the hydrophobicity prevents the bacteria from contacting the pyrite surface. Indeed, there is considerable controversy as to whether bacterial contact is actually required for bacterially mediated oxidation to proceed since it is typically mediated through the generation of Fe^{3+}.

7.5.1.2   Inert coatings

Georgopoulou et al.[31] examined the possibility of creating an iron phosphate coating on pyrrhotite to prevent oxidation. Using hydrogen peroxide to oxidise the pyrrhotite surface, then potassium orthophosphate to provide the phosphate for the coating and sodium acetate to buffer the pH, the researchers successfully formed a coherent iron phosphate coating on the pyrrhotite in small-scale column experiments. The resistance of the coating to oxidation was tested using hydrogen peroxide as the oxidising agent and was shown to be substantially better than control samples. Further research is necessary to detail the economic costs and technical constraints of larger-scale applications, but this approach does look promising, possibly as a means of treating segregated high-sulphide wastes prior to disposal. Other coating agents have also been suggested (e.g., humic acid, oxalic acid, sodium silicate) but remain unproven.[32,33]

7.5.1.3   Bactericides

Bactericides are normally based on detergents that break down the greasy film at the surface of the bacterial cell wall. The film normally protects the bacteria from the surrounding acid environment and once it is removed, the bacteria cannot survive the acid conditions. Although bactericides can temporarily disrupt the activity of acidophilic bacteria such as *Thiobacillus ferrooxidans* and *Thiobacillus thiooxidans*,[8] if the bactericide is removed or becomes depleted, bacteria quickly repopulate the local environment. Slow-release formulations are commercially available that reduce acid generation up to 10 years. Ideally, this is long enough to establish stable soil and vegetative covers that prevent reacidification (see Engineered covers, Section 7.5.2.2). Bactericides have proven to be most effective in delaying or preventing acidification in highly pyritic ARD sources.[34]

Alternative agents based on thiol-blocking agents have also been suggested for the inhibition of neutrophilic *Thiobacillus* species that are able to oxidise various sulphur compounds under neutral pH conditions.[35]

7.5.2   PREVENTION AND CONTROL OF OXYGEN AND/OR WATER INFILTRATION AND MIGRATION

7.5.2.1   Seals, grouting, interception trenches, and subsurface barriers

Sealing mine openings, adits, tunnels, and adits can prevent the infiltration of water into and the migration of ARD out of underground workings. Preventing the movement of water through such workings can minimise sulphide oxidation even if the workings are flooded, as static water will quickly become anoxic as oxygen is consumed by chemical and biological reactions. Cementitious grouts can also be applied to prevent the infiltration of oxygenated water.[36] One of the main concerns relating to the use of seals (other than cost) is that of sudden failure and the consequences of a sudden and massive ARD release (e.g., the release of over 10 million gallons of highly contaminated ARD into the Carnon River from Wheal Jane mine, Cornwall, U.K. in 1992 following the failure of the plug sealing Nangiles Adit). The risk of sudden failures in now being reduced by the fitting of relief valves to allow excess water to be drained off (and treated) during high flow periods,[37] thereby controlling the pressure on the seal and surrounding host rock.

Interception trenches have also been used to prevent the migration of ARD. For example, in Montana at the Clark Fork site (containing four contiguous Superfund sites), the sites' owner Atlantic Richfield Corporation prepared an intercept trench to force the contaminated water to the surface, where it could be treated by a passive system (see Mueller et al.[38] for details of the passive treatment system). This was possible due to the unusually shallow water table.[37] At other sites, impermeable slurry walls have been used to impede the flow of contaminated sub-surface water.[37]

7.5.2.2 Engineered covers

Although waste dump geometry can be important in defining surface area exposure and air infil-tration rates,[7,39] engineered covers are more effective at controlling oxygen and water infiltration and can be classified as oxygen barriers, oxygen consumers, or reaction inhibitors.[84] Soils and geofabrics (e.g., PVC and high-density polyethylene) have been used to prevent and control water and oxygen infiltration into ARD-generating wastes. Geofabrics are expensive, but if applied properly (i.e., to avoid punctures and rips) they are likely to have useful working lives in excess of 100 years.[8] Clays have often been used because of their minimal permeability and relatively low cost; however, there is a danger that if the clay cover dries, deep cracks can occur, allowing the rapid ingress of water and oxygen.

Engineered cover systems often include layers that promote lateral rather than vertical move-ment of water, a substrate for vegetation and protective layers between the geofabric (if used) and the waste to reduce the risk of physical damage. A free-draining layer of closely sized coarse material is sometimes used to ensure that a capillary water column cannot form between the waste and overlying soil, thereby preventing the upward movement of contaminants. This coarse layer also helps to prevent the mixing of wastes and upper soil layers by burrowing animals, worms and other soil invertebrates. In general, the greater the number of layers, the greater the protection offered against ARD, but the greater the cost of application (composite soil covers and plastic liners are generally the most expensive options).[40]

Additional layers may include sewage sludge and composted municipal waste to promote oxygen consumption. One potential drawback to this approach has been highlighted by recent research, which has shown that ferric oxyhydroxides present in weathered tailings dissolve when in contact with organic acids originating in carbon-rich oxygen consuming covers.[41] The dissolution of the oxyhydroxide phase can result in the release of adsorbed or coprecipitated non-ferrous metals into the aqueous phase. Therefore, it is possible that in some cases, the use of oxygen-consuming layers may exacerbate the problem that they are designed to prevent,[41] and that the inclusion of carbon-rich oxygen consuming layers should be viewed with caution until further research has been conducted.

7.5.2.3 Sub-aqueous disposal

Sub-aqueous disposal can take the form of dumping into, and subsequent flooding of, open pit transfer to flooded subsurface workings, man-made and natural lakes or impoundments,[8] or into the marine environment.[6] Whichever option is chosen as the most practicable, cost-effective, and environmentally sound, the immediate disposal of reactive wastes underwater is preferable to storage (and oxidation) of the wastes prior to disposal. Storage can extend significantly the period of acid generation and metal release.[42]

Public and regulatory concerns over impacts on benthic life (due to close contact of such species with both the water and sediment phases)[43] and water quality may restrict disposal at natural lake and marine sites. Given its controversial nature and the lack of public and regulatory support,[44] it is essential that scientific justification is made for this approach, possibly on a site-specific basis.

Underground disposal and flooding requires a thorough understanding of local and regional hydrology and hydrogeology: tailings and waste rock should not be exposed to anything but stagnant (anoxic) water. Back filling into open pits or underground workings can also be expensive, as it involves double handling of materials.

Disposal in purpose-built constructed impoundments may be the most acceptable approach, although this implies that the capacity to *plan* ahead must be in place before this option is chosen: retrospective engineering of existing impoundments to allow adequate flooding may be prohibitively expensive. Disposal underwater in constructed sites is less of an issue.[45] However, in some cases, topographical constraints, the absence of an elevated water table, and inappropriate climate can limit the application of flooded surface disposal sites.

Despite these limitations, according to Filipek et al.,[8] subaqueous disposal has the greatest promise in controlling ARD generation from solid wastes, as the diffusion rate of oxygen through water is four orders of magnitude less than in air. Therefore, the supply of oxygen becomes the rate-limiting step in ARD generation, and significant generation ceases. However, research over a three-year period at the Noranda Technology Centre has shown that although water covers can reduce the rate of acid generation by 99.7%,[46] the concentration of metals in the surface water can still exceed regulatory limits.[47] The inclusion of a biologically active organic layer on top of tailings has been suggested as a means of further reducing contact with oxygen and acting to confine metal contaminants.[48] Oxygen infiltration of the tailings is controlled by its consumption in the organic layer (e.g., via conversion to carbon dioxide and water), while metal diffusion into the water can be further controlled by inoculation of the organic layer with sulphate-reducing bacteria, capable of precipitating the metals as sulphides.

7.5.2.4   Hardpan formation

Hardpans normally form where the oxidation of pyrite and pyrrhotite leads to the cementation of mineral particles at the surface by a matrix of iron oxyhydrate. They may also form where solubilised ferrous and ferric iron reprecipitates at depth (typically at the interface between oxic and anoxic layers). Although natural hardpans are neither particularly chemically stable or physically robust, it has been suggested that the deliberate construction of hardpans (using electrochemical methods) may help control infiltration by water and thus reduce or prevent the generation of ARD.[49] However, this work has so far only proved possible under laboratory conditions.

7.5.3   Isolation, Segregation, and Blending

Partitioning of wastes into sulphide-rich and sulphide-depleted fractions offers the chance to expand waste management options. In theory, the low-volume sulphide-rich fraction can be disposed of at a highly engineered disposal site or at least isolated as buried "cells" within the bulk waste, while the high volume sulphide-depleted fraction can be disposed of as an inert waste. However, the USEPA considers that the containment of sulphide-rich fractions may pose considerable engineering challenges.[50] Pre-treatment prior to disposal may be required (e.g., surface coating or co-disposal with neutralising agents). Rather than adopting this practice as an "end-of-pipe" technique, it should be possible and indeed preferable to integrate it with the production process at the feasibility stage prior to permitting.

In a recent paper, Humber[51] analysed the relationship between sulphide elimination, acid generation, and operating/capital cost estimates. Several mineral processing techniques were examined for their capacity to separate acid-generating sulphides from mill tailings, thereby generating a sulphide-depleted fraction, and a low volume sulphide-rich fraction. Methods examined included gravity separation (e.g., centrifugal concentrator, shaking table, spiral concentrator), flotation, magnetic separation, and cyclone classification. These were tested on samples from three mines. The sulphide concentrates were also examined to see if they had any commercial value. However, it is worth noting that there has been a substitution of elemental sulphur for pyrite in the production of sulphuric acid,[52] showing that waste minimisation can be a two-edged sword: the production of a pyrite concentrate might not be as commercially viable or environmentally acceptable as it was a few years ago given this fact. The move to more expensive sulphur reflects the need to treat the wastewaters resulting from the use of pyrite to remove dissolved metals, and also the problems associated with stockpiles of pyrite and pyrite residues (e.g., ARD). However, research by Zouboulis et al.,[53] shows that pyrite may have some use in the removal of dissolved arsenic species by adsorption.

The application of magnetic separation looked promising in the cases where pyrrhotite was present in significant amounts.

For each of the samples, as the sulphide concentration increased, so did the capacity to generate acidity. The sulphide minerals were well liberated in each waste, and present at sufficiently low concentrations to produce, theoretically, a low-volume sulphide concentrate that could be handled separately. None of the gravity methods attempted produced non-reactive tailings (i.e., one with no net capacity for acid generation). Flotation was more successful, although to a certain extent this reflected the simple nature of the mineralogy and the fine size distribution of the sulphides. In addition to the benefits of reduced ARD generation potential, the concentration of other environmentally significant metals (such as cadmium, copper, and zinc) was also reduced in the sulphide-depleted fraction.

Based on the test conditions used to generate these optimum NNP values, the operating and capital costs associated with implementation of the process changes required were also estimated. Estimates were based on recent quotes for installation of similar equipment, and took into account existing plants that could be used or adapted. Total capital costs ranged from US$130,000 to US$1,275,000. Operating costs estimates ranged from US$0.50 t^{-1} to US$1.35 t^{-1}.

Unfortunately, the paper says nothing about existing operating or capital costs (to place the additional costs in context), or about the likely feasibility of full-scale implementation. However, there are examples of actual implementation at plant-scale of this approach, although not necessarily driven by environmental concerns. The Magma Copper Company has produced pyrite products at its Superior Mine operation by passing tailings from the copper circuit through an additional flotation circuit, thereby generating a less reactive tailings product, plus a fine, virtually pure, pyrite product, and a coarse pyrite concentrate (45 to 47% Fe, 48 to 50% S).[50] Approximately 500 t per month of pyrite products were sold in 1994, probably 90 to 95% of the U.S. market.[54]

The production of saleable pyrite products was possible because the ore (a) contained up to 25% pyrite and (b) had few impurities. These factors may make the deposit relatively unique and the transfer of this approach to other operations more difficult than it might first appear. As noted above, the driver behind pyrite production was not strictly environmental (although the company did recognise the benefits accruing from reduced ARD generation), but rather *demand* for the product — when there was no demand, there was no recovery of pyrite. This emphasises the difficulty of dealing with materials that have little or no market niche or value.

Blending of acid-generating and non-acid generating wastes can also be used to prevent or control ARD. However, as the latter is often exempt from permitting due to its inert nature, blending can sometimes result in the permitting of a much larger volume. This approach may therefore be environmentally attractive, but constrained by an operator's reticence to extend or further complicate the permitting process.

7.6 TREATMENT OF ARD

While the aim of best practice must be to prevent or control ARD generation and migration at *all* sites using techniques such as those outlined above, in reality there will always be cases where remedial treatment is required either as a temporary measure, or on a more permanent basis. However, best practice should in theory reduce the significance attached to treatment as a component part of the closure and post-closure process.

There are a number of overlapping approaches to the treatment of ARD, which are nominally categorised here as (a) active; (b) passive; and (c) active-passive hybrid systems. A fourth approach — resource (e.g., metal) recovery — cuts across these three categories, but is treated separately here as it is likely to become a more feasible option in the future as technology continues to advance. Typically, active treatment systems require less land than passive systems, but have higher operational and chemical costs. Capital costs *may* be lower than for passive systems, although this is not always true.

The use of neutralising reagents is the principal active treatment method, and lime (calcium oxide or hydroxide) is the most commonly deployed chemical.

Passive methods use chemical and biological processes to reduce dissolved metal concentrations and neutralise acidity. Constructed wetlands are now the principal passive treatment method, although this method is still in its infancy, particularly in the context of high-volume ARD from metal mining operations.

Innovative active-passive hybrid techniques are being developed in order to incorporate the best features of both, while minimising operational and performance limitations. Examples include microbial mat systems and bioreactors.

7.6.1 ACTIVE TREATMENT SYSTEMS — LIMING AND RELATED TECHNOLOGIES

In conventional lime treatment, there are five basic steps following collection of the ARD: equalisation of the treatment plant feed to minimise variations in water quality; aeration to oxidise Fe^{2+} to the less soluble Fe^{3+} (normally using air, but hydrogen peroxide, ozone, and biological oxidation have also been investigated;[55] neutralisation to increase pH to precipitate metals as hydroxides; sedimentation to separate water and solids, often involving the use of chemical coagulants and flocculents; and finally, sludge disposal. Combining certain steps can make financial savings. For example, the U.S. Bureau of Mines In-Line System aerates and neutralises the water simultaneously,[56] reducing capital and operating costs, particularly in cases where the ARD has a high oxygen demand.

Although the most commonly used neutralising agent is hydrated lime ($Ca(OH)_2$), other agents that have been applied (normally in the treatment of low volumes of water) include ammonia, magnesium hydroxide, soda ash, caustic soda, and limestone. A small number of proprietary reagents are also available, which normally compete with standard reagents on environmental performance and/or direct cost (e.g., KB-1™, KB-SEA™ and META-LOCK™ from KEECO Inc., Lynwood, U.S.,* which are currently being assessed by the EPA at trial sites in the U.S.). Unfortunately, few, if any, higher environmental performance proprietary reagents are currently being used on a commercial basis and liming continues to dominate, based on its relatively low price and wide availability.

Although the cost of lime varies depending on volume bought, transport costs, and other site-specific contract details between user and supplier, it is normally considered the most cost-effective technique where prevention or control has failed, particularly where an immediate response is required. However, this is likely to change, as alternative proprietary reagents such as those developed by KEECO become increasingly cost-competitive.

Lime is normally introduced into the system as 5 to 20% (by weight) water-based slurry, although it is sometimes applied as a dry powder when the water volume requiring treatment is low. Its addition consumes acidity, forming water and insoluble calcium sulphate (if the sulphate concentration is sufficiently high), and causes the precipitation of dissolved metals. The precipitate may then be settled or otherwise separated from the "clean" water. The principal reactions are summarised in Equations 5 to 7. For more detailed analysis of the reactions between lime and acidic, metal-contaminated effluents, see Marchant.[57]

$$Ca(OH)_2 + H_2SO_4 \quad \rightarrow \quad Ca^{2+} + SO_4^{2-} + 2\,H_2O \qquad \text{Eq. 5}$$

$$Ca(OH)_2 + FeSO_4 \quad \rightarrow \quad Fe(OH)_2 + Ca^{2+} + SO_4^{2-} \qquad \text{Eq. 6}$$

---

* KEECO Inc., 19029 36th Avenue West, Lynnwood, WA 98037, U.S. and http://www.keeco.com

$$3\,Ca(OH)_2 + Fe_2(SO_4)_3 \quad \rightarrow \quad 2\,Fe(OH)_3 + 3Ca^{2+} + 3SO_4^{2-} \qquad \text{Eq. 7}$$

Iron and other metals may precipitate as hydroxides, basic sulphates, or other compounds depending on the variety and relative concentration of anions and cations initially present in the aqueous phase. The stability of such precipitates in acid and alkaline environments varies considerably. Other metals such as aluminium, copper, and zinc may precipitate as hydroxides (Eq. 2 and Eq. 3). However, more complex species may also form, depending on the speciation and relative concentration of anions and cations. For example, arsenic (which has oxyanionic chemistry) may precipitate as a variety of ferric arsenate species, each having a different solubility. The precipitated sludge arising from liming operations should not, therefore, be considered a simple hydroxide precipitate, but a complex mixture of chemical species, for which the long-term stability may be extremely difficult to predict.

According to Kuyucak,[58] there are a number of factors that can affect the efficiency of treatment and the nature of the sludge produced, including:

- Rate of neutralisation (e.g., high rate result in gelatinous and voluminous sludges).
- Aeration and the oxidation of ferrous iron — ferrous hydroxide, is soluble over a broader pH range than ferric hydroxide; therefore, aeration is often employed to ensure iron is present in the ferric form prior to precipitation. However, the ratio of ferrous to ferric iron can also influence the sludge density, with the highest densities being achieved for high ferrous iron waters.
- Mixing rate of lime and ARD — rapid mixing (e.g., in high shear mixers), can detrimentally affect sludge density.
- Sludge recycling — sludge recycling has been shown to improve sludge density and settling characteristics while reducing lime consumption (see High Density Solids and NTC processes, below).

In order to compare liming with some of the alternative approaches, the benefits and drawbacks need to be examined.

7.6.1.1 Benefits

- Plant occupies a limited amount of space (although large areas may be required for sludge settling and disposal).
- Unexpected variations in water quality and flow can be accommodated by relatively easy adjustment of the operating parameters.
- Treatment is largely unaffected by temperature, and efficiency throughout winter is maintained.
- Effective in the treatment of highly acidic waters.
- A proven technology — mechanisms of metal removal and acid neutralisation are well documented and understood.

7.6.1.2 Drawbacks

- Equipment maintenance is relatively high due to the periodic buildup of calcium carbonate and calcium sulphate "scale."
- High pH is required to remove metals such as manganese. However, based on the author's own experience, certain metal hydroxides (e.g., $Al(OH)_3$) may redissolve in the highly alkaline solutions required to complete metal precipitation, necessitating a multistage treatment to reduce all metals to acceptable concentrations.

- Sludges derived from the liming of ARD are chemically unstable and will partially redissolve if exposed to a sufficiently acidic (or in some cases, alkaline) environment. Therefore, the chemical nature of the sludge poses a possible long-term, post-disposal hazard. For example, in the U.S., due to this fact, some sludge is treated as hazardous wastes, dramatically increasing the cost of disposal.
- Sludges are composed of fine particles, and normally have a low solids content. Consequently, the sludges are difficult and expensive to handle, and require some form of dewatering prior to disposal. Settling lagoons (the cheapest dewatering option) sterilise large areas of land, while mechanical dewatering may be prohibitively expensive.
- The sludges normally have no commercial value, and reprocessing to extract the metal content is uneconomic with existing technologies, due to the large excesses of lime that are often used during treatment to ensure complete precipitation.

The long-term disposal of liming-derived sludges and liabilities associated with the disposal site are probably the principal concerns of operators using this method, particularly in regulatory regimes that allow retrospective prosecution (e.g., the U.S.). Landfill is almost always prohibitively expensive, unless the site is owned by the operator and located near or on the mine site. The alternatives, such as on-site reclamation (including stabilisation, isolation and/or capping) are also likely to incur major engineering costs. As on-site disposal represents a long-term liability for the operator, the engineering standards for site reclamation will be extremely demanding, further elevating the cost. Underground (deep mine) disposal of liming-derived sludge has been proposed by some researchers. However, it is currently accepted that the long-term environmental impacts of this disposal scenario have not yet been properly assessed and that further research is required.

To address some of these drawbacks, a number of refinements to standard liming treatments have been developed in recent years. Two such examples include the High Density Sludge (HDS) process and the patented NTC process developed at the Noranda Technology Centre; these are being applied increasingly within the industry (see cited references below for examples).

7.6.1.3   High Density Sludge (HDS) process

The HDS process produces a more compact and higher density sludge than standard liming treatment (<2% solids); for example, 10 to 30% solids, rising to 40 to 50% with time in the impoundment area.[59] Treatment is carried out in aerated reactors and part of the settled sludge is recycled to the beginning of the process, where it is mixed with the lime slurry. The sludge acts as an adsorbent for hydroxide anions, which are then gradually rereleased into the untreated ARD. This prevents the rapid neutralisation of ARD (which can lead to a voluminous, gelatinous precipitate with low solids content and poor settling characteristics) and produces a denser sludge with improved settling performance.[58] Capital and operating costs of the HDS process were compared with those of precipitation plus centrifugation or filtration, biochemical extraction, and sulphate reducing bacteria at Wheal Jane.[5] Capital costs varied from £25 million for biochemical extraction down to £4.4 million for precipitation with either centrifugation or filtration. Operating costs ranged from £2 million per year for biochemical extraction down to £0.64 million per year for the HDS process. Given that the latter produced a sludge with a solids content of 45% (compared to 25 to 30% for the other treatments), this was the preferred choice at the site, despite the slightly higher capital cost.

7.6.1.4   NTC process (US Patent Number 5,427,691)

This is a two-stage process. The pH of the influent is raised to between 4 and 5 by sludge recycling (with no aeration) in the first reactor. In the second reactor, the pH is adjusted to between 9 and 10 by the addition of further lime/recycled sludge (with aeration). Sludge density and settling rates are improved relative to the HDS process, with sludges containing 30 to 50% solids being achievable.[58,60]

Despite the commercial availability of processes such as HDS and NTC, increasingly stringent legislation is likely to drive mine operators to look beyond the use of lime to avoid incurring growing disposal costs, and to avoid the possibility of future liability and litigation. One alternative to which serious consideration has been given is the use of constructed wetlands.

7.6.2 Passive Treatment Systems—Constructed Wetlands

In general terms, wetlands can be described as areas flooded or saturated by surface water or groundwater often or long enough to support those types of vegetation and aquatic life that have specifically adapted to saturated soil conditions. Constructed or engineered wetlands attempt to duplicate natural systems, and use chemical and biological processes to reduce dissolved metal concentrations and neutralise acidity. Compared with conventional active chemical treatment (e.g., liming), passive methods generally require greater land area, but use cheaper materials (sometimes waste products) to support the chemical and biological processes, and require less operational attention and maintenance. However, they are not "walk-away" solutions but, theoretically, low-maintenance, low-energy systems.[5]

Passive systems encompass a number of discrete unit processes, typically in the following sequence:[85]

- *Anoxic ponds* reduce the dissolved oxygen in the water and marginally increase alkalinity. Reducing the oxygen content reduces deposition of metals within anoxic limestone drains (the next treatment process). However, anoxic ponds are not yet considered a proven technology. Generally, these ponds are engineered to drain downward, forcing the water to flow through an organic-rich (oxygen consuming) layer.
- *Anoxic limestone drains* consist of an enclosed layer of limestone that adds alkalinity and thereby reduces the size of aerobic cells (the next treatment process). These drains are typically placed to intercept acidic, anoxic groundwater. ARD that contains significant dissolved oxygen must first be deoxygenated (e.g., using an anoxic pond).
- *Aerobic cells* are planted with suitable wetland plant species (such as broad-leaved cattail (*Typha latifolia*)). The water flows across the surface of the cell, promoting contact with oxygen in the air. The cells remove iron and aluminium as hydroxides. Arsenic and a fraction of the other dissolved metals are removed by adsorption onto the surface of iron hydroxide particles. Plants aerate the substrate and produce alkalinity by passing carbon dioxide through the root system. However, the cells are net producers of acidity, as sulphuric acid is produced as a by-product of iron precipitation.
- *Anaerobic cells* are organic matter-rich, and may or may not be planted. Planting allows the regeneration of the organic matter as the plants die and decompose, and helps filter suspended solids.[8] Water flows vertically down through the organic matter, minimising contact between the water and oxygen in the air. Bacteria in the organic substrate produce hydrogen sulphide from sulphates, which then reacts with dissolved metals to form insoluble metal sulphide precipitates. The cells remove cadmium, zinc, copper, some iron and sulphate, and produce net alkalinity.
- *Rock filters* provide a large surface area on which algae and manganese-oxidising bacteria grow. The algae generate a high pH that allows manganese to be removed from the water. Algae also require manganese as a micronutrient, and can accumulate large amounts (up to 5.6% by weight based on dry tissue matter). Manganese removal is only significant at pH > 6 and low iron concentrations, so these filters are generally only effective in the later stages of treatment.

While the general principles of operation are understood, many of the actual reaction pathways and mechanisms by which contaminants are removed from the water are poorly defined.

Equation 8 shows the main reaction occurring in the aerobic cells.

$$2\ Fe^{2+} + \tfrac{1}{4}O_2 + 1\tfrac{1}{2}H_2O \quad \rightarrow \quad FeOOH_{(solid)} + 2\ H^+ \qquad\qquad \text{Eq. 8}$$

In Equation 8, FeOOH represents several possible iron precipitates, the exact nature of which depends on a range of environmental factors such as pH.

The main reactions occurring in the anaerobic cells are shown in Equations 9 and 10.

$$2\ CH_2O + SO_4^{2-} \quad \rightarrow \quad H_2S + 2\ HCO_3^- \qquad\qquad \text{Eq. 9}$$

$$Zn^{2+} + H_2S \quad \rightarrow \quad ZnS_{(solid)} + 2\ H^+ \qquad\qquad \text{Eq. 10}$$

In Equation 9, bacteria utilise organic compounds (represented by CH_2O) and in the process increase alkalinity and reduce sulphate to hydrogen sulphide. This hydrogen sulphide reacts with dissolved metals (such as zinc shown in Equation 10) to form insoluble sulphides, which are then physically entrapped in the organic substrate.

7.6.2.1 Benefits

- Capital costs may be relatively low if the volume of water to be treated is low.
- Operational and maintenance costs appear to be relatively low.
- Generally self-maintaining and requires little or no operator supervision.
- Anaerobic cells may be able to utilise waste organic materials (such as composted sewage sludge, farmyard manure, and sawdust) as suitable substrates, thus mitigating one waste problem in the treatment of another.
- Wetlands provide a wildlife habitat and replace a valuable natural resource that has been significantly damaged by past human activity.

7.6.2.2 Drawbacks

- Wetlands can require large areas of flat or gently sloping land (measured in tens of hectares). Even if suitable land is available, purchase for conversion to wetland may be prohibitively expensive.
- As a result of the reliance on biological processes, treatment during winter months declines due to reduced bacterial activity at lower temperatures.[8] Unfortunately, this often coincides with periods of peak flow of contaminated water.
- Long periods are required for system equilibrium to be reached (possibly measured in decades) due to the biological nature of the treatment system. Consequently, the long-term efficacy of wetlands, particularly in the treatment of ARD from metal mining operations, remains uncertain as few, if any, systems have yet reached equilibrium.
- Aerobic cells are more efficient at removing metals from less acidic waters and can be prone to freezing during winter months.
- Anoxic ponds as a pre-treatment step for anoxic limestone drains are a recent development and relatively unproven.
- When acidic water containing any ferric iron or aluminium contacts limestone in the anoxic drain, metal hydroxide precipitates form, "armouring" the limestone,[8] thereby reducing the generation of further alkalinity and impeding flow through the drain. Any

decrease in alkalinity produced by the anoxic limestone drain will increase the required area of aerobic cells.

- Passive treatment systems cannot be expected to perform indefinitely. In the long term, the conditions that facilitate treatment will be compromised.
- The impact of metal-contaminated substrates on the local and regional food chain is largely unknown.
- Unexpected or unpredictable variations in water quality and flow may cause the metal removal capability of the wetland to be exceeded, causing the release of untreated water or introducing the need for additional back-up treatment systems. Flow-through of untreated water may have a detrimental impact on established plants in the latter stages of the wetland treatment process.
- The percentage of carbon available for actual use may only be a small fraction of the total present in the organic substrate. This would indicate that substrates would have to be replaced more frequently than previous estimates, [61] and that periodic maintenance will also be essential.[62]

Many of the concerns relating to liming-derived sludges also apply to engineered wetland treatments. Two principal questions have yet to be answered: (a) is the substrate, or will the substrate become, a hazardous waste, and (b) what are the options for disposal or use of the spent substrate? Until these questions are answered, constructed wetlands must be considered a potential future liability in much the same way as liming sludge disposal sites.

Once the operational capacity of the wetland treatment system has been reached, not only do the problems of substrate treatment and/or disposal have to be addressed, but also the costs of reinstatement. In early research on constructed wetlands, it was assumed that they would be capable of working indefinitely. Estimates for the working life have since been revised numerous times, and the average now quoted is around 25 years. Consequently, multiple phases of capital expenditure may be expected if ARD treatment is required for a longer period. This might dramatically affect the economic viability of wetlands in many cases, in particular where the volume of water to be treated (and hence size of wetland required) is large. Taking into account these points, constructed wetlands may be more useful for the treatment of low, long-term flows containing low concentrations of contaminants,[63] rather than high-volume, highly contaminated ARD. The more extreme cases may be better treated using active-passive hybrid systems or, in some cases, represent the opportunity to attempt resource recovery (e.g., metals, metal salts, and other by-products).

7.6.3 PASSIVE *IN SITU* TREATMENT SYSTEMS

Porous reactive walls have been proposed as a means of treating sub-surface ARD *in situ*.[64,65] By excavating a portion of the aquifer body ahead of the contaminated plume and replacing it with a permeable reactive mixture, the ARD can be treated *in situ*. It has been suggested that the reactive mixture should be based on organic carbon to promote the action of sulphate-reducing bacteria and the consequent precipitation of metal sulphides (c.f., Bioreactors, below). This approach has only been attempted on a limited scale, although the preliminary results appear promising in terms of both economic and technical performance.[64]

The use of sulphate-reducing bacteria in open pits or underground workings has also been suggested as a means of treating ARD *in situ*.[66] This approach might be suitable for low-load scenarios, where a suitable organic substrate for bacterial growth is locally and cheaply available. However, efficiency might be compromised by a single addition of substrate if the substrate is so deep that mass transfer is detrimentally affected. The operational temperature at depth is also an issue, as this will influence bacterial activity (c.f., performance of constructed wetlands during colder winter months).

7.6.4   ACTIVE-PASSIVE HYBRID SYSTEMS

Hybrid systems often use biological processes to neutralise acidity and remove dissolved metals from ARD, but normally in purpose-built reactors rather than in natural system analogues. Current examples include microbial mats and bioreactors based on sulphate reducing and other bacteria.

7.6.4.1   Microbial mats

Microbial mats are a heterotrophic and autotrophic bacterial community dominated by cyanobacteria and held together by the slimy secretions of various microbial groups.[67] The mats are tolerant of high levels of toxic metals, and are capable of rapidly removing metals from solution via a number of processes, including the release of metal flocculating polymers. Various oxidative, reductive, and precipitative reactions occur through the mat profile, which displays a steep redox gradient (passing from an oxic to anoxic environment). Based on small-scale field trials, the mats are durable, low-cost, and can be adapted to include specific microbial communities. This opens up the possibility of development of mats for specific applications (e.g., depending on type of effluent and suite of metals present). However, further research to scale up the processes is required.

7.6.4.2   Bioreactors and other biologically-based treatments

Bioreactors have been suggested as alternatives to constructed wetlands. Johnson[68] analysed the possible applications of iron and sulphate-reducing bacteria in the treatment of ARD using a series of bioreactors that would simulate anaerobic wetland cells. In the first stage, iron would be reduced from ferric to ferrous state; in the second, metals would be precipitated as sulphides. By reducing iron to its ferrous form, the freshly precipitated sulphides are protected from the oxidising power of ferric iron. This approach is at a "concept" stage and requires fundamental research to determine such basic parameters as the retention time required in the bioreactors and to identify suitable bacterial strains.

Another approach is to produce the hydrogen sulphide (H_2S) in a bioreactor that is isolated from the ARD. Pilot- and bench-scale tests in Canada,[69,70] and the U.S.[71] have been very encouraging. Depending on the water quality, it appears possible to produce high-grade sulphide concentrates that, when shipped to a smelter, can offset the operating costs of a water treatment plant. Other advantages are that: (a) potentially toxic metals are removed separately from the non-toxic metal waste (e.g., aluminium, iron, and manganese), greatly decreasing sludge disposal costs, and (b) sulphate concentrations are significantly reduced. Field tests are in progress in Utah, where contaminated groundwater is being treated,[72] and are planned for the Berkeley Pit Superfund site in Montana.[73] A similar dual biological-chemical system ("Biosulphide") has been piloted at the Britannia Mine (Vancouver, Canada) by NTBC Research Corporation (Richmond, Canada). Treated water is used as a source of sulphate for the bioreactor, where H_2S is produced. The hydrogen sulphide is then used to precipitate metals in the chemical circuit.[70,74] Pilot trials indicate that the capital costs of the system are less than for an equivalent liming plant, and that an operating *profit* would arise from the sale of zinc and copper sulphide concentrates (at this particular site). Other companies have also been involved in extensive research on similar applications of biological systems (e.g., Noranda Technology Centre).[75]

Bacteria and fungi have been used for the direct biosorption of metals. This is defined as uptake by biomass (living or dead) via physico-chemical processes such as adsorption and ion exchange.[76] Fixed (immobilised) biomass can compare favourably with synthetic ion exchange resins in terms of loading capacity, selectivity and re-usability. Fixed metals can be concentrated via elution (typically with a mineral acid) and recovered by standard downstream processes. However, currently available fixed biomass systems are unlikely to be capable of dealing with high volume ARD flows. Fixed biofilms (e.g., rotating disc contactors) probably offer greater opportunities in terms of high-volume throughputs (e.g., the biological treatment process applied at Homestake Mine to degrade cyanide, thiocyanate, ammonia, and biosorb metals simultaneously).[76]

7.6.5 RESOURCE RECOVERY

Current large-scale technologies used in the mining sector (e.g., electrowinning) are unable to recover metals directly or economically from "raw" ARD; consequently, ARD remains a major environmental problem and a financial drain on the industry and often, by default, the taxpayer. To circumvent this constraint, two innovative responses appear feasible. Firstly, the development of alternative metal recovery processes that are capable of directly treating ARD; and secondly, the development of an efficient and economic means of isolating and concentrating the component contaminants (metals) so that they may be recovered using existing technologies. Both these routes have merit and have been explored by a number of government, industry, and academic researchers (e.g., the Resource Recovery Project in the U.S., the MEND programme in Canada, and the now-defunct U.S. Bureau of Mines) using precipitation, solvent extraction, ion exchange, bio-adsorption, and electrowinning. In particular, the Resource Recovery Project in the U.S. (sponsored by the U.S. Department of Energy Office of Science and Technology) demonstrates and evaluates technologies for the recovery of marketable metal products and clean water from water contaminated with heavy metals, and uses the Berkeley Pit in Butte (Montana) as a test bed for all such technology demonstrations.[77]

Reagent costs and energy requirements can often adversely affect the economic viability of such approaches.[58] To date, selective ion exchange and liquid emulsion membrane separation have appeared the most promising options. However, a recent review by Canadian researchers indicated that current ion exchange resins are limited in their capacity to recover selectively specific metals, and further research is required to develop more effective alternatives.[78] Natural ion exchange materials such as zeolites, which are widely available at relatively low cost, have also been investigated (e.g., Zamzow et al.,[86] Mondale et al.[79]), although there are substantial problems in terms of ensuring consistent quality for what are largely unprocessed industrial minerals. Liquid emulsion membrane separation of copper was being developed by the U.S. Bureau of Mines prior to its demise in 1996,[80] and had reached the pre-commercial demonstration unit stage. According to cost evaluations by Nilsen et al.,[81] copper recovery using this approach should be economical for water with at least 100 mg Cu L^{-1}. Research is now being continued as part of the U.S. Department of Energy's research programme.[34]

Researchers at McGill University (Montreal) have been conducting laboratory-scale research on the use of magnetic ferrites formed from controlled treatment of ARD with sodium hydroxide as a means of recovering metals from ARD. As the ferrites form, they incorporate other metals into their structure. The ferrites can then be recovered using a magnetic filter.[88] The major advance has been the capacity to form ferrites at ambient temperatures rather than 60°C as previously required. However, sodium hydroxide is more expensive than lime, so further refinements are required before the process would become economically viable.

7.7 CONCLUSIONS: GAPS IN OUR UNDERSTANDING AND FUTURE RESEARCH REQUIREMENTS

Todd and Struhsacker[87] contend that the past performance of the mining industry as typified by old sites and abandoned operations does not represent what will happen at new and modern sites. This point is certainly borne out by the results of their survey; however, it is interesting that a notable absentee from the mines which they considered was Summitville. While it is unjustified to use Summitville as a "stick" with which to beat the industry, neither is it sensible to ignore those modern sites where significant failures have occurred.

Real and substantial advances have been made in the prediction of ARD from non-ferrous metal mining wastes in the past five years.[20] Appropriate predictive testing along with waste characterisation and scientific interpretation of the data are essential if proper waste management practices are to be developed, disseminated, and sustained. There is a need for flexibility in the way that predictive techniques are applied within the regulatory framework. Prescriptive measures will further confuse the processes of planning and permitting by promoting the generation of meaningless and unreliable data, with severe consequences for operators and regulators alike. In the same way that regulation regarding contaminated land is moving away from total contaminant levels to an approach based on ecotoxicology, so coarse predictive techniques are likely to become more refined and reliable in the future. However, as this occurs, there will be an inevitable move away from generic standards, which may prove to be a considerable headache for the regulators. Consequently, field validation of predictive technology is critical.

As our understanding of the processes that are responsible for the generation of acidity from sulphide-bearing wastes improves, so we will be able to assess the acid-generating potential of wastes at ever earlier stages in the operation's life cycle (e.g., during the exploration phase). Simulation and modelling of processes as affected by different physical, chemical, and biological parameters will allow advance estimation of acid generation potential for different disposal regimes and strategies.[19]

Steps are being taken to deal proactively with ARD throughout the life cycle of a mine (from exploration through to closure). These incorporate technical, scientific, engineering, and managerial facets. As an example, Placer Dome's Environmental Engineering Department classified company properties into four general categories,[24] with an implicit understanding that operations at varying stages of their life cycle represent very different challenges in the context of ARD (e.g., future mines versus existing and operating mines versus closed mines).

The company has proposed some generic principles, which are largely substantiated by the literature reviewed in this chapter:

- Each mine site is unique and this fact handicaps the application of prescriptive methods.
- As ARD may occur several years after sulphides have been exposed, it is never too late to assess ARD potential at existing mines.
- Prediction and prevention should be the primary objective.
- Data from predictive trials may justify changes in mining and tailings disposal to reduce the potential for ARD.
- Predictive trials should be applied at an early stage to facilitate planning for further testwork and alternative waste disposal options.
- The success or otherwise of preventative approaches should be monitored to facilitate the future development of improved techniques.
- Further work is required to properly understand the geochemical processes associated with the formation and migration of ARD.

Beyond the gaps in our understanding of the biogeochemical complexities of ARD generation and migration, there are other obstacles to the development of "best practice." The U.S. Federal Advisory Committee to Develop On-Site Innovative Technologies created the Abandoned Mine Waste Working Group to specifically address the barriers to the development, deployment, and commercialisation of innovative technologies to remediate abandoned mine waste.[82] Part of the Group's conclusions was that technology development should be market-driven, based on the needs of those involved in the cleanup. In this way, the scientific and non-scientific site-specific requirements are met by allowing a flexible response in terms of local site conditions, desired level of remediation, cost and so on. To facilitate this, a *lead organisation* existing as a partnership between public and private bodies is required. This organisation would clarify the technology requirements in mine waste remediation, develop a priority list for technological solutions, and develop and promote a mechanism for technology transfer.[82] Although this conclusion was developed in a U.S. context, the concept is sound and broadly applicable to scenarios in developed and developing countries alike.

The Acid Drainage Technical Initiative (ADTI) has recently been established in the U.S. and has as its objective "to identify, evaluate, and develop cost-effective and practical acid drainage technologies which will facilitate decision-making and subsequent compliance with permit conditions. As a policy development program ADTI aims to identify and develop the best science available in the field of acid drainage." ADTI will coordinate with the EPA and ASTM to ensure that this objective is met and hopes to build upon the advances generated within the MEND programme, bearing in mind that climatic differences between the U.S. and Canada represent a significant variation between the two countries.

Industry is conservative when investing its funds in unproven technologies or approaches. This reflects, in part, the low profit margin of many companies. However, new technology must be incorporated into mine design and closure plans as it becomes available. One way to promote new technologies would be to ensure that companies fully internalise their environmental costs. Promoting increasingly stringent constraints on acceptable discharges can do this. Another route would be improved funding for pilot-scale trials of technologies that have been developed at laboratory-scale. This is a point at which central government could intervene, although the argument that industry should pay for near-market research may have some validity. If discharge constraints were tightened, and pollution prevention and control promoted over remediation, operators may well discover the incentive to invest in pilot-scale research. Independent verification of technologies (as per the USEPA's Environmental Technology Verification Program) will make available credible and independent performance data,[83] and remove some of the risk element of investing in new technologies.

ACKNOWLEDGEMENTS

The following are gratefully acknowledged for their informative, timely, and positive contributions to this chapter: Prof. Joe Barbour (University of Bath); Mr. Philip Gray (Fellow, Royal Academy of Engineering); Dr. Bob Kleinmann (Director, Environmental Science and Technology Division, Federal Energy Technology Center, U.S. Department of Energy, Pittsburgh); Dr. Nural Kuyucak (Golder Associates, Ottawa); Dr. Jacek Kostuch (ECC International R&D, St Austell); Dr. David Barr (Río Tinto, Melbourne), Dr. Peter Eaglen (Rio Tinto, Melbourne), Dr. Ian Firth (Río Tinto, Melbourne) and Dr. Luke Danielson (Mining Policy Research Initiative, Montevideo, Uruguay). Finally, the author would like to thank Ms. Liz Smith and Mrs. Yvette Haine for their assistance in the preparation of the final manuscript.

REFERENCES

1. Doyle, F. M., Acid mine drainage from sulphide ore deposits, in *Sulphide Deposits — Their Origin and Processing*, Gray, P. M. J., Bowyer, G. J., Castle, J. F., Vaughan, D. J., and Warner, N. A., Eds., The Institution of Mining and Metallurgy, London, 1990, 301.
2. Noranda, Environment, Health and Safety Report, Noranda, Inc., Toronto, Canada, 1995.
3. García, G. G., The Río Tinto Mines, Huelva, Spain, *Mineralogical Record*, 27, 275, 1996.
4. N.R.A., *Abandoned Mines and the Water Environment*, National Rivers Authority, Water Quality Series No.14, Her Majesty's Stationary Office, London, 1994.
5. Cambridge, M., Wheal Jane Minewater Project. The development of a treatment strategy for the acid mine drainage, in *Minerals, Metals and the Environment II*, The Institution of Mining and Metallurgy, London, 1996, 293.
6. British Columbia AMD Task Force, Draft Acid Rock Drainage Technical Guide Volume 1, Province of British Columbia AMD Task Force, Canada, 1989.

7. U.S. Environmental Protection Agency, Acid Mine Drainage Prediction, U.S. Environmental Protection Agency, Office of Solid Waste, Special Wastes Branch (Washington, D.C.), EPA 530-R-94-036, December 1994.

8. Filipek, L., Kirk, A., and Schafer, W., Control technologies for ARD, *Mining Environmental Management*, 4, 4, 1996.

9. Bowell, R. J., Fuge, R., Connelly, R., and Sadler, P. J. K., Controls on ochre chemistry and precipitation from coal and metal mines, in *Minerals, Metals and the Environment II*, The Institution of Mining and Metallurgy, London, 1996, 293.

10. Sherlock, E. J., Lawrence, R. W., and Poulin, R., On the neutralization of acid rock drainage by carbonate and silicate minerals, *Environmental Geology*, 25, 43, 1995.

11. Gray, N. F., Influence of secondary sulphate mineral formation on the impact of acid mine drainage to surface waters, Water Technology Research Technical Report: 16, University of Dublin, Trinity College, 1995.

12. Salomons, W., Environmental impact of metals derived from mining activities: processes, predictions, prevention, *Journal of Geochemical Exploration*, 52, 5, 1995.

13. Sengupta, M., *Environmental Impacts of Mining: Monitoring, Restoration and Control*, Lewis Publishers, Boca Raton, 1993.

14. Nordstrom, D. K. and Alpers, C. N., Remedial investigations, decisions, and geochemical consequences at Iron Mountain Mine, California, Proc. Sudbury '95 Conf. on Mining and the Environment, Sudbury, Ontario, 1995, 633.

15. Singer, P. C. and Stumm, W., Acidic mine drainage: the rate determining step, *Science*, 167, 1121, 1970.

16. Schwertmann, U. and Taylor, R. M., Iron oxides, in *Minerals in Soil Environments (2nd Edition)*, Dixon, J. B. and Weed, S. B., Eds., Soil Science Society of America Book Series 1, Madison, 1989, 379.

17. Herbert, R. B., Precipitation of Fe oxyhydroxides and jarosite from acidic groundwater, *GFF*, 117, 81, 1995.

18. Sullivan, P. J. and Yelton, J. L., An evaluation of trace element release associated with acid mine drainage, *Environ. Geol. Sci.*, 12, 181, 1988.

19. Orava, D. A. and Swider, R. C., Inhibiting acid mine drainage throughout the mine life cycle, *CIM Bulletin*, 89, 52, 1996.

20. Durkin, T. V. and Herrmann, J. G., Focusing on the problem of mining wastes: an introduction to acid mine drainage, in *Managing Environmental Problems at Inactive and Abandoned Metals Mine Sites*, USEPA Seminar Publication No. EPA/625/R-95/007, 1995.

21. Struhsacker, D. W., The importance of waste characterization in effective environmental planning, project design and reclamation, in *New Remediation Technology in the Changing Environmental Arena*, Scheiner, B. J., Chatwin, T. D., El-Shall, H., Kawatra, S. K., and Torma, A. E., Eds., Society for Mining, Metallurgy, and Exploration, Inc., Littleton, Colorado, 1995, 19.

22. Mitchell, P. B., The application of industrial minerals in the control of pollution emanating from metalliferous mine waste, Ph.D. thesis, Camborne School of Mines, University of Exeter, U.K., 1991.

23. Mitchell, P. B. and Atkinson, K., The novel use of ion exchange materials as an aid to reclaiming derelict mining land, *Minerals Engineering*, 4, 1091, 1991.

24. Robertson, J. D. and Ferguson, K. D., Predicting acid mine drainage, *Mining Environmental Management*, 3, 4, 1995.

25. Mine Environment Neutral Drainage programme (MEND), New methods for the determination of key mineral species in acid generation prediction by acid-base accounting, MEND Project 1.16.1c, Ottawa, April, 1991.

26. Pratt, A. R., Nesbitt, H. W., and Mycroft, J. R., The increased reactivity of pyrrhotite and magnetite phases in sulphide mine tailings, *Journal of Geochemical Exploration*, 56, 1, 1996.

27. Smith, M. L., The role of geostatistical characterization in the remediation of mine wastes, in *New Remediation Technology in the Changing Environmental Arena*, Scheiner, B. J., Chatwin, T. D., El-Shall, H., Kawatra, S. K., and Torma, A. E., Eds., Society for Mining, Metallurgy, and Exploration, Inc., Littleton, Colorado, 1995, 7.

28. Mine Environment Neutral Drainage programme (MEND), Guide for predicting water chemistry from waste rock piles, MEND Project 1.27.1a, Ottawa, July, 1996.

29. White, W. W., Trujillo, E. M., and Lin, C. K., Chemical predictive modelling of acid mine drainage from waste rock: model development and comparison of modelled output to experimental data, Proc. Int. Land Reclamation and Mine Drainage Conference and the 3rd Int. Conf. on the Abatement of Acidic Drainage (Vol. 1), Pittsburgh, April 24–29, 1994, 157.

30. Nyavor, K., Egiebor, N. O., and Fedorak, P. M., Suppression of microbial pyrite oxidation by fatty acid amine treatment, *The Science of the Total Environment*, 182, 75, 1996.

31. Georgopoulou, Z. J., Fytas, K., Soto, H., and Evangelou, B., Feasibility and cost of creating an iron-phosphate coating on pyrrhotite to prevent oxidation, *Environmental Geology*, 28, 61, 1996.

32. Maki, S., Belzile, N., and Goldsack, D., Inhibition of pyrite oxidation by surface treatment. Proc. Sudbury '95 Conf. on Mining and the Environment, Sudbury, Ontario, 1995, 1.

33. Mitchell, P. B. and Atkinson, K., The treatment of acid rock drainage: a preliminary study of enhanced lime treatment by the co-application of soluble sodium silicate, Proc. Sudbury '95 Conf. on Mining and the Environment, Sudbury, Ontario, 1995, 467.

34. Kleinmann, R. L. P., personal communication, 1997.

35. Stichbury, M., Béchard, G., Lortie, L., and Gould, W.D., Use of inhibitors to prevent acid mine drainage, Proc. Sudbury '95 Conf. on Mining and the Environment, Sudbury, Ontario, 1995, 613.

36. Scheetz, B. E., Silsbee, M. R., and Schuek, J., Field applications of cementitious grouts to address the formation of acid mine drainage, Proc. Sudbury '95 Conf. on Mining and the Environment, Sudbury, Ontario, 1995, 935.

37. Mineral Policy Center, *Burden of Gilt*, Mineral Policy Center, Washington, D.C., 1993.

38. Mueller, R. F., Sinkbeil, D. E., Pantano, J., Drury, W., Diebold, F., Chatham, W., Jonas, J., Pawluk, D., and Figueira, J., Treatment of metal contaminated groundwater in passive systems: a demonstration study, Proc. 1996 National Meeting of the American Society for Surface Mining and Reclamation, Knoxville, Tennessee, 1996, 590.

39. Rastogi, V., Scharer, J. M., and Pettit, C. M., Factors affecting ARD production: size and configuration of waste rock piles, Proc. Sudbury '95 Conf. on Mining and the Environment, Sudbury, Ontario, 1995, 597.

40. Mine Environment Neutral Drainage programme (MEND), Economic evaluations of acid mine drainage technologies, MEND Report 5.8.1, Ottawa, January, 1995.

41. Ribet, I., Ptacek, C. J., Blowes, D. W., and Jambor, J. L., The potential for metal release by reductive dissolution of weathered mine tailings, *Journal of Contaminant Hydrology*, 17, 239, 1995.

42. St-Arnaud, L., Water covers for the decommissioning of sulfidic mine tailings impoundments, Proc. Int. Land Reclamation and Mine Drainage Conference and the 3rd Int. Conf. on the Abatement of Acidic Drainage (Vol. 1), Pittsburgh, 1994, 279.

43. Mine Environment Neutral Drainage programme (MEND), Literature review report: possible means of evaluating the biological effects of sub-aqueous disposal of mine tailings, MEND Project 2.11.2a, Ottawa, March, 1993.

44. Fraser, W. W. and Robertson, J. D., Subaqueous disposal of reactive mine waste: an overview and update of case studies—MEND/Canada, Proc. Int. Land Reclamation and Mine Drainage Conference and the 3rd Int. Conf. on the Abatement of Acidic Drainage (Vol. 1), Pittsburgh, 1994, 250.

45. Price, W. A. and Errington, J. C., ARD policy for mine sites in British Columbia, Proc. Int. Land Reclamation and Mine Drainage Conference and the 3rd Int. Conf. on the Abatement of Acidic Drainage (Vol. 3), Pittsburgh, 1994, 285.

46. Payant, S., St-Arnaud, L. C., and Yanful, E., Evaluation of techniques for preventing acidic rock drainage, Proc. Sudbury '95 Conf. on Mining and the Environment, Sudbury, Ontario, 1995, 485.

47. Aubé, B. C., St-Arnaud, L. C., Payant, S. C., and Yanful, E. K., Laboratory evaluation of the effectiveness of water covers for preventing acid generation from pyritic rock, Proc. Sudbury '95 Conf. on Mining and the Environment, Sudbury, Ontario, 1995, 495.

48. St.-Germain, P., and Kuyucak, N., Biological water covers — a preliminary assessment, *Extractive Metallurgy and Mineral Processing Journal* (in press).

49. Ahmed, S. M., Chemistry of pyrrhotite hardpan formation, Proc. Sudbury '95 Conf. on Mining and the Environment, Sudbury, Ontario, 1995, 171.

50. U.S. Environmental Protection Agency, Innovative Methods of Managing Environmental Releases at Mine Sites, U.S. Environmental Protection Agency, Office of Solid Waste, Special Wastes Branch (Washington, D.C.), OSW Doc. 530-R-94-012, April, 1994.

51. Humber, A. J., Separation of sulphide minerals from mill tailings, Proc. Sudbury '95 Conf. on Mining and the Environment, Sudbury, Ontario, 1995, 149.

52. Berkowitz, J. B., Environmental cost considerations and waste minimisation in new plant design and process optimization, in *Hazardous Waste: Detection, Control, Treatment*, Abbou, R., Ed., Elsevier Science Publishers B.V., Amsterdam, 1988, 1727.

53. Zouboulis, A. I., Kydros, K. A., and Matis, K. A., Arsenic(III) and arsenic(V) removal from solutions by pyrite fines, *Separation Science and Technology*, 28, 2449, 1993.

54. U.S. Environmental Protection Agency, Treatment of Cyanide Heap Leaches and Tailings, U.S. Environmental Protection Agency, Office of Solid Waste, Special Wastes Branch (Washington, D.C.), EPA 530-R-94-037, September, 1994.

55. Rao, S. R., Finch, J. A., and Kuyucak, N., Ferrous-ferric oxidation in acidic mineral process effluents: comparison of methods, *Minerals Engineering*, 8, 905, 1995.

56. Ackman, T. and Kleinmann, R. L. P., An in-line treatment system for treatment of mine water, *International Mine Waste Management News*, 1, 1, 1991.

57. Marchant, P. B., Cation precipitation from complex acidic sulphate solutions using lime, in *Proc. Int. Symp. on Crystallization and Precipitation*, Strathdee, G. L., Klein, M. O., and Melis, L. A., Eds., Pergamon Press, 1987, 211.

58. Kuyucak, N., Conventional and new methods for treating acid mine drainage, Proc. CAMI '95 Symposium, Montreal, Canada, 1995, 1.

59. Murdock, D. J., Fox, J. R. W., and Bensley, J. G., Treatment of acid mine drainage by the high density sludge process, Proc. Sudbury '95 Conf. on Mining and the Environment, Sudbury, Ontario, 1995, 431.

60. Kuyucak, N., Payant, S., and Sheremata, T., Improved lime neutralisation process, Proc. Sudbury '95 Conf. on Mining and the Environment, Sudbury, Ontario, 1995, 129.

61. Eger, P. and Wagner, J., Sulfate reduction for the treatment of acid mine drainage; long term solution or short term fix?, Proc. Sudbury '95 Conf. on Mining and the Environment, Sudbury, Ontario, 1995, 515.

62. Eger, P., Wagner, J. R., Kassa, Z., and Melchert, G. D., Metal removal in wetland treatment systems, Proc. Int. Land Reclamation and Mine Drainage Conference and the 3rd Int. Conf. on the Abatement of Acidic Drainage (Vol. 1), Pittsburgh, 1994, 80.

63. Gusek, J. J. and Wildeman, T. R., New development in passive treatment of acid rock drainage, in *Pollution Prevention for Process Engineering*, Richardson, P. E., Scheiner, B. J., and Lanzetta, F., Eds., Engineering Foundation, New York, 1995, 29.

64. Blowes, D. W., Ptacek, C. J., Waybrant, K. R., and Bain, J. G., In situ treatment of mine drainage water using porous reactive walls, in *Proc. 11th Annual General Meeting of BIOMINET: Biotechnology and the Mining Environment*, Lortie, L., Gould, W. D., and Rajan, S., Eds., Natural Resources Canada, Ottawa, 1995, 119.

65. Waybrant, K. R., Blowes, D. W., and Ptacek, C. J., Selection of reactive mixtures for the prevention of acid mine drainage using porous reactive walls, Proc. Sudbury '95 Conf. on Mining and the Environment, Sudbury, Ontario, 1995, 945.

66. Kuyucak, N., and St.-Germain, P., *In situ* treatment of acid mine drainage by sulphate reducing bacteria in open pits: scale-up experiences, Proc. Int. Land Reclamation and Mine Drainage Conference and the 3rd Int. Conf. on the Abatement of Acidic Drainage (Vol. 2), Pittsburgh, 1994, 303-310.

67. Bender, J., Lee, R. F., and Phillips, P., Uptake and transformation of metals and metalloids by microbial mats and their use in bioremediation, *Journal of Industrial Microbiology*, 14, 113, 1995.

68. Johnson, D. B., Acidophilic microbial communities: candidates for bioremediation of acidic mine effluents, *International Biodeterioration & Biodegradation*, 35, 41, 1995.

69. Rao, S. R., Kuyucak, N., Sheremata, T., Leroux, M., Finch, J. A., and Wheeland, K. G., Prospect of metal recovery/recycle from acid mine drainage, Proc. Int. Land Reclamation and Mine Drainage Conference and the 3rd Int. Conf. on the Abatement of Acidic Drainage (Vol. 1), Pittsburgh, 1994, 223.

70. Rowley, M. V., Warkentin, D. D., Yan, V. T., and Piroshco, B. M., The Biosulphide process: integrated biological/chemical acid mine drainage treatment — results of laboratory piloting, Proc. Int. Land Reclamation and Mine Drainage Conference and the 3rd Int. Conf. on the Abatement of Acidic Drainage (Vol. 1), Pittsburgh, 1994, 205.

71. Hammack, R. W., Dvorak, D. H., and Edenborn, H. M., Bench-scale test to selectively recover metals from metal mine drainage using biogenic H_2S, Proc. Int. Land Reclamation and Mine Drainage Conference and the 3rd Int. Conf. on the Abatement of Acidic Drainage (Vol. 1), Pittsburgh, 1994, 214.

72. de Vegt, A. L., Bayer, H. G., and Buisman, C. J., Biological sulfate removal and metal recovery from mine waters, SME Annual Meeting Preprint, Denver, Colorado, 1997.

73. Hammack, R. W., Edenborn, H. M., and de Vegt, A. L., Bacterial sulfate reduction treatment of mining-related wastewaters: pilot plant results, unpublished paper (in review).

74. Rowley, M. V., Warkentin, D. D., and Sicotte, V. S., Treatment of acidic drainage from the Britannia Mine with the Biosulphide Process — results of a 10m^3 on-site pilot project, Proc. 13th Annual BIOMINET Meeting, January 13, Ottawa, Canada, 1993.

75. Kuyucak, N. and St.-Germain, P., Possible options for *in situ* treatment of acid mine drainage seepages, Proc. Int. Land Reclamation and Mine Drainage Conference and the 3rd Int. Conf. on the Abatement of Acidic Drainage (Vol. 2), Pittsburgh, 1994, 311.

76. Gadd, G. M. and White, C., Microbial treatment of metal pollution - a working biotechnology, *Trends in Biotechnology*, 11, 353, 1993.

77. Fletcher, A., Environmental breakthrough: the Resource Recovery Project, MERN Research Bulletin and Newsletter, University of Bath, U.K., June 1994, 31.

78. Riveros, P., Ion Exchange Extraction of Metals from Acid Mine Drainage, CANMET Division Report MSL 94-41 (CR), Ottawa, Canada, 1994.

79. Mondale, K. D., Carland, R. M., and Aplan, F. F., The comparative ion exchange capacities of natural sedimentary and synthetic zeolites, *Minerals Engineering*, 8, 535, 1995.

80. Wright, J. B., Nilsen, D. N., Hundley, G., and Galvan, G. J., Field test of liquid emulsion membrane technique for copper recovery from mine solutions, *Minerals Engineering*, 8, 549, 1995.

81. Nilsen, D. N., Hundley, G. L., Galvan, G. J., and Wright, J. B., Field testing of a liquid-emulsion membrane system for copper recovery from mine solutions, in *Chemical Separations with Liquid Membranes*, Bartsch, R. A. and Way, J. D., Eds., American Chemical Society (ACS Symposium Series 642), Washington, D. C., 1996, 329.

82. AMWWG, Abandoned Mine Waste Working Group — Final Report, Federal Advisory Committee to Develop On-site Innovative Technologies, Western Governors' Association, June 1996.

83. U.S. Environmental Protection Agency, EPA issues Environmental Technology Verification Strategy, USEPA Press Release, March 21, 1997.

84. Mine Environment Neutral Drainage programme (MEND), Evaluation of alternate dry covers for the inhibition of acid mine drainage from tailings, MEND Project 2.20.1, Ottawa, March, 1994.

85. N.R.A., Wheal Jane — A Clear Way Forward, National Rivers Authority South Western Region, Exeter, PR3/94, 1994.

86. Zamzow, M. J., Eichbaum, B. R., Sandgren, K. R., and Shanks, D. E., Removal of heavy metals and other cations from wastewater using zeolites, *Separation Science and Technology*, 25, 1555, 1990.

87. Todd, J. W. and Struhsacker, D. W., Environmentally responsible mining: results and thoughts regarding a survey of North American metallic mineral mines, Proc. Environmentally Responsible Mining: The Technology, the People, the Commitment, Milwaukee, Wisconsin, U.S., February 17–18, 1997.

88. Wang, W., Xu, Z., and Finch, J., Fundamental study of an ambient temperature ferrite process in the treatment of acid mine drainage, *Environmental Science & Technology*, 30, 2604, 1996.

8 Planning for Waste Management and Disposal in Minerals Processing: A Life Cycle Perspective

Mary Stewart and Jim Petrie

CONTENTS

ABSTRACT

Good environmental performance in mining and minerals processing operations relies on effective waste management. In turn, the generation of waste, both its quality and quantity, is a direct function of technology choice and is limited further by the thermodynamic constraints under which the industry operates. It is only over the former that operating companies can exercise any control. We argue that improvements in environmental performance are realised primarily by changes in technology — not simply hardware choice, but also operating and management practices. Whilst such changes are driven by environmental concerns expressed by society (often in the form of legislative guidelines), it is operating companies alone which can effect such change. Current environmental management systems offer little in the way of technical guidance for mining and minerals processing companies intent on improving their environmental performance. They tend to reflect a static operating condition, in which the domain of influence is seen as external to the mining or minerals process. Generally, opportunities for improved process development and design are overlooked. There is little attempt to relate environmental impacts from, for example, waste generation, to the processes that generate these wastes. This "closed box" view of technology is limiting in another way also. It restricts opportunities to see environmental performance in the same context as economic performance. The latter is driven by measures of efficiency — capital, labour, and technology. So too is the former. The ability to reconcile these two aspects of company performance, and so make explicit any trade-offs between the two that are required as part of operational strategy, arises naturally when a causative link is made between technology choice/operation and waste generation. This becomes clearer when it is recognised that "avoidable" wastes represent process inefficiency, which directly affects economic performance. In this instance, an "avoidable" waste can be defined as the difference between the optimal and the actual waste generated by the technology in place.

We have developed a systemic model of mining and mineral processing which reflects all environmental issues arising from resource extraction and mineral beneficiation, including waste generation. The model is dynamic and process-based; critical resource consumption and waste flows are linked directly to unit operations. We have adopted the methodology of Life Cycle Assessment (LCA) to relate resource and waste issues to environmental impacts. This reverse-mapping exercise, in which impacts can be traced back to particular wastes or emissions, which, in turn, are related to specific unit technologies, allows companies to target strategic improvements in environmental performance by reviewing their process design and operation philosophies. This approach is consistent with proactive "planning for closure."

In this way, issues of waste management and disposal are brought to the fore. We propose that this approach will assist operating companies in identifying their long-term liability associated with waste disposal. Rehabilitation of waste deposits on cessation of mining and minerals processing can be guided by the above and implemented cost effectively. The model provides a basis upon which to explore in detail the relationship between income generation and waste generation potential for a given process requirement, specifically in terms of technology selection and operation. In this way, it may be possible to identify constraints on the part of operating companies to fund closure management practices. Equally, the model gives some guidance on how best to apply regulatory mechanisms to closure planning.

We have explored the use of this generic modelling approach to the South African minerals industry, which has been analysed on a sectoral basis *vis-à-vis* gold, coal, base metals, platinum group metals, beach sands, and ferro-alloys. Case studies from the base metals sector have been used to demonstrate how structural features of the industry dictate systemic environmental performance. The value of the model in an expanded environmental management programme framework is highlighted in the context of the prevailing regulatory regime.

8.1   INTRODUCTION

The design of mining and minerals processing plants has been driven historically more by techno-logical issues relating to resource extraction and minerals beneficiation than it has been by waste management and disposal considerations. As such, it has proved difficult to develop strategic assessment tools for process design which can provide objective guidance to plant operators concerning their potential environmental liability — both as a consequence of "day-to-day" oper-ations, and that which might arise post-closure. We argue in this chapter that planning for closure requires an understanding of all aspects of the technologies in place over the complete life cycle of the project — from design concept, then through commissioning, operation, closure, and post-closure. Here we are not concerned exclusively with the engineering or process performance of individual unit technologies, but equally with such issues as the degree to which their operations have been integrated, the management structure within which they function, and the social milieu in which the operation is placed—after all, the issue of environmental liability will always have a subjective element to it, informed by social preferences. With that in mind, our objective is to develop an understanding of mining and mineral processing projects that will facilitate the closure process in a cost-effective and environmentally responsible manner, and be acceptable to all stakeholders.

This gets to the heart of the "role of waste management and prevention in planning for closure" debate which has been covered ably by Mitchell in Chapter 14. Whilst policy makers are duty bound to reflect in their deliberations all societal concerns regarding the industry's environmental performance, it is to the industry itself we look to improve its performance within a given regulatory framework. The challenge is to provide industry with as much flexibility as possible to meet this objective without compromising overall environmental performance or sacrificing economic advan-tage. The question which comes to mind immediately is whether this can be achieved through a "business as usual" scenario, which focuses on waste management, or whether there is a need for a fundamental shift in thinking about the way in which mining-related activities are defined — in particular the social, spatial, and temporal domains over which the influence of these operations should be considered. It is our belief that current practices within the industry do not reflect these perspectives adequately, and we devote much of the remainder of this chapter to informing this expanded view.

Our point of departure from accepted practice in the industry (which tends to focus on waste once it is waste), is that there are many opportunities to reduce the risks inherent in closure management by focusing on the formative stages of the life cycle, the process synthesis, and design stages, for it is here that technologies are selected, and here where overall process flexibility is constrained. As the design process itself unfolds, the number of technological options is reduced, and individual units are integrated in a particular way to ensure that the overall design objective is met. Typically, this has been economic only. Waste generation and management practices are defined at this stage. No matter how effective treatment and disposal options may be, the environmental risk attendant on the project is dependent largely on these elements of project planning. The promotion of cleaner technologies within the industry requires that we develop a strategy to force project planners to recognise this view. In terms of planning for closure, this means that the role of the process design stage in defining overall environmental performance, over both short and long time scales, must be accentuated.

However, the tools of process synthesis and optimisation in mining and mineral processing are relatively new, are concerned with economic optimisation alone and are, more often than not, based on heuristic arguments. Recent attempts at process synthesis in the general process engineering context with an explicit waste minimisation objective include those of Sarigiannis[1] and Douglas.[2] In mineral processing and extractive metallurgy, only the work of Reuter et al.[3,4] is notable. It has to be argued, however, that minimising waste is not the same as minimising environmental impact.[5]

For this reason, simply optimising processes as far as waste generation is concerned is not sufficient. Process optimisation for environmental benefit needs to consider the complete suite of impacts associated with the process.

We propose that Life Cycle Assessment (LCA) provides a structure wherein the linkages between wastes and impacts can be quantified objectively. The LCA approach is an all-encompassing one in which the impacts associated with the entire life cycle of the product are accounted for, from resource extraction, through production and use of the product, to final disposal.[6,7] Also included are the impacts of all the materials used, and the impacts of the wastes generated. It is this last aspect that is central to the approach which we develop in this paper. In addition, the outputs from an LCA exercise reflect in a transparent manner the trade-offs between environment and economic performance that arise from a particular technological action, and should thus be meaningful to all stakeholders. A discussion of LCA, its structure, and value is presented below.

There is a need for the above information in order to provide guidance to mining and minerals processing operating companies around the environmental dimension of their business. In particular, our primary objective is to provide such information in a suitable format, and with adequate detail, to assist in the design and operation of process technologies, which will lead to environmental improvement. Given that LCA starts from a rigorous input-output analysis, in terms of mass and energy balances, it is consistent with the approach taken by process design engineers, and thus is well positioned to inform technology choices. We have demonstrated that it is possible, by modelling minerals-related activities as a series of structured flow sheets, to link waste arisings to specific unit operations/technologies. In this way, process improvements can be targeted to provide greatest effect.[8] It is encouraging that developments in process synthesis and design are making increasing mention of the need to accommodate environmental performance as part of a hierarchical design approach.[2] We foresee the direct use of LCA in this context.

It has been suggested already that LCA can provide links between waste generation and waste management. This being so (and this position needs still to be supported), it should be possible to apply an LCA perspective to an assessment of waste disposal and closure practices within the industry. Both these aspects of company performance are driven largely by prevailing legislation. In this paper, we provide an account of this interaction for the South African case, and discuss how LCA can be used to inform the legislative process.

8.2 ANALYSIS OF THE PROBLEMS AND CHALLENGES

The *physical structure* of the minerals processing industry is such that it poses a number of problems as far as "planning for closure" is concerned. The industry processes low concentration raw materials into higher concentration products with significant energy input. Table 8.1 shows average feed concentrations for the industry. As a result, significant quantities of waste materials are unavoidable. This is true both for the initial beneficiation stages where metal concentrates are produced, and for the various value-added stages (e.g., purification of metals by electrowinning). Thus, adopting a waste minimisation approach to waste management problems within the minerals industry will not in of itself present a comprehensive solution.

A further problem posed by the physical nature of the industry is the fact that the feed materials are thermodynamically stable. In order to impose increased order on the system (by decreasing the disorder represented by the dispersion of the metals within the ore), significant energy input is required. This thermodynamic constraint goes beyond energy provision when opportunities for waste recovery/reprocessing are entertained. There is a need to consider explicitly "second law" inputs in terms of exergy (combined energy and entropy). This analysis is particularly relevant when examining material cascades within the minerals economy, e.g., ferro-alloys to steel to scrap to ferro-alloys.

The wastes themselves are usually very diverse in nature. The waste stream from a single mineral processing plant can contain a variety of metals as well as water, sulphur, aqueous salts,

---

TABLE 8.1
Average Feed Concentrations

Sector		Average Concentration (mass weighted)
Gold		4 g t^{-1}
PGMs	Noble metals	5 g t^{-1}
Base Metals	Copper	1%
	Lead	6%
	Phosphate	8%
	Zinc	3%
Ferro-Alloys	Chrome	40%
	Iron	60%
	Manganese	40%
	Vanadium$^+$	2%
Beach Sands		1%

$^+$ (as V_2O_5 in Fe-rich ores)

---

organic, and inert materials. Environmental liability arises from the discharge/disposal of these wastes. It is, however, not a simple matter to quantify this liability, which is determined to a great extent by fate and transport considerations. The liability associated with a waste is dependent also on the ability of the local environment to absorb the impacts associated with the waste, its so-called "assimilative capacity." It is not only the amount of material discharged to the environment which is important in these calculations, but also the rate of such discharge, and, in the case of metals, the rate at which these are mobilised within their environment. These factors together determine the total quantity of waste that the environment is able to absorb before a threshold level is crossed and irreparable damage is done. For these reasons, the geographic location of the process is important as are waste management regimes adopted.

The previous discussion highlights the fact that it is a difficult task to assign strict environmental liability to waste management practices within the industry. It needs to be recognised also that the ability to improve the environmental performance of minerals processing activities rests principally with the companies involved. The challenge, therefore, is to develop a strategy for quantifying environmental liability in a manner that provides a direct link to waste-generating processes. By doing so, critical points of waste generation can be identified insofar as they relate to process equipment choice, operational strategy and resource selection, which together define the ambit of technology selection. By making this express linkage, we afford operating companies the opportunity to bring an awareness of environmental liability directly into the decision-making process. This should ensure that the only limitation on the environmental performance of the process is thermodynamic rather than "artificial," as would result from inappropriate technology choice. An artificial waste can be defined as the difference between the best achievable performance of available technology and the design performance level of the selected technology. The concepts of "artificial" and "avoidable" inefficiencies are illustrated in Figure 8.1. In this figure, two technologies are compared; neither achieves the thermodynamic performance though Technology B is clearly better than Technology A. Our design objective is to move to a situation where the operating efficiency of our selected technology achieves the thermodynamic limit for the system. In this context, the thermodynamic limit (which will always be less than 100%) reflects the maximum attainable conversion efficiency of primary ore to product.

These distinctions are important in the minerals processing industry, where valuable products are dispersed in the ore matrix. This can be described as a situation with high entropy. The recovery of product from the dispersed phase requires significant energy input to concentrate the material.

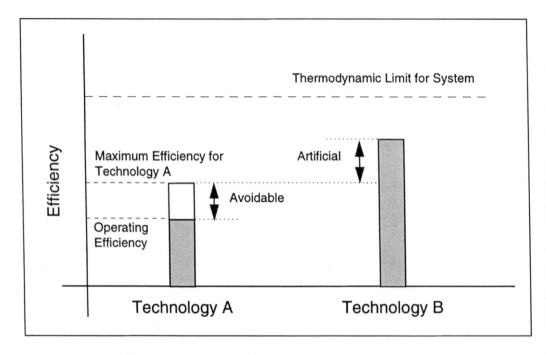

FIGURE 8.1   Thermodynamic limits to process efficiency.

This results in a decrease in the entropy of the product, but an increase in the entropy of the overall system due to dissipate processes including waste generation. The challenge in environmental management is to develop a systematic approach that will allow:

- Thermodynamic performance limit for the whole mineral life cycle to be identified.
- An objective comparison of competing technologies to be carried out on a thermodynamic basis.

We argue here that this strategy requires a detailed process model of the industry, which embodies all material and energy resource flows. This model should be developed in a structured manner and be sufficiently detailed to make explicit the linkages between unit operations within a single process and between unit operations and waste generation. Furthermore it should have generic qualities in order to facilitate comparisons across different geographic locations and between disparate resource uses. In this way, the model supports the decision-making process in the broader sense (spatial, temporal, and economic).

The second problem facing the industry results from its *management structures*. The size of the minerals industry has given rise to two basic corporate structures, one in which decisions are devolved to operating plant level, and the other in which the holding company retains the right to direct decisions. Quite often these are in conflict, and this is most readily apparent with respect to environmental performance. Delegation of environmental competency to an appropriate level within operating structures is often misdirected — a case in point here is where multinational operations fail to appreciate local conditions. The ability of either of these structures to inform the decision-making process around planning for closure is not clear.

Management structures are complicated further by company networks that link suppliers and markets. There are many cases in the minerals processing industry where the economic viability of the operation is dependent upon identification of markets for various waste streams. Optimal performance of any one process (i.e., node within the network) will not necessarily ensure optimal systemic performance. Any model developed to assist in the quantification of environmental performance needs to recognise this network structure.

The previous discussion has highlighted the complexity of operational structures within the industry. The flexibility of operating companies to implement environmental improvements to any process relates to the point within the project life cycle at which such improvements are introduced. In order to minimise the environmental risk associated with minerals processing, it is important to ask the right questions at the right time, and ensure that decisions/actions are implemented at the right time.

One method which attempts to take account of this in the decision-making framework is the so-called "phased demand matrix" approach, which links decisions to the technology life cycle, from project conception to closure.[9] In this context, the phased demand matrix links technology choice to decisions required at each stage of the life cycle and traces the effects of these decisions.

We suggest in Figure 8.2 a way to illustrate, qualitatively, the interrelation between economics, production, and environmental impact of a mineral processing activity. There are three parts to this figure. Part A details the situation if there is no mitigation of environmental impact at any stage of the life of the project. In Part B, the decision to remediate has been taken at some arbitrary point X during the production phase of the project. This is a common occurrence with projects; legislative or social pressures are brought to bear and a "cleanup" campaign begins. Part C illustrates the effects of including mitigation strategies in the initial stages of the project. In this case, graphical trends reflect a precautionary approach; in other words, the mitigation methods included in the planning stage of the process are sufficient to foresee any changes in legislative or social requirements and to cater for these. In this scenario, no additional mitigation is necessary over the life of the project.

The value of **e** in each case is an indication of the differential environmental impact associated with the project. In the case of Part C, this is attributed only to thermodynamic limitations, i.e., all technologically possible interventions have been implemented. It is clear that planned mitigation results in decreased environmental impact. The difference between the profit line and the environmental impact line gives an indication of the ability of the process to fund post-closure mitigation. The value of **P** in each case is an indication of the total net profit. The challenge is to drive P_C towards P_A and to reduce the size of trough in the profit curve.

This figure serves to illustrate the likely trade-off between economics and environment and shows how the point at which decisions are made can affect the environmental liability associated with a process. The phased demand matrix approach overlays a matrix of impacts onto this picture in order to make transparent the effects of decisions taken at each point within the technology cycle. It highlights also the opportunities for technology intervention and, equally, quantifies the cost of delaying the recognition of environmental impacts as well as indicating the costs associated with delaying the implementation of a waste management/site remediation strategy.

In addition to the quantitative elements of environmental risk identified above, there is a qualitative component relating to *social perceptions* of environmental impact, which is driven by social processes. A number of systemic tools have been developed with the intention of reconciling the social and technical dimensions of environmental impact of any project. These include Strategic Environmental Assessments (SEAs), Environmental Impact Assessments (EIAs), Environmental Management Plans (EMPs), and so on. Though different in scope and intent, each assessment tool has a positive role to play. However, none of these tools is sufficient in itself to address the environmental impacts associated with all the phases of a project life cycle. Their main shortcoming is that they tend to be static reflections of perceived impacts associated with a particular suite of technology choices up to point **Z** in Figure 8.2. They take no account of the evolutionary nature of technology and the effect that this has on cumulative environmental impact.

It was with a view to addressing the various challenges presented here that the methodology described in the rest of this section has been developed. It couples a chemical engineer's view of processes with a LCA view of environmental impacts to produce a quantifiable evaluation of

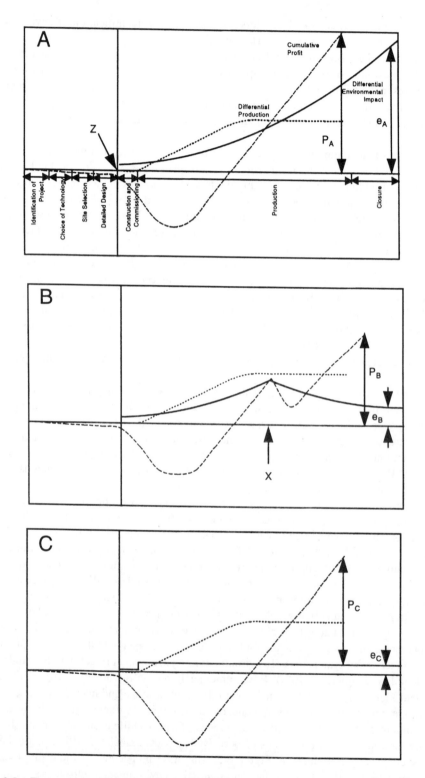

FIGURE 8.2 Economic, environmental, and production details for the life cycle of a technology.

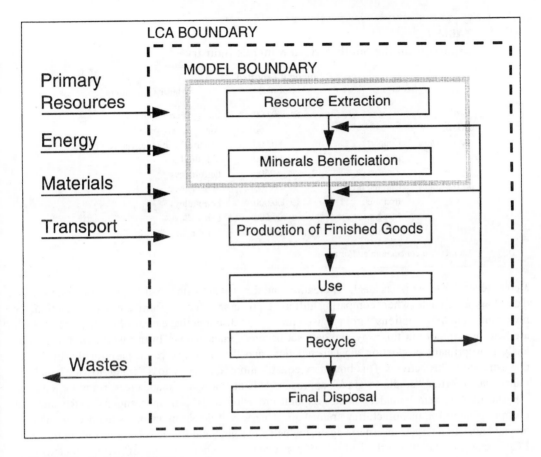

FIGURE 8.3   Definition of LCA boundary.

environmental impact. It is generic in nature, and can be used to compare disparate technologies in different locations. It can be made plant-specific as well as being used to describe the overall performance of an entire industrial network. It is also dynamic and grows with process development to reflect the environmental and economic effects of changes within the process. One method by which social criteria are included in the decision-making process is through the environmental legislation framework and its role as a driver for improved performance. Our methodology takes explicit account of these mechanisms.

8.2.1   LIFE CYCLE ASSESSMENT (LCA)

Life Cycle Assessment (LCA) can be thought of as a form of environmental systems analysis.[10] It is based on rigorous mass and energy balances calculated by modelling the flow sheets of processes. These balances are used to evaluate the resource consumption and waste generation inventories of the process, called the Life Cycle Inventories (LCI). Life Cycle Assessment links these inventories to recognised environmental impacts in an objective manner.

There are four stages in an LCA. The first is the *definition of the system and its boundaries*. Life Cycle Assessment is a holistic approach to the quantification of environmental impact as it is based on a cradle-to-grave view of a product or process. In order to achieve this, it expands the boundaries of the system under scrutiny to include the impacts of resource extraction, plus the use and reuse of the product through to final disposal.[6,7] It includes all impacts associated with the manufacture and transport of the product. Such a boundary is illustrated in Figure 8.3. Whilst such

TABLE 8.2
Exports from the South African Minerals Industry[32]

Sector		Amount Exported (tonnes)	Major Consumers
Coal		54,600,000	Energy, Ferro-alloys
PGMs		160	Autocatalysts, jewellery
Base Metals	Copper[+]	80,000	Electrical and industrial equipment
	Lead[+]	90,000	Industrial
	Zinc[+]	2500	Ferro-alloys
Ferro-Alloys	Chrome ore	900,000	Ferro-alloys
	Iron ore	19,000,000	Ferro-alloys
	Manganese ore	1,300,000	Ferro-alloys
	Vanadium	26,000	Ferro-alloys

[+] a metal in concentrate and/or metal

a comprehensive view is an ideal, its creation and assessment often lead to unwieldy amounts of data. It is acceptable when comparing different processes that perform the same function, or reviewing technology options within a given process, to discount those parts of the life cycle which are held in common. In this way, the data set becomes manageable. In our work, we restrict our analysis to primary production via beneficiation, and therefore ignore all parts of the life cycle downstream of this activity. This boundary equates more accurately to a "cradle-to-gate" analysis rather than a formal "cradle-to-grave" approach. This distinction is made clear in Figure 8.3.

The LCA system boundary makes explicit provision for social interventions in the material chain outside of the manufacturing framework. One way it does this involves the use of environmental legislation targets or criteria against which environmental impacts are assessed. In this way, LCA recognises the oft-overlooked role of consumers and other elements of society in generating environmental burdens. LCA views manufacturing processes as an integral step in the provision of services valued by society.[11]

It is this expansion of the system boundary which allows for the linking of resource extraction economies with resource consumption economies, and thus can be used to place liability for the impacts associated with a product in a global framework. An example of such a coupling is the linking of minerals beneficiation in a country like South Africa with the downstream manufacture of metal products in the OECD countries, Table 8.2 includes information on the amount of material exported by South Africa in the 1995 financial year. The LCA can be used to apportion the environmental impacts associated with a process to both the producers and the users of a product.[12,13] In this way, accountability for environmental impacts is brought to the attention of consumers of the product who otherwise are likely to be oblivious to impacts of mining and minerals processing. This should not be perceived as dilution of environmental responsibility, but rather as a way to bring on board everyone involved in the management of the material chain.

The second stage of LCA is the establishment of *waste inventories*. These require mass and energy balances for the service/process. Typical LCA inventories involve large sets of data that are manipulated using conventional process systems tools.[11,14]

The *assessment of impacts* of the process forms the third stage of LCA. Here the relationship between wastes and impacts is quantified. This is not an easy exercise due to the number of components in a waste inventory. In LCA, wastes are linked to a greatly reduced number of well-recognised environmental impacts. This is done by aggregating the information according to a specific set of priorities which, in the case of minerals processing, should allow for a regional/local perspective to be developed. In this way, it is easy to envisage a mechanism whereby people most directly affected by the impacts of a particular process can be directly involved in ranking these

TABLE 8.3
Criteria Used to Determine Unit Operation Boundaries

Criteria	Details
Common Function	If units perform the same function within a process, they are integrated into one unit. This was the case with the entire initial beneficiation process, Figure 8.4A.
Mass Flow Rate	Units which have a high flow rate of any input or output should be kept separate in order that their effect on the waste stream will not be diluted .
Hazardous Waste	If a unit is either the entry or exit point of a hazardous material, then it must be kept separate in order that the origin of the hazardous material can be pinpointed.
Energy Intensity	Energy is of importance within LCA, and units with a high energy consumption must be kept separate.
Common Waste Generation	The unit from which a waste leaves the system is not necessarily the unit that dictates waste quality. It is important to link waste generation with its point of exit, as in the linking of leaching with filtration in the zinc flow sheet, Figure 8.5C.

impacts. Before opportunities for technological improvements can be identified, it is important to rank the array of impacts according to a set of pre-defined criteria. This exercise requires decision-makers to declare their preferences. The manner in which this is done is invariably subjective, and it is necessary to make this process as transparent as possible to avoid the distortion of information in the LCI. This is where a robust decision-making framework has value.

The final stage of the LCA is that of *improvement analysis*. It is at this stage that the flow-sheeting basis of the LCA becomes important, as it is possible to target the specific units within the production process which require improvement. Improvements can be iterative; the benefits of process changes can be reflected in a new LCI and further process improvements suggested. This is repeated until the process is optimised within the production, economic, and environmental constraints.

8.2.2   GENERALISED METHODOLOGY FOR FLOW SHEET STRUCTURE

The flow sheets were constructed to deliver information that is consistent with the LCA framework. For this reason, flow sheets were developed with a focus on waste type as well as on the origin of the waste within the process. Table 8.3 details the five criteria used when deciding on unit operation boundaries within the process.

This generalised methodology for flow sheet development was used to model all of the processes used in the South African minerals beneficiation industry. These models were then used to evaluate the waste inventory for the industry. The waste inventories include information on the component composition of each stream.

8.2.3   APPLICATION OF THE METHODOLOGY—THE SOUTH AFRICAN MINERALS BENEFICIATION INDUSTRY

We have elected to demonstrate this modelling approach in the context of the South African minerals processing industry. The motivation for this comes from the role of the industry in this country's economy. In addition to this, the minerals industry is also a major contributor to the waste stream in the country. For the purpose of this study, the South African minerals beneficiation industry was divided into six subsectors: gold, coal, base metals, platinum group metals, ferro-alloys, and beach sands (which are washed to produce a titanium-rich product). These groupings were chosen as they divide the sector along lines of common process routes.[8] They are also the divisions of various other structures that feed into the model of the industry, for example, the stock exchange, corporate configuration, and national statistics. Generalised flow diagrams of all these sectors are included in Figure 8.4. It is not our intention in this paper to describe the totality of the information contained

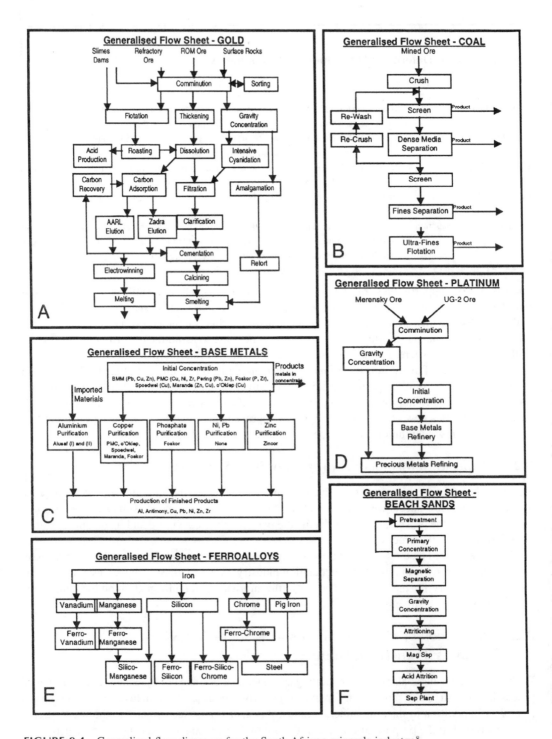

FIGURE 8.4   Generalised flow diagrams for the South African minerals industry.[8]

in this larger document.[15] Rather, we have chosen to explore different aspects of our approach with specific reference to individual subsectors.

The base metals industry in South Africa is used to illustrate the generalised approach to flow sheet development. The flow sheets identify unit operations within the industry. Each flow sheet is

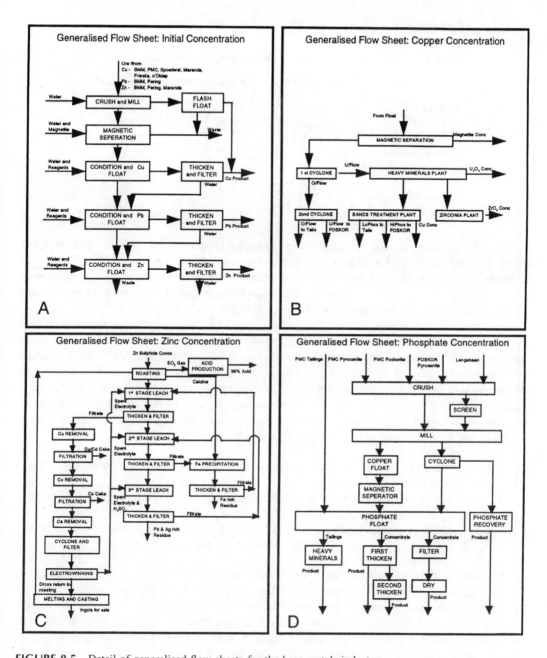

FIGURE 8.5   Detail of generalised flow sheets for the base metals industry.

translated to a quantitative model, which includes detailed mass and energy balances for each unit within the flow sheet. These balances by their nature include a complete waste inventory for the industry, as well as the input-output balance required by LCA. Figure 8.4C shows the generalised flow sheet for the South African base metals industry.

Each of the units in Figure 8.4C is further detailed by a separate flow sheet. These flow sheets are included in Figure 8.5. To simplify matters, the aluminium flow sheet has not been included, as there is no bauxite mining in South Africa. All raw materials are imported. Though it is not part of the base metals sector, phosphate production has been included therein, as there is a strong link between the processing of copper and the processing of phosphate in South Africa.

TABLE 8.4
Sources of Information Used in Developing the Models

	Technical Staff	Environmental Staff	Other Staff
Academic	9	1	2
Consultants	24	9	6
Corporate	58	7	2
Sector Associates		4	4
Government Departments	11	8	1
NGOs		8	
Research Organisations	18	6	2
TOTALS	120	43	17

The mass balances for the flow sheets were prepared in spreadsheet format. There were two sets of information required in order to ensure that the mass balances were accurate. The first set of information governs the operation of each unit within the process. The second set of information governs the stream splits between the units. These had to be evaluated with care in order to ensure that the performance for the industry as a whole was being calculated.

Once the mass balances had been quantified for each of the generalised flow sheets, they were taken to the South African industry as a whole for verification both of the accuracy of the generalised flow sheets and the assumptions governing the unit operations and stream splits. Table 8.4 reflects the various constituencies that have been canvassed and their effective contribution to the data validation exercise. We do not pretend that this distribution is either optimal or can be translated to similar exercises in other countries. We are encouraged, though, by the level of commitment we received and believe this attests to the validity of our approach and to the structure of the models.

There are many advantages to this type of approach:

- It pinpoints which units are responsible for generating which wastes and thus gives guidance to operator intervention.
- Improvement can be continuous, as it is possible to evaluate the performance of different technologies *in situ*.
- Changing the assumptions that govern unit operation performance gives an understanding of the sensitivity of the process to the operating conditions of a single unit.

Although this model structure does not provide an explicit account of waste management practice within the industry, it does provide guidance on how to manipulate processes to minimise the impact which waste generation has on the environment. In many cases, changing the process can decrease its waste management requirements. It is very important that all groups dealing with the environmental performance of an industry be appraised of the fact that it is the process that gives rise to the wastes, and thus that there are limitations to the extent to which waste management practice alone can ensure environmental protection. We would argue that this cause is best served by process engineering intervention, rather than by placing unwarranted confidence in the ability of management systems to deliver the same. This sea change in operational practice is consistent with the tenets of clean technology outlined in Chapter 14 by Mitchell.

8.3   ANALYSIS OF THE PROBLEMS AND CHALLENGES FACING THE SOUTH AFRICAN MINERALS INDUSTRY

Figure 8.6 contains details of the total inputs to the industry. It shows the vast amount of water consumed by the industry. The category "reagents" includes all chemicals as well as fluxes and reductants required by the various processes.

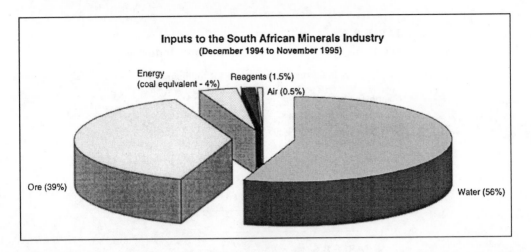

FIGURE 8.6   Total inputs to the South African minerals industry.

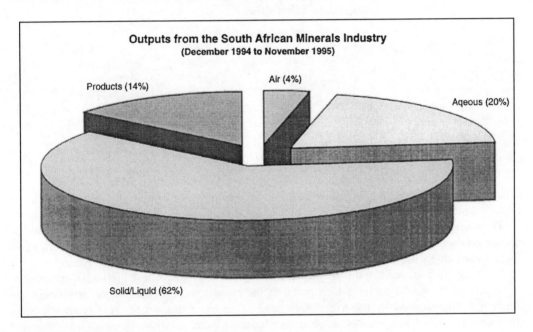

FIGURE 8.7   Total outputs from the South African minerals industry.

The input of energy includes both the coal used in provision of direct energy, as well as the electricity requirement of the industry, on a coal equivalent basis (0.54 kg of coal per kWh).[16] This is a fair estimate, as more than 90% of South Africa's energy comes from coal-based power generation.[17]

Also included is Figure 8.7, which details the total outputs from the industry. The wastes from the industry have been divided between contributions to airborne waste, aqueous waste, and to the solid/liquid stream.

Figure 8.7 shows the expected high contribution to the wastewater stream. The total product mass from the industry is small in comparison to the mass of waste generated during production.

The environment includes the economy that it supports and thus the two cannot be considered in isolation. Table 8.5 shows the importance of the industry to the South African economy. The industry plays a significant role in the generation of foreign exchange and a smaller part in GDP

TABLE 8.5

Economic Contribution of the Minerals Industry

Indicator	R (million)	%
Contribution to GDP	30,150	8.4
Contribution to state revenue*	1102	1.5
State aid to the mining industry	25	
Contribution to exports	38,118	48.7
Gold's contribution		29.7
Employment	617,147	4.1

* Company tax only (excludes income tax)

TABLE 8.6

Waste Generation Normalised with Respect to Sales Value

Subsector	Waste generated (kt) per sales value (million Rand)
Gold	18
Coal	5
Base Metals	8
PGMs	12
Ferro-Alloys	4
Sands	14

of the country as well as in contributions to state tax income. Employment figures are disproportionately low.

The waste inventories for the six subsectors have been normalised with respect to sales value of their products. This gives an indication of their ability to fund closure management practices. These values are included in Table 8.6.

It is interesting to note that the industries with high-value products have high waste generation. In the case of the gold industry, which produces the highest waste figure per ton of product, the situation is compounded by the hazardous nature of much of this waste. This is an important consideration — our analysis has attached significance to various waste streams on the basis of their generation rate. There needs to be an explicit account of the hazardous nature of these waste streams defined in terms of both rigorous quantitative measures and more subjective qualitative values. Whilst LCA provides an indication of hazard potential, the normalised economic analysis in the table has ignored this for the time being. We will return to this as part of our consideration of legislative structures.

Other interesting trends are also highlighted in this table. For example, another process with high waste generation is the beach sands industry. This is due to the low concentration of ilmenite and rutile in the sands. In contrast to gold, however, most of this waste is benign. Table 8.6 also shows that it is the ferro-alloys industry that is most likely to be able to fund closure than the other processes.[18]

Many companies within the industry mitigate their poor environmental performance by highlighting their role in job creation. Whilst Table 8.5 gives an indication of the overall industry record with respect to employment, Table 8.7 provides some data on specific sectors where employment figures have been normalised against the amount of waste generated. These values give an indication

TABLE 8.7
Waste Generated per Job Offered

Subsector	Waste Generated (t) per Job offered
Gold	658
Coal	624
PGMs	703
Base Metals (excl. Phosphate)	1152

TABLE 8.8
Average Water Quality Indicators for South African Gold Mines

Determinand	Rock Dump	Slimes Dam
pH	3.7	2.5
SO_4 (mg L^{-1})	3,100	1,700
TDS (mg L^{-1})	2,000	45,000

of the positive contribution that selected subsectors make to society through the provision of jobs, whilst contrasting this with their environmental impact.

This information would suggest that employment growth within the coal industry would impact least negatively on the environment. However, in South Africa a major downstream user of coal is the power generation sector, using 57% of domestic coal.[19] The environmental impact of this sector is substantial.[20] Once again, this emphasises the need for an LCA perspective.

Leaving the macro-scale evaluation of the processes, it is possible to pinpoint various points of concern within the industry on a subsectoral basis.

8.3.1 GOLD

- The first problem is the lack of control of leachates from the dumps. Many of the dumps belong to existing companies, but the dumps are extremely old and they have been inadequately engineered to contain runoff and leachate. Recent experience suggests that the management of slimes dams and tailings ponds is also inadequate. Most notable here is the case of the Harmony slimes dam which collapsed, killing 17 people. This disaster is discussed later in this chapter.
- Cyanide in the waste stream from the process is of concern. The notion that cyanide is broken down within the dumps at a sufficiently fast rate to ensure that surface runoff and leachate discharged are benign is open to debate.[21] However, the perception that cyanide presents the greatest environmental hazard is a false one. The reality is that sulphates and other acid-forming constituents contained in the waste give rise to most of the water quality problems associated with the processing of gold. Table 8.8 includes typical leachate concentrations for the industry. These high sulphate levels result in high salinity in ground and surface water systems.
- Low water prices favour inefficient use of this very scarce resource. Many water circuits operate in open loop because there is little pressure to do otherwise. This practice aggravates the problem of increasing salinity in surface and ground water. We recognise this problem as that of greatest concern to the maintenance of environmental welfare within the country.

- Gold mining at significant depth (the deepest mine is South Africa extends to more than 4 km below the surface) requires continuous pumping of underground water of high salinity to surface impoundments.

8.3.2 Coal

- The main problem in the coal industry is associated with the increasing stockpiles of fine and discard coal, which arise from beneficiation to support the export market. In 1988, the total discard produced by the coal industry was over 36 million tonnes.[22] These stockpiles are a significant source of acid mine drainage and, because of their potential to ignite through low temperature oxidation, a major contributor to regional air pollution.
- Near surface mining as well as deep level long-wall mining results in major surface subsidence, thereby prejudicing post-closure land-use.

8.3.3 Platinum Group Metals (PGMs)

- The main problem, as in the gold industry, is with the use of water. However, the platinum processes are newer than the gold processes and thus technology and engineering are more modern, resulting in a more efficient use of this resource.
- Platinum ores are high in iron and other metals that are discharged as wastes typically in slurry form, leading to the possibility of water contamination. The ores of the Merensky Reef contain typically 30% chromium trioxide and iron occurs in a ratio of 1.33 to 1 with the chrome.
- Slimes dams in this industry are huge. Our many attempts to quantify both tonnages of rock contained therein and the land surface area occupied by tailings and rock dumps have met without any success. These dumps will require a very innovative solution for remediation on closure of the mine.

8.3.4 Base Metals

- Water again is a problem, although these mines have a far less effect than, for example, the gold mines, merely by virtue of their size.
- Opencast mining poses problems at time of closure.
- Most base metal concentrates are sulphidic. Smelting technologies in place generate sulphur dioxide which, with few exceptions, is released into the atmosphere untreated. Air and water quality suffers directly as a result.

8.3.5 Ferro-Alloys

- The industry is extremely energy intensive. Electricity pricing structures have discouraged investment in more energy-efficient technologies.
- There is a legacy of mis- or non-management of slag dumps in this industry.
- Acid pickling and other surface treatments of the primary steel products generate large quantities of acidic waste streams of high salinity.

8.3.6 Beach Sands

- Most of the ilmenite and rutile deposits are in scenic areas of the country, specifically in coastal areas and areas with delicate habitats in the Western Cape and Kwa-Zulu Natal provinces. The challenge on closure is to return these sites to as near a pristine condition as possible. Significant effort has been made to ensure that this is done.

TABLE 8.9
Waste Streams for the Purification of Copper and Zinc

Technology	Component	Waste Generated (kt per year)	Generation to		
			Air	Water	Solid/Liquid
Copper	Water	21,000			
Smelting	Ore	15,000			
	Air	2800			
	Copper	100			
	Carbon	100			
	Flux	85			
	Reverts	66			
	SO_2	42			
	Oxygen	11			
	Nitric Acid	6			
	Total	**39,210**	**2853**	**0**	**36,357**
Zinc	Ore	190			
Electrowinning	Water	120			
	H_2SO_4	60			
	Oxygen	47			
	Steam	33			
	SO_3	15			
	SO_2	12			
	Fe	6			
	Total	**483**	**107**	**156**	**220**

8.4 A TECHNOLOGY ASSESSMENT OF ENVIRONMENTAL PERFORMANCE

The previous discussion is restricted to an overall assessment of the industry. We recognise that improved environmental performance requires an awareness of the potential for introducing cleaner technologies at subsector level. We attempt in what follows to demonstrate an approach to technology assessment based on our overall flow sheet models.

Case studies are included with the intention of showing the ability of processes to fund waste management both during the life of the processes and after mine/plant closure.

8.4.1 COMPARISON OF PURIFICATION TECHNOLOGIES

There are two types of purification technology used in the South African base metals industry: hydrometallurgical purification, as in the leaching and electrowinning of zinc (Figure 8.5C); and pyrometallurgical purification, as in the smelting of copper (Figure 8.5B). The two technologies perform the same service, that of purifying a concentrated ore. However, they do this in two very different ways and thus have very different waste inventories.

The most notable components in the waste streams from the two technologies are detailed in Table 8.9. These components are either those with the highest flow rate or the most hazardous components.

It is evident that the streams are very different in both nature and quantity. The stream from the pyrometallurgical purification of copper has emissions to both air and the solid/liquid stream. The greatest impacts are the contribution of the process to acidification and the leaching of material from the slag heaps.

TABLE 8.10

Comparison of Waste and Income Generation Information

Technology	Product: Waste (t t^{-1})	Sales Income: Waste ($ t^{-1})	Sales Income: Hazardous Waste ($ t^{-1})
Copper smelting	0.002	90	1300
Zinc electrowinning	0.030	1800	10,500

On the other hand, the electrowinning of zinc is an aqueous process with acidic emissions to the water stream being the greatest impact from the process. Both processes are energy intensive—the purification of copper requires energy for smelting, and the purification of zinc requires energy in the electrowinning stage.

The comparison of these two processes would usually now move onto the third stage of LCA, that of the impact analysis. The impacts have been identified and they must now be "rated" in some order of importance. This is not always an easy task. In order for the decision-making process to be better informed, the information detailed in Table 8.10 needs to be included.

It is evident that the purification of copper has a far greater waste generation potential than the electrowinning of zinc. By normalising each waste stream with respect to the income generated by the process from which it originates, taking into account different commodity process, an even clearer picture is obtained. These figures indicate that an electrowinning process generates 30 times more income per tonne of waste than does a smelting process, making it easier to fund closure in the case of the hydrometallurgical process — discounting any measure of "tractability" or intrinsic hazard.

It is also possible to give an indication of the ability of a process to pay for the impacts for which it is responsible. It is up to the decision-makers to decide whether an acid rain potential coupled with very low income relative to waste generated is a better proposition than a process which generates an aqueous waste, but has a far better income:waste ratio.

8.4.2 PROCESSING A WASTE STREAM

The further processing of a waste stream to recover valuable materials has often been proposed as a method of waste minimisation. In the base metals industry, there is an interesting link between a copper process and a phosphate process. The copper process sells a waste stream to the phosphate producer. The phosphate producer removes the phosphate present in this stream and returns the water to the copper process. This is a very elegant example of waste reprocessing and cooperation between industries.

However, in order to quantify the environmental impact of this further processing of the stream from the copper process, the two processes need to be separated and their waste inventories quantified independently. This is simple with the methodology developed; the assumptions governing the stream-split need to be altered, after which the copper and phosphate processes can be modelled independently.

Figure 8.8 shows the change in the waste stream for the uncoupling of the two processes. Note that this figure is plotted on a log scale in order to give an idea of which components in the waste stream are changing. In order to detail the extent and magnitude of this change, Figure 8.9 and Table 8.11 have been included. Figure 8.9 shows the percentage change of the various components in the waste stream and Table 8.11 gives the magnitude of these changes.

There is an increase in the waste stream of 620,000 tonnes per month when the processes are linked. The economic effect of linking the processes can be summarised by quantifying the increase

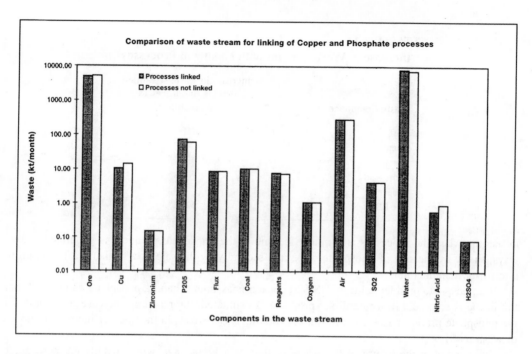

FIGURE 8.8   Comparison of waste streams for the linking and unlinking of the copper and phosphate processes.

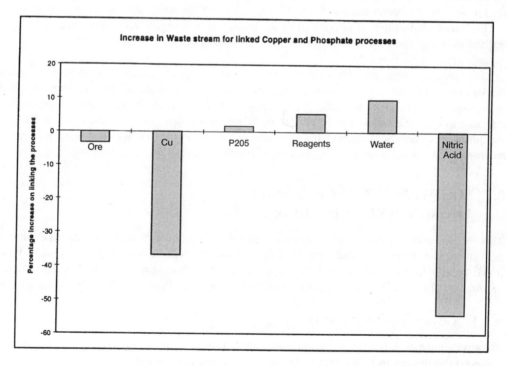

FIGURE 8.9   Percentage increase in the waste stream for unlinking the copper and phosphate processes.

TABLE 8.11
Increase in Waste Stream with Linking of Processes

Waste Component	Increase in waste stream on linking the processes (kt month^{-1})
Ore	−160
Copper	−4.0
Water	790
Nitric acid	−0.3
Reagents	0.4
Net Effect	**626.1**

in income when the processes are linked. Income to the phosphate producer is generated in two ways; firstly by the sale of phosphate rock, and secondly by the sale of a further waste stream to a fertiliser manufacturer.

If a phosphate product value of US $20 per ton is assumed,[23] and a phosphate value of $9 per ton in the stream sold to the fertiliser producer is estimated, the resulting increase in income for the phosphate producer is $1.04 million per month. The increase in income per increased kiloton of waste is $1,600.

The change in the nature of the waste is shown in Figure 8.9, which details the percentage change in various components. It shows a decrease in the amount of nitric acid in the waste stream but an increase in the reagents used in phosphate production. The question which decision-makers need to answer is whether the additional $1,600 can cover the additional one kiloton of waste.

Both case studies hint at opportunities for improved environmental performance that can be realised only through an awareness of the physical and management structures that exist within the industry (see Section 8.2).

The two case studies have shown how a generalised flow sheet approach, as has been described, can be used to identify constraints on the part of operators to fund closure management practices. In essence, we have identified an explicit linkage between economics and environment without resorting to a normalisation which reduces environmental impact to a monetary value. The next section of this paper will address the role of current legislative practice in driving responsible environmental management practice, specifically with respect to planning for closure.

8.5   SOUTH AFRICAN MINING AND ENVIRONMENT LEGISLATION

This section reviews the role of the various government departments that control the environmental performance of the minerals industry. This reflects another type of structural management which impacts the industry's performance. We follow this with a discussion of prevailing legislation in an attempt to highlight the link between it and the LCA approach identified in the previous sections.

8.5.1 GOVERNMENT DEPARTMENTS

The environmental legislation that governs the minerals industry in South Africa is very fragmented, falling under the jurisdiction of the following government departments:

- Department of Mineral and Energy Affairs (DMEA): this department legislates over all mines in the South Africa. It is responsible for conditions in the workplace. It also issues closure certificates. Such a certificate is issued only when the DMEA is satisfied that the closure plans drawn up have been carried out during the life of the mine. Once a closure certificate has been issued, the liability for wastes generated and disposed of becomes the responsibility of the government.

There is a conflict inherent in the Minerals Act of 1991. This act not only puts closure management in the hands of the DMEA, but also charges it with the international marketing of the industry. This places the DMEA in the role of both public custodian and industry promoter — an untenable position. The environment is often seen to be in conflict with the profitability of the industry. It is easy to envisage a situation in which issues of public welfare and environmental protection are subjugated to corporate gain.

- Department of Water Affairs and Forestry (DWAF): the Water Act considers both aqueous and solid wastes. The DWAF issues emission certificates that determine the maximum permissible concentrations of various contaminants in the water leaving a process. It is responsible also for permits that determine the site and structure of slimes dams and rock dumps.
- Department of Environment Affairs and Tourism (DEAT): the DEAT is responsible for air emissions. The Atmospheric Pollution Act of 1965 has a confidentiality clause that prohibits the public from accessing all air emission certificates. Though this infringes constitutional issues, it has not been tested in court.

8.5.2 ENVIRONMENTAL LEGISLATION CONCERNING MINE CLOSURE

There is one overriding piece of legislation governing the closure planning for mines, which forms part of the Minerals Act, Act 50 of 1991. This act is built on three pillars:

- Optimal processing of minerals.
- Safety and health of workers.
- Rehabilitation of land surface during and after mining operations.

The Minerals Act requires that the operating company draw up an Environmental Management Programme Report (EMPR). Included in the EMPR is the Environmental Management Programme (EMP). This programme includes details company plans to ameliorate environmental impacts during the life of the process as well as post-closure. Central to this process is the requirement for companies to set aside funds to be used at time of closure. The exact amount that is required for this purpose is determined at the time of acceptance of the EMPR. On closure, liability is transferred to the state. Should the accumulated funds be inadequate for site rehabilitation, state agencies may choose to curtail the rehabilitation process or make up the shortfall from treasury funds. This practice has not been tested as yet. Most companies have chosen to centralise the closure funds at a corporate level. These contributions are tax deductible as running costs.

A detailed discussion of remediation bonds is beyond the scope of this work. Suffice it to say that the practice in South Africa has not evolved to a level where there is clarity over or consistent application of the concept.

The EMPRs have been drawn up according to an Aide Memoire issued by the DMEA.[24] This document evolved from extensive consultations between government departments, the mining industry, and the agricultural sector. There are two main problems with the Aide Memoire, these being that:

- Companies have complained that reports they are required to submit are repetitive in the extreme.
- The reports submitted have varied in the quality and comprehensiveness of the information they contain.

The main structural elements of the EMPR are:

- Executive summary.
- Project description.
- Description of pre-mining environment.
- Project motivation.
- Detailed description of the process.
- Description of the environmental impact of the process.
- Environmental management programme.

It is this last element which is the legally binding component of the report. Provided the management programme, whose structure and content are based on the other information contained within the EMPR as well as a collective departmental view of site conditions, is adhered to, the award of a certificate of closure is assured.

As the documents are drawn up by the industry, it can be seen that this is a self-regulatory approach to environmental control. The DMEA had approved over 2000 of the more than 3000 EMPRs to the end of August 1996.[25]

8.5.3   EMPR Pros and Cons

The EMPR has both positive and negative aspects.[26] The *positive* aspects are:

- For the first time, information on the environmental performance of the industry is being collected in a methodical manner. The EMPs require the monitoring of water quality in individual catchment areas for the first time.
- Environmental issues have been brought to the fore. The EMPR structure is designed to encourage the active participation of all stakeholders in the maintenance of environmental quality. In particular, plant-operating personnel are immediately brought into the direct decision process, as their input is critical to the preparation and review of the EMPR.
- The consultative process involving all relevant government departments has identified opportunities for process intervention as well as improved waste management. This is consistent with the philosophy of clean technology. This has been made possible by engendering a conciliatory atmosphere of cooperation rather than confrontation. The people dealing with the effects of pollution were talking to the people at the source of the pollution. This cooperative effort is time-consuming but very necessary.

The *negative* aspects of the EMPR process are:

- The reports are disparate in nature mainly because they reflect a number of different views. The authors of the reports range from plant metallurgists, through environmental and technical consultants from the corporate level, to environmental consultants from outside the industry. For this reason, it has taken far more time to ratify the reports than was estimated initially.
- The EMPR directive does not cover the entire minerals industry, only processes that are within the boundaries of the mine property. For example, the base metals refining stage of the platinum industry is not included, as these plants are at sites remote from the mine. This is also the reason for most of the ferro-alloys industry not being covered by this legislation.
- The EMP is built on the philosophy of Integrated Environmental Management (IEM). There is one critical difference, however. The IEM process provides for social intervention in the planning process. The EMP makes no such provision. Once a new project

has been proposed, there is no way in which the people affected by that process can stop it from proceeding.

8.5.4   LCA AND EMPR

It is proposed that LCA could make a meaningful contribution to the EMPR process. Specifically, its "systems" approach to the quantification of impacts generates information which can be directly incorporated into the EMPR, and in doing so, ensures that the quantification of impact is as comprehensive and objective as possible. This would make the reports more uniform and easier to evaluate. Should this suggestion be adopted, it would be possible to propose the integration of discrete EMPRs into a unified document that would provide a countrywide perspective.

On another level, LCA could be used to evaluate the proposed EMP, to quantify the effect of the programme and to compare the efficacy of the programme with other feasible programmes. It would thus be possible to ensure that the programme being put in place is optimal. This would be to the advantage of both the government and the operating company.

The general approach to flow sheet design outlined at the beginning of this paper could be used to ensure that the process description included in the EMPR is sufficient to describe both production and environmental performance. The linking of wastes to impacts using LCA would also make it possible for the people involved in the EMPR evaluation process to decide whether it is more efficient to change the operation of a process instead of making provision for amelioration of the environmental effects of the process, be these in the short- or long-term.

8.6   ANALYSIS OF BEST PRACTICE AND FUTURE TRENDS BASED ON EVIDENCE FROM MINE SITES AND PLANNED PROJECTS

The previous discussion has attempted to identify the links between process operations and waste generation, and we have hinted that this is a necessary first step towards improving the overall efficiency of the process. Measures of economic performance under these conditions give an indication of the ability of an operation to fund its waste management practices. We have not, thus far, explored the relationship between waste generation and waste disposal that, after all, will be the focus of any closure management plan.

It is not our intention to devote any discussion to a consideration of engineering aspects of waste disposal insofar as this is considered under planning for closure. We consider it more important to be able to quantify the nature of waste streams committed to permanent disposal, as this will give an indication of the long-term liability to which operators are exposed. This characterisation is made easier when the role of specific unit operations within a given process in waste generation is known.

The above points notwithstanding, there is merit in reviewing current waste management practices in the South African minerals industry to provide backdrop against which to assess the potential for implementing improved practices in the light of our previous discussion.

Remediation of gold and coal mine dumps has been ongoing for many years. This includes the grassing of dumps as well as the planting of trees at the base of dumps to stabilise the soil. Collection and containment of surface runoff is common. However, the effect of leachate generation and mobility on groundwater quality is largely unknown. Groundwater management practices differ from site to site and have evolved largely in an uncoordinated manner. This situation is compounded by the large number of abandoned mine dumps in existence around the country. Monitoring of these particular sites is left to the discretion of the DMEA, which does not attach a particularly high significance to these problems.

On the positive side, many of the old gold dumps are being reworked in order to recover their gold value. Residues from this operation are re-deposited to current compliance standards.

South Africa has been at the forefront of the development of mining technologies and the design of waste deposits has also been part of this. Dumps and dams being constructed at the moment are being built to some of the strictest standards in the world. One of the main motivating factors for this was the Merriespruit disaster in February 1994, in which 17 people were killed when a slimes dam collapsed. In addition, 600 people were injured and $18 million damage caused to property. The legal case that followed made South African history. Though certain individuals involved were indicted on charges of culpable homicide, the case did not come to court. Six people were fined a total of $22,000 and the companies involved in both the running of the mine and the construction of the dam were fined a further $60,000. Though these fines were paltry, the case has served to heighten the awareness of personal liability associated with environmental risk.

Other problems which exist within the industry as a whole are the excessive use of water and electricity. Legislation governing discharge of aqueous effluents requires compliance with "zero effluent" standards. Whilst this is an ideal, some plants have managed to approach this, and are experiencing a reduced running cost as a result — a clear example of best practice. However, the industry is reluctant to publish these results, as they are not achieved uniformly. The same has been the case with electricity where, as a consequence of increased auditing, energy consumption has been reduced, to the economic advantage of the companies concerned.

As already stated, the EMP enforces the establishment of a company-controlled fund to pay for amelioration of the environmental effects of waste disposal. This requirement has encouraged companies to investigate the best methods for offsetting their impacts. At present, research is being carried out in many areas. Three will be discussed here:

8.6.1   NEUTRALISING THE EFFECTS OF ACID MINE DRAINAGE

The main drive here has been towards development of passive treatment systems. Currently, there is a project being managed by the Water Research Commission, which is evaluating the use of an integrated system of lime beds and wetlands to neutralise the acid and immobilise the metals.

8.6.2   EXTINGUISHING BURNING COAL DUMPS

Most attempts here have been directed at starving the dumps of oxygen by covering them with impermeable layers of soil. There is also research into the conditions under which coal spontaneously combusts, and on other methods to alleviate this condition.

8.6.3   MANAGEMENT OF LEACHATES FROM SLAG DUMPS

Projects are underway at present to quantify the leach rate of metals and salts from these dumps. Whilst international trends suggest that thermal treatment processes are to be favoured, there is significant potential for the integration of hydrometallurgical treatment of various slag and dust products. This is being explored to some degree in the ferro-alloys industry.

8.7   ANALYSIS OF CONSTRAINTS AND OPPORTUNITIES FOR TRANSFERRING BEST PRACTICE

The main constraint to the transfer of best practice, as we perceive it, is the industry view that productivity, employment, and economic advantage are in direct conflict with the environmental view. Whilst this at first glance seems a simplistic notion, it has been reinforced time and time again by both management and labour.[27,28] In effect, environmental management is seen as an add-on cost. There is an apparent inability on the part of management to incorporate environmental best practice into its economic framework, despite the proliferation of company documentation attesting to a commitment to responsible environmental management. The labour view sees clean technology equating to capital intensive technology, thereby posing a direct threat to employment. This is one rare example where both management and labour are comfortable with the same view, being the perception that a commitment to the environment will detract from overall profitability, thereby compromising the balance between labour and capital.

The first environmental statement was included in a company report as recently as 1990. Since that time, the quality and detail of company reporting has improved considerably, driven mainly by public relations initiatives. There is as yet no legal imperative for companies to disclose their record of environmental performance. Indeed, there are sectors within the industry where there is apparent collusion amongst operating companies to keep such information out of the public domain. A commonly held view within the industry is that a self-regulatory approach is optimal — the EMP process being one example of this. Due recognition should be given to those companies which have developed environmental auditing structures as a consequence of the EMP process.[29] However, much of this information is not communicated to a wider audience.

Current legislative practice is being led by this industry view. Whilst it can be argued that improvements to environmental performance are best identified and managed by the industry itself, it is difficult to see how industry can prescribe guidelines for compliance which retain a sufficient degree of objectivity. Industry should be encouraged, rather, to define its operational mode within a regulatory framework that recognises that there is an economic dimension to the ideal notion of sustainable development. It is the requirement of government to define such a framework in consultation with all stakeholders.

Company reporting structures, which should be compiled in an objective and transparent manner, could ably assist the above process. We would propose that the environmental section of annual reports includes a register of the material and energy resources consumed, wastes and emissions generated, and waste management practices. Such information is consistent with the LCA methodology.

Historically, the minerals industry has been slow to adopt new technologies. In part, this is due to thermodynamic constraints as well as the high capital cost of critical unit operations. However, the industry is a conservative one, containing much inertia. An example of this is the time it took the mining section of the industry to adopt the new hydraulic drills currently being phased in for use in underground workings (previously, the drills were driven by compressed air). There is a cost savings in using water, as less energy is required for the hydraulic drills because they use the available water head. In spite of this cost savings, it has taken 15 years from the initial development of the technology for this development to gain significant penetration in the industry. Other instances of similarly long lag times include the Carbon-in-Pulp gold process, bioleaching of refractory ores, and direct reduction furnaces in ferro-alloys. Taking these as examples, it could be inferred that the ability of the industry to adopt new technologies, even if they are cheaper to run, is not good. How, then, can the industry be encouraged to take a more proactive stance with respect to its environmental performance?

Two pressures on the industry to improve its position are:

- International markets.
- Local interest groups.

International markets are having an immediate impact. Recent large-scale capital investment in the South African Minerals Industry (for example, Alusaf Hillside Smelter, Columbus Joint Venture for ferro-alloys, and the Namakwa Titania sands projects) reflects a corporate commitment to levels of environmental performance consistent with international norms even where these are more stringent than the prevailing local legislation.

Local interest groups, whose common-law rights are entrenched in the new Constitution, now have sufficient muscle to influence decisions surrounding development within the industry. Two cases in point are Saldanha Steel and the St. Lucia beach sands project, where issues related to natural resource management dominated the impact assessment process.[30,31]

Both the above pressure groups have succeeded to some degree in their objectives, albeit by putting industry in a reactive mode. We would argue that greater all-round benefit could be achieved by appraising industry of the direct synergy between sound environmental practice and good economic planning identified in our preceding discussion around the role of LCA. Success in this objective, we believe, would elicit a proactive response from industry. Given that the nature of this interaction is dynamic, this should make industry more robust and better able to respond to changing pressures in the environmental domain.

8.8   CONCLUSIONS AND REVIEW OF AREAS FOR FURTHER RESEARCH

It is reasonable to expect that the EMP process will remain the basis upon which corporate environmental performance is assessed. However, there needs to be an objective set of guidelines within which the EMP operates. These cannot be prescribed by the industry itself but should evolve out of a wide-ranging consultative process involving all interested parties. The EMP makes explicit provision for closure management, and identifies funding structures to achieve this. What it lacks is the ability to guide industry on how best to manage its day-to-day operations, both to minimise long-term impacts and to ensure that profitability remains high enough to generate the necessary closure funds.

Industry needs to be encouraged to adopt a proactive stance with respect to environmental management. One way to achieve this is to promote the perspective that optimal economic efficiency requires sound environmental practice, given that avoidable wastes are indicators of process inefficiency. LCA, in the context of a multi-objective optimisation regime, can be used to link environmental and economic performance. We argue that the generalised approach to developing flow sheets for the LCA process can be easily merged into the EMPR process. In so doing, the philosophies of environmental thinking embodied within LCA will form an integral part of the legislative process.

We do not wish to give the impression that LCA is a "catch-all" which can or should replace all other environmental management tools. Where it does offer significant advantages is that its assessment module is linked directly to a detailed understanding of industrial processes. For this reason, it is consistent with process engineering tools for technology assessment and optimisation, and thus can be easily adopted by industry.

There are two areas in which the LCA methodology requires further development. Both of these impact directly on mining and minerals processing, and specifically closure management practices. They are

- The effect of resource provision.
- The long-term effects of waste disposal.

Each is deserving of significant research attention. In the case of the former, changing ore grades will, to some extent, dictate both technology choices and waste management practices. Furthermore, energy provision to specific mining activities needs to be looked at as part of a national energy network, and water consumption in terms of other uses for this resource. In the case of waste disposal, LCA cannot accommodate a temporal dimension, which is required in order to assess long-term risks associated with such practices. This is a major limitation of the approach. The identification of appropriate measures for closure planning requires some quantitative measure of the intrinsic risks of waste disposal. This is a difficult task, requiring multi-disciplinary input. The challenge to the mining community, having minimised waste formation as far as possible, is to ensure that waste residuals are benign, thereby making the previous task much easier.

REFERENCES

1. Sarigiannis, D. A., Computer-aided design for environment in the process industries, *Computers in Chemical Engineering*, 20, Supplement, S1407, 1996.
2. Douglas, J. M., Process synthesis for waste minimisation, *Industrial Engineering Chemical Research*, 31, 238, 1992.
3. Reuter, M. A., Sudhoter, S., Kruger, J., and Koller, S., Synthesis of processes for the production of an environmentally clean zinc, *Minerals Engineering*, 8, 201, 1995.
4. Reuter, M. A., Sudhoter, S., and Kruger, J., Some criteria for the selection of environmentally acceptable processes for the processing of lead and zinc containing flue dusts, *SAIMM Journal*, 97, 27, 1997.
5. Rossiter, A. P., Process integration for pollution prevention, *AIChE Symposium Series*, 90, 12, 1994.
6. Clift, R. and Longley, A., Clean Technology and the Environment, in *Introduction to Clean Technology*, Kirkwood, R. I. and Longley, A. J., Eds., Blackie Press, Glasgow, 1995, Ch. 4.
7. Welford, R. and Gouldson, A., *Environmental Management and Business Strategy*, Pitman Publishing, London, 1993.
8. Stewart, M. and Petrie, J. G., Life cycle assessment for process design — the case of minerals processing, in *Clean Technology for the Mining Industry*, Sanchez, M. A., Vergara, F., and Castro, S.H., Eds., University of Concepcion, Chile, 1996.
9. Coates, J. F., Anticipating the environmental effects of technology — a primer and workbook, prepared for UNEP, The Kanawha Institute, Washington, D.C., 1995.
10. Petrie, J. G. and Clift, R., Life cycle assessment, *MERN Research Bulletin*, University of Bath, U.K., December, 1994, 9.
11. Boustead, I., General principles for life cycle assessment databases, *J. Cleaner Production*, 1, 167, 1993.
12. Ayres, R. U. and Simonis, U. E., *Industrial Metabolism — Restructuring for Sustainable Development*, United Nations University Press, New York, 1994.
13. Ayres, R. U., Industrial Metabolism, in *Technology and Environment*, Ausubel, J. H. and Sladovich, H. E., Eds., National Academy Press, Washington, D.C., 1989, 23.
14. Society for Environmental Toxicology and Chemistry (SETAC), *A Conceptual Framework for Life Cycle Impact Assessment*, SETAC, Brussels, Belgium, 1993.
15. Stewart, M. and Petrie, J. G., The waste generation potential of the South African minerals industry, unpublished report, 1996.
16. Eskom, Statistical year book, Eskom, Johannesburg, South Africa, 1993.
17. Van Horen, C., *Counting the Social Costs — Electricity and Externalities in South Africa*, Elan Press and UCT Press, South Africa, 1996.
18. Petrie, J. G., Petersen, J., Stewart, M., Dustan, A. C., and Cohen, B., Rehabilitation of ferro-alloy waste deposits, presented at Sixth Workshop of the Mining and Environment Research Network, Harare, Zimbabwe, August 1996.
19. Department of Mineral and Energy Affairs (DMEA), South Africa's Mineral Industry 1994/95, South African Minerals Bureau, Pretoria, 1995.
20. Petrie, J. G., Burns, Y. M., and Bray, W., Air Pollution, in *Environmental Management in South Africa*, Fuggle, R. F. and Rabie, M. A., Eds., Juta and Co. Ltd., Cape Town, 1992, Ch. 17.
21. Smith, A. and Mudder, T., Cyanide — dispelling the myths, *Mining Environmental Management*, 3, 4, 1995.
22. Department of Mineral and Energy Affairs (DMEA), Discard and unsold duff coal in South Africa — 1988, South African Minerals Bureau, Pretoria, 1991.
23. Department of Mineral and Energy Affairs (DMEA), South Africa's Mineral Industry 1993/94, South African Minerals Bureau, Pretoria, 1994.
24. Department of Mineral and Energy Affairs (DMEA), Aide Memoire for the Preparation of Environmental Management Programme Reports for Prospecting and Mining Projects, Government Press, Pretoria, 1992.
25. Department of Mineral and Energy Affairs (DMEA), personal communication, 1996.
26. Ayres, T., The environmental management programme report — a critique. South African Institute of Mining and Metallurgy Colloquium on Environmental Management in the Mining, Minerals and Metals Industry, Mintek, Randburg, South Africa, 1996.

27. Keeton, M., Business and the Environment, South African Institute of Mining and Metallurgy (SAIMM) Colloquium on Environmental Management in the Mining, Minerals and Metals Industry, SAIMM, Mintek, Randburg, South Africa, May, 1996.

28. Industrial Development Research Corporation (IDRC), The Environment and the Reconstruction and Development Programme, IDRC, Midrand, South Africa, 1996.

29. Industrial Environment Forum (IEF), Environmental Self-Assessment Programme, IEF, Johannesburg, South Africa, 1992.

30. Council for Scientific and Industrial Research (CSIR), Final Environmental Impact Report — Saldanha Steel Project Phase 2, CSIR, Pretoria, South Africa, 1992.

31. Environmental Evaluation Unit (EEU) and Council for Scientific and Industrial Research (CSIR), Environmental Impact Assessment for a 466,000 tpa aluminium smelter, EEU, Cape Town, South Africa, 1992.

32. Department of Mineral and Energy Affairs (DMEA), personal communication, 1995.

9 The Decommissioning of Leach Dumps and Protection of Water Quality: Lessons for Best Practice from Copper and Gold Leaching Operations in the United States

Gavin Bridge

CONTENTS

1-56670-365-4/00/$0.00+$.50
© 2000 by CRC Press LLC

ABSTRACT

This chapter discusses the challenges associated with the protection of water quality during the operation and closure of leaching facilities. It assesses best-practice approaches to decommissioning and reviews the changing public policy framework within which closure takes place by drawing on the experience of copper and gold leaching operations in the United States. The practice of leaching is defined, and an increase in the number of leach operations over the last 15 years is identified and explained. The potential for leaching operations to contaminate groundwater poses a technological and managerial challenge for leach facility operators. The high costs of groundwater remediation, long-term monitoring, and the possibility of incurring environmental and health liabilities provide a strong argument for planning closure from the outset by designing and managing leach facilities to prevent contamination. In response to increased public concern for environmental protection in general, and groundwater quality in particular, many states are pursuing regulatory strategies for the prevention and containment of contamination. The chapter highlights innovative initiatives in the framework of legislation regulating leaching practices, outlines emerging best-practice management and technological approaches, and discusses obstacles to their effective transfer and diffusion.

9.1 LEACHING: DEFINITION AND SCOPE

Leaching refers to the chemical and biochemical processes by which solutions transfer metal compounds from a solid phase to a liquid phase (usually an aqueous solution) from which the valuable metal component can then be recovered through downstream processing. Although leaching is used widely for the production of many metals (for example, virtually all aluminium derived from bauxite is produced by the Bayer caustic leaching process which takes place at elevated temperatures and pressures), leaching from open heaps and dumps at ambient temperatures and pressures is confined at the commercial scale to the production of gold and copper. However, ongoing research points to future commercial applications of heap and dump leaching in the cleaning of high-sulphur coals and the remediation of contaminated soils, including land-farming of organically contaminated materials.[1-4]

This chapter focuses on the specific environmental issues associated with the operation and closure of open heap and dump leaching operations, and draws on the experiences of copper and gold operations in the U.S.

Commercial leach production is typically undertaken in one of four formats — vat, heap, dump, or *in situ* leaching—of which heap and dump leaching are the focus of this chapter. *Vat leaching* occurs in large tanks where concentrates are mixed with process solutions, leached for a short period, removed, and replaced with fresh material. Vats enable key environmental parameters such as temperature, pressure, acidity, and the availability of oxygen to be closely monitored and controlled to optimise metal recovery. Vats may also be agitated to enhance the effectiveness of chemical reactions and improve metal recovery rates. *In situ* leaching refers to the leaching of ore that has not been actively mined and is still in the ground. Process solutions are injected into an orebody and recovered from the lowest point through pumping. The process was first developed for the extraction of uranium from shallow deposits, but has since been widely applied for copper. There are currently 18 *in situ* copper operations in the U.S. that leach rubblised ore in underground workings, and in 1995 Magma (now BHP) announced plans for the first commercial-scale operation that does not depend on preexisting underground workings at Poston Butte near Florence, Arizona.

ASARCO and Freeport McMoRan have collaborated since 1988 in a joint venture with the U.S. Bureau of Mines (and, since the closure of the Bureau in 1995, the U.S. Bureau of Reclamation) to investigate the leaching of deep copper oxides using wells at the Santa Cruz project near Casa Grande, Arizona. Pilot scale testing began in February 1996. Inmet of Canada announced in May

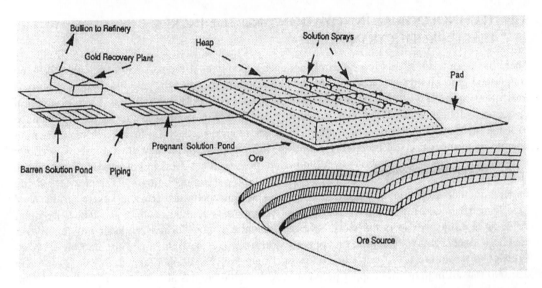

FIGURE 9.1   Heap leaching process — gold.[50]

1996 that it would conduct pilot scale testing of *in situ* techniques at the Copper Range facility in Michigan. The experience obtained from studies conducted over the last 25 years suggests that *in situ* leaching can be commercially viable although key technical questions remain to be resolved. The principal obstacle to large-scale commercial *in situ* leaching, however, is political and centres on the environmental acceptability of introducing and containing potential contaminants in groundwater at a time of heightened concern for the quality of groundwater supplies.[5]

Heap leaching involves the application of solutions to relatively homogenous and well-defined heaps of ore that have been placed on pre-prepared surfaces. These leach pads are designed to minimise the loss of leachate in an attempt to optimise leachate recovery and reduce environmental impacts. At some "on-off" heap leaching facilities, leached ore may be removed and replaced with fresh ore after a period of time, whereas at others new ore is simply added to the heap that grows in size. *Dump leaching* refers to the leaching of large, amorphous, and heterogeneous waste dumps or stockpiles which have accumulated over the life of the mine. Leach dumps are typically massive and are sited directly on existing topography rather than on pre-prepared surfaces.

As used here, the term dump leaching refers to the purposive, actively managed practice of extracting metals in solution from constructed piles of mineral-rich material. "Dump" is used, therefore, in a generic fashion to refer to both ore which has been piled in heaps for the sole purpose of leaching (distinguished in the literature as "heap leaching"), and to ore and waste rock which has formerly been discarded in a waste dump or stockpile but which subsequently has been incorporated as part of a leach operation ("dump leaching"). As will be shown, the distinction between heaps and dumps can be of value in a discussion of appropriate management techniques for decommissioning. Where no distinction is made in the text, however, the term dump leaching should be understood as a generic term covering both heaps and dumps. The discussion in this chapter does not address the specific concerns of *in situ* leaching operations or practices for the disposal, control or treatment of tailings from milling or vat leach operations, although many of the ecological issues surrounding the protection of water quality can be substantially similar.

Figure 9.1 illustrates a typical heap leaching format for gold. Low-grade gold ores are piled on a pre-prepared leach pad and irrigated with a solution of sodium cyanide. The solution percolates through the ore, dissolving the gold to form a gold-cyanide complex in solution which is then collected at the base of the heap. This "pregnant leach solution" is sent to a gold recovery plant where the gold is stripped from the cyanide and processed, and the barren cyanide solution recycled to the heap.

9.2 TECHNOLOGICAL INNOVATION AND THE INCREASE IN LEACH PRODUCTION

There has been a substantial increase in heap leach capacity over the past 15 years within the U.S. copper and gold mining industries. Cyanide heap leaching capacity for gold has increased 16-fold over the period 1972–1992 (Figure 9.2). Over the same period, dump and heap leach production of copper increased from 2% of U.S. output to over 30% (Figure 9.3). The increased significance of leach production in the U.S. is matched by accelerated investment in heap and dump leaching operations in other mineral-producing countries. For example, of the new investment planned for Latin American mining operations over the period 1995–2000, 52% is in gold and 32% in copper, and much of this is associated with large-scale low-grade leach operations.[6] Examples of specific projects include Newmont's cyanide leach project at Yanacocha in Peru, the largest producer of gold from heap leaching in the world; the joint venture between Outokumpu and Placer Dome at Zaldivar in Chile, which is the world's largest greenfield mine-for-leach copper project; BHP's proposed leach/SX-EW operation on copper oxide stockpiles at Tintaya in Peru; and the Cyprus-Codelco joint venture at El Abra, Chile, the world's largest SX-EW operation, capable of producing 225,000 tons of high quality copper per year.[7,8] Although there has also been increased investment in non-leach facilities (such as Phelps Dodge's investment at La Candelaria, Chile), the increase in leaching reflects its ability to cost-effectively treat low-grade ores and has been made possible as the result of innovations in engineering and mineral processing. Increasing economies of scale in earthmoving equipment over the past 30 years (many of them dependent on innovations in rubber technology) have made it economically feasible to construct large heaps and dumps. For example, the average size of mechanical shovels has increased four-fold over the period 1960–1990, the capacity of haul trucks eight-fold, and the size of copper flotation cells ten-fold.[9] Innovations in hydrometallurgical process control technology in the last 15 years have facilitated the production of a consistent, premium-quality product. Pioneering work by the U.S. Bureau of Mines in the 1970s on carbon-in-pulp cyanidation, for example, made possible the leaching of the ultra-low grade gold deposits of the Carlin Trend in northeastern Nevada, which were too low in grade to be economically treated by gravimetric or physical means. In the 1980s, these deposits underpinned the greatest gold boom in the history of the U.S. as production surged from 30 tonnes per year in 1980 to over 300 tonnes in 1992 (Figure 9.2).

Investment in the acid leaching of copper oxides and bioleaching of copper sulphides increased dramatically in the 1980s, following improvements to the specificity and tolerance of salicylal-doxime and ketoxime solvent-extraction reagents which made it possible to link low-grade leach production with established electrowinning technology (SX-EW). By the late 1970s the solvent-extraction and electrowinning processes had been improved to the extent that high quality cathode (99.999% pure) could be produced from dilute dump leach solutions which required no further refining.[10] Although large-scale, commercial dump leaching has been practised in the U.S. for over 50 years, increased investment in dump leaching during the 1980s enabled the U.S. copper industry to survive its deepest recession since the Great Depression of the 1930s. As part of a comprehensive restructuring of their production practices, many copper producers sought to reduce production costs by acid leaching waste oxide material which was not amenable to smelting, and which had been stripped and dumped in the course of mining sulphides for conventional production, and by the bio-leaching of low-grade sulphide ores that fell below the cutoff grade for the concentrator. Since these sources bore no mine costs, copper could be produced by leaching "wastes" for as little as $0.20/lb, at a time when production costs averaged over $0.60/lb. Investment in oxide and sulphide leaching has been sustained beyond the initial opportunity to turn accumulated waste dumps into productive assets: a significant proportion of new investment by U.S. copper and gold producers during the last five years in both domestic and overseas markets has been in leach capacity, including investment in mine-for-leach operations.

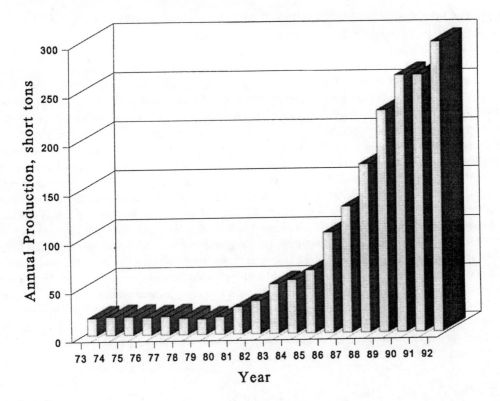

FIGURE 9.2   U.S. gold production by cyanidation, 1973–1992.[50]

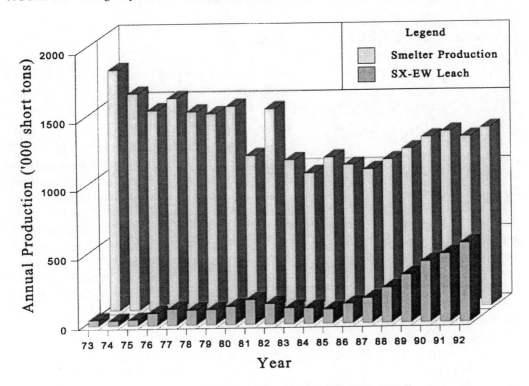

FIGURE 9.3   U.S. copper production by smelting and leach/SX-EW, 1973–1992.[50]

Further increases in leaching capacity are likely to result from one of the most significant developments in leaching in recent years: the commercial harnessing of the bacterially assisted oxidation of sulphide ores.[11] Commercial bioleaching processes for gold, copper, and nickel have been developed that capitalise on the ability of a group of bacteria (*obligate chemolithoautotrophs*) to obtain energy through the oxidation of insoluble inorganic sulphides or ferric iron.[12] Although bacterial oxidation and the leaching of metals occurs naturally and has a long history of productive utilisation, it is only in the last two decades that process parameters have been sufficiently understood and process control technology developed to the point that a consistently high quality product can be produced at a low cost. In the late 1970s, for example, the South African mining firm Gencor pioneered a biological oxidation process for refractory gold deposits. Patented as BIOX, the process went into full-scale commercial production at the Fairview mine in South Africa in 1986 and has since been licensed to six other producers, including the largest installation at Ashanti Goldfields's Sansu mine in Ghana in 1994.[13] The production of gold from refractory deposits — those that are typically pyritic, arsenopyritic, or carbonaceous and do not respond well to simple cyanidation — is likely to increase as many near-surface oxidised deposits are depleted and investment focuses on deeper sulphides which often underlie the oxide deposits. As a consequence, pre-treatments such as biooxidation leaching will become increasingly important. Newmont Gold has combined a bio-oxidation pre-treatment with ammonium thiosulphate heap leaching as an alternative to conventional cyanidation. The technique extends the range of grades that can be profitably processed, and makes viable the re-mining of waste dumps. Gencor has also pioneered the bioleaching of nickel sulphide concentrates, making possible the development of complex nickel sulphide ores (such as those with high concentrations of arsenic) which cannot be processed with conventional pyrometallurgical technology.[14,15]

9.3 DECOMMISSIONING LEACH DUMPS: THE CHALLENGE OF GROUNDWATER PROTECTION

Dump leaching poses a number of regulatory and managerial challenges for the protection of environmental quality during operation, decommissioning, and post-closure. These challenges arise from the nature of the leaching process itself (which liberates metals in solution from ore dumps located in close proximity to the natural environment); the recent construction of many leach operations and the associated lack of accumulated long-term experience with successful decommissioning techniques for large leach dumps; and the rapidly evolving regulatory framework.

9.3.1 PRESERVATION OF WATER QUALITY AND QUANTITY

Potential impacts from leaching operations on the environment are most likely to be experienced as changes to surface and groundwater quality. This is for two reasons: first, since leaching involves the extraction of metals in solution, a process failure will lead to the introduction of contaminants to the environment in a liquid form; and second, those feed materials and waste products which are solids rather than liquids are generally too coarse to enter the atmosphere, but are susceptible to chemical breakdown on contact with precipitation. Since many dump leaching operations are located in arid or semi-arid environments where surface water is limited, the environmental medium most often affected is groundwater (e.g., copper leach operations in the southwestern U.S., the Sonoran desert of Mexico, the Atacama desert of Chile, or gold heap leaching in Nevada or Western Australia). The principal pathways by which leach contaminants can enter groundwater are: leakage or spills from storage ponds, leach pad liners, or conveyances during the operation of the facility; entrainment of solvent-extraction reagents into heaps or dumps and subsequent leaching to groundwater; stormwater run-on/runoff; and uncontrolled leaching from heaps and dumps following closure.

Potential contaminants associated with leaching include process reagents and the products of reactions which occur as part of the leaching process. In addition to impacts on groundwater of process reagents, such as sodium cyanide or dilute sulphuric acid, process reagents can initiate or facilitate the mobilisation of metals which, if not contained, can generate increased metal concentrations in groundwater. The bacterial oxidation of sulphide minerals which generates acid mine drainage, for example, can liberate other metals whose solubility is inversely related to pH, including iron, cadmium and nickel (see Mitchell, Chapter 7). Entrainment (carry-over) of reagents from downstream processing can result in organic compounds entering and accumulating in the dump. In the solvent extraction — electrowinning (SX-EW) of copper, for example, an organic reagent (a hydroxy phenol oxime diluted in kerosene) is used to extract the copper from the pregnant leach solution. After copper is stripped from this organic reagent, the reagent is then recycled to perform its extractive function. The efficiency with which the reagent is separated and captured, however, represents a tradeoff with the metal recovery rate, and some reagent is lost from the system and is returned to the heap. If the reagent is not attenuated or evaporated in the dump, it can be flushed into the groundwater, where over time it may become a significant source of contamination.

Available evidence suggests that local groundwater contamination at leach sites may be commonplace.[16,17] A lack of comprehensive groundwater data, however, complicates a systematic assessment since, until recently, many prominent mining states did not require systematic groundwater monitoring. For example, groundwater monitoring in Arizona, a state which produces nearly two thirds of U.S. copper, and which led the way in the application of hydrometallurgy to the red metal throughout the 1970s and 1980s, was only achieved with the implementation of Aquifer Protection Permitting following the passage of the Environmental Quality Act in 1986.

Where groundwater data does exist, the spatial scale at which it is collected has made it difficult to isolate contamination due to leaching operations from other mine activities (e.g., tailings disposal) and naturally high background levels. Empirical research at two large copper leach operations in the southwestern U.S., however, found elevated levels of groundwater contaminants that were chemically and hydrogeologically consistent with that of leaching operations. At one site, for example, leaching began in early 1990, and by the summer seepage from the leach dump threatened contamination of the regional aquifer to the point that a network of over 300 pump-back and monitoring wells had to be installed to dewater a series of small canyons underlying the dump in order to contain the contaminant plume. At another site, an investigation by the state groundwater agency shortly after operations began reported that the chemical composition of groundwater in the vicinity of the dumps was virtually indistinguishable from undiluted leach solution.[18]

Mine operators, regulatory agencies, and environmental groups in the U.S. are increasingly embroiled in litigation over contaminated groundwater at active and abandoned leaching operations. The Environmental Protection Agency's (EPA's) Mine Waste Program has found a number of operations at which water quality has been degraded as a result of leaching activities, and has expressed concern over the potential for release of metals and acidic drainage from the 182 million tons of copper leach waste produced each year.[19] At the Ray complex in Arizona, for example, process solutions from dump leaching and from an electrowinning facility entered groundwater and emerged at the surface, exceeding state water quality standards for cadmium and copper, although the ecological impact from these exceedances is unclear. Other examples include the Bluebird mine at Miami, Arizona where increased copper concentrations in groundwater have been recorded downgradient from the leach facilities, and the Bagdad Complex and the Copper Cities site in Arizona where dump solutions entered the groundwater and surfaced in nearby creeks.[20]

Fluctuations in the water table caused by high rates of water extraction may be issues of local concern. Although many mines use water extremely intensively, collecting and recycling water where possible, leaching requires large volumes of water to compensate for losses by evaporation. Pit dewatering techniques can lead to drawdown of local water tables so that springs and wells no longer flow. The potential for political conflict over scarce water supplies is exemplified by recent events in the Crescent Valley of northeast Nevada. The boom in gold production during the 1980s brought state-of-the-art leach production to the Valley's isolated ranch lands that are claimed by the Western Shoshone, a Native American tribe. The Shoshone's dependence upon springs and wells for watering their cattle and for their spiritual significance has galvanised a movement in opposition to further leach developments in the area, which are seen as threatening long-term water supplies.[21]

9.3.2 PLANNING FOR POLLUTION PREVENTION

The high expense and uncertain success of remediating contaminated groundwater at poorly designed heaps and dumps highlights the economic and environmental benefits to be gained from building closure into the initial design and permitting phases. Groundwater may be the example *par excellence* of the cost-savings to be made from proactive environmental management, which aims to prevent contamination rather than remediate after the fact. Preventative planning can significantly reduce the costs associated with the need for extensive remediation on closure and the likelihood of post-closure liabilities arising from groundwater contamination and impairment of use. Best-practice approaches to heap and dump leaching therefore require the implementation of management practices and control technologies to prevent groundwater contamination from the initial design and permitting stage throughout the life cycle of the operation. These practices will be designed to protect groundwater from the introduction of potential pollutants such as cyanide, acidic drainage, and organic reagents, control natural bio-oxidation processes and the mobilisation of heavy metals in the heaps and dumps, and reduce the adverse impacts on local communities from fluctuations in the water table.

9.3.3 REGULATING THE TECHNOLOGICAL FRONTIER

Preventative planning and design may be a common sense goal, yet regulatory tools to facilitate the achievement of this goal are only just becoming available. Leaching processes for the production of metals have been actively used for centuries. For example, the active use of the natural processes of bacterial oxidation has a long history: the Romans utilised naturally occurring solutions emanating from copper sulphide deposits; the natural oxidation, leaching, and precipitation of copper on iron were observed during the East Han Dynasty in China in the first century A.D.; copper-bearing solutions were recycled through waste heaps in Hungary in the mid-16th century; and acid solutions were actively applied to large heaps of oxide copper ores at Rio Tinto, Spain in the 1750s.[5,22,23]

However, both the scale of leach operations and levels of public concern for environmental protection in general (and groundwater quality in particular) in the U.S. have undergone a substantial transformation in the last two decades. The large-scale application of heap and dump leaching in the U.S. gold and copper mining industries described above has occurred at a time of heightened public awareness of the potential human health and ecological impacts of pollution and an increased social valuation of environmental quality. Not only has the level of concern amongst the public as a whole increased, but many mining regions in particular have undergone an economic and social transformation. Once isolated from contemporary environmental concerns, traditional mining regions (such as the many mining towns of Colorado, Montana, and Idaho) are becoming increasingly integrated with the social and cultural norms of the metropolitan mainstream as a result of improved transportation and communication systems and far-reaching demographic and economic changes. In many mining regions, the transition in the economic base away from primary products (agriculture, mining, lumber) towards manufacturing and service provision and the associated rise of an urban middle class has accelerated demands for environmental preservation from both the public and an increasingly powerful business community whose profitability is dependent on high environmental quality (for example, the real estate sector and tourist interests).

Translating this growing public support for environmental protection into effective legislation for the preservation of groundwater quality, however, has yet to be fully accomplished. The application of heap and dump leaching technologies for the recovery of copper and gold in the

U.S. occurred largely in a regulatory vacuum as far as the regulation of groundwater quality was concerned. While there have been extensive and highly publicised regulations governing air quality (e.g., Clean Air Act of 1970 and subsequent amendments), endangered species (e.g., Endangered Species Act of 1973), permitting and environmental impact statements (e.g., National Environmental Policy Act of 1969) in the U.S. since the 1970s, regulations governing groundwater quality and the reclamation of mining wastes (including heaps and dumps) have been developed in many states only in the last ten years. This gap between the evolution of leaching technology and evolution of groundwater regulations is rapidly closing as state regulatory agencies seek to address the potential for groundwater deterioration arising from mining operations. Each state, however, has acted independently with little federal coordination. The variation among states in terms of design and performance standards, permitting and monitoring requirements, and enforcement capacities has amplified uncertainty and concern within the mining industry at the increasing regulation of mine operations.

9.4 ENVIRONMENTAL REGULATORY TRENDS IN LEACH DUMP DECOMMISSIONING

The closure of leach dumps which have ceased to be productive is a normal part of mine operations and has therefore been standard operating procedure for many years. In the last decade, however, closure procedures have been required to take account of additional environmental considerations. This recent trend towards active decommissioning to minimise environmental degradation is regulatory-driven. The widespread commercial application of hydrometallurgical processes in the 1970s and 1980s occurred largely in a regulatory vacuum, since groundwater and mine waste reclamation were unexplored regulatory frontiers in many of the states where leaching was adopted. The last decade, however, has seen regulatory agencies increasingly focus on groundwater quality and the reclamation of mine wastes. The EPA's regulatory determination in 1986 upholding the Bevill Amendment (exempting wastes generated during the extraction and beneficiation of ores and minerals from regulation as hazardous wastes) created the prospect of new federal regulations for mine waste under Subtitle D of the Resource Conservation and Recovery Act. The Bevill Amendment (Section 3001(b)(3)(A) to the Resource Conservation and Recovery Act (RCRA)) excluded wastes and was passed in 1980 (for the 1986 determination, see Federal Register 24496: July 6, 1986).

Possible regulatory approaches to mine waste (which specifically addressed heap and dump leaching) were drawn up by the EPA in 1988 and 1990 (the so-called "Strawman I and II" proposals) and were circulated for review and consultation by interested parties (including industry) as a working document (EPA 1990). The Strawman draft reflected the EPA's recognition of the gaps in existing legislation and their identification of the need for a multimedia, risk-based approach within which the principal objective would be pollution prevention. For example, in Strawman II the EPA identified the potential benefits of a regulatory approach which did not introduce regulatory oversight only on closure (when heaps and dumps officially became mine wastes under existing legislation) but which provided for regulatory oversight from the initial design phase, through operation, to decommissioning: "the rationale for regulating heap and dump leaching operations prior to closure is to ensure that design and operating requirements prevent releases from leaching units while those units are operational, and to detect and remedy any releases that do occur."[24]

Although Strawman energised a debate over the appropriate regulatory tools for promoting environmental best practice in the U.S. mining industry, the debate has not yet led to new legislative initiatives at the federal level. The Strawman proposals are currently in a state of suspended animation and are unlikely to be developed further until the Resource Conservation and Recovery Act comes up for reauthorization. The EPA has, however, recently developed a framework document,[25] which seeks to improve the environmental management of hardrock mining operations but

TABLE 9.1

Recent State-Level Regulatory Initiatives Affecting the Operation and/or Closure of Leaching Facilities

STATE	REGULATORY INITIATIVES
Arizona	Environmental Quality Act 1986
	BADCT Guidelines 1990
	Mined Land Reclamation Act 1994
California	Reclamation Standards 1992
Colorado	Mined Land Reclamation Act, Amendments 1993
Idaho	Cyanidation Regulations 1988
Montana	Custom Milling and Reprocessing Regulations 1990
Nevada	Water Pollution Control Zero Discharge Program 1989
	Mine Land Reclamation Act 1990
New Mexico	Mining Act 1993
Oregon	Mined Land Reclamation Act, Amendments
	re. chemical processing 1992
South Dakota	Mined Land Reclamation Act, Amendments 1992

by using existing legislation and regulatory authority. According to the EPA, the framework is a strategic plan to guide agency activities, and aims to identify more flexible alternatives to the standard regulatory tools from within existing policy.[26] Hardrock mines were not included in the comprehensive regulatory programme of the 1977 Surface Mine Control and Reclamation Act, which was restricted to coal operations. The major industry-specific piece of federal legislation affecting hardrock operations — the General Mining Law of 1872 — contains no environmental provisions, a fact that contributed to a strong, but unsuccessful reform effort in 1994. The lack of industry-specific environmental regulation at the federal level does not mean, however, that hardrock operations have been unregulated: on federally-held lands, mines are subject to the environmental guidelines and procedures of the Bureau of Land Management[27] (BLM) and Forest Service (USFS), to various environmental provisions at the state level, and like all other economic activity, to media-specific regulations such as the Federal Clean Air and Water Acts.

9.4.1 REGULATORY INITIATIVES AFFECTING OPERATION AND DECOMMISSIONING OF LEACH DUMPS

The prospect of new federal legislation, particularly the comprehensive and potentially costly programme which Strawman was seen to represent, proved to be a sufficient incentive for those states without mine waste programs (such as Arizona and New Mexico) to introduce state-level regulations in an attempt to head off any Federal regulation. Those states which had already passed mine waste legislation, such as Colorado, South Dakota, and Idaho, have subsequently passed amendments extending the scope and detail of existing legislative frameworks. Table 9.1 illustrates examples of recent state regulatory initiatives that have had a direct impact on the operation and closure of leach facilities. The close relationship between decommissioning and water quality is reflected in the increasing involvement of both mining and environmental regulatory agencies in the closure process. Whereas mine closure was formerly the province of *either* a reclamation-based program (e.g., Colorado, Montana, South Dakota) *or* a water quality program (Arizona, Nevada, New Mexico), recent regulatory changes dovetail reclamation laws with water quality regulations. New Mexico's Mining Act, for example, reserves a prominent role for the Environment Department's Groundwater Bureau, although the Energy, Minerals and Natural Resources Department is its administrator. In Nevada, reclamation and water quality issues are fully integrated under the administration of a single unit within the Division of Environmental Protection.[29] This integration can provide for streamlined and consistent regulatory policy which eases the administrative burden on operators. Not only does it facilitate one-stop permitting and reporting but, by charging a single agency with responsibility for all environmental media, administrative integration can help achieve effective integrated waste management and pollution control.

9.4.2 PERFORMANCE STANDARDS, TECHNOLOGICAL STANDARDS, AND ZERO-DISCHARGE REQUIREMENTS

The operation and decommissioning of leach facilities is currently regulated by the states rather than the federal government. Regulatory authority for the closure and reclamation of heap and dump leach facilities and the protection of groundwater quality is primarily vested in environmental agencies at the state level. The federal government (in the form of the EPA) has some limited authority for groundwater quality (via the Wellhead Protection and Underground Injection Programs of the Safe Drinking Water Act), but most mine-related groundwater issues are overseen by the states. Federal initiatives for mine waste management have been reviewed above.

State programmes vary in the extent to which they have opted for performance standards versus prescriptive technology standards. Reflecting a collective wariness of prescriptive federal regulation, most states have chosen to adopt performance standards which require that water quality contaminant concentrations not be exceeded at designated "points of compliance." Companies are free to propose decommissioning procedures that will be adequate to meet the performance standards and the specific choice of technology or management practice is negotiated as a site-specific agreement between the operating company and the regulatory agency. Arizona's Aquifer Protection Permit, for example, allows companies to discharge to local groundwater as long as water quality standards are met at a point of compliance on the perimeter of the property. Performance standards are increasingly preferred by regulatory agencies on the grounds of both financial efficiency and environmental effectiveness. Performance standards not only provide maximum flexibility to operators over the life of the mine and encourage the search for lower-cost methods of achieving standards of environmental performance, but if properly implemented they can also lead to better environmental practice, since they do not assume that a certain level of environmental performance is being met simply because a particular technology is in place. To be effective, however, performance standards require significant monitoring capacity (to ensure that agreed performance levels are being met) and the technological expertise and resources necessary to evaluate each site on a case-by-case basis.

To facilitate the selection of management techniques which will ensure that performance standards are met, some states define a set of best-management practices for the control of leachate from heaps and dumps. In Arizona, for example, the Aquifer Protection Permit provisions authorise the Director of the Arizona Department of Environmental Quality to prepare Best Available Demonstrated Control Technology (BADCT) guidelines for a number of different industries. BADCT guidelines for mining were released in 1990 (revised in 1995) and include a guidance manual for heap and dump leach operations. Colorado was recently (1998–1999) undertaking review and revision of cyanide guidelines and acid rock drainage guidelines to determine, and encourage the diffusion and replication of, best-management practices while providing sufficient incentive for the innovation of new approaches.

The Arizona guidelines describe a range of optimal pre-approved technologies and practices, but also present an alternative suite of options, which may be used under certain specific conditions. The revised 1995 edition of the manual indicates a retreat from the detailed prescription of 1990, with increased opportunity for firms to propose techniques and demonstrate their effectiveness.[28] New Mexico's Mining Act is similarly based on performance standards, but encourages the application of the "most appropriate technology and best management practices" during operation and reclamation. Since appropriate technology is decided on a case-by-case basis and best-management practices are currently undefined, individual operations are free to propose techniques and practices to meet the general requirement of the Act that "measures must be taken to reduce, to the extent practicable, the formation of acid and other toxic drainage that may otherwise occur following closure to prevent releases that cause federal or state standards to be exceeded."

Prescriptive technology-based standards are more common for cyanidation leaching than for base metals. The state of Nevada, for example, has developed minimum design criteria for leach pads, liners, ponds, and vats associated with gold cyanidation. These establish minimum contaminant control technologies and include a zero-discharge requirement in areas of the state where annual evaporation exceeds annual precipitation. Technology standards for leach pads, for example, specify an engineered liner system that "provides containment equal to or greater than that provided by a synthetic liner placed on top of a prepared sub-base of 12 inches of ...soil," where synthetic liners must meet a technology standard of resistance to leachate throughflow (e.g., equal to a coefficient of permeability of 1×10^{-11} cm sec^{-1}). The application of these minimum design criteria does not, however, exempt a leach operator from compliance with performance-based water quality standards. This blending of performance and design standards in most state regulations is in part a pragmatic response to the lack of a definitive set of proven management practices. It is also reflective of a political compromise between the demands of industry and environmental groups during the rule-making process. Industrial interests (including mining) were keen on solely performance-based standards, leaving them free to meet standards at the point of compliance through the selection of the most appropriate technique, and allowing discharge to groundwater as long as contamination was contained on-site. In contrast, environmental groups advocated zero-discharge to groundwater through the use of rigorous design standards and procedures for heap construction, monitoring during operations, and during decommissioning.

Zero-discharge requirements are often regarded as unworkable legalisms that ignore the fact that leaching is not a closed loop process but produces wastes that require disposal. If appropriately defined as a target objective and limited in application to the operational life of the mine, zero-discharge standards can serve as an effective tool for the protection of groundwater quality and can serve as an integral part of a policy of preventative design and planning. In areas of high-quality groundwater, tightly defined zero-discharge standards for leach pad design, for example, can facilitate pollution prevention at the outset and be a cost-effective alternative to containment-in-perpetuity or remediation after-the-fact.

It must be recognised, however, that zero-discharge from leach pads to the subsurface vadose zone (the soil region situated above the water table) is unlikely to be achieved in practice, as a degree of seepage from leach pads and storage facilities is inevitable, given the volumes of material concerned and the containment technologies currently available. However, in some situations zero-discharge *to groundwater* can be achieved during the operational life of the mine through the installation of secondary containment and pump-back systems, adequate monitoring and contingency planning, and the use of the local buffering and/or attenuation capacity of the vadose zone itself. A requirement to achieve zero-discharge during the post-closure phase and *in perpetuity*, however, is likely to be unattainable. The absence of active water management and maintenance techniques following closure, together with the infinite period of time in the post-closure phase, suggest that in most situations a measure of seepage to groundwater is likely to occur from even the best-planned operation. It is important to recognise, therefore, that technologies and management practices that can make zero-discharge a practicable objective during the relatively short operational life of a mine, are not necessarily "technological fixes" that will endure over the longer term. As a consequence, zero-discharge — or other best-practice objectives — should not be considered a substitute for public debate and collective decision-making over the impacts of mineral extraction on the environment in the longer term.

A policy of zero-discharge during the operational life of the mine may not be applicable in all situations. In areas of relatively poor background groundwater quality and/or where there is no projected future domestic use, or where local conditions provide adequate buffering capacity, a

TABLE 9.2
Examples of Regulatory Requirements to Plan for Decommissioning

Arizona	Aquifer Protection Permitting requires development of a conceptual *"post-closure plan"* which will *"eliminate, to the greatest extent practicable, any reasonable probability of further discharge from the facility and of exceeding aquifer water quality standards at the applicable point of compliance."* Mined Land Reclamation Act requires the submission of a closure and reclamation plan prior to conducting new operations. The plan must include a *"tentative schedule"* of the timing of disturbance and reclamation activities.
Colorado	Mined Land Reclamation Act: *"any operator proposing to engage in a new mining operation must first obtain... a reclamation permit"* which requires the submission of a reclamation plan. Reclamation plans should indicate how reclamation is to proceed in 5-year phases, and should minimise *"disturbances to the prevailing hydrologic balance... and the quality and quantity of water in surface and groundwater systems both during and after the mining operation and during reclamation."*
Montana	Hard Rock Mining Reclamation Act: *"In a reclamation plan accompanying an application for operating permit, the applicant shall provide the board with sufficiently detailed information regarding method(s) of disposal of mining debris...,"* and *"reclamation shall be as concurrent with development or mining operations as feasible."*
New Mexico	Mining Act: all new mines must submit in their permit application a reclamation plan, which includes a description of *"how all waste, waste management units, pits, heaps, pad and any other storage piles will be designed, sited and constructed in a manner that facilitates, to the maximum extent practicable, contemporaneous reclamation and are (sic) consistent with an approved reclamation plan."*
Nevada	Mining Reclamation Act of 1990: all operations must acquire a permit, which is conditional on the submission of a reclamation plan. Specific reclamation activities *"must be conducted simultaneously with the mining operations to the extent practicable..."*

policy of zero-degradation may be applicable. This would allow a minimum discharge of potentially polluting effluent to groundwater, but would prohibit discharges that caused a further deterioration in groundwater quality. Climatic variation, particularly the relationship between rainfall and potential evaporation, is especially important in determining the feasibility of zero-discharge operations since it determines the water balance at the mine. Zero-discharge is most easily implemented in areas where evaporation exceeds rainfall (as codified in the Nevada statutes), although even here seasonal rainfall events can temporarily overwhelm evaporative capacity and promote seepage to groundwater. In these arid or semi-arid regions, excess process water is often recycled as part of mine operations or can be evaporated to leave a solid residue for disposal. In areas where rainfall exceeds potential evaporation, zero-discharge is still possible but may require construction to facilitate water storage during seasonal periods of excess supply, the redesign of tailings pond and impoundments to increase evaporative loss, treatment of contaminated water, and waste disposal.[30]

9.4.3 REGULATORY APPROACHES TO PLANNING FOR CLOSURE

Driven by public concern over the transfer of cleanup costs from operators to the public at a few, high-profile cases (e.g., Summitville), recent regulatory initiatives at the state level indicate a trend towards better planning for closure. Table 9.2 illustrates regulatory language from several states requiring companies to plan the closure of their leach dumps at the permitting phase. These regulations are recognition of the ecological and financial benefits of planning heap decommissioning from the outset of operations. Attempts at closing heaps and dumps that were poorly designed have proved lengthy and resource-intensive, highlighting the economic and environmental gains from building closure into the initial design. Concern over the long-term economic and environmental impacts of unreclaimed gold mining operations in the Black Hills of South Dakota forced a three-year moratorium on new mining operations in 1989. As a result, new legislation was signed in 1992, which strengthened post-closure requirements for care and monitoring and, most significantly, required the provision of a detailed closure plan at the initial application stage.

The specific experiences of the EPA and the State of Colorado with an abandoned open-pit gold mine and heap leach facility at Summitville in the San Juan Mountains of southern Colorado has highlighted the high financial costs and potential for significant ecological degradation which can result from inadequate mine planning and a failure to incorporate closure considerations into the initial design phase. A combination of insufficient water management capacity, poorly installed heap leach facilities, and a lack of contingency planning culminated in the abandonment of the site in 1992 at a time when rising water levels in the heap leach area were rapidly increasing the risk of a sudden and extensive pollution event. Since December 1992, when the EPA became involved with the site, reclamation and rehabilitation costs for detoxifying and containing leach wastes, neutralising acidic drainage, removing and encapsulating waste piles, and re-grading the site to engineer a net negative water balance around the mine are estimated to reach between $150 and $200 million.[51,52] The experience of Summitville has driven new regulation to promote increased oversight and closure planning of mining operations in general (and heap and dump leach facilities in particular) not only in Colorado (the Mined Land Reclamation Act of 1993 was a response to the perceived regulatory failure at Summitville), but also in other states, the EPA, and in other countries. For example, the experience of Summitville was influential in the debate over the design of a mine reclamation law in New Mexico during public hearings prior to the passage of the New Mexico Mining Act in 1993.

Regulatory requirements to build closure plans into the design and permitting phases of a mining operation are supported by incentives to promote concurrent reclamation, such as the provision for phased bond release, enabling companies to recover some of their capital as they complete reclamation work. For example, Colorado's Mined Land Reclamation Act encourages concurrent reclamation by requiring lower reclamation bonds from operators if less land is disturbed at any one time. Concurrent reclamation provides an opportunity to reduce long-term liability, bonding requirements, and other costs and strengthens the permitting position of the mine site. The San Luis heap leach facility operated by Battle Mountain Gold in southern Colorado has been conducting concurrent reclamation of its open pits by backfilling the initial pit with overburden and waste excavated during later production. Not only has this enabled the operation to minimise the environmental footprint of the mine both spatially and temporally, but it has also reduced the total bonding costs for the site and may ease future permitting actions for other sites (such as the company's proposed Crown Jewel Project in Washington state) by demonstrating a best-practice approach.[31] The Florida Canyon Mine in northwest Nevada began concurrent reclamation of its 242-acre heap leach in 1993, and estimates that it has saved "hundreds of thousands" of dollars on total reclamation costs by expensing them while the operation is generating revenue.[32]

The objective of planning for closure is also being met through regulations that require a full characterisation of the mine site as a condition of permitting. Many states now require the submission of closure plans as part of the permit application for all new mines and are gradually addressing suitable closure options for existing mines. In Colorado, for example, permitting requires the prospective operator to undertake a full characterisation of the site and its waste products, involving a baseline study of mineralogy, acid-base accounting, and an assessment of the long-term weathering potential of materials through the use of humidity cell tests, or other appropriate static and dynamic testing procedures. Over the past two years the EPA (via the Hardrock Mining Workgroup) has been developing a new regulatory approach to environmental management at mines, which stresses the significance of detailed preventative planning at the permit stage. A principal objective is to address the Achilles heel of current closure planning by ensuring that proposed closure and remediation techniques are not only technically feasible, but also economically viable. The case of Summitville illustrated very clearly how the traditional approach to regulation not only made fairly heroic assumptions about the performance of environmental control technologies (e.g., assumptions that liners are 100% effective or that acid mine drainage can be controlled and contained) but also that the remedial actions proposed in the original permit to deal with an unforeseen pollution event were not always economically viable in practice. Thus, when a problem arose, there were insufficient funds available to fix it, and an operator could cite economic hardship in order to avoid meeting remedial action requirements. Recognising that remedying problems after the fact is expensive and often inherently risky, the EPA is now moving towards an approach which evaluates the technologies and practices of closure and remediation on a case-by-case basis at the permitting stage, i.e., during the development of an Environmental Impact Statement, as required by the National Environmental Policy Act of 1969 (NEPA). If the proposed closure actions are not demonstrated to be economically viable (even though they may be technically feasible) the operator must reengineer the process design or mine plan to ensure that the proposed closure procedures can be undertaken with the financial resources available. In essence, this approach attempts to reduce the risk of uncontrolled environmental contamination during operation and on closure by building viable prevention strategies into the design stage.

In practice, this approach involves quantifying the performance standard to take account of site-specific conditions, and then engineering a technological design to meet this standard. The prospective operator has the responsibility for developing a process able to meet this quantified performance standard, and for demonstrating the economic viability of the proposed closure and remedial actions. By requiring closure procedures to be viable, this approach is leading some operators at new mines to actively reengineer the production process and reassess the proposed operating plan by, for example, altering mill and mine design, incorporating post-mining land uses into the planning stage, and increasing the extent of backfill. EPA's approach contrasts with that of other federal agencies such as the Bureau of Land Management and the Forest Service which in general continue to focus on technological approaches to single environmental issues such as, for example, cyanide at gold leach operations.[27] The more integrated and flexible approach to pollution control at mining operations developed by the EPA addresses a broader range of target issues such as metals mobility and nutrient and salt loadings, and attempts to capture the highly site-specific nature of environmental issues at mine sites. It acknowledges that the suite of target issues will not be the same for all sites, but will vary with the specific physical and technological conditions encountered in each mining project.

9.4.4 EMERGING BEST-PRACTICE TECHNIQUES FOR LEACH DUMP DECOMMISSIONING

Leach dump decommissioning often occurs as part of a broader process of mine closure, requiring the removal of infrastructure, grading, contouring and revegetation, and the establishment of a post-mining land use. Since these issues are dealt with at length in other chapters, the focus here is on the protection of water quality during the closure of leaching facilities. The decommissioning of leach facilities is an evolving technical and regulatory frontier as companies, regulators, and affected communities attempt to find ways of containing and controlling effluent from active and abandoned heaps and dumps. A number of gold leach operations have been successfully closed in the U.S., but as yet no large copper leach operations have been fully decommissioned. The state-of-the-art, therefore, reflects current regulatory thinking rather than proven technologies or management practices. The aim of decommissioning is to reduce the potential of the facility to impact water quality into the future. This typically involves action in four areas: neutralisation of spent ore and waste residues; elimination of free liquids; geotechnical, erosional, and chemical stabilisation of heap and dump materials; and recontouring/covering to reduce erosion. There is no generally accepted means of achieving these goals and in practice a range of techniques have been applied with differing degrees of success. Site-specific conditions play a key role in deter-mining both the selection of the technique (or combination of techniques) and the extent of its success. Two broad categories of treatment techniques can be identified: those that seek to reduce pollution by actively treating potential contaminants, and those that seek to reduce the potential for contamination of the wider environment by controlling the migration of potential pollutants and isolating the source.

9.5 POLLUTION TREATMENT TECHNIQUES

9.5.1 RINSING AND NEUTRALISATION

Rinsing can be used to purge reagents and other potential contaminants from heaps prior to their closure. The aim is to detoxify the heaps by removing potential contaminants from the heap through flushing with multiple pore volumes of freshwater and/or the application of neutralising agents. The contaminant-rich effluent coming off the heaps is then treated, and heap material can either be left on-site or removed. Rinsing is prescribed as a decommissioning technique by several states and it has been used successfully to fulfil closure requirements at over 25 gold heap leach operations.[33] Rinsing can take anywhere from a few months to several years, with speed and success heavily dependent on-site-specific factors. Schafer and Van Zyl[34] report that, based on a study of heap decommissioning at the Zortman Landusky gold heap leach operation in Montana, up to 80% of cyanide can be removed in a 10-day freshwater rinse. Rinsing of the heap facility at Summitville, however, has involved the circulation of over 100 pore-volumes of water over a 3-year period to achieve a reduction of around 90% total cyanide. Despite the wide variation in the time taken to satisfactorily rinse a heap, rinsing is still significantly quicker than the process of unaided degradation (primarily via volatilisation and photodegradation) following passive abandonment which, according to Schafer and Van Zyl,[34] can take between 6 to 10 years.

Although freshwater rinsing is widely practised, rinsing is more commonly undertaken by recirculating untreated process solutions and/or neutralising agents through the heap. A number of techniques are available to detoxify residual cyanide, including peroxidation, chlorination, acidification, and biological treatment (see Table 9.3). A small, abandoned heap leach facility in Toole County, Utah, for example, was decommissioned by saturating the heap with a solution of hydrogen peroxide, then bulldozing, contouring, and hydromulching the heap to encourage vegetative cover.[53] A similar process has been used successfully for the reclamation of the Annie Creek Mine at Lead, South Dakota, and is currently being used at the Brewer and Barite Hill Mines in South Carolina. In addition to its application to cyanide facilities, rinsing has also been proposed for the closure of a number of copper oxide heaps. Outline decommissioning procedures for copper oxide leach heaps at the Sanchez Copper Project in Arizona, for example, involve rinsing the heaps with freshwater or neutralised raffinate using the existing irrigation system (with an option to undertake additional liming if necessary) until the effluent demonstrates a consistent pH of around 6. Proposals for decommissioning at San Manuel, Arizona involve rinsing the heaps with "natural meteoric events" over a number of years, and sending the effluent to the SX-EW plant for metals recovery.[35] Rinsing (or "sweeping") has also been used for the decommissioning of *in situ* uranium leaching operations.

Rinsing has been successful at several facilities, but it is not, however, a panacea. The process is often unable to access the total volume of materials in the heap due to the presence of preferential pathways and channels, and it can leave large areas of the heap retaining potential contaminants. Seasonal or altitudinal low temperatures can also reduce the effectiveness of the rinsing process. Incomplete rinsing of the leach material due to winter freezing at the Ortiz Gold Mine in New Mexico, for example, led over several years to the development of a groundwater contaminant plume as the result of the release of cyanide from tailings during the seasonal spring thaw. More fundamentally, many ores — most notably sulphides — have the potential to continue to generate contaminants even after rinsing. Although rinsing may be suitable for copper oxide ores, it is not possible to rid a copper sulphide stockpile of its potential for acid mine drainage simply by rinsing it, since the sulphides are an inherent part of the ore and will continue to oxidise and leach naturally long after rinsing has stopped. For non-rinsable ores and wastes, an encapsulation procedure and/or

TABLE 9.3
Neutralisation Procedures at Gold Heap Leach Operations[33]

Operation	Heap Size (million tons)	Method of Detoxification	Comments
American Girl (California)	2.2	Rinsed with process solution and freshwater.	Reduced free cyanide levels from 710 mg L^{-1} to less than 1 mg L^{-1} over 22 months. Closure approved.
Carson Hill (California)	3.6	Freshwater rinsing with carbon absorption to scavenge metals, and ultraviolet degradation of cyanide.	WAD cyanide levels in effluent reduced from 10 mg L^{-1} to 0.5 mg L^{-1}. Closure approved after 10 months.
Morning Star (California)	1.4	Rinsing with process solution and freshwater, carbon absorption, followed by addition of hypochlorite.	Closure approved after 48 months. Calcium and sodium hypochlorite added to oxidise cyanide.
Yellow Pine (Idaho)	1.5	Rinsing with process solution, biological detoxification, and freshwater rinsing.	99.4% reduction in heap cyanide after 7 months of biological treatment, freshwater rinsing was designed to reduce pH of effluent.
Landusky (Montana)	N/A	Natural degradation and limited active rinsing.	Idling of the heap over a decade reduced cyanide levels to within regulatory levels.
Golden Eagle (Nevada)	0.4	Freshwater rinsing.	Closure approved, with a variance allowed for high mercury levels.
Illipah (Nevada)	1.2	Alternating rinse-rest cycles using process solution and freshwater.	Closure approved, with variance for high WAD cyanide and metal concentrations.
Brewer (South Carolina)	3.8	Rinsed with process solution then peroxide neutralisation.	Closure ongoing.

active treatment may be necessary, although preventative approaches such as selective mining, waste segregation, and buffering acidic wastes by blending with alkaline wastes (where available) may be more effective strategies at new mines.

9.5.2 Active Treatment

A range of techniques is currently available for the treatment of contaminated effluent, such as that emanating from heaps and dumps. Much of the research in this area has been conducted in the context of acid mine drainage caused by the oxidation of sulphides. The addition of neutralising additives (e.g., limestone, lime, sodium hydroxide) to acid drainage at coal mines has proved an effective buffering technique, although to maintain neutral drainage additives must be added repeatedly, representing a long-term commitment on the part of the operator. The construction of artificial wetlands to remove sulphates and metals and neutralise acid drainage has also been successful on a small scale, but constructed wetlands are used more as a polishing process than a primary treatment technique. Work has been conducted on metal-tolerant plants that are able to accumulate and fix significant amounts of metals from contaminated drainage waters. Recent research indicates that the sequestering of heavy metals by plants, soils, and fungi can be highly effective in removing aquatic contaminants such as copper and zinc.[36,37] If plant communities are to continue to be effective in removing and fixing contaminants, however, active ecological management may be required.

A number of biochemical processes have been proposed as a treatment for acid mine drainage. The application of bactericides to sulphide ores can slow the rate of acid generation by creating an environment hostile to the reproduction of oxidising bacteria. Benzoates, sorbates, or anionic surfactants can effectively destroy the conditions for bacterial growth but can only reduce rather than eliminate acidic drainage. They are therefore most effective if used in conjunction with other control techniques such as capping and covering.[38]

Bacterial processes have also been used with success as part of the neutralisation of cyanide facilities. The biochemical decomposition of cyanide is based on the use of cyanide as the carbon and/or nitrogen source for the metabolic processes of native bacterial populations and consists of two oxidation steps. In the first step, cyanide and thiocyanate are oxidised to produce carbonate, ammonia, and sulphate, and in the second step the ammonia is then nitrified to produce nitrates. Metals liberated by the oxidation of the metal-cyanide complexes can be precipitated out. Decommissioning of the Copperstone gold heap leach in Arizona and Yellow Pine in Idaho, for example, was achieved by neutralising residual cyanide with non-pathogenic bacterial solutions. The pregnant leach solution pond was converted into a bioreactor in which bacterial populations were incubated prior to their application to the heap. Bore and rinsate samples verified that detoxification was effective, and since the heap was deemed to have no potential for acid generation, it was regraded to promote long-term stability, and seeded and planted in appropriate areas. Neutralisation of the 1.3 million-ton heap at Yellow Pine was successfully achieved in less than 6 months, in contrast to the two years estimated for chemical treatments, and the site was awarded the Pacific Northwest Pollution Control Association's Industrial Pollution Control Award for its innovative approach to prevention of water contamination. Once successfully neutralised, pea-sized gravel material from the heap at Copperstone was used as a capping for the remediation of tailings from a vat leach operation on the site.[39–41] Bioremediation is now a proven technique for attaining closure status at gold cyanidation heaps and is seeing increasing application in the U.S. For example, Pintail Systems of Aurora, Colorado has demonstrated the commercial viability of bacterial decomposition of cyanide at heap leach facilities and was involved in the Yellow Pine, Copperstone, and Summitville projects described above. Pintail reports that not only can bioremediation reduce the cost of detoxification, but that in some instances (e.g., Yellow Pine) it can even be a net generator of income by increasing gold recoveries towards the end of productive operations.[42]

The EPA also experimented with biological oxidation for the detoxification of the heap leach facility at Summitville. Although rinsing and hydrogen peroxide treatment were ultimately selected because of the need for a proven, low-risk approach, the results of bioremediation trials indicated its ability to rapidly and cost-effectively reduce total cyanide concentrations to well within statutory limits.[43] Bacterial processes are also being explored for application in passive treatment systems for the long-term treatment of draindown at heap leach facilities.[44] This "biopass" system is similar to the passive wetland systems developed for the treatment of acid mine drainage and consists of an emptied and lined solution pond containing organic matter which is connected to the leachate collection systems of the heap. Following the cessation of production, the heap is stabilised and vegetated to reduce percolation, and drain-down solutions are collected at the base of the heap and fed to the anaerobic reactor. Residual cyanide in the leachate is degraded by bacterial action, residual nitrates and sulphates are reduced, and precipitated metals are fixed within the organic layer. While this system is largely unproven at the commercial scale, laboratory evaluations have demonstrated considerable potential. The technique provides an opportunity to avoid handling the large volumes of solution generated by rinsing, and to reduce the time taken to achieve effluent standards for closure.

9.5.3 WASTE SEGREGATION AND SITING CRITERIA

The problems and liabilities associated with acid-generating material in heaps and dumps have led several operators to explore preventative strategies. These include reviewing environmental parameters in the initial site selection; the avoidance of unnecessary disturbance to acid-generating material to prevent the ingress of water and oxygen; and the selective mining and blending of waste piles. The initial location of heaps, dumps, and waste units can be one of the most simple yet most effective factors in enhancing the successful long-term protection of groundwater quality. Unlike many manufacturing activities which produce relatively small volumes of waste that can be economically transported to specific disposal facilities, mining produces large volumes of waste that cannot be cost-effectively transported over long distances. This means that the location of waste units is constrained to sites proximate to the site of extraction. Within these constraints, however, planning the siting and design of a leach facility is perhaps the foremost opportunity to promote pollution prevention since the stability, geochemistry, and hydrology of the area can be selected to maximise natural protective and assimilative capacity, and minimise the need for excessive engineering intervention. This is particularly the case with leach heaps and dumps, which at many sites have historically been sited in creek beds or in close contact with alluvial systems. At the McLaughlin Mine in California, engineers screened over 30 potential sites for a tailings disposal facility, rejecting those with technical problems and environmental sensitivities. The selected site exhibited natural protective features, such as high groundwater in the surrounding upland ridges, providing a barrier to lateral migration and adequate downstream space to collect seepage, monitor groundwater, and implement remedial containment if necessary.[45]

On-site tests can determine the acid-potential of rock prior to it being mined and plans for heap and dump preparation can be prepared to minimise acid generation. The development of an acid-base accounting procedure for the mine facilitates the blending of acid-consuming carbonate wastes with acidic wastes to provide effective buffering. The Hudson Bay Mining and Smelting Company, for example, has adopted a policy at its copper and zinc mines of adding a third category to the usual divisions of ore and waste rock, by subdividing waste into stable-waste and acid-generating waste. According to their environmental director, "the additional money spent during design, development, and operations on a proper waste rock programme can mean economic returns many times greater at the time of closure."[46] The new Carlota heap leach development in Arizona has developed a continuous sampling programme for waste rock, enabling the blending of acid-generating material with wastes possessing excess buffering capacity. In addition, operators are proposing to segregate oxide and sulphides in their heap leach. Although undertaken primarily to enhance operating efficiency, segregation will enable the selective use of the most effective decommissioning technique. Oxide sections can be decommissioned by rinsing and land application disposal, while sulphides can be capped in place. Although law does not mandate selective waste separation, legislation in several states requires a characterisation of mine waste as a condition of permitting. For example, California's Water Quality Act requires mining waste to be classified according to, amongst other things, the waste's potential to generate acid drainage and establishes different engineering standards for the disposal of wastes in each category.

9.5.4 POLLUTION PREVENTION THROUGH ISOLATION

Key to the prevention of contamination is the isolation of the leaching process and solutions from the wider environment through process controls and the use of barrier technologies. At the Carlota copper project in Arizona, for example, the decommissioning plan for the heap leach facility is to reclaim the leach area so that it functions as an isolated unit with minimal connection to the external environment. Since the Carlota project has been strongly contested and has undergone several revisions in response to concerns over potential environmental impacts, it is arguably a very good example of current best-practice technological and planning approaches towards environmental management at heap leach operations. The Environmental Impact Statement for the heap leach pads summarises the policy of isolation as follows: "the reclaimed leach pad is designed to function as an isolated unit. The pad would be sealed from below by a synthetic liner. The pad closure is designed to promote rainfall runoff by heap surface soil sealing, and to use evapotranspiration from the vegetation to prevent the remaining water from entering the heap."

Widely used techniques include surface water management, engineered barrier systems (such as liners) and the manipulation of hydrological conditions. The management of surface water conditions through site selection and engineering represents a "low-tech" yet very effective means of reducing the potential for contamination through the initial design process. A number of new mines have addressed the long-term pollution potential of heaps, dumps, and tailings ponds by designing them to achieve a post-mine water balance which maximises the opportunity for passive water control and minimises the formation of acidic drainage. Designing for positive drainage on closure can reduce water management problems such as those experienced at Summitville in Colorado, where excess water flooded both the heap leach facility and Cropsy Waste Pile, and surface topography accelerated the formation of acidic drainage by channelling runoff directly into the pit. Remedial action at Summitville has involved extensive regrading of the surface topography and rechannelling creeks to divert runoff from the site and minimise the contact with leach dumps. Several new heap leach operations in the U.S. illustrate this proactive approach to water management. The Sanchez copper project in Arizona, for example, has avoided the need for permitting under Section 402 of the Clean Water Act (the National Permit Discharge Elimination System) by diverting surface water around the site to prevent run-on, and operating the mine as a zero-discharge facility by controlling waters inside the mine boundary through a central collection system designed to accommodate a 100-year flood event. The Lisbon Valley copper project which is currently undergoing permitting in Utah proposes to operate a 266-acre heap leach as a zero-discharge facility by diverting run-on and designing the facility to contain both process water and direct precipitation. The closure plan includes provisions to decommission the leach heap by capping and revegetating it in place, and grading the top and benches to establish positive drainage.

9.5.4.1   Engineered barrier systems

Barrier systems are designed to prevent regional groundwater contamination by isolating leaching processes from the wider environment. The most commonly used barrier devices are liners, the design of which is a rapidly evolving field. Liners reduce the permeability at the base of the leach heap and consist of materials selected for low hydraulic conductivity and leakage control, such as compacted earth, bentonite, high-density polyethylene (HDPE), polyvinylchloride (PVC), or a combination of these. Liners are a regulatory requirement for precious metal heap operations and at most base metal heap facilities, although several states do not require that synthetic, or multiple-layer "liner systems" be used. Under Arizona law, for example, mine operators need not use liners if they are able to demonstrate that natural topographic and geomorphological conditions are such that they make lining unnecessary. State-of-the-art liner systems comprise several layers of liner material and incorporate leak detection and recovery devices. The Sanchez copper project in Arizona, for example, was initially approved for permitting with a compacted clay liner but, following concerns over groundwater protection, permitting was subsequently made conditional on the use of a synthetic liner with an additional clay layer and supported by a comprehensive network of vadose zone monitoring wells. Oregon's new standards for heap leaching, adopted in 1994, require a three-foot thick layer of stable, low permeability soil/clay to be covered with a layer of 80-mm thick synthetic sheeting. This is followed by a layer of high permeability material (sand) in which are embedded leak detection devices such as pipes, which channel leakage to the side of the heap where it can be collected and monitored. The sand layer is overlain by a further synthetic liner, on top of which a cushion of fine material (usually crushed ore) is placed as a base for the ore heap.[47] The use of a bentonite clay layer beneath the synthetic layer provides a self-sealing mechanism should the synthetic layer become torn. Contact with percolating solutions causes the clay to swell, plugging the leak. In addition, clays can attenuate materials and facilitate the removal of contaminants.

Field experience with liners demonstrates that all liners leak to some degree, and that the primary function of liners is not to prevent discharge, but to significantly reduce the rate at which discharge occurs. Synthetic liners are fragile, their installation relatively exacting, and they are susceptible to ultraviolet photodegradation over time. Liner failure at a gold leach operation near

Elk City, Idaho led to extensive cyanide contamination of domestic water supplies.[48] The Summitville gold heap leach operation in the San Juan Mountains of Colorado experienced liner failure soon after the application of leach solution in 1986, due to hasty installation on frozen ground. The failure of the leach pad liners has contributed significantly to the extensive acidic and metal-containing drainage experienced at Summitville, and to its listing in 1994 as a Superfund site. To counter the difficulties with conventional liners, a number of innovations in liner technology have been developed in recent years. The Robertson Barrier Liner system, for example, consists of a sandwich of two synthetic liners, sealed into a series of cells, each of which has a partial vacuum drawn across it. Advantages include easy leak detection by monitoring any losses of vacuum pressure, and the ability to test the system prior to the addition of material to the heap. In addition, the cellular construction enables the isolation of leaks, speeding up any remedial work.

The use of liners to isolate leaching processes from the environment is not possible for leach dumps, or for those heaps that began operating prior to regulations requiring liners. The size of these facilities makes it prohibitively expensive to retrofit liners or other physical barriers: dumps at Chino, New Mexico, for example, cover 6400 acres and contain over 4360 million tonnes of rock. The use of dumps as feedstocks for leaching has proved especially problematic for groundwater protection and for instituting an adequate closure program. By their very nature, dumps were not designed as leach facilities but as waste piles and very few have any sort of engineered liner. Many predate concerns over environmental impacts of mining operations and the locational decision were made primarily to minimise haul distances. Many are located in valley washes or on flood plains with a high potential to leach contaminants to groundwater. Since physical barrier techniques are not suitable control techniques, most dump leaching operations rely on pit dewatering (and/or localised pumping) to maintain favourable hydraulic gradients in order to prevent contaminants from migrating offsite (see below). In this respect, BADCT for dump leaching are very similar to those for *in situ* mining, which revolve around control of the leach solution through groundwater modelling and pilot-scale testing, controlled recovery by overpumping, and aquifer restoration.

9.5.4.2 Encapsulation

Effective isolation of the heap or dump from the surrounding environment requires that surface percolation and groundwater infiltration to the heap be controlled. By controlling the rate at which groundwater, surface water, rainwater, and snowmelt enter the heap, the rate of leachate discharge and acid generation can be significantly reduced. Berms and riprapped diversion channels to control stormwater run-on to heaps and dumps are simple yet very effective means of reducing the volume of effluent discharged from the heap, and once established require minimal maintenance. Kennecott has used check dams at its Bingham Canyon operation to capture runoff prior to its entering the mine and divert it around waste dumps and ore piles. An 11-mile canal rings the mine area collecting runoff and leachate seepage from the mine area and dumps, and diverting it to the company's reservoir.

Several facilities use low permeability interfaces (clays or synthetic cappings) to isolate acid-generating materials from infiltration. Since acid generation requires the presence of sulphides, air, · and moisture, acid generation can be reduced by restricting the availability of any one of these inputs. The decommissioning of the abandoned heap facility at Summitville, for example, involves reducing water access to the heap material to a minimum by grading the heaps to achieve a 4H:1V slope, covering the top and sides of the heap with a geosynthetic clay liner (two layers of geotextile with a layer of bentonite clay between), and doming the heap surface to facilitate runoff. The objective is to keep the localised water table within the heap relatively static over time by effectively isolating the heap from run-on and runoff. Subaqueous storage of sulphide ores through flooding of open pits or underground workings has proved successful at some locations. Sulphide heaps, which are not readily detoxified for land application disposal, can alternatively be bulldozed and backfilled in old open pit workings where flooding can reduce the rate of acid formation to within statutory levels. Battle Mountain Gold's San Luis mine in Colorado, for example, is undertaking closure activities by backfilling open pits with overburden and waste and allowing the water table to rise to within 30 feet of the surface. Together with the natural buffering capacity provided by the mineralogy of the host rock, this practice of submersion will reduce the potential for the formation of acidic drainage. A layer of locally available carbonaceous, alkaline material (which had been segregated and stockpiled during mine development) is then placed on top of the backfilled waste to provide additional buffering capacity above the water table.[31] Where flooding is not an option, a favoured technique is to cap heap and dump material in place to reduce the availability of moisture. The new Carlota heap leach project in Arizona is proposing to decommission some of its heaps, using a cap and cover procedure. Rinsing is infeasible for the sulphide heaps and capping will reduce the rate of acid formation by impeding the entry of water. In the absence of subaqueous disposal options, this method of isolation can be the most effective tool for the decommissioning of sulphide heaps that continue to bioleach over a long period of time. Decommissioning of Barrick Goldstrike's leach facility in Nevada used a rinsing process followed by the capping of the heap with waste rock, the surface of which was then graded to reduce the infiltration of rainwater.[38] Closure plans for the extensive sulphide dumps at Chino, New Mexico combine a capping technique with the placing of a "shell" of non-acid generating material around the perimeter of the leach units to improve the quality of stormwater runoff from the heaps.

9.5.4.3   Hydraulic gradients

The reality of dump leaching is that in the absence of a liner, groundwater in the vicinity of the dump will often be contaminated as a result of leaching activities. Many state regulations recognise this fact by not specifying zero-discharge requirements at dump or existing heap operations. The Arizona Aquifer Protection Permit, for example, does not preclude contamination of local groundwater through discharge from leaching operations, but aims to prevent more widespread contamination through the requirement that aquifer quality standards be met at designated points of compliance. One of the most widely used methods of controlling the migration of contaminated groundwater is the manipulation of local hydraulic gradients. For *in situ* leaching, hydrological control is often the only management technique. The hydrosink principle is acknowledged by several states as a possible long-term water management strategy and many mines utilise the hydrological cone of depression caused by pit dewatering as a contamination control technique. The Arizona Aquifer Protection Permit application for Magma's (now BHP's) San Manuel facility, for instance, states that "as long as Magma maintains this cone of depression, it is hydrologically impossible for ambient groundwater outside the mine to become impacted by mine operations."[35] In addition to the hydraulic sink created by the pit, pumped wells can be used to control the direction of groundwater flow. Pumping creates a localised cone of depression in the groundwater table, so that contaminants in the vicinity of the pump well flow towards the well. Interceptor wells are used at Chino, New Mexico to prevent migration of contaminants into localised, perched aquifers, and at Tyrone, New Mexico over 350 wells were drilled as part of a remediation effort to contain regional aquifer contamination from a newly commissioned leach dump.

Once productive operations come to an end, however, the costs of continuing to pump water are prohibitive. Without the cone of depression, groundwater flow resumes its natural hydrological pattern and potential contaminants can impact local and regional aquifers. Although groundwater contamination may not pose much of an immediate threat to water supplies off-site during the life of the mine, long-term reversals of the hydraulic gradient on closure may make containment problematic. Recognising the uncertain financial obligations of continued pumping, regulatory agencies are requiring new facilities to minimise their reliance on active intervention, and are encouraging existing facilities to progressively reduce dependency on active methods where possible. New Mexico's Mining Act, for example, states that new mines "will be designed to meet *without perpetual care* all applicable environmental requirements of the Act, these Rules and other laws following closure" (emphasis added), while the 1995 draft Closeout Guidelines require that acid rock drainage controls "minimise the need for post-closure technologies such as chemical treatment of leachate."

In semi-arid areas, high rates of evaporative loss relative to the rate of groundwater recharge may render the pit an effective hydraulic sink without the aid of pumping. In such situations, the pit provides a passive containment mechanism that will prevent the off-site migration of contaminants even after closure. Many mining operations in Arizona are currently looking to passive containment as a decommissioning technique following its listing in the Aquifer Protection Permit statute as an acceptable control technology. Arizona defines passive containment as "natural or engineered topographical, geological, or hydrological control measures that can operate *without maintenance*" (emphasis added). The success of passive containment, however, is highly dependent on local climatological and hydrological conditions, and there is evidence to suggest that it will not be applicable at all those operations that propose to use it. In Arizona, for example, there may be only one mine site in the state (Silver Bell) where the hydrosink principle can be used in perpetuity. Silver Bell is located on a catchment divide and therefore has very little inflow and, unlike many other sites, it is not tied into alluvial deposits. Preliminary indications are that most of the state's other copper properties will not qualify.[40] Several of the operations currently proposing to use passive containment were inundated by flash floods during 1992 and 1993 (indicating that rates of recharge can massively exceed evaporative loss, at least in the short term) leading to deterioration in groundwater quality as a result of reversals in the hydraulic gradient.

9.6 TRANSFERRING BEST-PRACTICE: CONSTRAINTS AND OPPORTUNITIES

Perhaps the most significant obstacle to the transfer of best practice is the fact that best practice is not reducible to a set of techniques and technologies that can be implemented *carte blanche*. Rather, best practice consists of the technical and managerial skills necessary to optimise and assure the maintenance of high-quality environmental performance. The effective transfer of best-practice environmental management for leach dump decommissioning therefore requires the acquisition of the capacity to effect and sustain the technological and managerial processes that promote high environmental quality throughout the project life cycle. Environmental management systems can be an effective tool in the pursuit of best practice, since they provide the opportunity for an iterative learning process which incorporates all levels of the management hierarchy, combining the observations and insights of those charged with the day-to-day implementation of environmental protection with the decision-making capacity of management. Joint ventures and other strategic partnerships between established leaders in environmental management and firms in emerging markets could serve as a conduit for this process of skill acquisition from one firm to another. The recent influx of foreign mineral investment to developing countries, associated with the liberalisation of investment regimes for mining around the world since 1989, has the potential to improve the environmental management practices of developing country mineral producers. The process of information transfer is two-way: local firms provide detailed knowledge and experience of local environmental and political conditions, while the established multinational can provide the capital and managerial resources for effective environmental management.[49]

There are a number of other constraints on the transfer of best practice. Recent regulatory initiatives indicate a move towards flexibility for site-specificity and an effort to avoid the costly rigidities of a blanket, one-size-fits-all approach such as that enshrined in the Surface Mining Control and Reclamation Act of 1977 (SMCRA), the federal regulation of coal mine reclamation. While this has the potential to improve environmental protection and reduce the costs of compliance by enabling firms to select the most appropriate environmental control technique, management practice, or post-closure use for the area, the introduction of site-specific elements requires that agency staff be highly qualified and have sufficient time and resources at their disposal to provide adequate review. Where these educational and financial resources are not available, applications for permitting that depart from prescriptive standards and propose site-specific procedures will not receive adequate review. To be effective, therefore, regulatory policy must reflect local institutional capacity. Where resources are limited, design standards for some of the more standardised elements of leaching facilities (such as liners and containment systems) may have some merit.

Many best-practice techniques rely on planning closure at the initial permitting stage and building preventative features into the design in an effort to minimise the final costs of decommissioning. Although planning for closure may be the most effective means of reducing the potential for contamination at new heap leach facilities, many existing dump leach operations do not have that option. The large size of most leaching operations precludes their resiting or the retrofitting of liners. At existing operations the range of decommissioning strategies is limited by pre-existing design. Where groundwater contamination has already occurred, closure procedures may have to rely on active control technologies such as dewatering or remediation through biotreatment. New regulations recognise this, and make a distinction between existing and new mines. Requirements for new operations are typically more stringent and focus on front-end pollution prevention rather than pollution control. Attempts to transfer best practice techniques and successful regulatory initiatives need to recognise that existing and new facilities pose different sets of issues and that different standards and procedures for decommissioning will often be appropriate.

9.7 CONCLUSIONS AND AREAS FOR FURTHER RESEARCH

Innovations in mining and mineral processing techniques making it possible to produce high quality output at relatively low cost from low-grade ores promise to ensure that the heap and dump leaching of copper, gold, and other metal ores will continue to be a vitally important part of U.S. metals production. The demonstrated success of heap and dump leaching will continue to encourage its widespread application at new operations outside of the U.S. It is equally certain that regulatory oversight of leach operations can be expected to continue and increase as public and commercial interest in the preservation of environmental quality in mining regions grows. Although interest in large-scale commercial leaching in the U.S. occurred initially in the absence of legislation protecting groundwater, state regulatory agencies have moved to establish requirements for the protection of groundwater quality and the closure and reclamation of mining projects, including heap and dump leach facilities.

In this climate of increased regulatory and public scrutiny of mining projects, operators seeking to reduce their exposure to financial risk and long-term liability are increasingly realising the merits of planning for closure by preventing groundwater contamination through initial design and operational practices. States have tended to eschew prescriptive technology-based standards and have opted for water quality performance standards augmented by technical guidance documents which identify a range of possible best-management practices. The variation in standards from state to state can be expected to decrease over time as dynamic operators demonstrate success with particular techniques in one jurisdiction and seek to replicate it in others. Although design and performance standards still illustrate variation, the diffusion and gradual homogenisation of best practice amongst the states is beginning to occur. For example, demonstrable success with compound liners on heap leach facilities has led several states to require such liners for leaching facilities with monitoring and secondary containment systems. If current public concerns for environmental protection continue, this levelling of the regulatory playing field between the states is likely to take place in one direction only: as innovative mining firms demonstrate that high environmental performance can be attained as part of a cost-competitive and highly productive operation, requirements for environmental protection will be standardised up rather than down.

Few regulatory agencies prescribe specific management practices for the closure of leach facilities, reflecting the relatively short collective experience of industry and regulators with leach dump decommissioning. There is, however, increasing interest on the part of the regulators and the regulated in planning closure from the outset. Experience with poorly designed facilities, complex remedial actions, and expensive lawsuits has demonstrated that the financial efficiency and ecological impacts of closure can be significantly improved if decommissioning requirements are made integral to the initial design process and operation of the facility. The promulgation of zero-discharge requirements for precious metal heap leaching in several states is part of this trend towards initial prevention rather than containment-in-perpetuity and/or remediation. Specific preventative techniques, however, will not be common to all circumstances. The different challenges posed by new heaps and existing dumps, the varying institutional capacities of regulatory agencies in different states, and the limited range of conditions under which zero-discharge requirements are appropriate demonstrate the need for local flexibility in the approach to groundwater protection. The selection of technique should consider local mineralogical and geohydrological conditions, existing groundwater quality, and the current and projected future uses of groundwater at the site.

The decommissioning of leach dumps is an evolving technological, managerial, and regulatory frontier and as such there are many possibilities for further research. Over the next few years, a number of large dumps will be decommissioned, providing an opportunity to conduct a comparative study of the long-term success and relative costs of different closure techniques. The limitations of passive containment — currently proposed as a decommissioning technique at several large copper properties — as a long-term, permanent method of dealing with contamination need to be more fully investigated. There is also the need for risk-based research on the link between groundwater degradation, ecological impacts, and public health, so that the regulatory policies adopted to prevent groundwater contamination at leaching operations can more fully incorporate an assessment of the risk to the public.

ACKNOWLEDGEMENTS

The author would like to express his grateful appreciation for comments and constructive criticism on earlier drafts of this chapter to the following: Philip Gray, Paul Mitchell of KEECO (U.K.) Limited, and Alyson Warhurst with the Mining and Environment Research Network, School of Management, University of Bath; Carl Mount of the Division of Minerals and Geology, State of Colorado; and two other anonymous reviewers. Any errors of omission, commission, or interpretation remain the responsibility of the author alone. Some of the fieldwork for this chapter was conducted as part of a research project within the Mining and Environment Research Network to investigate Technology Transfer and the Diffusion of Clean Technology in the Minerals Industry. The Global Environmental Change Programme of the U.K. Economic and Social Research Council supported this project.

REFERENCES

1. Bartlett, R. W., Soil decontamination by percolation leaching, in *Extraction and Processing for the Treatment and Minimization of Wastes*, Hager, J.P. Ed.,. The Minerals, Metals, and Materials Society, San Francisco, California, 1993, 411.
2. Pommier, L., Extraction of organic compounds and metals from hazardous soils, in *Extraction and Processing for the Treatment and Minimization of Wastes*, Hager, J. P., Ed., The Minerals, Metals, and Materials Society, San Francisco, California, 1993, 425.
3. Hanson, A., Dwyer, B., Samani, Z., and York, D., Remediation of chromium-containing soils by heap leaching, *Journal of Environmental Engineering*, 119, 825, 1993.
4. Hymanth, D., Hammack, R., Finseth, D., and Rhee, K., Biologically-mediated heap leaching for coal depyritisation, presented at the First International Symposium on Biological Processing of Coal, Orlando, Florida, May 1–3, 1990.

5. Hiskey, J., In situ leaching recovery of copper: what's next, in *Hydrometallurgy '94,* Institution of Mining and Metallurgy and Chapman and Hall, Cambridge, England, 1994, 43.

6. Hinde, C., An analysis of trends in the mining industry, presented at Financing Mining Projects Conference, London, 16–18 April, 1996.

7. Harris, L., Conway, T., Santa Cruz, C., Schwalb, F., Diaz, M., Arguelles, L., Cotts, N., Villaneuva, M., Meza, F., Guerra, F., Orams, P., and Chang, J., Newmont's Yanacocha project: the joint venture three years later, *Mining Engineering,* 48, 41, 1996.

8. Hickson, R., El Abra: the world's largest SX-EW mine on track to join copper mining elite, *Mining Engineering,* 48, 35, 1996.

9. Andrews, C., Mineral sector technologies: policy implications for developing countries, *Natural Resources Forum,* 16, 212, 1992.

10. Burniston, T. and Severs, K., Advances in copper solvent extraction by improved reagent technology, in *Proceedings of the First International Conference on Hydrometallurgy,* Yulian, Z. and Jiazhong, X., Eds., International Academic Publishers, Beijing, 1988, 231.

11. Warhurst, A., Metals biotechnology for developing countries and case studies from the Andean Group, Chile, and Canada, *Resources Policy,* 17, 54, 1991.

12. Brewis, T., Bacterial oxidation, *Mining Magazine,* 173, 197, 1995.

13. Marais, H., Gencor (consulting metallurgist), Johannesburg, personal communication, 1996.

14. Warhurst, A. and Bridge, G., Improving environmental performance through innovation: recent trends in the Mining Industry, *Minerals Engineering,* 9, 907, 1996.

15. Haines, A., Environmental impact of increasing production of gold from hydrothermal resources, in *Hydrometallurgy '94,* Institution of Mining and Metallurgy and Chapman and Hall, Cambridge, England, 1994, 27.

16. Clawson, R., Enforcement Officer, Mining Task Force, EPA Region 9, U.S. Environmental Protection Agency, personal communication, 1995.

17. Pond, E., Arizona Department of Environmental Quality (Aquifer Protection Permit Section), personal communication, 1995.

18. Bridge, G., Restructuring Nature: the Ecological Origins and Environmental Implications of Restructuring in the U.S. Copper Industry, unpublished doctoral dissertation, Clark University, Massachusetts, 1997.

19. U.S. Environmental Protection Agency, Report to Congress: Wastes from the Extraction and Beneficiation of Metallic Ores, Phosphate Rock, Asbestos, Overburden from Uranium Mining, and Oil Shale, Environmental Protection Agency, Washington, D.C., 1985.

20. Eccles, L., Office of Research and Development, Las Vegas, Nevada, U.S. Environmental Protection Agency, personal communication, 1995.

21. Sewall, C., Western Shoshone Defence Project (Research Officer), Nevada, personal communication, 1995.

22. Li, T. and Snedecker, M., Current status and future trends, in *Surface Mining,* Kennedy, B., Ed., Society of Mining Engineers, Littlewood, Colorado, 1990, 8.

23. Fletcher, J., In situ Leaching at Miami, in *Mineral Processing Handbook,* Weiss, N., Ed., Society of Mining Engineers, Littlewood, Colorado, 1985, 14D-7.

24. U.S. Environmental Protection Agency, Strawman II: Recommendations for a Regulatory Program for Mining Waste and Materials under Subtitle D of the Resource Conservation and Recovery Act, Office of Solid Waste, Washington, D.C., 1990.

25. U.S. Environmental Protection Agency, Draft Final Hardrock Mining Framework, EPA Hardrock Mining Workgroup, Washington, D.C., 1996.

26. Walline, R., Mining Division, EPA Region 8, Denver, Colorado, U.S. Environmental Protection Agency, personal communciation, 1996.

27. U.S. Department of the Interior, Bureau of Land Management, Cyanide Management Policy for Activities Authorized Under 43 CFR 3802/3809, Instruction Memorandum 90-566 from the Director to all State Directors, Bureau of Land Management, Washington, D.C., 1990.

28. Environmental Law Institute, *Hard Rock Mining: State Approaches to Environmental Protection,* ELI, Washington, D.C., 1996.

29. Environmental Law Institute, *State Mining Waste Regulation: Current State of the Art,* ELI, Washington, D.C., 1992.

30. Cale, S., Towards zero-discharge mining: minimization of water outflow, in *Hydrometallurgy '94*, Institution of Mining and Metallurgy and Chapman and Hall, Cambridge, England, 1994, 949.
31. Battle Mountain Gold, interview with Sally Hayes, San Luis Mine, Colorado, 1996.
32. Pegasus Gold, Florida Canyon's innovative concurrent reclamation program, in *Environmental Success Stories: Good News About Gold*, Echo Bay Mines and the Gold Institute, Denver, Colorado, 1994.
33. U.S. Environmental Protection Agency, Current Methods of Cyanide Heap Detoxification, Draft Final Report, EPA Contract 68-W4-0030, Office of Solid Waste, Washington, D.C., 1995.
34. Schafer, W. and Van Zyl, D., Cyanide degradation field study of spent heap-leach ore at the Landusky Mine, *Heap and Dump Leaching*, 6, 1991.
35. U.S. Environmental Protection Agency, Technical Resource Document (EPA 530-R-94-031) on the Extraction and Beneficiation of Ores and Minerals Volume 4: Copper, National Technical Information Service PB94-200979, Washington, D.C., 1994.
36. Kastning-Gulp, N. and Lockwood, J., Development of High Mountain Plant Communities as Wetland Mitigation Systems for Copper Mine Effluent, Final Report, Department of Plant, Soil, and Insect Science, University of Wyoming, 1993.
37. Morrey, D., Using metal-tolerant plants to reclaim mining wastes, *Mining Engineering*, 47, 247, 1995.
38. Sengupta, M., *Environmental Impacts of Mining: Monitoring, Restoration, and Control*, Lewis Publishers, Boca Raton, 1993.
39. Burchard, G., Arizona Department of Environmental Quality (Mining Unit of the Aquifer Protection Permit Division), personal communication, 1996.
40. Finton, P., Arizona Department of Environmental Quality (Mining Division), personal communication, 1996.
41. Hecla Mining Company, Using bioremediation for reclamation at Yellow Pine, in *Environmental Success Stories: Good News About Gold*, Echo Bay Mines and the Gold Institute, Denver, Colorado, 1994.
42. Pintail Systems, Inc., *The Bioremediation Report*, Aurora, Colorado, 1992.
43. Ketallapper, V., Project Manager, Summitville Reclamation Project, EPA Region 8, U.S. Environmental Protection Agency, personal communication, 1996.
44. Mudder, T., Miller, S., Cox, A., McWharter, D., and Russell, L., Lab evaluation of an alternative heap-leach pad closure method, *Mining Engineering*, 47, 1007, 1995.
45. Homestake Mining Company, Designing a new waste management unit at McLaughlin Mine, in *Environmental Success Stories: Good News About Gold*, Echo Bay Mines and the Gold Institute, Denver, Colorado, 1994.
46. Fraser, W., Mine decommissioning: an HBM&S case study, *Minerals Industry International*, 1025, 24, 1995.
47. Reece, B., Leaks and Liners 101, *Clementine*, Mineral Policy Center, Washington, D.C., Summer, 3, 1995.
48. Elliott, R. and Thomssen, D., Restoration of water resources after cyanide contamination: a case study, presented at the Northwestern Groundwater Issues Conference, Portland, Oregon, May 5–7, 1987.
49. Warhurst, A., and Bridge, G., Economic liberalisation, innovation, and technology transfer: opportunities for cleaner production in the minerals industry, *Natural Resources Forum*, 21, 1, 1997.
50. U.S. Bureau of Mines, *Minerals Yearbooks*, Government Printing Office, Washington, D.C., 1973–1993.
51. Anonymous, Summitville employees face prison, *Mining Engineering*, 50, 19, 1998
52. Ketallaper V. L. and Christiansen, J. W., The effectiveness of acid rock drainage control strategies at the Summitville mine, in Proc. 1998 Annual Meeting of the American Society for Surface Mining and Reclamation, St. Louis, Missouri, May 16–21, 1998.
53. Brooks, S. Reclamation of the Timberline heap leach. U.S. Department of the Interior, Bureau of Land Management, Technical Note 386/BLM/SC/PT-92/002+3024, 1992.

10 Recent Developments in the Treatment and Utilisation of Metallurgical Residues

C. P. Broadbent and N. J. Coppin

CONTENTS

ABSTRACT

Metallurgical residues are often disposed of as waste products. Growing environmental and feed constraints coupled with changing perceptions of acceptability question this attitude. There is a need to consider the most appropriate chemical form for eventual disposal. Often, a reprocessing step is required to achieve this, making it possible to consider the residues as resources. The suitability of silicate slag as a disposal medium for heavy metals is discussed. Substantial amounts of energy are lost with the disposal of molten slag from smelting operations. New schemes are considered which offer the potential to recover the greater part of this energy. A number of extractive industry residues are considered with a view to recovering valuable by-products. The results of recent research in which mercury is recovered from natural gas production sludges, alumina from colliery spoil, and metal and alloy from nickel smelter slags are discussed. It is concluded that these schemes can be realised commercially within a short time frame.

10.1 INTRODUCTION

Metals may sometimes exist in the Earth's crust in their native or elemental form. However, by far the most important states, with respect to quantity and occurrence, are the sulphides and oxides. Ores are concentrations of the valuable metal compounds associated with other (usually) unwanted minerals (gangue), which are often silicates. The choice of extraction route employed for a particular orebody will depend largely on the cost per tonne of metal extracted, though increasingly, notice must be made of environmental factors such as pollution control and safe disposal of residues.

Historically, pyrometallurgical processes have provided the main routes for the extraction of metals, owing to the relative abundance of "low-cost" fossil fuels and ability to achieve high production rates. In smelting processes, two immiscible liquid layers are formed; a liquid matte or metal layer containing the valuable metal and a liquid slag formed of the undesirable metals and gangue. Traditionally, this slag phase has been discarded without any attempt at further processing to recover (valuable) by-products or recover the intrinsic energy content. Often, disposal took place with little or no regard to the long-term environmental stability.

The amount of slag produced in different processes can vary tremendously. An iron blast furnace produces typically 1 tonne of slag per 3 tonnes of metal, whereas in the non-ferrous extraction industries this ratio can be considerably higher (e.g., nickel laterite smelting often produces 10 tonnes of slag per tonne of metal).

A background of growing environmental and feed constraints and changing perceptions of acceptability means that it is no longer necessary, or indeed possible to accept the inevitability of disposal of these waste products. At the very least, policies should be adopted which ensure disposal of residues in the most appropriate form and in an environmentally acceptable manner. In many cases, especially where some form of reprocessing is required to convert the residue to an alternative form, it is often possible to consider the waste by-product as a resource in its own right. Metals (and/or alloys) of saleable purity can be removed and recovered, enabling the economic utilisation of environmentally safe metallurgical residues.

Each orebody and associated metallurgical process offers its own unique set of opportunities for utilisation of residues. This paper will illustrate some of the possibilities by drawing on specific examples. Individual countries have developed their own regulations concerning disposal and reuse of metallurgical wastes. Reference will be made to procedures proposed in the Netherlands, as these represent some of the most stringent in western Europe, and serve as an indication of likely long-term trends in environmental policy. Further sections will include a review of the current status of research initiatives directed towards profitable and safe utilisation of metallurgical residues, including energy recovery, utilisation as a product (with and without further reprocessing), recovery of (valuable) metals and/or alloys, and use of silicate slag as an effective immobilisation medium for toxic residues.

10.2 REGULATIONS CONCERNING DISPOSAL OF EXTRACTIVE INDUSTRY RESIDUES

Traditionally metallurgical residues, such as slags, ashes, sinters, calcines, and sludges have been disposed of by tipping. Many of the environmental concerns associated with these practices have only been identified relatively recently and a number of countries are developing regulations for the control of the disposal of hazardous wastes. Indeed, a practical problem, and one which should not be overlooked, is the definition of hazardous waste and the selection of criteria that should be used for regulatory purposes.

In western Europe, the Netherlands has been at the forefront of environmental regulation policy. A concept proposal has been made, referred to as the "bouwstoffenbesluit — construction material decree" in order to control the deposition and reuse of metallurgical residues in the best way possible.[1] The aim is to ensure that as many waste products as possible can be reused, but always in a way such that there is no damage to the environment. This proposal would govern the disposal and/or use of all metallurgical residues. Background values for Dutch soil, groundwater, and surface water are given in Table 10.1. These have been considered as target values for the Netherlands and have been used in attempting to define acceptable criteria for use in the construction material decree.

A combination of chemical composition and leaching behaviour is proposed in the conceptual protocol. The limits for metals are shown in Table 10.2.

TABLE 10.1
Background Concentrations in Unpolluted Soil in the Netherlands

Substance	Ground (mg kg^{-1})	Groundwater (mg L^{-1})	Surface water (mg L^{-1})
Cr	100	1	5
Co	20	20	NA
Ni	35	15	9
Cu	36	15	3
Zn	140	65	9
As	29	10	5
Mo	10	5	NA
Cd	0.8	0.4	0.05
Sb	(2.6)	NA	NA
Se	(1)	NA	(10)
Sn	20	10	NA
Ba	200	50	(200)
Hg	0.3	0.05	0.02
Pb	85	15	4
V	(68)	NA	NA
F	500	500	1500
CN—complex	5	10	NA
CN—free	1	5	(50)
SO$_4$	(500)	150,000	100,000
Br	20	300	8000
Cl	(200)	100,000	200,000

If chemical composition values do not exceed the S1 unit, no restrictions apply. If residues exceed the limit S1, leaching tests have to be performed on the material. These tests are based on a combination of diffusion and column tests.[2] Where the leaching behaviour is such that the metal values in solution are lower than the U1 limit, once again no restrictions apply; if they are within the U2 category, the material can be used with certain restrictions, such as:

- Maximum of 5000 m^3 slag in one discrete engineering unit.
- Inspection, maintenance (or repair) of the isolation layer.
- Inspection of adjacent soil once every 5 years for possible contamination.
- A financial bond given for the liability for damage as a result of (possible) contamination of the soil.

Material failing to meet these standards has to be disposed of as industrial waste (weakly contaminated) or *chemical waste* (severely contaminated). In the Netherlands, a typical dumping charge for the worst category of chemical waste is US $280 t^{-1} which reduces to $75 t^{-1} if the material can be classified as industrial waste.[3] Thus, in many cases there is an economic incentive to reprocess wastes, removing (and recovering) the deleterious element(s) and thereby reducing the disposal costs.

It should be noted that the current Dutch proposals exist only in concept form and are liable to alteration before implementation. They serve, however, as a useful guide to likely trends with respect to environmental legislation. The regulatory aspects appertaining to specific metals will be discussed in some of the following case studies.

In the United Kingdom, however, the disposal of wastes from extractive industries is generally not considered separately from the disposal of other wastes. These materials are classified as

TABLE 10.2
Maximum Composition (S1) and Leaching Limits (U1 and U2) for Residues. Interim Values from Dutch Construction Material Decree

Material	Granular products		
	Leaching		Composition (interim values)
Level	U1 mg kg^{-1}	U2 mg kg^{-1}	S1 mg kg^{-1}
Chromium	1.0	10	250
Cobalt	0.2	2	50
Nickel	0.35	4	50
(Copper)	0.35	4	75
Zinc	1.4	14	250
Arsenic	0.3	3	30
Molybdenum	0.05	0.5	25
Cadmium	0.01	0.1	2
Antimony	0.03	0.3	10
Selenium	0.02	0.2	10
Tin	0.2	2	50
Barium	4	40	1000
Mercury	0.005	0.05	1
Lead	0.8	8	250
Vanadium	0.7	7	250

"controlled wastes," and as such are covered by the provisions of the Environmental Protection Act, 1990[4] and the subsequent Waste Management Licensing Regulations.[5]

A number of specific exemptions are given by the Regulations to the wastes of extractive industries which mean that, in some cases, waste licensing is not required for their disposal.

Nevertheless, the U.K. has adopted a general policy regarding the disposal of controlled wastes as follows, specified at Section 33 of the above Act:

> "*A person shall not (c) treat, keep or dispose of controlled waste in a manner likely to cause pollution of the environment or harm to human health.*"

Thus, rather than the somewhat prescriptive stance adopted in, for example, the Netherlands, the U.K.'s general policy statement allows the various regulatory bodies involved with waste disposal practices to interpret the provisions of the Act as they see appropriate.

In common with all European Community Countries, the U.K. and the Netherlands must ultimately aim to implement waste disposal regulations/directives produced at European level (e.g., the Waste Framework Directive).[6]

10.3 ENERGY RECOVERY

Substantial amounts of energy are required in melting the gangue minerals remaining in feeds to conventional smelters. For example, a typical iron blast furnace slag is produced at approximately 1823 K with a heat content of approximately 1200 J g^{-1}. This slag flow alone represents an energy resource of 7-35 MW. Usually, this slag is discarded with no attempt at energy recovery. Where water granulation is employed to provide, for example, aggregate material, additional electrical

TABLE 10.3
Energy Recovery from Slag: Potential Value of Generated Electricity
at 3.5 cent kWh^{-1} (after Broadbent and Warner)[8]

Scenario	Calculated Values
a) Large Iron Blast Furnace	
Hot metal rate	10,000 tonnes per day
Mass of slag at 300 kg slag t^{-1} hot metal	1.09 million tonnes per year
Potential value of electricity generated	$3.73 million per year
b) Typical Nickel Smelter	
Max. projected throughout of concentrate with O$_2$ enrichment	750,000 tonnes per year
Probable slag production	733,000 tonnes per year
Potential for electricity generation	9 MW
Potential value	$2.5 million per year

energy is employed in pumping the granulation water (water:slag volume ratio of 20:1 is often used). Furthermore, if the slag product is sold, subsequent drying costs are estimated at between $2-5 t^{-1}.

Implementation of dry granulation techniques[7,8] offers scope for energy recovery whilst maintaining a granulated product which is amenable to easy disposal or utilisation as a slag product. Indeed, where *lightweight* slag products are required in, for example, slag cement production or more novel utilisation, such as sound insulation material, it has been suggested that granulated slag produced by a dry granulation process, such as impinging jets, may represent a premium over traditional water-granulated slag.[9]

Energy could be recovered as either hot gas, for reuse in the metallurgical process or used for the generation of electricity. Table 10.3 shows the theoretical amounts of energy recoverable from typical ferrous and non-ferrous operations.

At present, energy recovery from slag has been performed at bench- and pilot-scales. However, demonstration facilities of dry granulation devices, inherent in energy recovery schemes (see Figure 10.1), have been commissioned at a number of iron and steel works. There must be hope of regarding energy recovery systems as being viable in the relatively short term. Schemes recovering energy in the form of (high-grade) hot gases are likely to be able to recover a greater part (>75%) of the energy content than those generating electricity (<30%). However, electricity generation is achieved without the coproduction of greenhouse gases, which may be a pertinent factor if carbon taxes are ever introduced.

10.4 PROCESSING METALLURGICAL RESIDUES FOR RE-USE/RECOVERY OF VALUABLE BY-PRODUCTS

10.4.1 USE OF SLAG PRODUCTS

Use of iron- and steel-making slags varies from country to country. Molten slag is either granulated and cooled by a stream of high-pressure water, or the slag flows into cooling pits where it solidifies as a coherent mass by air cooling alone or with additional spraying of water. Most U.K. blast furnace slag (BFS) relies on air cooling and almost all this material is sold to supply slag-based products to the construction industry.[10] In countries such as the Netherlands, almost the entire production of BFS is water granulated.[11] Conventional uses for slag products include use as landfill, a raw material for use in blast furnace cement production and use in road-making and civil engineering applications (e.g., coastal protection/river protection). It is interesting to note that a review of slag utilisation in

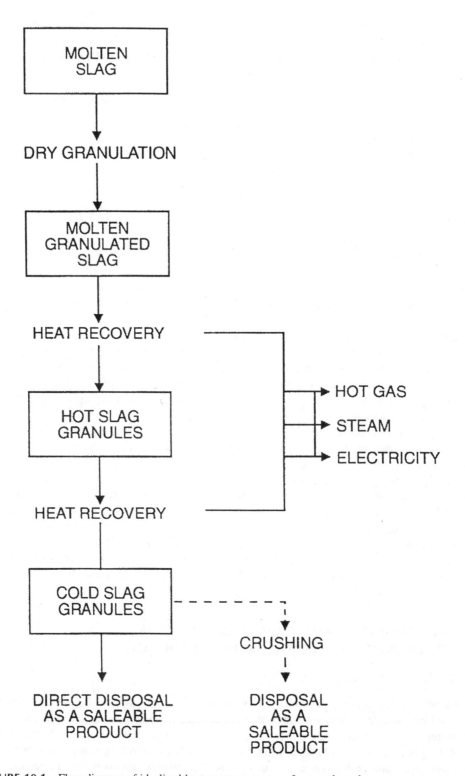

FIGURE 10.1   Flow diagram of idealised heat recovery process from molten slag.

TABLE 10.4
Review of Potential Uses of Metallurgical Slags

Metallurgical Process	Slag Treatment Route	Use	Notes
Iron blast furnace	Water Granulation	Slag cement	$CaO:SiO_2$ ratio
	Air cooling/grinding	Slag products to construction industry	Compressive strength chemistry
Lead-zinc blast furnace	Water granulation	Difficult due to heavy metal content	Granulation for ease of handling
	Vacuum de-zincing	Metal recovery and use of slag as grit blast etc.	
		Architectural bricks	Suggested further tests needed
Secondary copper smelting	Water granulation	Grit blast	
Nickel smelting	Oxidative pressure leaching	Recovery of base metals leach residues used as pozzolanic colour	
	Water granulation	Soil-less potting medium	Depends on trace metal content
General smelting	Water granulation	"Foamed" products used for insulating properties	Chemical composition

the Netherlands[12] concludes that water granulated slag is preferred as it limits the emission of sulphur-containing compounds, released during the production of air cooled slags.

Slag from the secondary copper smelter at James Bridge (U.K.) is all water granulated and sold as a *grit blasting* material. Other uses suggested for slags (especially non-ferrous slag) include sound insulation material, soil-less potting mediums, refractory furnace linings, and even architectural brick manufacture. This latter use has been proposed as extremely dense black bricks can be produced from many slags which are suitable for decorative use on buildings. A summary of some of the uses, other than landfill, for metallurgical slags is provided in Table 10.4.

Where the chemistry of the residue precludes utilisation without further treatment, many metallurgical residues can be considered as a resource for recovery of (valuable) by-products. These treatment routes are by necessity dependent upon the overall chemistry of the process. Three examples will be provided to illustrate the potential.

10.4.2   Recovery of Mercury from Natural Gas Production Sludges

Natural gas often contains traces of mercury. In the Groniagen field, the single largest supplier of gas for the Netherlands, the presence of mercury was first recognised in 1965. The major part of the mercury is condensed in the Low Temperature Separation gas treatment facilities, resulting in a number of mercury-bearing waste streams, such as sludges, sand and filter material. Shell Research in the Netherlands has developed a High Temperature Oxidation (HTO) process capable of removing and recovering mercury from all the various wastes, which are converted to an inert material which can be used as an aggregate in the construction industry or disposed of as industrial waste. In order to satisfy the limits of industrial waste in the Netherlands, material must contain mercury at less than 50 mg kg^{-1}. Detailed descriptions of pilot plant testing have been given elsewhere,[13,14] but during this process mercury is collected in a metallic form, of sufficient purity for sale (if a market exists) or for safe storage, and all calcine residues (with respect to their mercury content) are inert materials suitable for use as an aggregate in the construction industry.

In addition to natural gas production sludges, the process has been demonstrated using contaminated soils/sands, used filters, sludges from oil refineries, spent catalysts, mercury-loaded

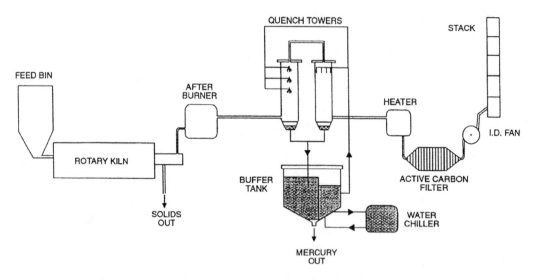

FIGURE 10.2   HTO process flow scheme.

carbon filters and mercury-contaminated sand blast grit. The proposed process flow scheme is illustrated in Figure 10.2.

10.4.3   The Extraction of Alumina From Colliery Spoil Materials by Acid Leaching

Vast amounts of colliery spoil materials are available in (former) coal mining regions. A number of investigations have been performed, which suggest that, in the case of a country such as the U.K., total alumina demand could be more than satisfied by recovering the alumina content from spoil tips. A large number of processes for the recovery of alumina from non-bauxitic materials have been proposed and described in a number of excellent reviews.[15,16] Generally, for the extraction of aluminium from clay minerals, acid leaching processes are preferred. Figure 10.3 shows a flowsheet proposed by Mahi and Bailey[17] for alumina recovery from colliery tailings, using a fluidised bed calcining stage. The process chemistry has been adequately demonstrated, although no pilot plant test work has been performed. At present, such schemes produce alumina more expensively than the Bayer process. However, significant reduction of the capital and operating costs can be expected by optimising the flowsheet.

10.4.4   Utilisation of Nickel Smelter Slags

Nickel smelter operations in the Sudbury district of Ontario, Canada produce slags containing approximately 0.136% nickel, 0.1% cobalt and 0.29% copper. Cobalt occurs mostly in the oxidic phases whilst nickel is found, both as sulphides as well as oxidic phases, and copper is present mostly in sulphide phases. Heinrich[18] demonstrated that oxidative pressure leaching with sulphuric acid of the finer fractions effectively recovered the entrained base metals, allowing normal utilisation of the coarse fractions (significantly depleted in base metals) and thereby alleviating potential environmental problems which are encountered if nickel-containing dusts are used in the construction industry.[19] Furthermore, the fine-grained residues from the leaching process could be used as a pozzolanic cement colour or, if no saleable outlet exists, could be treated by addition of calcium oxide to form a stable, cement-like material.

Reviews of the physical chemistry of nickel laterite smelting slags have been published previously.[20] Figure 10.4 presents the mass balance for a typical electric arc furnace — rotary kiln process producing ferronickel. The bulk chemistry of slag produced during ferronickel smelting

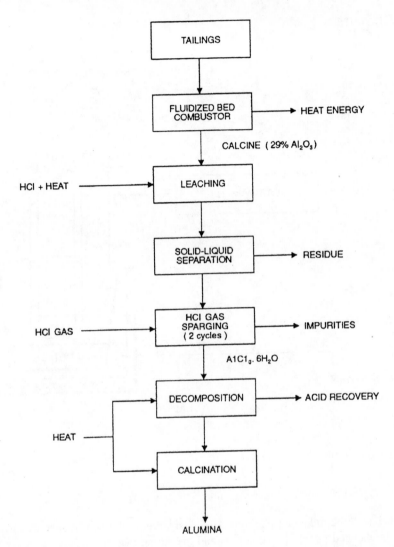

FIGURE 10.3 Proposed flowsheet for alumina recovery from colliery tailings, using a fluidised bed calcination stage.

can be regarded as a potential silicon or ferrosilicon source. Thermodynamics predicts that iron oxide and silica will be reduced before magnesia, suggesting that by careful control of the addition of reductant the purity of ferrosilicon can be controlled, leaving a final discard slag containing all the magnesium oxide, together with some of the iron oxide and silica. Oxides such as chromium trioxide will be amongst the first to be reduced and (if present) it is possible that the first alloy to be formed will be a ferrochrome, followed at a later stage by formation of ferrosilicon. If present in sufficient quantity, tapping of separate ferrochrome and ferrosilicon phases may be commercially viable.

The prevention of magnesium volatilisation would be of practical importance a) to limit the overall energy requirement for the process, and b) perhaps more importantly, to prevent the associated handling difficulties. A single step process, represented schematically in Figure 10.5, could be expected to show a modest (<25%) saving in energy requirements over traditional ferrosilicon production routes (due to the credit gained from the sensible heat of the already molten slag). This process could be envisaged to produce ferrosilicons containing 50% Si°. Production of higher-grade ferrosilicons, containing in excess of 50% Si° could be contemplated by addition of

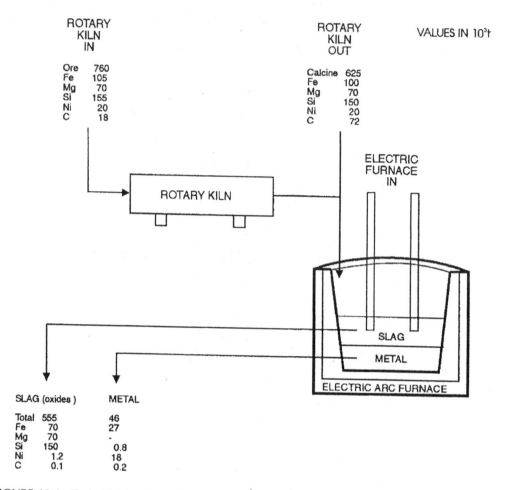

FIGURE 10.4   Typical Ni-laterite smelter mass flows.

silicon together with the reductant (carbon). Obviously, the production of specific ferroalloy from the slag will be dictated largely by economic and marketing factors.

10.5   USE OF SILICATE SLAGS AS A DISPOSAL MEDIUM FOR HEAVY METALS

Many countries are implementing stringent legislation governing the disposal of hazardous wastes containing heavy metals such as arsenic, antimony, lead, cadmium, or mercury. These wastes cannot be dumped without providing clear guarantees that the material concerned is environmentally safe and will not leach toxic elements to the surroundings. A survey of proposed regulations in various countries for arsenical wastes indicate quite some differences between individual countries, as indicated below:

- U.K. — co-deposition in landfill may not release more than 20 mg L^{-1} arsenic.
- U.S. — arsenic concentration following U.S. Environmental Protection Agency's buffered acetic acid leach test (Toxicity Characteristic Leaching Procedure) is set at 5 mg L^{-1} (100 times the World Health Organisation's drinking water standard).
- Netherlands — leaching according to the column test NVN2508 has a limit of 3 mg kg^{-1} recommended, recalculated in terms of mg kg^{-1} solids leached at a liquid:solid ratio of 10.

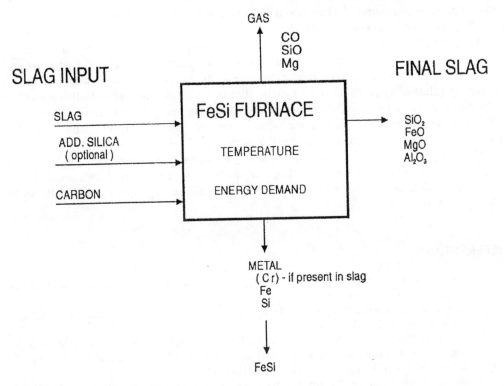

FIGURE 10.5 Schematic flowsheet showing production of FeSi from Ni-laterite slag using a one-stage process.

Arsenical residues are often produced as by-products in the non-ferrous metallurgical industry. Pyrometallurgical operations (usually) produce arsenical residues in the form of arsenic trioxide-containing flue dust and speisses. If the flue dust is of high purity, it can be sold for use as an active agent in wood preservatives, although even this is coming under challenge from the use of safer substitutes. Often the dust quality is not amenable to upgrading and no uses can be considered.

Normal practice is to dissolve the dusts in acid prior to precipitation as calcium or, more usually, ferric arsenate. Recent work has questioned the stability of ferric arsenate and has highlighted concerns over the suitability of this process, especially as adoption of a hydrometallurgical process may be seen as inappropriate by an established pyrometallurgical plant if it creates additional aqueous effluents. Hence, alternative methods of disposal have been considered.

Test work has shown that up to 10% arsenic (by weight) can be incorporated into glassy silicate slag with low levels of arsenic leaching.[21] Comparative leaching tests performed on a number of potential arsenical residues[22] have demonstrated that arsenic-containing glassy slag displays leaching characteristics superior to those of calcium arsenate and as good as the best ferric arsenate species. Furthermore, the "slag" option appeared to be more flexible than the ferric arsenate route towards accommodating changes in chemical composition. More effort is required to determine the overall long-term stability of arsenic-containing glassy slags; however, incorporation of arsenical residues in molten silicate slags with subsequent water-quenching would seem to represent the most environmentally acceptable route for the disposal of many arsenical residues.

10.6 CONCLUSIONS

Metallurgical residues, whether they be slags, calcines, sludges, and so on, are often regarded as "waste products" with little or no intrinsic value. Against a background of growing environmental and feed material constraints, the time is rapidly approaching where this traditional attitude towards metallurgical residue disposal must be challenged.

As responsible environmental scientists, there is an obligation on us to ensure that, at the very least, residues are in the most appropriate chemical and physical form for disposal. Further treatment of these residues offers exciting possibilities for investigating the use of plant residues as a resource, recovering valuable by-products (e.g., metals, alloys, or energy content) and allowing effective utilisation (e.g., aggregate, cement additive, and grit blast material) of the remaining residue. Indeed, if the metallurgical residues can be regarded as a resource, many of the environmental problems associated with disposal of the untreated residues can be solved.

Many of the examples given in this paper are recent developments and are based on the results of bench- or pilot-scale investigations. However, the technologies involved are such that we can be optimistic in believing that many of the schemes could be exploited commercially within a relative short time frame.

REFERENCES

1. Ontwerp Bouwstoffenbesluit, ISBN 90.39.90104X, Department of the Environment, Government Printing, The Hague, 1991.
2. Nederlandse Noornorm, Bepaling van de uitloog karakteristiken van kolenstoffen NVN, 2508, Government Printing, The Hague, 1988.
3. Broadbent, C. P., Machingawuta, N. C., van der Kerkhof, A. R. F., and Bongers, E. A. M. Heavy metal immobilisation in silicate slags, in *Pyrometallurgy for Complex Materials and Wastes*, Nilmani, M., Lehnerand, T., and Rankin, W. J., Eds., TMS, 1994, 93.
4. *Environmental Protection Act*, Her Majesty's Stationery Office, London, ISBN 0-10-544390-5, 1990.
5. *The Waste Management Licensing Regulations*, Her Majesty's Stationery Office, London, ISBN 0-11-044056-0, 1994.
6. Council Directive on Waste (75/442/EEC) as amended by Council Directive 91/156/EEC and 91/692/EEC, OJ No. L194, 25.7.1975 p39, OJ No. L 078 26.3.1991, p32, and OJ No. L 377, 31.12.1991, p 48.
7. Pickering, S. J., Recovery and utilisation of the energy in molten slags, unpublished doctoral dissertation, University of Nottingham, U.K., 1983.
8. Broadbent, C. P. and Warner, N. A., A novel approach to energy recovery from slag, *Iron and Steel Int.*, 57, 101, 1984.
9. Broadbent, C. P., Internal structure of solidified iron — blast furnace slag (bfs) droplets formed by self-impinging jets, *J. of Materials Science Let.*, 6, 1264, 1987.
10. Thomas, G. H., The utilisation of blast furnace and steelmaking slags, presented at the Technology of Reclamation Conference, University of Birmingham, April 7 to 11, 1975, 10.
11. Scholes, S., New building and enginering materials from metallurgical slags, presented at the Technology of Reclamation Conference, University of Birmingham, April 7 to 11, 1975, 9.
12. Schuur, H. M. L. and Broadbent, C. P., Slag utilisation in the Netherlands, presented at Minerals Industry Research Organisation Conference on the Economic Treatment of Mineral Wastes, Bath, U.K., June 21–22, 1990, 321.
13. Broadbent, C. P. and Rots, K. J., Decontamination of mercury-containing wastes, in *International Minerals & Metals Technology*, Roberts, N. J., Ed., Sterling Publishers Limited, 1994, 163.
14. Broadbent, C.P., Berkenhagen, F. P., Aurich, V. G., and Rots, K. J., High temperature oxidation process for decontamination of mercury containing wastes, *Aufbereitungs Technik*, 35, 299, 1994.
15. Christie, P. and Derry, R., Aluminium from indigenous U.K. sources — a review of possibilities, Warren Spring Report L R 219 (ME), Warren Springs, U.K.,1976.
16. Hammer, C.A., Acid extraction processes for non-bauxite alumina materials, CANMET Report 77–54, CANMET, Ottawa, 1977.
17. Mahi, P. and Bailey, N. T., The extraction of alumina from colliery spoil materials by acid leaching, presented at Symposium on the Reclamation, Treatment and Utilisation of Coal Mining Wastes, Durham, U.K., September, 1984, 44.1.

18. Heinrich, G., On the utilization of nickel smelter slags, *CIM Bulletin*, 82, 87, 1989.
19. Whitby, L. M., Stokes, P. M., Hutchinson, T. C., and Myslik, G., Ecological consequences of acidic and heavy-metal discharges from the Sudbury smelters, *Can. Mineralogist*, 14, 47, 1976.
20. Broadbent, C. P. and Machingawuta, N. C., The physical chemistry of nickel laterite smelting slags, Proc. Paul Quineau Symp., Denver, 1993, 3.
21. Machingawuta, N. C. and Broadbent, C. P., Incorporation of arsenic in silicate slags as a disposal option, *Transactions of the Institution of Mining and Metallurgy*, 103, C1, 1994.
22. Broadbent, C. P., Machingawuta, N. C., vant Sant, J. P., and van der Kerkhof, A. R. F., Leaching behaviour of arsenical materials, in *Impurity Control and Disposal in Hydrometallurgical Processes*, 24th Annual Hydrometallurgical Meeting, Proceedings of International Symposium, Toronto, Canada, August 21–24, Harris, B. and Krause, E., Eds., 1994, 125.

11 Management of Ferro-Alloy Wastes

Jochen Petersen, Mary Stewart, and Jim Petrie

CONTENTS

ABSTRACT

The ferro-alloy industry in South Africa can be divided into two groups of processes: the first concerned with the mining and initial beneficiation of ores, and the second with the production of various metallic products, ranging from pig iron to various ferro-alloys and stainless steel.

Waste production from these processes is considerable, with the majority of waste products in the form of solids, such as furnace slags, emission control dusts, flotation tailings, and leach residues, to name but a few. Traditionally, these waste materials have been disposed of to landfill, usually in the form of large-scale heaps or dams. It is only recently that the management of ferro-alloy wastes has been governed by concerns about the long-term environmental stability of such deposits, which are associated with the potential leachability of salts and heavy metal species contained therein.

The concept of clean technology is beginning to influence process design options in the sector with the ultimate aim of reducing material discarded as waste, and to improve the environmental acceptability of these wastes. This includes not only the installation of new process technology, but also process-integrated recycling of waste materials and new downstream processes that use waste products as raw materials. Examples of such options are discussed in the context of stainless steel production.

These new technologies notwithstanding, waste materials that ultimately arise from any given process will have to be managed in such a way as to ensure minimal impact on the receiving environment. We argue in this paper that operator liability with respect to waste management practices can be assessed only with the aid of a comprehensive model of the physical and chemical processes that affect leach resistance within waste deposits. The point of departure from other work in this field is our interest in conditions within the landfill (in chemical engineering terms) as these relate to technology choice and operating conditions within the process(es) which generated the waste. The development of such a model is discussed, and a methodology that combines laboratory waste characterisation with the model to provide a predictive tool for the assessment of leachate generation and mobility is proposed.

11.1   WASTES IN THE FERRO-ALLOY INDUSTRY

The South African ferro-alloy industry represents a well-established, comprehensive array of processes. It can be broadly divided into two different types of processes. The first represents the mining and initial beneficiation of ores, mainly those containing manganese, vanadium, chrome, iron, and silicon. These are extended to include the production of any chemical compound forms of the metals (e.g., salts or other compounds). The second group of processes covers the production of various metallic products, ranging from pig iron to various types of ferro-alloys and stainless steel.

Table 11.1 shows South Africa's ranking with respect to ore reserves, production, and exports for the primary ferro-alloys sector. These figures make it plain that not only is this a sector with a very long projected life, but that there is immense room for growth.

What is less obvious from these figures is the generation of large amounts of wastes during the beneficiation of ores and the production of ferro-alloy metals. While most of the beneficiated product leaves the country through export, the associated wastes remain behind and the local environment most immediately feels any potential impacts of these.

In terms of a waste inventory for the South African ferro-alloy sector, there exists a large discrepancy between data available from official studies,[1,2] and those calculated from a rigorous flowsheet-based approach,[3] incorporating the production figures quoted in Table 1. This comparison is given in Table 11.2. The discrepancies are particularly significant for gaseous and aqueous effluents, which can be attributed to poor monitoring and rapid dispersal of these streams, whereas the paths of solid wastes are generally more easy to follow.

The majority of waste materials in the ferro-alloy sector arise as solids, such as furnace slags, emission control dusts, flotation tailings, and leach residues, to name but a few. Traditionally, these wastes are disposed of to landfill, usually in form of large-scale heaps if the solids are deposited dry, or slimes dams where solids are disposed of as slurries. It is particularly the latter disposal method that gives rise for concern regarding the uncontrolled migration of an aqueous phase through the waste deposit before entering the underlying substrata. Not only does this constitute a loss of valuable process water, but creates an aqueous environment throughout the deposit, which can potentially support the release and migration of hazardous heavy metal species contained in the waste solids.

It is only recently that operators of ferro-alloys processes in South Africa have become aware of the potential environmental hazards associated with current waste management strategies, and that they will have to accept the liability for environmentally sound disposal practice and monitoring of the disposal site, which may even extend beyond closure. However, responsible waste management within an industrial process is not confined to disposal of the waste as generated, but also extends to the process design stage, where the choice of appropriate technology can effect a reduction not only of the amounts of wastes generated, but also the environmental hazards associated with the wastes.

11.2   WASTE MINIMISATION AND INTEGRATED WASTE MANAGEMENT IN STAINLESS STEEL PRODUCTION

The concept of clean technology is beginning to influence process design options in the sector, with the ultimate aim of reducing material discarded as waste, and to improve the environmental acceptability of these wastes. This includes not only the installation of new process technology,

TABLE 11.1

Reserves, Production, and Exports of the South African Ferro-Alloys Sector

Product	Reserve Base Mt	Rank (% of world reserves)		Production kt/annum	Rank (% world production)		Export kt/annum	Rank (% of world exports)	
Chrome	3200	1	(72.0%)	2826.70	1	(33.5%)	983.70	1	(37.1%)
Iron	5900	7	(5.8%)	29,400.00	8	(3.1%)	19,000.00	6	(4.8%)
Manganese	3992	1	(83.2%)	2506.70	3	(11.7%)	1661.50	2	(25.0%)
Silicon				34.50	7	(5.4%)	32.20	4	(11.7%)
Vanadium	12.5	1	(44.7%)	26.87	1	(46.4%)	21.34	1	(79.2%)

TABLE 11.2

Comparison of Modelled Waste Inventory with Data from Literature

Process/Wastes	Calculated (kt per annum)	CSIR Data (kt per annum)	Ratio (Calculated/CSIR)
Steelmaking (solids)	8000	2826	2.8
FeCr and FeSi (solids)	4700	1658	2.8
FeMn (solids)	2300	370	6.2
Gas emissions (sector)	2300	13	170
Wastewater (sector)	9700	16	600
Water excluding initial beneficiation	380	16	24
Contaminants in water	21	16	1.3

TABLE 11.3

Normalised Waste Generation Potential of Steel Production Technologies

Technology	Waste Generation (kiloton per kiloton product)
Blast Furnace	66
Basic Oxygen Furnace	0.3
Corex	5.6
Corex/DRI	0.1

but also process-integrated recycling of waste materials and new downstream processes that use waste products as raw materials.

Particularly in the steel industry, many such technology shifts have already taken place and continue to be developed. For example, development of furnace technology has resulted in a simultaneous improvement in the waste generation per tonne of product (Table 11.3). The first option is blast furnace technology, the oldest technology employed at present; then, the basic oxygen furnace (BOF), which was the next step in technological development; finally, the Corex process and the coupling of a Corex reactor with a downstream direct reduction process (DRI), which uses the CO- and CO_2-rich off-gas from the Corex process in the reduction of iron.

As can be seen in Table 11.3, the environmental performance of the blast furnace is the worst as it generates the most waste per tonne of product. When the Corex process is run without a downstream DRI section, the environmental performance is also fairly poor. Coupling a DRI and a Corex process gives the best environmental performance of the four technologies compared here.

At a different level, developing process technology has focused on improving process-integrated recycle and recycling of waste materials as raw materials for new processes, thus reducing the amount of material that leaves the process cycle as waste. Finally, modifications in the process can also be made to ensure that the waste materials which are ultimately disposed represent a reduced hazard for the environment.

An integrated waste management approach, which takes on board all three of these options, has been devised for a combined ferro-chromium and stainless steel facility.[4] The underlying production process entails the following main steps:

- Production of ferrochrome metal from chromite ore, reductants, and various fluxes, typically in a submerged arc furnace.
- Remelting of this product with additional scrap and other alloys in an electric arc furnace (EAF), followed by addition of other alloys.
- Decarburizing of the melted product in a refining vessel by selective blowing with argon and oxygen.
- Casting of molten steel into slabs.
- Reheating, rolling, and annealing.
- Surface treatment of rolled product (i.e., acid pickling).

A rigorous waste minimisation assessment for the integrated stainless steel production facility has identified the following main waste streams:

- Bag filter dusts—ferrochrome and stainless steel.
- Slag product—ferrochrome and stainless steel.
- Acid wastewater—stainless steel surface treatment.
- Landfill leachate—from all solid residues.

Solid wastes from all furnaces (fine bag filter dusts and coarser slag) tend to contain constituents that have been identified as hazardous, such as chromium, nickel, lead, cadmium and other heavy metals. Liquid wastes include acidic rinses from annealing and pickling operations, gas scrubber liquors, process water used for slurry transport, as well as leachates recovered from dump sites. These liquid streams contain heavy metals in solution, as well as high salt levels and some organics. The ability to treat effectively the ongoing production of these wastes, whilst at the same time remediating waste materials already accumulated in waste depositories from previous production, is demonstrated in the process shown in the process block diagram in Figure 11.1. Details of this integrated process have been described elsewhere,[4] and only the fundamental concepts are reiterated here.

Acid waste waters and any other process waters requiring treatment are mixed with the solid dust and fine slag materials, both from ongoing production and from recovered waste deposits, to effect additional leaching of heavy metal species contained in these. The residual solids, from which a significant amount of the heavy metals have been removed in the leach step are separated, dewatered, and solidified prior to landfill disposal. The leachate in turn is reduced and all dissolved metals are precipitated as hydroxides. These are again separated, dewatered, and subsequently briquetted to be returned to the process (i.e., as part of the charge to the EAF furnace). The residual leachate, containing mainly dissolved salts which were unaffected by the precipitation process, is clarified and charged to an evaporator/crystalliser unit which yields distilled water for return to the process and salt crystals for further downstream refining.

The particular attractiveness in this process lies in the concept of treating waste with waste to yield solid products which can be recycled to the process, used as a raw material for downstream

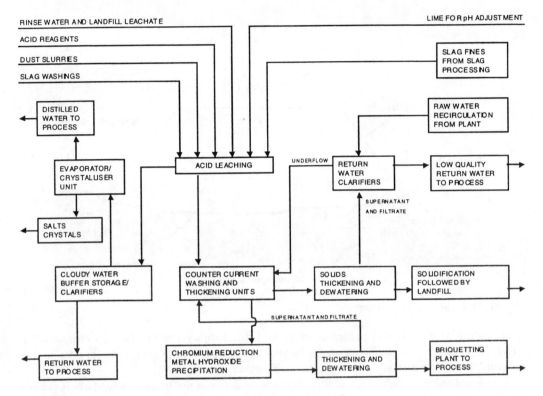

FIGURE 11.1 Integrated treatment of solid wastes and liquid effluent.

processing or as waste products significantly reduced in their hazardous constituent content. Moreover, the process has no liquid effluent and all process waters are — after treatment — returned to various stages in the overall process. Although developed for the specific situation found in the combined ferro-chrome/stainless steel operation, the fundamental ideas of this process — and clean technology in general — should be equally applicable to other operations in the ferro-alloy sector and facilitate waste minimisation and amelioration.

11.3 PREDICTION OF LEACHATE GENERATION POTENTIAL

The potential for waste reduction and treatment by choice of appropriate technology in the ferro-alloy industry has been demonstrated above. What is less apparent, however, is how to objectively assess the long-term risks associated with residual waste deposits. The long-term environmental stability of such deposits is linked directly to the potential for leachate generation, and the subsequent mobility of leached constituents into associated ground water.

The presence of trace amounts of various heavy metals in many solid minerals processing waste materials constitutes a considerable environmental risk for this form of disposal. Rainwater percolating through a deposit can induce a variety of chemical processes that can potentially lead to the release of heavy metal species, their transport through the deposit, and further dispersion. The ecotoxicity of dissolved heavy metals, even at relatively low concentrations, is reasonably well established.[5] Thus the gradual generation and transport of an environmentally harmful leachate from minerals processing waste deposits can turn these into long-term environmental hazards.

The hazard potential of this type of industrial waste is now acknowledged generally and, increasingly, more attention is being given to environmentally sound disposal of the waste materials where any forms of the minimisation or treatment technologies discussed above are not feasible. There is a clear trend in legislation to place the environmental liability for wastes with the operators

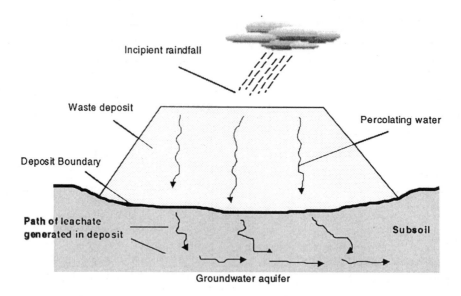

FIGURE 11.2   Leachate dispersion.

of the industrial process that generates them, and this liability remains beyond closure of the operation. Thus there should be an urgent interest by the operators of such disposal operations in assessing the environmental risk associated with their operation.

The quantitative measurement of this environmental risk is not trivial, particularly if it has to be achieved *a priori*. It requires a sufficiently rigorous model of the leachate generation and transport processes likely to take place within a deposit based on a meaningful characterisation of the waste materials it contains. Current waste characterization and classification procedures, such as the well-known Toxicity Characteristic Leaching Procedure (TCLP)[14] devised by the U.S. Environmental Protection Agency (USEPA), at best offer indicators to the hazard of wastes, but are insufficient to allow prediction of leachate generation as it may occur over the operational life span of a disposal operation and beyond.

The focus of the work presented in the following is the formulation of a mechanistic model which describes the transport of an aqueous phase through the pores of a granular solid matrix and the chemical interactions it can enter into at the particle level. This model should provide a vehicle with which the release and transport of chemical species, particularly heavy metals, from a full-scale deposit can be effectively modelled over extended periods of time.

11.3.1   CONCEPTUAL LEACHATE GENERATION PROCESS

Before embarking on the development of such a modeling tool, it is useful to first take a conceptual look at the leachate generation process within the bed of landfill deposit. Essentially, the landfills considered here are heaps of granular waste particles that gradually develop shape and form in accordance with ongoing production. These are usually engineered to promote structural integrity and easy runoff of incipient rainfall and water draining from deposited slurries. Nonetheless, a significant fraction of incipient water will continuously enter the granular matrix of the deposit and percolate through it towards its base, where it can enter the soil underneath the deposit — if a suitable liner and leachate collection system is not in place (Figure 11.2).

This percolating water creates an aqueous environment in the interstices between the particles deposited in the waste bed (Figure 11.3). Irrespective of whether there is bulk movement through the bed or not, particles are continuously in contact with water and hydro-chemical interactions

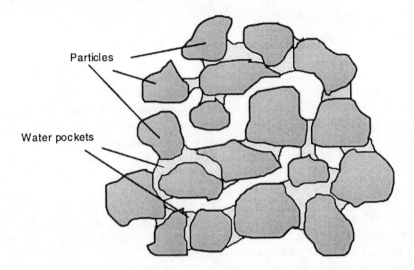

FIGURE 11.3   Development of interstitial water.

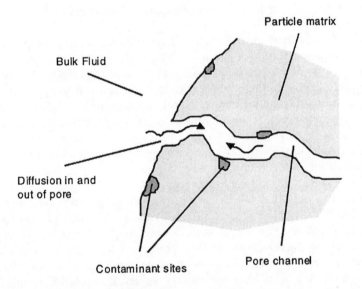

FIGURE 11.4   Particles and diffusional transport of contaminants.

with the solids, such as dissolution, can take place. These interactions are not confined to the wetted particle surfaces.

Water can migrate into particle pores and also create an aqueous environment there that can interact with the solid throughout the particle. Because this pore water is in contact with the bulk fluid on the outside, it offers a medium through which dissolved chemical species can enter and leave the pores by diffusional transport (Figure 11.4). Thus even contaminants located on the inside of particles can be mobilised potentially and be transported to the particle surface and further into the bulk fluid, where they can move more widely by bulk transport.

Consequently, the movement of a reactive aqueous phase through the granular matrix of porous particles in a waste deposit can be likened to processes taking place in a packed bed chemical reactor. Thus successful modeling of this process can extensively draw from chemical engineering concepts and theory.

11.3.2　Model Development

Modeling of contaminant transport through an aqueous phase within a porous solid matrix per se is not new. Geohydrology is concerned with groundwater movement and there is a whole host of hydro-geochemical models which incorporate transport of both water and dissolved contaminant species.[6] At a chemical level, such models concern themselves mainly with adsorptive attenuation of dissolved species onto solid surfaces. However, they provide little assistance in terms of describing chemical release from the solids by reactions such as dissolution and leaching, nor do they take account of kinetic effects. In many modeling scenarios, the leachate composition is in effect assumed as given and only the fate of this leachate as it is transported away from its source through soils and aquifers is considered.[7] In this way, the usefulness of such models to describe leachate generation within waste deposits is limited.

Heap leaching, which is essentially a low-cost minerals processing option to extract metals from low-grade ores, exhibits a striking similarity to the conceptual leach process described in Section 11.3.1. Here, a heap of ore particles is continuously sprayed with a leach liquor, which percolates through the bed, releases the desired metals from the particles by chemical reaction at the particle level and transports them to the base of the deposit, where the pregnant liquor is collected for stripping of the mineral values. Dixon and Hendrix[8,9] have proposed a fundamental model to describe the heap leach process. Their approach essentially divides the process into two distinct regions:

- At the level of the individual particle, the model, by assuming the particle as porous spheres, considers diffusion of reactant into the particle pores, reaction within the pores, and diffusion out of the particle of the dissolved reaction products (Figure 11.4).
- On the overall bulk level, the model describes convective flow of the fluid phase in a vertical direction through the particle matrix and migration of dissolved species in and out of the particle pores, accounting for a distribution of particle sizes (Figure 11.5).

The use of the continuity equation for mass transport on both levels ensures a mathematically rigorous approach and the authors have demonstrated successful application of the model with data obtained from laboratory-scale column experiments.

Thus the Dixon model offers an excellent starting point for the development of a model to describe the transport and reaction processes likely to take place within a heap of particulate waste material. Certain extensions to the model were, however, required in order to account for the less specific reaction and transport conditions likely to prevail in a waste deposit. These include multiple reaction mechanisms, such as dissolution, adsorption, and leaching, governed by both kinetics and thermodynamic equilibrium. The Dixon model only considers kinetic leach reaction by one principal reactant, which is valid for a heap leach scenario, but insufficient for waste deposits, where generally no specific reaction dominates over all others.

- Combined convective and diffusive bulk transport. The Dixon model only considers convection, since flow rates in heap leach scenarios are usually high enough to render bulk diffusion effects insignificant. In waste deposits, infiltration of rainwater is seasonal and normally proceeds at much slower rates, resulting in an often stagnant or slow-moving phase, in which diffusion effects are much more pronounced.

These extensions to the Dixon model have resulted in the formulation of a set of fundamental waste heap model equations, a detailed discussion of which is beyond the scope of this paper, but is available elsewhere.[10,11] The significant features of the model are the separation of particle and bulk level processes and the fact that all solid-fluid reactive interactions are modelled on the particle level alone. Parameters for the model, particularly those characterizing reactions, need to be obtained mainly by experimental waste characterisation methods.

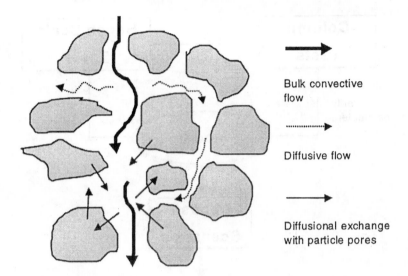

Bulk convective
flow

Diffusive flow

Diffusional exchange
with particle pores

FIGURE 11.5   Flow regimes linking particles and the external environment.

11.3.3 Waste Characterisation

There is a multitude of solid waste characterisation methods that are commonly practiced. These include pH and redox controlled batch leach tests; sequential leach tests,[12] in which waste is contacted with increasingly aggressive leach media; serial batch leach tests,[13] in which the same sample of waste is repeatedly washed in fresh leach liquor (normally water); and column leach studies at various scales, where leach liquor is percolated through a bed of particles. While column studies generally tend to better reproduce conditions of waste leaching as they occur in the field, they are often tedious and costly to conduct. Laboratory bench scale experiments, on the other hand, are quick and simple, but generally limited by their highly specific experimental conditions and results obtained can seldom be directly applied to simulate the time-dependent generation of leachate in a deposit scenario. Jackson et al.[13] have attempted to correlate results from serial batch leach tests with those obtained from lysimeter studies, but only with limited success.

In the absence of a more meaningful method, many waste characterisation methods merely serve to estimate the worst-case leach potential of a particular material, which is then used to classify waste as hazardous or non-hazardous. Prominent among these is the USEPA's TCLP,[14] which enjoys widespread use throughout the world. Tests such as this do little to assist the estimation of leachate generation over time from a particular waste material.

As indicated in Section 11.3.2, an important feature of the model is the distinction between reaction-diffusion at the level of the individual particle and bulk transport effects through the deposit. The particle level model remains valid even if the bulk conditions change (e.g., different bed geometry or a mixed batch reactor scenario). The advantage of this aspect is that particle level parameters, particularly all relevant chemical reaction parameters, can be established from suitable bench-scale batch experiments on small samples of a particular waste material. These parameters remain valid irrespective of the bulk conditions in which the particles are placed. Thus the model can serve as a vehicle to allow meaningful correlation of laboratory data to simulate the time-dependent leach behaviour in a deposit situation.

Laboratory experiments must be suitably designed in order to obtain data that is relevant in the particular chemical environment likely to prevail in the deposit scenario. In this sense, worst case assessment methods, such as the TCLP, which employ relatively aggressive leach conditions, are not well suited to generate data on a waste material, which in a deposit situation is exposed to a much less aggressive environment. An initial "feel" for likely deposit conditions can be obtained

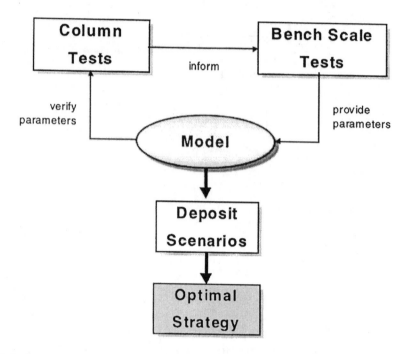

FIGURE 11.6   Proposed assessment methodology.

from a small-scale lysimeter column experiment. Interactions under these conditions can then be more specifically investigated in batch experiments; for example, leaching of various species at the prevailing pH values, or adsorption of relevant species in the prevailing concentration ranges. The degree to which individual reactions are discerned and quantified by experiments is governed by the degree of rigour required for a specific assessment as well as by time and cost restraints of the investigation. Generally, however, for most waste materials there is only a small number of principal species of concern and reactions they are likely to be involved with. A reasonably good assessment is possible on the basis of data for these, even if side reactions and less significant species are ignored.

11.3.4   ASSESSMENT METHODOLOGY

Drawing model and experimental assessment techniques together, one can formulate a powerful waste assessment methodology which allows the prediction of leachate generation over time of a particular waste material in a particular deposit situation on the basis of relatively simple bench-scale laboratory tests. Moreover, since particle level parameters obtained from such tests are obtained independently from the deposit scenario that is being modelled, no undue empiricism is introduced which would limit the scaling of results. This way a number of different deposit scenarios could easily be simulated in order to identify an optimal disposal strategy, which would minimise the risk of long-term leachate generation with the associated environmental hazards.

A block diagram for the proposed assessment methodology is given in Figure 11.6. Column tests identify the prevailing chemical conditions in a deposit situation and thus inform the selection of suitable batch experiments to quantify these in more detail. Experimental results are either used directly or after optimisation — in combination with the model adapted to the batch leach scenario — to obtain the relevant particle level reaction and transport parameters. The parameter set can be verified by applying the model to the column experiment and simulating the initial leach curves. If this is successful, the model can be applied with reasonable confidence to a full-scale heap scenario. Repeating this for a number of scenarios should help to identify the optimal disposal strategy for a particular waste material with which the lowest environmental risk is associated.

11.4 CONCLUSION

The ferro-alloy industry sector in South Africa as a whole offers potential for growth in the foreseeable future. However, mining and primary beneficiation of the various ores as well as secondary processing to ferro-alloy products generate large volumes of waste materials. The majority of these wastes is solid and is usually stored in landfill deposits, with little regard given to their long-term environmental stability. From this perspective, it is desirable to reduce both the amount of waste generated and the potential toxicity thereof. There is tremendous potential to achieve both these objectives in the sector through the installation of suitably chosen process technologies and innovative steps to achieve process-integrated recycling of wastes, treating wastes with other wastes and passing a waste as raw material to a downstream process. As was shown for the example of stainless steel production, such technologies can be easily integrated into the existing process flowsheet and ongoing operation.

However, while such technological measures can significantly contribute to a reduction of the waste load from the sector, there remains a liability with the operators of an industrial process to ensure responsible management of whatever wastes are ultimately generated from the process. In terms of solid waste deposits, this liability is directly associated with the potential for leachate generation. The *a priori* assessment of waste stability in this context requires a thorough methodology based on both a rigorous mathematical model of the leachate generation process and laboratory characterisation methods of the waste materials in question. It is this requirement which the proposed methodology is aimed to fulfil.

Our motivation for this step is rooted in the belief that, by characterising the stability of specific waste products, we can begin to focus attention backwards in the material chain to the waste-generation processes themselves, and to identify opportunities for process improvements to minimise waste formation and to render wastes environmentally benign.

REFERENCES

1. Council for Scientific and Industrial Research, The Situation of Waste Management and Pollution Control in South Africa, Report CPE 1/91, Pretoria, 1991.
2. Noble, R. G., Ed., Hazardous Waste in South Africa, Department of Environmental Affairs, Council for Scientific and Industrial Research, Pretoria, 1992.
3. Stewart, M. and Petrie, J. G., The waste generation potential of the South African minerals industry, unpublished report, University of Cape Town, 1996.
4. Petrie, J. G. and Paxton, R. G., Integrated technologies for treating liquid and solid wastes from ferrochrome and stainless steel production, Pollution Prevention for Process Engineering, Proceedings of the Conference on Technical Solutions for Pollution Prevention in the Mining and Minerals Processing Industries, Palm Coast, Florida, January 1995.
5. Alloway, B. J. and Ayres, D. C., Chemical Principles of Environmental Pollution, Blackie Academic & Professional, London, 1992.
6. Mangold, D. C. and Tsang, C., A summary of subsurface hydrological and hydrochemical models, Review of Geophysics, 29, 51, 1991.
7. Rowe, R. K. and Booker, J. R., Pollutant migration through liner underlain by fractured soil, J. Geotech. Eng., 117, 1902, 1991.
8. Dixon, D. G. and Hendrix, J. L., A general model for leaching of one or more solid reactants from porous ore particles, Metallurgical Transactions B, 24, 157, 1993.
9. Dixon, D. G. and Hendrix, J. L., A mathematical model for heap leaching of one or more solid reactants from porous ore pellets, Metallurgical Transactions B, 24, 1087, 1993.
10. Petersen, J., Assessment and modeling of chromium release in minerals processing waste deposits, unpublished doctoral dissertation, University of Cape Town, South Africa, 1997.
11. Petersen, J. and Petrie, J. G., A modelling strategy to predict leachate generation within minerals processing waste deposits, presented at XX International Minerals Processing Congress, Aachen, Germany, September, 1997.

12. Stegemann, J. A. and Cote, P. L., Investigation of test methods for solidified waste evaluation — a cooperative program, EPS 3/HA/8, Environment Canada — Wastewater Technology Centre, Ontario, 1991.
13. Jackson, D. R., Garrett, B. C., and Bishop, T. A., Comparison of batch and column methods for assessing leachability of hazardous waste, *Environ. Sci. Technol.,* 18, 668, 1984.
14. U.S. Environmental Protection Agency, The Toxicity Characteristic Leaching Procedure, Code of Federal Regulations, 40 (part 261, app I), 1992.

12 Sustainable Rehabilitation and Revegetation: The Identification of After-Use Options for Mines and Quarries Using a Land Suitability Classification Involving Nature Conservation*

N. J. Coppin and J. Box

CONTENTS

ABSTRACT

The principles of sustainability require that land utilised for mineral extraction, waste disposal, and processing should be returned to some beneficial use once the operations are finished. Returning land to beneficial use — either the original state or some new purpose — involves rehabilitation and revegetation.

It is crucial to recognise that both rehabilitation and revegetation are processes rather than specific events involving just the return of stored soils and the introduction of vegetation. As a process, careful definition of the output (the final land-use) will determine the type and duration of the inputs (materials or vegetation management/cultivation) needed to drive the overall process. Moreover, such processes can be monitored to determine the need for further inputs. The resultant

---

* Reprinted with amendments from Fox, H.R., Moore, H.M., and McIntosh, A.D. (eds), *Land Reclamation: Achieving sustainable benefits*. Proceedings of the 4th international conference, Nottingham, U.K., 7-11 September 1998. 560 pp., EUR75.00//U.S. $88.00/GBP 53.00. A A Balkema, P.O. Box 1675, Rotterdam, Netherlands.

BOX 12.1
Definitions

Reclamation is a widely used term to describe the process of creating a land-use, which may be hard (industrial, commercial) or soft (agricultural, amenity), on a site where mining and quarrying operations have finished.

Rehabilitation has a broadly similar meaning to reclamation, though it implies that the after-uses are related to the land-use on the site prior to mining and quarrying.

Restoration is often used to mean restoring the original land-use or vegetation, or even the same landform. It is also applied to active mineral operations where after-use is developed by the mineral operator as part of the site activities, rather than starting with an abandoned or derelict site.

Revegetation is the process of vegetation establishment and aftercare undertaken as part of reclamation, rehabilitation, or restoration.

Recultivation is not widely used in English-speaking countries, but is often the literal translation of equivalent terms in other languages. It generally applies to the agronomic and ecological aspects of reclamation, rehabilitation, or restoration.

After-use means a land-use to which a site is returned, which should be beneficial although not necessarily economic.

Aftercare describes the crucial process of managing the soils and the vegetation systems after the initial revegetation or recultivation in order to ensure that the desired after-use is attained within a reasonable time period. The process would involve soil amelioration and vegetation management that is more intensive than normally associated with land in that particular use.

---

feedback will ensure that the necessary resources are deployed to maintain progress towards the desired after-use.

The ability of a site to achieve a given after-use depends on the viability of the soil-plant-animal systems that develop. This development is governed by a number of factors including slope and erosion, soils and soil-forming materials, climate and exposure, vegetation management/cultivation requirements, and the effects of associated animals on the developing vegetation. These factors can be linked with various potential land-uses through use of a *land suitability classification.* Such a classification can be used to suggest final land-uses, to examine the amelioration required to achieve a given land-use, to define site design and management parameters for a given land-use, and to compare the economic and environmental costs and benefits of alternative land-uses.

There is increasing interest in the application of ecological principles to the revegetation of abandoned mineral workings by allowing natural colonisation and natural succession to proceed towards climax vegetation such as woodland. Rehabilitation and revegetation can also be geared towards replacing or recreating natural vegetation communities that have been lost to agricultural intensification and industrial development. The creation of complex wildlife habitats has implications in terms of both the long-term ownership and the provision and funding of wildlife management.

12.1   INTRODUCTION

Mining and quarrying are only temporary uses of land, albeit very long-term in some cases. The principles of sustainability require that land utilised for mineral extraction, waste disposal, and processing should be returned to some beneficial use once the operations are finished. Returning land to such beneficial use, either the original use or some new land-use, involves rehabilitation and revegetation. Other terms used to describe the process of returning land to a beneficial after-use include reclamation, restoration, and recultivation. It is confusing that all these terms are used somewhat interchangeably to mean much the same thing. However, there are some differences between the terms that are worth noting (see Box 12.1).

It is important to recognise that rehabilitation and revegetation are ongoing processes rather than specific events. Both processes involve a series of activities, including planning, design, engineering a new landform, soil preparation, vegetation establishment, aftercare, and vegetation management. The requirements in any particular situation will be site-specific, depending on the objectives, the available resources, and the constraints that have to be overcome. Careful definition of the output (the final land-use or *after-use*) will determine the type and duration of the inputs needed to drive the process of rehabilitation or revegetation. Moreover, such processes can be monitored to determine the need for further inputs, whether of materials and/or vegetation management and cultivation. The resultant feedback needs to be directed to ensure that the necessary resources are used to maintain progress towards the desired after-use.

12.2   THE PRINCIPLES OF A REHABILITATION PLAN

As a general rule, prevention is better than cure. In terms of sustainable development, it is much better if the closure and decommissioning of a mine or quarry is undertaken, and paid for, as part of the mineral operation. It is now becoming widely accepted that project proposals and investment plans should include provision for full decommissioning. For example, the Ontario Ministry of Northern Development and Mines in Canada produced a technical document "Rehabilitation of Mines: Guidelines for Proponents,"1 which sets out principles, objectives, and criteria for a mine closure and rehabilitation plan. Mining and quarrying companies in many countries now produce plans which apply these or similar guidelines to both existing mineral operations and to new proposals.

The purpose of a decommissioning plan is to ensure that, after the mining or quarrying operation is finished, the site does not impose a hazard to public health and safety as a result of physical or chemical deterioration. It is important that a mineral operation, as a temporary use of the land, does not impose any permanent constraints on the options for future beneficial use of the site and adjacent land, nor has any permanent effects on the local water resources, biodiversity and overall landscape quality. In terms of the long-term acceptability of a rehabilitated site, it is more appropriate to consider a design life span measured in centuries rather than in decades.

A mine or quarry closure plan should, therefore, include a long-term assessment which covers:

- *Physical stability* of structures and workings that remain after closure and rehabilitation.
- *Chemical stability*, particularly mobilisation and dispersal of pollutants into the environment.
- *Land-use* and long-term compatibility of the rehabilitated site with the surrounding land.

The *Ontario Guidelines* identify three categories of rehabilitation in terms of the after-use that is sustainable in the long-term:[1]

- A *walk-away* status, where there are no residual constraints on the future use of the land remaining after rehabilitation has been carried out and where there are no additional monitoring or maintenance requirements.
- *Passive care*, where there is minimal need for monitoring and infrequent maintenance of non-critical structures.
- *Active care*, requiring regular operations, monitoring and maintenance of the site that is not typical of normal land management practices; there may be permanent constraints on the beneficial use of the land, such as high metal concentrations.

In principle, a site can only be considered properly rehabilitated if it has achieved a "walk away" or "passive care" status. In the long term, "active care" will not be suitable without the provision of considerable resources. The aim should be to rehabilitate the land to a use that is beneficial, whereby the resources required to maintain it are consistent with the benefits derived from the land. Such benefits may be economic, such as industrial use and crop production, or intangible, such as amenity or wildlife value. In either case, the long-term management burden on the occupier/owner must not be such as to prohibit the maintenance of the after-use, otherwise the site may become neglected. Planning for disposal of residues and spoils has to consider the long-term ability of the land to be rehabilitated or restored to a beneficial after-use. In other words, it

TABLE 12.1
Relative Benefits of After-Use Options in Temperate Climates

Use		Economic	Landscape	Amenity/Wildlife
Agriculture	Arable.	Moderate to high, depending on local agricultural conditions. Takes time to establish.	High where field boundaries are retained; restored land can provide a variety of colour and texture. Moderate where prairie farming practical.	Very low unless low input/organic farming techniques are used.
	Pasture.	Moderate to high for dairy and beef production, depending on local agricultural conditions.	High where traditional field boundaries are retained.	Low unless low input/organic farming techniques are used.
	Rough grazing.	Low.	Moderate to high, depending on vegetation.	Moderate, but grazing intensity may cause damage to vegetation.
Forestry	Timber or pulp.	Moderate, but requires long-term investment.	Moderate when properly planned.	Low when pure monoculture. Moderate to high in mixed planting.
Energy Crops	Biomass, woody or non-woody.	Moderate.	Moderate when properly planned.	Low to moderate, dependant on the crop.
Nature Conservation	Biodiversity/ wildlife habitats.	Low, unless as part of wider attractions.	High, especially with a variety of habitats or extensive areas.	High.
	Industrial heritage	Low, unless as part of wider attractions.	High, if areas maintained as part of local landscape.	Moderate to high.
Recreation and Tourism	Intensive uses (golf, fishing).	High. Can be maximised through tourism.	Low to moderate depending on the activities.	Low to moderate depending on the activities.
	Extensive uses (walking, riding).	Moderate where membership scheme can be implemented.	Moderate to high.	Moderate to high.
Development and Building	Industry and commerce.	High, provided that the cost of clean up is not onerous.	Low to moderate, depending on development type.	Low.
	Housing.			

is necessary to adopt an after-use led approach with the land being considered as a potential asset with a long-term future rather than a liability to be discharged as quickly and cheaply as possible.

12.3 AFTER-USE OPTIONS

A selection of typical after-use options and their relative benefits are given in Table 12.1. For each situation and site, the feasibility of the available rehabilitation or restoration options has to be examined, taking into account factors such as: the site and its access, the soil-forming materials that are available, the supply of water, contamination, climate, and the management resources which would

TABLE 12.2*
A Land Suitability Classification for China Clay Wastes*

Site Factor	Degree of Limitation	Land Suitability Class
Gradient	<11°	1
	11° to 18°	2
	18° to 25°	3
	>25°	5
Exposure	Very sheltered	1
	Moderately sheltered	1
	Moderately exposed	2
	Very exposed	3
	Severely exposed	4
Type of soil-forming material	Ameliorated sands and mica	1
	Mica	2
	Sand	3
	Stent/overburden	3
Depth of soil-forming material	>300mm	1
	50 to 300mm	2
	<50mm	3
Stone content of soil-forming material	<10%	1
	10 to 35%	2
	>35%	3

TABLE 12.3*
Potential After-Uses and Land Suitability Classification of China Clay Wastes*

After-use		Land Suitability Class				
		1	2	3	4	5
Agriculture	Arable	●				
	Pasture	●	●			
	Rough grazing			●	●	○
Forestry	Timber	●	●	●		
	Firewood	●	●	●	○	○
Nature Conservation	Woodland	●	●	●	○	○
	Semi-natural scrub	●	●	●	●	●
	Moor and heath			●	●	○
Recreation	Intensive	●	●	●	●	●
	Extensive			●	●	●

● Suitable uses based on present knowledge
○ Potential uses given further research and development
* Tables 12.2 and 12.3 Crown Copyright. Reproduced with the permission of the Controller of Her Majesty's Stationery Office.

be required. The suitability for different potential after-uses can then be considered, alongside the engineering feasibility, the land area available, and any local factors affecting the choice of land-use.

12.4 LAND SUITABILITY CLASSIFICATION

The ability of a site to be rehabilitated or revegetated depends on the viability of the soil-plant-animal systems that develop. This development is governed by a number of factors including slope and erosion, soil-forming materials (e.g. type, depth, fertility, pH, water-holding capacity, cation exchange capacity, and drainage), climate and exposure, vegetation management requirements, and the effects of associated soil invertebrates and grazing animals on the developing vegetation. These factors can be linked with various possible after-uses through a *land suitability classification*.

A land suitability classification developed for the reclamation of china clay wastes in the southwest of England[2] is summarised in Tables 12.2 and 12.3. This is based on the principles of *land capability classification*,[3] and *agricultural land classification*,[4] which are used, respectively, by land-use and agricultural planners to classify land according to its potential for different uses.

The various degrees of limitation of five critical site factors for china clay wastes — the gradient, the exposure, and the type, depth, and stone content of the soil-forming material — are used in Table 12.2 (other factors such as drainage, pH, or soil fertility could also be used if appropriate). Five levels of land suitability classification are shown in Table 12.2. Land suitable for agricultural use is covered by classes 1 to 3: class 1 has few limitations and class 3 has moderately severe limitations. Classes 4 and 5 have severe limitations and include land that is only suitable for non-agricultural use. The land suitability classification for the whole of a site, or for part of a site, is determined by the site factor imposing the most severe limitation. For example, a site with classes of 1 or 2 for all the site factors except gradient, where the class is 5, will have an overall land suitability classification of 5.

The potential after-uses for sites with land suitability classifications of 1 to 5 are given in Table 12.3. Agriculture and forestry are restricted to china clay wastes with land suitability classifications 1 to 3, whereas nature conservation is an after-use option for all levels of land suitability classification.

A land suitability classification can be used as a design tool whose purpose is to define the required land characteristics for it to be suitable for a particular after-use. Such a land suitability classification can be used to:

- Suggest suitable after-uses for an existing site.
- Examine the possibilities for amelioration to achieve a given after-use.
- Define the parameters of site design and management that are required by a given after-use.
- Compare economic and environmental costs and benefits of alternative after-uses.

12.5 LESSONS FROM NATURE

Mineral workings are often perceived as a threat to nature conservation due to destruction of wildlife habitats, alterations to the hydrology of adjacent habitats, changes to the sediment burden or chemistry of adjacent watercourses, or disturbance of wildlife sensitive to human activity. Wildlife habitats have, however, become established through natural colonisation and succession within abandoned mineral workings or have been deliberately created as part of a planned restoration programme.[5-7]

There is increasing interest in the application of ecological principles to the revegetation of mineral workings by allowing natural colonisation and natural succession to proceed, perhaps to climax vegetation such as woodland. Revegetation can be geared towards replacing or recreating natural vegetation communities that have been lost due to past agricultural intensification or industrial development.

Recently, the contribution of past mineral workings to nature conservation in England has been assessed in a research project undertaken by Wardell Armstrong.[8] Six counties were chosen to provide a geographical spread and a range of types of mineral workings. Only sites with significant nature conservation designations in the six counties were included in the study:

TABLE 12.4
Proportion of SSSIs, LNRs and WTNRs
Associated with Mineral Workings

	SSSIs	LNRs	WTNRs
Cambridgeshire	21%	33%	23%
Derbyshire	36%	27%	9%
Devon	30%	7%	8%
Durham	38%	100%	19%
Essex	17%	0%	9%
Staffordshire	20%	100%	15%
TOTAL	27%	23%	13%

Notes: a) Table based on 1993/1994 data.
  b) Those LNRs and WTNRs that are also
     SSSIs have been included in both the rel-
     evant sections.

TABLE 12.5
Proportion of SSSIs Associated with
Mineral Workings

	Geological SSSIs	Biological SSSIs	Mixed Interest SSSIs
Cambridgeshire	70%	16%	0%
Derbyshire	48%	23%	53%
Devon	65%	4%	17%
Durham	75%	25%	60%
Essex	59%	7%	0%
Staffordshire	87%	7%	20%
TOTAL	64%	13%	27%

Note: Table based on 1993/1994 data.

- *Sites of Special Scientific Interest* (SSSIs)—statutory sites in public or private ownership which are notified by English Nature (the statutory nature conservation agency in England) under the Wildlife and Countryside Act 1981.
- *Local Nature Reserves* (LNRs)—statutory sites which are declared by Local Authorities under the National Parks and Access to the Countryside Act 1949 (some of these sites are also SSSIs).
- *Wildlife Trust Nature Reserves* (WTNRs)—sites which are owned and managed for nature conservation by County Wildlife Trusts (some of these sites are also SSSIs and/or LNRs).

Second-tier sites such as *Sites of Importance for Nature Conservation* (SINCs) were not included due to the large number of such sites and the relatively incomplete information available on these sites compared to that available for SSSIs, LNRs and WTNRs.

The results for the six counties (Table 12.4) show that mineral workings are included within the site boundary of 27% of the SSSIs, 23% of the LNRs and 13% of the WTNRs. Mineral workingsassociated with SSSIs, LNRs or WTNRs can be sites where the nature conservation interest of the site derives wholly from the mineral workings (for example, geological exposures or natural

BOX 12.2
Oxlow Rake — A Case Study

Oxlow Rake occupies the site of a former lead deposit at the northern end of the Derbyshire orefield. The mine workings and dumping of waste minerals have created topography of hummocks and hollows of varying lead content. These toxic and inhospitable soils have only a thin cover of a vegetation that includes plants which have adapted to the environmental conditions (metallophytes). The most notable of these metallophytes is the spring sandwort, or leadwort, which is confined to heavy metal soils throughout nearly all of its British distribution. Oxlow Rake is of considerable research interest and has contributed to the understanding of how plant species adapt to stressed environments.

BOX 12.3
Wingate Quarry— A Case Study

Wingate Quarry (Durham) was worked for magnesian limestone between the mid-18th century and the 1930s. The large quarry complex consists of several connecting quarries that have been abandoned for varying lengths of time. Consequently, the vegetation in each quarry has reached different stages in the process of natural colonisation and succession that range from open vegetation to young ash woodland over hawthorn scrub. The most important habitats are the tall herb communities, which are particularly rich in insects, and the magnesian limestone grassland, which is of great nature conservation value due to its restricted distribution in Britain.

BOX 12.4
Little Paxton Pits — A Case Study

Little Paxton Pits (Cambridgeshire) is an extensive area of flooded gravel workings of varied ages with correspondingly diverse vegetation. The site is of national importance for wintering wildfowl and is an important stopping point for migrating birds. The invertebrate fauna associated with the pits is extremely diverse and includes a number of national rarities. Woodland, scrub, and areas of grassland with hedges enhance the wildlife value of the aquatic and marsh vegetation. Positive management of the site by the gravel companies who own the site over the years since gravel extraction ceased has been undertaken in conjunction with Huntingdonshire District Council which established a Local Nature Reserve in 1989. The result is that Little Paxton Pits is making a major contribution to nature conservation in the Ouse Valley.

recolonisation within a quarry). There are also SSSIs, LNRs, and WTNRs where an area of mineral working is included within a site boundary that includes other biological or geological features not related to mineral working. More information was available for the SSSIs from which it was possible to distinguish those SSSIs that were either biological or geological or a mixture of the two (mixed-interest sites). As would be expected, two thirds of the geological SSSIs (64%) and over one-quarter (27%) of the mixed-interest SSSIs in the six counties were associated with mineral workings (Table 12.5).

Importantly, mineral workings were associated with some 13% of all the biological SSSIs in the six counties. The proportion ranged from 4% in Devon to as high as 25% in Durham where the grasslands which develop on magnesian limestone workings are very valuable for nature conservation. Natural colonisation and succession can create a variety of wildlife habitats on abandoned mineral workings, which in some cases are of particular importance for nature conservation either in their own right (Box 12.2) or as part of a larger area which includes the area of mineral workings.

The development of wildlife habitats takes a long time in limestone or chalk quarries where there are severe environmental stresses due to lack of water and nutrients. The biological communities that eventually develop on these habitats are often of particular value for nature conservation due to the rarity of exposed limestone and chalk as habitats and the species diversity of the vegetation that develops. Chalk and limestone workings are usually associated with a range of terrestrial habitats demonstrating stages along the natural succession from grassland to woodland (Box 12.3). Barnack Hills and Holes (Cambridgeshire), for example, is famous as the source of the Jurassic limestone from which Peterborough Cathedral was constructed in medieval times, as well as for the species-rich limestone grassland which has developed over the old mine workings. Clay and sand/gravel workings will support a variety of grasslands, scrub and woodland communities. However, wet sand and gravel workings and clay workings with high or perched water tables will naturally develop into a wide range of wetland habitats (Box 12.4). Mining subsidence tends to result in a range of wetland habitats such as open water, reedswamps, and marshes.

Operational mineral workings can also be associated with wildlife. Badgers, dragonflies, peregrine falcons, and sand martins spring to mind as being able to take advantage of these havens to which the public does not have access.

These results were followed up by the identification of issues that those involved in the minerals industry considered to be important in the restoration of mineral workings for nature conservation. A survey of 44 Mineral Planning Authorities (MPAs) and 18 mineral operators throughout England was undertaken using a questionnaire which contained ten questions relating to the nature conservation value of mineral workings. Further correspondence with the MPAs and the mineral operators provided information to supplement the analysis of the questionnaires.

Replies were obtained from 36 of the 44 Mineral Planning Authorities (82%) and 13 of the 18 mineral operators (72%). These high returns within the very short time-scale imposed by the project indicate that the subject of nature conservation and mineral workings is one of real interest to those involved in the minerals industry. The results are given in Table 12.6.

The majority of the Mineral Planning Authorities (83%) and the mineral operators (85%) have policies that encompass the nature conservation after-use of mineral workings (Q1). There is, however, a variable response in both groups to the emphasis given to nature conservation as an after-use (Q2); it is usually combined with another after-use, with recreation, education or landfill being the most popular (Q3/Q4). Provisions are made by the majority of the mineral operators for the conservation of interesting ecological or geological features (Q7/Q8) with more mineral operators making provisions, albeit qualified in many cases, for ecological features (92% of mineral operators) than geological features (69% of mineral operators).

There was a wide range of views on the use of the natural development of vegetation as a suitable method for establishing sites of ecological interest on mineral workings (Q6). There are concerns about allowing the natural processes of colonisation and succession to provide the vegetation required by a restoration scheme. These concerns undoubtedly relate to the terms of the restoration conditions. For example, a longer time-scale is required for natural processes compared to the quick greening that results from the use of fertiliser and a seed-mixture of amenity grasses.

The majority of both the mineral operators (77%) and the Mineral Planning Authorities (86%) would like to be able to predict the potential of a site to develop an ecological or a geological interest (Q10). Only 54% of the mineral operators, however, maintain a record of the nature conservation interest that has developed on mineral workings (Q9) in their ownership.

It is clear from this study that both mineral operators and Mineral Planning Authorities appreciate the role that mineral workings can play in creating wildlife habitats and geological exposures. Both groups have made positive commitments to nature conservation through the development of appropriate policies. Nature conservation can be an end-use in its own right, often in combination with recreation, education or even landfill.

12.6 THE WAY FORWARD

It is significant that a high proportion of both mineral operators and Mineral Planning Authorities want to be able to predict whether a given site will develop ecological or geological features of importance for nature conservation. Any such predictions must recognise the role played by chance

— particularly in relation to natural colonisation and succession. The time-scale involved in such natural biological processes is often long and is related to the type of mineral involved and the method of working, as well as to the source of seeds and spores.

The scale of modern quarries and current methods of excavation produce relatively straight working faces with flat floors and benches free of stone waste. These workings are in contrast to the older, small-scale workings, which are often associated with important wildlife habitats. Blasting with black powder left shattered faces and large quantities of waste fragments heaped up into spoil banks that provide the varied topography and microclimates so characteristic of these older sites. A similar situation is found in the sand and gravel industry, where modern workings tend to have less variation in slope, aspect, and shoreline, thus restricting the opportunities for nature to colonise.

Modern restoration schemes have been developed for these modern working methods. Flat quarry floors and benches are restored to become bland areas of grass with planted trees. A prevalent sense of tidiness reinforces this general approach. Are restoration schemes going too far? Could better use be made of natural processes and more varied features and landforms? Should there be more untidiness?

The integration of restoration proposals with the extraction programme can help to ensure that the final mineral workings are designed for nature conservation where this is an agreed end-use.[9–11] Techniques are available for creating more natural landforms from the faces left in quarries,[12] and for designing sand and gravel quarries to take maximum advantage of high water-tables once extraction and pumping have ceased.[13,14] Such techniques provide a wide range of habitats for plants and animals to colonise and can bring significant environmental benefits.

It is clear from the concerns expressed during this study that the current system of restoration conditions needs to be looked at in relation to the development of wildlife habitats. The results of the questionnaire survey indicate the wide range of views amongst both mineral operators and Mineral Planning Authorities over whether the natural development of vegetation is a suitable method to establish wildlife habitats in mineral workings. Wildlife habitats that develop naturally are much more complex and species-rich than the vegetation that develops from mixtures of sown grasses and wildflowers. But natural processes "green up" a site more slowly than enhanced processes involving sowing and planting.

There are, however, amenity, landscape, and local visibility issues that have to be taken into account. The current five-year period for aftercare means that, in practice, there is a presumption in favour of quick and reliable methods of establishing vegetation on a site. The objectives for the restoration of abandoned mineral workings need to be carefully thought through. Appropriate methods are needed which take account of the needs of nature conservation and biodiversity, which may conflict with the needs of amenity and reduced local visual impact. Zoning may help to resolve some of these conflicts: the external landscape may have a different purpose (amenity) from that within the restored site (nature conservation). Local communities may be willing to tolerate a relatively poor visual appearance in the short term if the time-scale and the purpose of a restoration to nature conservation are explained.

Long-term management aims and procedures need to be developed which allow for natural colonisation and succession.[15] The benefits will be the development of a greater wildlife interest through such natural processes. Such management techniques will undoubtedly include an element of what might be called "doing nothing" or "benign neglect." It is often harder to justify doing nothing than to be seen to be actively managing a site.

Two issues were raised many times during the study: how to finance the long-term management of a restored site and who should own a restored site once the mineral operator has no further economic need for the site. The two are clearly related. Restoration to a nature conservation after-use may not be compatible with the transfer of the site to the local farmer. There are mechanisms available for addressing these issues — for example, Section 106 Agreements involving planning obligations or management agreements with voluntary bodies or local author-

TABLE 12.6
Responses to Questionnaire Sent to Mineral Planning Authorities (MPA) and Mineral Operators (MO)

Q1	*Does the authority/operator have policies on nature conservation (ecological and geological) after-use of mineral workings?*					
	Yes	No				
MPA	83%	17%				
MO	85%	15%				

Q2	*Does the authority/operator consider nature conservation to be a high priority after-use?*					
	1	2	3	4	5	Depends on site
	(High)				(Low)	
MPA	28%	44%	19%	3%	-	6%
MO	23%	38.5%	38.5%	-	-	

Q3	*Is nature conservation combined with other after-uses?*					
	1	2	3	4	5	Depends on site
	(Always)				(Never)	
MPA	-	28%	61%	5.5%	-	5.5%
MO	15%	46%	31%	8%	-	

Q4	*Which after-uses are combined with nature conservation?*					
	Recreation	Education	Landfill	Industrial	Agriculture	Housing
MPA	92%	53%	47%	11%	39%	11%
MO	77%	23%	46%	15%	31%	8%

Q5	*Does the authority/operator advise on the type of mineral workings most suitable for a nature conservation after-use?*		
	Yes	No	Other
MPA	64%	27%	9%
MO	31%	46%	23%

Q6	*Does the authority/operator regard the natural development of vegetation as a suitable method to establish an ecological interest in mineral workings?*				
	1	2	3	4	5
	(Always)				(Never)
MPA*	5.5%	30.5%	42%	11%	8%
MO*	-	31%	38%	8%	8%

*The percentages do not total 100%, as some respondents did not reply to this question

Q7	*Does the authority/operator make provision for the conservation of interesting geological features exposed during the life of a mineral working?*		
	Yes	No	Other
MPA	83%	17%	-
MO	69%	15.5%	15.5%

Q8	*Does the authority/operator make provisions for the conservation of interesting ecological features developed during, and after, the life of a mineral working?*	
	Yes	No
MPA	89%	11%
MO	92%	8%

TABLE 12.6 (continued)
Responses to Questionnaire Sent to Mineral Planning Authorities (MPA) and Mineral Operators (MO)

Q9   *Does the authority/operator maintain a record of mineral working sites that have developed a nature conservation interest?*

	Yes	No	Other
MPA	47%	50%	3%
MO*	54%	15%	21%

*The percentages do not total to100%, as some respondents did not reply to this question

Q10   *Would you like to be able to predict the potential of a site to develop an ecological or geological interest?*

	Yes	No	Other
MPA	86%	6%	8%
MO	77%	8%	15%

ities. No easy general formula is available. Rather, the needs of each site have to be dealt with on a site-specific basis.

12.7   ISSUES FOR THE FUTURE

There is still a general perception that mineral workings are a threat to the environment, despite recognition of the link between mineral workings and nature conservation. Planners and decision-makers need to acknowledge the potential benefits of creating new wildlife habitats and geological features as an integral part of restoration schemes. Increasing land-use pressures mean that new mineral workings need to have significant environmental as well as economic benefits.[16] Restoration schemes cannot afford to be seen as "making the best of a bad job." A change in prevailing attitudes is needed in order that minerals can continue to be extracted and that interesting wildlife habitats and valuable geological features can be created.

Significant benefits to the minerals industry and to nature conservation would come from actions that would:

- Promote the contribution made by past mineral workings to nature conservation due to the creation of a wide variety of new wildlife habitats and geological features.
- Define the types of vegetation communities that have developed on a range of different mineral workings, as well as the scope for changes in these communities as a result of natural succession. This data would allow simple predictive models to be developed for the benefit of mineral operators and planners.
- Develop restoration approaches with an emphasis on the creation of natural landforms with a diversity of slopes, aspects, drainage, and physical features in order to maximise the number and variety of niches available for natural colonisation by plants and animals. Nature can do an excellent job of restoring abandoned mineral workings without the need for heavy-handed restoration schemes. Too much intervention and tidying up can reduce the potential of a site for wildlife.
- Investigate appropriate funding mechanisms for habitat management and aftercare, as well as arrangements for land tenure, which maximise the nature conservation resource provided by disused mineral workings. The usual five-year aftercare period (derived mainly for agricultural restoration) places undue emphasis on quick and reliable methods for "greening" a site.

ACKNOWLEDGEMENTS

The research project was carried out by John Mills during an industrial placement with Wardell Armstrong (mining, minerals, engineering, and environmental consultants) as part of an M.Sc. degree in Environmental Analysis and Assessment undertaken at Anglia Polytechnic University (Cambridge). The assistance of many people from Mineral Operators and Mineral Planning Authorities throughout England, as well as from Wildlife Trusts and English Nature in the six counties studied, is gratefully acknowledged. The results of this study were first presented in *Mineral Planning* and their use is with the permission of the editor (*Mineral Planning*, 2 The Greenways, Little Fencote, Northallerton, DL7 0TS, U.K.). Crown copyright in respect of Tables 12.2 and 12.3: reproduced with the permission of the Controller of Her Majesty's Stationery Office.

REFERENCES

1. Ontario Ministry of Northern Development and Mines, *Rehabilitation of Mines: Guidelines for Proponents*, Ministry of Northern Development and Mines, Sudbury, 1992.
2. Department of the Environment, *Landscaping and Revegetation of China Clay Wastes: Main Report*, Department of the Environment, London, 1993.
3. Bibby, J. S. and Mackney, D., *Land-Use Capability Classification*, Technical Monograph No. 1, Soil Survey of England and Wales, Harpenden, 1969.
4. Ministry of Agriculture, Fisheries and Food, *Agricultural Land Classification of England and Wales: Revised Guidelines and Criteria for Grading the Quality of Agricultural Land*. Ministry of Agriculture, Fisheries and Food, London, 1988.
5. Ratcliffe, D. A., Ecological effects of mineral exploitation in the United Kingdom and their significance to nature conservation, *Proceedings Of The Royal Society Of London*, Series A, 339, 355, 1974.
6. Gemmell, R. P., The origin and botanical importance of botanical habitats, *Urban Ecology*, Bornkamm, R., Lee. J. D., and Seaward, M. R. D., Eds., Blackwell Scientific Publications, Oxford, 1982, 33.
7. Box, J., Conservation or greening? The challenge of post-industrial landscapes, *British Wildlife*, 4, 273, 1993.
8. Box, J. D., Mills, J., and Coppin, N. J., Natural legacies: mineral workings and nature conservation, *Mineral Planning*, 68, 24, 1996.
9. Bradshaw, A. D. and Chadwick, M. J., *The Restoration of Land: The Ecology and Reclamation of Derelict and Degraded Land*, Blackwell Scientific Publications, Oxford, 1980.
10. Merritt, A., *Wetlands, Industry and Wildlife. A Manual of Principles and Practices*, The Wildfowl and Wetlands Trust, Slimbridge, 1994.
11. Department of the Environment, *Reclamation of Damaged Land for Nature Conservation*, Her Majesty's Stationery Office, London, 1996.
12. Department of the Environment, *Landform Replication as a Technique for the Reclamation of Limestone Quarries*, Her Majesty's Stationery Office, London, 1992.
13. Andrews, J. and Kinsman, D., *Gravel Pit Restoration for Wildlife: A Practical Manual*, Royal Society for the Protection of Birds, Sandy, 1990.
14. Littler, A., *Sand and Gravel Planning and Restoration*, The Institute of Quarrying, Nottingham, 1990.
15. Department of the Environment, *Cost Effective Management of Reclaimed Derelict Sites*, Her Majesty's Stationery Office, London, 1989.
16. Coppin, N. J. and Richards, I. G., *Use of Vegetation in Civil Engineering*, CIRIA/Butterworths, London, 1990.

13 Integrated Planning for Economic Environmental Management During Mining Operations and Mine Closure

David R. Morrey

CONTENTS

ABSTRACT

The decision to initiate and continue production at any mine operation is based upon economic criteria. Key determinants in the decision-making process include the current and predicted market value of the commodity, and the cost of production. Production costs necessarily include expenditures related to environmental management during operations. The costs of reclamation and closure are usually the most significant throughout the project life cycle. Consequently, the cost of environmental management during operations and at closure is considered at an early stage in mine planning, often during project feasibility analysis. This chapter presents the main issues of environmental cost management, which ones are assessed during feasibility studies, and sometimes as early as the pre-feasibility assessment. The process of reducing costs by incorporating sequential reclamation and phased closure concurrently with mining operations is advocated. The concepts of systematic reclamation and closure technology selection during early stages of project planning, and the cost benefit of designing for closure, are discussed. In particular, reference is made to uncertainty and risk as factors that must be included in closure design and cost management.

1-56670-365-4/00/$0.00+$.50
© 2000 by CRC Press LLC

13.1 INTRODUCTION

Within the U.S., the regulatory environment at both state and federal levels has exerted significant economic pressure on mine operators. In particular, increased stringency in standards for environmental protection on and around mines has a tendency to increase operating costs during production, and capital costs at closure. Also, as a result of increasing environmental awareness in the public sector, mine operators must be cognisant of contingent liabilities related to regulatory compliance and third party, common-law action. This awareness applies as much to the environments of developing countries as it does to established industrial nations.

Hutchison and Ellison[1] present evidence of the cost sensitivity of mining operations to changes in environmental regulation in California. For example, more stringent requirements for precious metal waste rock containment were estimated to increase rock dump management costs three-fold, if it should be necessary to introduce liner systems for groundwater protection. Undoubtedly, mining projects are cost-sensitive to environmental regulation, and can incur reduced financial rates of return, resulting from increased costs in environmental management during operations and at mine closure.

The most expensive closure components relate to physical rehabilitation of mine disturbance, and the elimination of long-term maintenance requirements and liabilities beyond closure. However, with careful planning, technology selection, and design, rehabilitation and closure costs can be significantly reduced, particularly if reclamation is performed concurrently with mining operations, and partial closures of exhausted mine components are completed within operational costs.[2]

13.2 MINING PROJECT DEVELOPMENT

Mining projects generally progress through a life cycle that can be characterised by the following phases:

- Exploration.
- Pre-feasibility assessment.
- Feasibility assessment.
- Construction.
- Production.
- Closure.
- Post-closure.

This progression is simplified, and does not indicate the levels of detail within each phase, or the interactions that exist between the phases. However, the model provides a useful basis upon which environmental management strategies can be conceptualised, finalised, and implemented.

Each project phase is associated with specific mining and environmental activities that are becoming integrated more frequently into a single plan of operations. The mining and environmental activities commonly associated with each phase of a mining project life cycle are summarised in Table 13.1.

Depending upon prevailing legislation, environmental planning can commence before a minerals exploration permit is granted. This may require the preparation of a brief environmental assessment and management plan that provides details of methods which will be implemented by the proponent to mitigate impacts caused mostly by the construction of access roads and core drilling.

Minerals exploration and pre-feasibility studies necessarily involve geochemical and mineralogical (petrological) analysis, and geostatistical and economic orebody evaluation. During this exploratory stage, preliminary environmental management work may include broad environmental characterisation, a succinct environmental assessment to determine potential impacts and liabilities, and a review of regulatory requirements. An initial evaluation of capital and operating costs for mine development, production, and minerals processing, relative to mineral yield and market price

TABLE 13.1
Principal Mining and Environmental Actions During Each Phase of Mine Development

Phase in Mine Project Development	Principal Mine Planning Action	Principal Environmental Management Action
Exploration	Exploration road construction Rock core drilling Geochemical analysis Geostatistical analysis Orebody evaluation	Environmental assessment Rehabilitation plan Exploration permit application
Pre-feasibility study	Initial mine and minerals process planning Facilities siting Scheduling Econometric analysis Initial technology selection	Environmental baseline study Environmental assessment "Fatal Flaw" analysis Initiation of permitting process
Feasibility study	Plan of operations Technology selection Conceptual to final designs Costing and cost benefit analysis Investment brokerage	Comprehensive EIA and review Mitigation planning Reclamation and closure planning Conceptual design for closure Reclamation and closure costing Closure fund design
Construction	Access and haul road development Site clearing and grubbing Earth moving and surface water management Mine dewatering Utilities installation Building and infrastructure construction	Installation of pollution control facilities General environmental management (air, water, land) Construction phase reclamation and closure
Production	Ore extraction Size reduction Minerals processing Smelting and refining Maintenance and upgrade	General environmental management Performance assessment/audit Monitoring Concurrent reclamation Final closure design Partial closure Partial bond release
Closure	Facilities decommissioning Dismantling Decontamination Burial Removal Asset recovery Recycling	Implementation of closure plan Site cleanup Final reclamation Final impact assessment Post-closure planning
Post-closure		Treatment Maintenance Monitoring Final bond release

projections, is also completed during the pre-feasibility phase. Estimates of the costs of permitting, concurrent environmental management, and final reclamation and closure are included in the analysis, as are the possible economic returns from asset recovery, and recycling or reuse of equipment and materials at closure. More recently, financial returns from the sale of privately owned mined land is being considered as a method of offsetting or transferring environmental liabilities.

The financial spreadsheet evolves to a more sophisticated, more accurate, and precise level of detail during the feasibility study. At this stage, all prospective stakeholders review detailed project economics, including the potential for environmental liability. Potentially interested parties include institutional and private investors, insurance companies, commodity purchasers, regulatory agencies, and public representatives.

Feasibility-level work is focused primarily upon econometric analysis and mine planning. An integrated approach will include descriptions of environmental management strategies and preferred mitigation technologies. Costs related to their implementation and maintenance are also presented. During this phase, mine plans, engineering designs — including reclamation and closure designs, and project costs are refined from a conceptual to a final level of detail. The process involves the selection of mining and environmental control technologies, and the selection of design elements, based upon performance and cost-benefit criteria. An integrated approach to this exercise allows optimisation of multi-task scheduling and cost distribution. It also forms the basis of an environmental economics model which provides sufficient capital for premature, enforced, or planned early closure. Physical and fiscal methods for managing operational and closure costs, and the reduction of regulatory pressure, are usually evaluated during the feasibility study, thus requiring an in-depth assessment of regulatory requirements, and initiation of the permitting process.

A series of regulatory-driven environmental tasks is triggered during the pre-feasibility study, for completion before project approval. The tasks include, but are not restricted to:

- Detailed environmental baseline studies.
- Environmental impact assessment.
- Impact mitigation planning and design.
- Reclamation and closure planning and design.

Whilst frequently viewed as an unwanted imposition by some mine proponents, such regulatory-driven tasks may be regarded more positively as opportunities to explore and evaluate innovative and economic alternatives for environmental management. They may also be viewed as vehicles to facilitate effective risk and liability management, reduced regulatory pressure, and efficient cost control during operations and beyond closure. In particular, scientifically meaningful and defensible baseline studies and impact assessments have proven to be invaluable tools to support the proponent when faced with litigation. It is often the case that environmental litigation involving alleged mining-related environmental impacts is initiated by third parties acting under common law privilege, and may be opportunistic. It is, therefore, arguable that the permitting process provides the proponent with an ideal foundation for effective risk management which, when integrated with the mine plan in a thoughtful way, facilitates "value added" environmental management. Additionally, the results of initial studies and planning provide the basis of an integrated plan of operations for minerals production and environmental management.

13.3   AIMS AND OBJECTIVES OF MINE CLOSURE

The aims of mine closure are generic, regardless of the regulatory arena within which a mine operates. In general, the closure of each mine component should comply with both national and local regulations, and should fulfil specific permit requirements. Mine managers plan to minimise environmental impacts and therefore reduce the need for impact mitigation, during all phases of the project life cycle. By advocating *a priori* design and construction for closure, the application of best technical practices, and the performance of phased reclamation and closure tasks during operations, environmental liabilities will be managed at an acceptable cost:risk ratio. The need for long-term, active maintenance after closure will also be minimised.

The fundamental aims and objectives of mine rehabilitation and closure planning are described in the Ontario Closure Plan Technology Guidelines.[3] The Ontario Guidelines summarise the main objectives of mine site rehabilitation as:

- Protection of public health and safety.
- Elimination or reduction of environmental impact.
- Restoration of disturbed land to its pre-mining condition, or an acceptable alternative.

In doing so, maintenance of air quality, the protection of surface and groundwater resources, and the reclamation of land for agricultural or ecological production are accommodated. Each of these objectives may be considered in terms of designing for physical stability, waste management, and acceptable land use. Within the context of physical stability, all facilities and structures that remain after closure should not pose any risk to human health and safety, or to the environment, as a consequence of failure or deterioration. All structures should remain stable under the influence of extreme events and perpetually disruptive forces. Similarly, wastes that contain potentially toxic or corrosive chemical components should not adversely affect health, safety and the environment. This category of wastes includes acid rock drainage, soluble metal salts, cyanide, and miscellaneous solvents and hydrocarbons. Mechanisms of release of these components to the environment include leaching, runoff, spillage, and erosion, the management of which must be addressed in the closure plan.

These concepts are widely accepted in the mining industry and have become standard frames of reference for rehabilitation and closure planning internationally. More recently, objectives related to socioeconomic factors and cultural resources have been included as key closure planning components for the mining industry. This trend is evident in the World Bank and Berlin Guidelines for operations closure, and several of the more advanced codes of corporate practice.[4,5]

Perhaps the least understood and often-neglected issue in closure planning is that of site reclamation for land-use development. A closure plan should describe rationale and methods for site rehabilitation (amelioration) and reclamation (recovery from a derelict or waste condition). The realistic objectives of land rehabilitation and reclamation may differ significantly from the idealistic goal of site restoration to a pre-mining condition, which is frequently cited as a permit or closure requirement by regulatory agencies. In the context of land reclamation for ecological purposes, the restoration of pre-mining conditions may be technically and economically difficult, if the original condition was an ecological climax system. Within a practical time scale, mined land may be reclaimed and rehabilitated to provide a capability for natural adjustment and succession towards an ecological endpoint that resembles the pre-disturbance condition. The resultant ecosystem is most likely to be a stable dis-climax or sub-climax. However, the long-term stability of the system depends upon the restoration of a complex of interactive components which allow ecologically rehabilitated land to be self-sustaining. As a general rule, the biological stability of reclaimed land, including vegetated mine wastes, increases with species diversity and habitat diversity. Long-term sustainability also depends upon the establishment of microbiological processes within the rooting zone. Therefore, the approach to successful rehabilitation should be holistic, and aimed at habitat reconstruction, soil development, and ecosystem restoration, rather than vegetation establishment alone. Usually, this is seen as a process of short-term management and long-term natural succession.

In some circumstances, the establishment of ecological or agronomic land capability on degraded land is infeasible as a consequence of the limiting factors that may prevail at a particular location. Therefore, principles of land-use potential, capability, and feasibility assessment should be included as critical components of the reclamation planning process. All land that has been disturbed by mining activities has potential for economic, recreational, or aesthetic after-use. One objective of reclamation planning is to identify the potentials of mined land, and to select technologies and design elements that will transform land-use potential into a sustainable capability. The practical feasibility of achieving this depends upon a suite of potentially limiting factors that include physical, social, economic, and regulatory variables.

Current trends in closure planning therefore advocate assessment of land capability and end-use at an early stage in project development and reclamation design. The application of guidelines for the use of contaminated land,[6] and reference to systems such as the U.S. Department of Agriculture criteria for land-use capability,[7] its British modification,[8] FAO guidelines for assessing land suitability,[9,10] or other internationally accepted protocols are recommended when planning the reclamation of mined land with particular end-uses in mind.

13.4   CLOSURE PLANNING AND DESIGN

Conventionally, closure plans present five critical categories of information, which may be summarised as:

- Environmental characterisation.
- Identification and description of operational components, infrastructure and utilities to be included in the closure plan.
- Potential closure issues and liabilities.
- Selection of closure technology alternatives.
- Schedule and costs.

The planning process is iterative and requires a resolution to the apparently mutually exclusive objectives of environmental and operations managers, and public sector and regulatory demands. Consequently, closure plans become dynamic entities which evolve throughout the project, from an initial statement of intent to a definitive set of final designs and technical specifications.

Technology selection and design for closure of each mine component can be performed systematically in a similar manner to that described in Figure 13.1. The evaluation of closure options is based upon assessments of effectiveness, risk of failure, long-term stability, and cost.

At present, trends in closure planning and design involve technical review and analyses of risk and cost benefit, in both engineering and environmental terms. Conceptual designs may be screened by means of quantitative performance assessments which are relevant to design evaluations during operational and closure phases. Such analyses assist the selection of the most appropriate design alternatives. Selection can be facilitated further by way of objective decision analysis.

Performance assessments incorporate elements of operations audit, as well as engineering and environmental risk assessments. The environmental risk assessments can be both human and ecological. The advantages of a risk-based approach to closure planning lie in the quantification of subjective factors and the analysis of uncertainty related to engineering design or environmental control performance and cost. The outcome of this type of assessment is either the selection and implementation of an appropriate technology, or risk management. Risk management may involve re-design, additional data collection, or more detailed modelling, leading to final selection. In doing so, mine management aims to reduce risk and uncertainty. A schematic flow diagram, which describes a performance assessment protocol, is shown in Figure 13.2.

13.5   ENGINEERING PERFORMANCE ASSESSMENT

An engineering assessment must be performed for all structures remaining on the mine site to evaluate their post-closure performance. The performance assessment should include an analysis of stability under extreme events, such as seismic loading, and under perpetually disruptive forces,

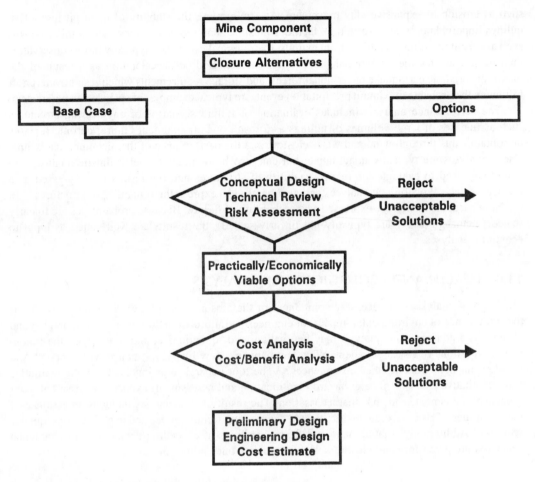

FIGURE 13.1   Selection of closure options.[17]

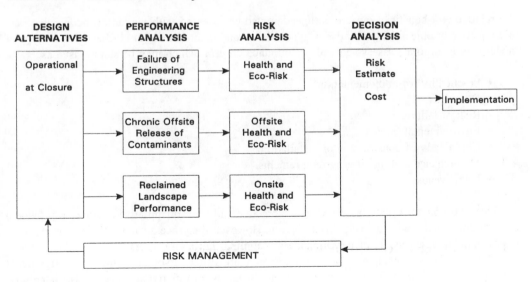

FIGURE 13.2   Performance assessment protocol.[18]

such as erosion or expansive clay movement. At a minimum, the stability of open pit highwalls, tailings impoundments, and waste rock facilities must be analysed. The forces used in this analysis, such as seismic loading or impacts from extreme hydrological events, are typically more conservative than those used for the operational life of the structure. Bedrock acceleration as a result of the maximum credible earthquake or erosional forces, and storage requirements which result from runoff following the probable maximum precipitation event, are typical examples of such design parameters.

The performance analysis includes evaluation of failure scenarios of engineering structures such as releases of large volumes of tailings as a result of impoundment failure; chronic releases of contaminants through wind and water erosion; and the performance of the reclaimed landscape. These analyses are typically quantitative; but can also be qualitative. Deterministic analyses are commonly used to evaluate the relative stability of structures and the results are expressed as a factor of safety. Probabilistic analyses can also be used to express the reliability of structures.[11] In the latter case, the results are expressed as a probability of failure. Because probabilities are linearly scaled (factors of safety are typically not linearly scaled), the results are well suited as input to economic analyses.

13.6   HEALTH AND ECOLOGICAL RISK ANALYSIS

This type of analysis is a necessary component of the closure technology screening procedure if the significance of environmental impacts of engineering failure, both during operations and beyond closure, is a concern. The analysis may be qualitative or quantitative, and relies upon the fundamental questions of risk assessment: What can happen? How likely is it that it will happen? And if it does happen, what are the consequences?[12] The four general steps involved in both qualitative and quantitative assessments are: hazard identification and assessment; exposure assessment; consequence assessment; and risk characterisation. The result of the analysis includes an estimate of the magnitude of risk associated with selected engineering failure modes and exposure/consequence scenarios. Additionally, contaminant-specific probabilities of exceeding prescribed environmental standards are given for each closure design alternative and failure mode.

13.7   COST-BENEFIT ANALYSIS

The relative cost-benefit of closure options can also be assessed with the aid of models that may be subjective or objective, qualitative or quantitative, and deterministic or stochastic. Each type of model may incorporate assessments of performance criteria, which may be summarised as:

- Practicality (of implementation).
- Durability.
- Risk of failure.
- Environmental impact.
- Complexity of construction or installation.
- Maintenance and monitoring requirements.
- Effectiveness.

These criteria are usually assessed in a subjective and semi-quantitative manner, to provide a cost per unit benefit ratio for each closure option. However, the process can be made more rigorous if probabilistic assessments of performance are applied. Difficulty arises, though, when reliable cost estimates for personnel, materials, and equipment are difficult to obtain, either due to significant variance in quotes, uncertainties of mass/volume estimates, or inflationary factors associated with long-term planning. This type of uncertainty in closure cost analysis can be managed using relatively simple statistical tools.[13]

13.8 ADDRESSING CLOSURE COST UNCERTAINTIES

The uncertainty or variability of parameters can be quantified using objective statistical approaches, or subjective decisions can be used to identify their variability. Typically, closure costs are calculated with the aid of a spreadsheet in which the quantities and the unit cost for the various items are listed and assumed to be deterministic. Two spreadsheet supplements are available which can describe any of the cells in a matrix as a variable. These additional software components are @RISK[14] and CrystalBall.[15] @RISK is compatible with the Lotus spreadsheet, while CrystalBall can be used with Lotus and Microsoft Excel software.

By defining the variable in any cell as a random variable, Monte Carlo or Latin Hypercube modelling can be performed to generate results expressed as a random variable. This will allow a large number of probability distributions to be simulated for the random variables. It is typical that estimates of the mean and variance about the mean are necessary to define the probability density function. In the case of some distributions, such as the β and γ distributions, it is also necessary to identify scale and/or shape parameters.

13.8.1 EXAMPLE OF MONTE CARLO SIMULATION

Consider a simplified example for closure costs of a mine waste dump. Assuming that there is no acid drainage present and that the waste dump can be reclaimed *in situ* by contouring the outer surface, placing growth medium, and vegetating the covered surface. Table 13.2 provides a typical cost estimate to perform the work. By assuming that all quantities and unit costs are random variables, the resulting closure cost based on expected values (mean values) can be estimated. Table 13.3 describes statistical distributions, quantities, and unit rates. For example, a triangular distribution is assumed for the quantity of earthworks required during surface recontouring. The triangular distribution has a minimum value of 180,000 m³, a maximum value of 220,000 m³ and a mean value of 200,000 m³. The unit cost is assumed to follow a normal distribution with a mean value of \$0.38 and a standard deviation of \$0.08. After performing a Monte Carlo simulation, the probabilistic results for the cost estimate shown in Figure 13.3 are obtained. These results show that when using expected values for all parameters, the projected value for closure costs is \$139,700 (the results shown in Table 13.2). However, there is a 10% probability that the closure cost will exceed \$163,700 or that it will be less than \$116,900. By using this approach, the upper, lower, and most likely costs can be estimated, and can be used by management as a decision-making tool for funds to be set aside for closure.

13.9 CLOSURE FUND ESTIMATION

Financial provision for closure may be planned by considering the following tasks:

- Estimation of the time to closure.
- Estimation of the period of post-closure reclamation and monitoring.
- Determination of the annual costs of post-closure reclamation.
- Determination of the value of fund required at closure.
- Forecast of the rates of return on the fund.
- Calculation of periodic payments.

Generally, closure costs may be reduced significantly if concurrent reclamation and partial closures of mine components are effected during the operational life of the mine. Such costs may be accounted for as ongoing operational costs and can be easily planned. However, the capital cost of final closure has several uncertainties attached to it.

TABLE 13.2
Reclamation Cost Example

Cost Item	Units	Quantity	Unit Cost	Cost
Recontouring	m³	200,000	$0.38	$76,000
Placement of growth medium	m³	28,000	$1.65	$46,200
Vegetative cover	hectares (ha)	35	$500	$17,500
TOTAL COST				$139,700

TABLE 13.3
Distribution Functions for Reclamation Example

Variable	Units	Distribution	Mean	Standard Deviation	Range
Recontouring	m³	triangular	200,000		180,000 to 220,000
	$/m³	normal	$0.38	$0.08	
Growth medium placement	m³	triangular	28,000		25,000 to 31,000
	$/m³	normal	$1.65	$0.17	
Vegetative cover	ha	triangular	35		30 to 40
	$/ha	normal	$500	$75	

Apart from the element of specific reclamation cost as illustrated above, mine management may wish to consider contingent costs associated with unforeseen environmental liabilities within the context of closure fund estimation. Exposure to such liability at and beyond closure is an uncertain variable, involving an uncertain cost. Often, the exposure arises as a consequence of design failure, changes in regulations, or litigation.

13.10   CAPITAL CLOSURE COST CALCULATION

Management would be able to include uncertain or risk-costs in the overall calculation of the final capital cost of closure, as shown in Equation 1, where

$$Q = \sum \frac{1}{(1+i)} \bullet \left(B_t - C_t - R_t \right) \qquad \text{Eq. 1}$$

Q = Net capital cost of closure.
B = Financial benefits (e.g., asset recovery or transfer, enhanced land value).
C = Final reclamation and closure (non-operational) costs including post-closure monitoring and maintenance.
R = Risk costs (costs × uncertainty).
i = Annual interest rate.
t = Time in years to achieve certified closure.

Note that financial benefits (B), maintenance or closure costs (C), and risk cost (R) can be expressed as probabilistic quantities. The example above for waste dump closure cost demonstrates how Q can be calculated on a probabilistic basis. A similar approach can be used to calculate the variability of B. The section below demonstrates an approach to calculating the expected value of R, which may be used to estimate the size of an environmental contingency fund.

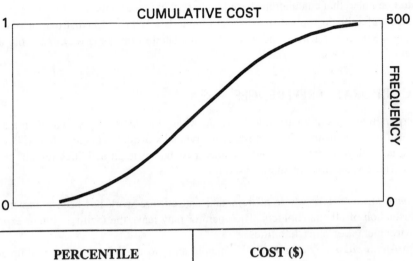

PERCENTILE	COST ($)
0%	92,462
10%	116,882
20%	123,567
30%	128,952
40%	133,688
50%	139,073
60%	144,087
70%	150,153
80%	155,229
90%	163,771
100%	203,882

FIGURE 13.3 Cost estimates.

13.11 RISK COST EXAMPLE

Risk cost (R) is a function of real cost and uncertainty. It is usually defined as:

$$R = p \times c \times \gamma \qquad\qquad \text{Eq. 2}$$

where p = Probability of cost occurring.
  c = Value of cost.
  γ = Management function.

The management function is a parameter that describes the overall behaviour of a system and/or its management, in terms of risk. Conservative, risk-averse management sets $\gamma < 1.0$; whereas in the case of poor management (whether identified by the company or during due diligence or regulatory review) the value of the management function is typically $\gamma > 1.0$.

By using Equation 1 above, it is possible to obtain the total expected cost Q on an annual or cumulative basis by using the expected costs from Monte Carlo analyses and by estimating the risk cost.

It should be noted that calculations based on expected values only are not fully "probabilistic." In reality, it is preferable to apply distributions of input data, to provide a distribution of contingency fund estimates. The contingency fund would be calculated in the same way, but using a statistical distribution of values for each variable.

13.12   CORPORATE PREFERENCES

Not all corporations have a clear philosophy on what objectives they should define as principal requirements for mine closure. The five scenarios typically available assume that all key mitigation and closure tasks have been completed satisfactorily, but residual activities remain as regulatory obligations. These scenarios are summarised as:

- *Compliant departure* where all remedial and closure work has been completed to the satisfaction of all stakeholders. The operator may leave the facility without exposure to environmental or other liabilities.
- *Short-term passive care* where all remedial and closure work has been completed, but monitoring and maintenance work will be required for a short period (e.g., vegetation establishment and short-term reclamation management).
- *Long-term passive care* where residual environmental impacts still need to be managed, or regular monitoring and maintenance work has to be performed for an indefinite period (e.g., limestone drains, wetland treatment systems, and chemical or geotechnical stability monitoring).
- *Short-term active care* where residual maintenance work has to be performed during a finite period (e.g., road maintenance, provision of critical utilities, and environmental management facilities).
- *Long-term active care* where long-term management, maintenance and monitoring work are required, at significant cost (e.g., collection and pumping systems, water treatment plant operations).

These scenarios are not mutually exclusive, and each can be applied to different operational units and support facilities at different times according to type, extent, and magnitude of environmental impact.

In the case of new mines, the choice should always be to develop a strategy that will allow for a final closure solution. There are a number of international jurisdictions where permits for new mines will not be issued if anything other than a final closure solution for compliant departure is proposed.

Because of poor experiences with active care, such as ongoing treatment of effluent from waste dumps, future corporate philosophy may be that only closure solutions which involve an economic exit strategy will be acceptable. However, if such solutions are developed, and the capital and operating costs are found to be prohibitive, it may be necessary to review the issue and consider closure designs with alternative objectives in mind. Alternatives may include underwritten or risk-managed transfer of liabilities and assets to third parties. The uncertainties associated with corporate preferences may also change with time.

13.13   DISCUSSION

Globally, the mining industry is tending towards an integrated approach to environmental and operations management and economic planning. This trend is particularly appropriate to new mining developments, and expansions of existing facilities. Also, environmental impact mitigation and reclamation are now implemented more frequently as integral components of the mining process.

This holistic strategy may also involve strategic decommissioning and closure of facilities, as they become redundant during the course of operations. The aim of phased closure during operations is to reduce effort and capital costs during the final decommissioning phase, whilst reducing environmental liabilities during operations. Indirect benefits of this type of strategy include reduced regulatory pressure, which is frequently associated with poor environmental compliance during mine development and production; and increased confidence amongst capital investors and other stakeholders. Overall, the advantages of measurably enhanced environmental performance are economic.

The relationship between production efficiency, economic efficiency and environmental performance of mining projects is described by Warhurst.[16] Analyses of these performance parameters indicate a cause-effect relationship which suggests that the environmental impact of a mining project varies inversely with technological efficiency and economic yield. A relationship between economic performance and prevailing regulatory conditions is also evident. Warhurst suggests that a key challenge in the resolution of the apparent conflict between minerals development and environmental protection will be for governments to ensure that mining operations remain sufficiently dynamic and profitable, so as to be able to afford efficient production technologies as well as effective environmental control systems during operations — and beyond closure.

Mining, minerals processing and environmental control technologies have developed over the last century to a high level of sophistication and efficiency. However, it is often possible to reduce the impacts from a site by 80% or more through the implementation of relatively simple and low-cost scientific and engineering technologies during the initial stages of project construction. Conversely, the cost of retroactive installation of environmental control technologies at later stages in the project life cycle is likely to be prohibitive. Therefore, judicious selection of control technologies and design elements to facilitate production efficiency and enhanced environmental performance is best undertaken at the feasibility stage of mine planning. If a minimalist and reactive environmental management philosophy is adopted during operations, the operator is frequently faced with an excessive level and extent of environmental degradation at closure, which may exceed the capacity of existing remedial technology. This scenario is likely to be accompanied by an excessive cost of long-term maintenance, treatment and monitoring which could be beyond the financial capacity of the company when production has ceased. Within the regulatory arena, sufficient flexibility must exist to allow for a degree of technological creativity and innovation to achieve cost-effective compliance with scientifically defensible environmental standards. The provision of financial, incentive-driven mechanisms to promote proactive mitigation efforts and concurrent remedial work, and to facilitate accrual of capital to cover final closure costs, would be particularly attractive to the minerals industry.

REFERENCES

1. Hutchison, I. P. and Ellison, R. D., Eds., *Mine Waste Management*, Lewis Publishers, Boca Raton, Florida, 1992.
2. Dahlstrand, A., Closure concerns at Sonora Mining's Jamestown Mine, *Mining Engineering*, 47, 236, 1995.
3. Robertson, A. M., Gadsby, J. W., Wiber, M. A., Brodie, J., and Solomon, W., Rehabilitation of Mines: Guidelines for Proponents, Version 1.1, Ontario Ministry of Northern Development and Mines, Toronto, Canada, 1991.
4. Campbell, J. M. and Emery, A. C., RTZ's approach to mine and smelter closure worldwide, Proc. Sudbury '95 Conf. on Mining and the Environment, Sudbury, Ontario, 1995, 8.
5. McKee, W. and McKenna, G., Improving environmental performance through mine closure planning, unpublished Syncrude Canada Ltd. Internal Corporate Position Paper, 1997.
6. ICRCL, *Guidance on the Assessment and Redevelopment of Contaminated Land*, Guidance Note 59/83, 2nd Edition, U.K. Interdepartmental Committee for the Redevelopment of Contaminated Land, Department of the Environment, London, 1983.

7. Klingbeal, A. A. and Montgomery, P. H., *Land Capability Classification*, U.S. Department of Agriculture Agricultural Handbook No. 210, U.S. Department of Agriculture, Washington, D.C., 1961.

8. Bibby, J. S. and MacKney, D., *Land Use Capability Classification, Soil Survey Technical Monograph No. 1*, Her Majesty's Stationary Office, London, 1969.

9. Riquier, J., *A New System of Soil Appraisals in Terms of Actual and Potential Production*, F.A.O., Rome, 1970.

10. F.A.O., *F.A.I. Soils Bulletin No. 32*, F.A.O., Rome, 1976.

11. Harr, M. E., *Reliability-Based Design in Civil Engineering*, McGraw-Hill, New York, 1987.

12. Kaplan, S. and Garrick, B. J., On the quantitative definition of risk, *Risk Analysis*, 1, 11, 1981.

13. Morrey, D. R. and Van Zyl, D., Including uncertainty in mine closure, Proc. 5th Western Regional Conference on Precious Metals, Coal and the Environment, Society for Mining, Metallurgy and Exploration, Black Hills, South Dakota, October 10, 1995.

14. Palisade, @RISK Risk Analysis and Modelling, Palisade Corporation, 1991.

15. Decisioneering, CrystalBall Version 3.0, Forecasting and Risk Analysis, Decisioneering Inc., Denver, Colorado, U.S., 1993.

16. Warhurst, A., Environmental best practice in metals production, in *Mining and Environmental Impact*, Hester, R. E. and Harrison, R. M., Eds., Royal Society of Chemistry, Cambridge, U.K., 1994, 133.

17. British Columbia AMD Task Force, Draft Acid Rock Drainage Technical Guide Volume 1, Province of British Columbia AMD Task Force, Canada, 1989.

18. Hamilton, H., Protocol for risk-based engineering and environmental performance assessment for oil sands mining, unpublished report, GAL, Calgary, 1994.

14 The Role of Waste Management and Prevention in Planning for Closure

Paul Mitchell

CONTENTS

ABSTRACT

This chapter primarily addresses the implications of clean mineral processing technologies for (a) efficient and cost-effective closure and (b) the continued use of waste management approaches. Clean (or sometimes "cleaner") technology is a generic term describing those processes that, overall, use fewer resources and cause less environmental damage than those with which they are economically competitive. Clean technology is normally implemented with the dual aim of reducing deliberate and fugitive waste emissions during operation and improving process economics through waste avoidance, reuse, recycling and recovery ("waste minimisation").

Emphasis is placed on new operations in the base and precious metal sectors, as it is there that some of the greatest opportunities for change exist, alongside some of the most intransigent and persistent problems. It has been developed within a framework based upon the recommendations of the 6th Workshop of the Mining and Environment Research Network, "Towards Integrated Environmental Management Planning for Closure from the Outset," in particular the delegates' comments pertaining to technological issues (see Warhurst, Chapter 29).

The chapter is presented in two sections, the first being an analysis of recent advances in the recovery of minerals and metals (including specific reagents and technologies) and their implications for closure and existing remedial waste management procedures. The second section looks in greater detail at two specific issues which cut across the interface between clean technology, waste management, and planning for closure, namely the use of cyanide in gold recovery and arsenic-contaminated waste treatment and disposal. Acid rock drainage (ARD) is another key issue that crosses this interface. However, since this has been addressed elsewhere in this book by the author (Chapter 7), it is not dealt with here as a separate issue.

While clean technologies have a clearly defined role in reducing environmental impacts during active operation, their potential to reduce the long-term hazards, risks, costs, and liabilities associated with waste disposal, acid rock drainage, post-closure revegetation, monitoring and other technical and engineering closure issues is less obvious, despite the existence of substantial and related literature. This chapter takes a lateral view of existing literature and draws inferences and conclusions from empirical and theoretical research originally conducted with little or no emphasis on closure issues and draws out the implications of waste prevention for the closure process.

14.1 THE ENVIRONMENTAL IMPACT OF MINING

The environmental impacts of base and precious metal mining activity have been well documented, particularly over the last two decades. For the purposes of this chapter, only a brief review is presented, based upon a number of key texts:

- *Environmental Effects of Mining*
  Ripley, E. A., Redmann, R. E., and Crowder, A. A., St. Lucie Press, Delray Beach, Florida, 1996.

- *Mining and Its Environmental Impact*
  Hester, R. and Harrison, R., Royal Society of Chemistry, Cambridge, 1994.

- *Environmental Management of Mine Sites*
  United Nations, United Nations Department for Development Support and Management Services and United Nations Environment Programme, Technical Report No. 30, 1994.

- *Environmental Impacts of Mining: Monitoring, Restoration and Control*
  Sengupta, M., Lewis Publishers, Boca Raton, 1992.

- *Minerals, Metals and the Environment*
  Institution of Mining and Metallurgy, Elsevier Science Publishers Ltd., London, 1992.

- *Environmental Aspects of Selected Non-Ferrous Metal Ore Mining: a technical guide*
  United Nations Environment Programme, Industry and Environment Programme Activity
  Centre, UNEP/IEPAC, 1991.

The mining life cycle can be represented by a number of linked phases, which may proceed in a linear or concurrent fashion. These include:[1]

- Exploration
- Site development
- Extraction
- Beneficiation
- Further processing
- Decommissioning

Although shown as the final stage, decommissioning (closure) can also occur throughout a site's life as a rolling programme of remediation and restoration, and there is now evidence to suggest that this approach is more cost effective (see, for example, Morrey, Chapter 13). The extent of remedial work required is a function of a number of variables including size of the site, method of ore extraction, ore mineralogy, processing and treatment routes, process efficiency, volume and nature of wastes, hydrology, hydrogeology, local environmental conditions, nature of the regulatory framework, and timing of remediation. Ideally, any operation should seek to minimise the work required through cost-effective preventative measures, particularly that portion of the work which is scheduled for the end of the site's working life (when the generation of revenue has diminished or ceased).

The sources of the major physical environmental impacts are ore extraction and processing: these cause deliberate (controlled) and fugitive solid, liquid, and gaseous releases to soil, air, and water. Arguably the most significant impact (with the possible exception of ARD) arises from the disposal of solid wastes, as these are the progenitor of many of the subsequent impacts, particularly in relation to soil and water contamination and pollution. In mining, as with most industries, less than 100% of the raw material reports to the valuable product or products. "Average" figures (based upon a survey of Canadian metal mines) indicate that 42% of the total mined material is rejected as waste rock; a further 52% from the mill as tailings; and an additional 4% from the smelter as slag, leaving as a valuable component only 2% of the originally mined tonnage.[2,3] At many gold operations, the concentration of valuable material is sufficiently low (e.g., less than 2 g t^{-1}) that effectively all of the mined ore is disposed of as waste. In effect, such sites are as much about waste disposal as they are about resource extraction.

The significance of this dual extraction/disposal role in the base and precious metal mining sectors is seen in global estimates for waste production. Gray[4] estimated that the annual generation of solid and gaseous wastes was 7.5×10^6 tonnes from primary copper production, 5.4×10^6 tonnes for zinc and 4.4×10^6 tonnes for lead, with these tonnages being dominated by the solid waste fraction. Although waste disposal directly sterilises a significant land area, it is the presence of potentially harmful elements, minerals, and other contaminants in the solid (and liquid) wastes generated during mining, mineral processing, and other downstream processes (e.g., waste rock, tailings, slags, smelter flue dusts, and precipitates from the chemical treatment of metal-contaminated liquid effluents) that determines the wider environmental impact. The contaminants may be organic (e.g., various flotation reagents) or inorganic (e.g., metals). Their original source may be the orebody itself or, alternatively, chemicals applied during extraction and processing of the ore. In broad terms, the contaminants may directly or indirectly cause a number of interrelated environmental impacts, including contamination and pollution of ground and surface waters and soil resources, sustained ecosystem degradation, and atmospheric suspension of respirable dusts. These can combine to extend the spatial and temporal "footprint" of a mine site, an obvious example being the long-term generation of ARD from surface wastes, and from abandoned and active underground and open pit workings.

ARD includes water originating in underground mines and runoff from open pit workings, waste rock dumps, mill tailings, and ore stockpiles, sometimes in flows measured in millions of gallons per day. While it is normally mines extracting coal or non-ferrous metals that suffer the greatest problems, similar problems may arise during construction or other engineering activity on certain rock types. It results from the exposure of pyrite (FeS_2), marcasite (FeS_2) and pyrrhotite (FeS_{1-x}) to oxygen and water, resulting in the formation of sulphuric acid, ferrous iron (Fe^{2+}) and ferric iron (Fe^{3+}). Ferric iron is a powerful oxidising agent, and oxidises insoluble metal sulphides to water-soluble sulphates. These may then be dissolved and transported in the acid medium. The rate of ferric iron and acid generation is increased dramatically by the presence of certain bacteria (e.g., *Thiobacillus ferrooxidans*, *Thiobacillus thiooxidans*). Metals commonly found in ARD include copper, zinc, nickel, cadmium, arsenic, aluminium, and manganese, although the exact speciation and relative concentrations are largely dependent on the mineralogy of the orebody, the surrounding host rock, and attenuation reactions. The nature of ARD originating from waste products is also influenced by process efficiency and economics, as these in turn partially determine the concentration and type of residual metal contaminants (present as sulphides or other species) in waste rock and tailings, and in the unexploited host rock.

14.2   MITIGATION OF ENVIRONMENTAL IMPACTS — A WASTE MANAGEMENT APPROACH

Although it is impossible to generalise across the industry, it is probably fair to say that waste management rather than waste prevention is the current norm, with "end-of-pipe" (remedial) approaches dominating. Current responses (examined elsewhere in this book) are normally chosen for reasons of cost rather than optimum environmental performance, and include the use of cover materials for land-based solid waste disposal sites, decontamination of wastes, and chemical or biological treatment of metal- and chemical-contaminated liquid effluents. While there are many examples where these have been successfully applied, there is a range of broader questions that need to be considered. For example, in the use of cover systems, "out of sight" is often synonymous with "out of mind" and there is a risk that wastes may become a "hidden" environmental liability for future generations and a financial liability for the site owner. Although it may seem absurd to consider that such waste disposal sites could become "hidden," one has only to compare this scenario with the "disappearance" of large numbers of highly contaminated abandoned gas works sites in urban areas around the U.K. during the 19th century and their subsequent "rediscovery" during the last two decades.

Drawbacks to other remedial approaches include the transfer of contaminants from one medium to another. An example of this is the transfer of metals from water to land via disposal of metal-contaminated water treatment sludges.

The dominance of waste management at the present time is perhaps a reflection of the prevailing attitude towards closure, namely that a civil engineering-style remedial approach is sufficient to properly decommission a site *and* avoid possible liability in the future. While this may have been true 20 years ago, technology, legislation, and the expectations of stakeholders have moved on. If there is a generic "thread" that links current waste management approaches, it is the possibility of future litigation and liability associated with what might be considered short-term (but relatively cheap) solutions to long-term environmental problems. The threat of incurring financial penalties in the future has helped to catalyse the industry into exploring alternatives to current practices; to establish "best practice" standards; and to move towards preventative action, particularly within the mineral processing arena. This is not to say that waste management has no role in optimising future environmental protection. However, in a legislative framework increasingly leaning towards preventative rather than remedial measures, any environmental benefits of remedial approaches relative to alternative approaches need to be quantified rather than relying on empirical and historical precedent.

14.3   LONG-TERM COSTS AND LIABILITIES

In many countries, there is a *legal* liability for the environmental damage associated with the operation of a site. From an operator's perspective, legislation forces the internalisation of environmental "costs" by establishing the concept of "polluter pays" as an underpinning principle in the regulator/operator relationship.

As poor or inefficient environmental performance can result in losses of metals to waste streams, so it can also increase a firm's exposure to liability,[5] through the mechanism of responsibility for the short-, medium-, and long-term environmental damage associated with such wastes. This, however, presupposes that an effective regulatory framework exists, which ensures that the full cost of any required remedial work is borne by the site operator, and that there is no option to "walk away" and create an orphan site. This can be accomplished, for example, by a system of bonds, levies, and/or other taxes that are applied during the active phase of the operation's life cycle (see Anderson, Chapter 15 and Danielson and Nixon, Chapter 17). Responsibility will certainly include an element of long-term monitoring, and most likely also entail some treatment of aqueous discharges and remedial steps at land-based disposal sites. Poor performance at one site may also prevent permitting at future sites and, therefore, liability at a particular site can have an effect on a company's economic performance far in excess of the hard-cash figure attached to that liability.

Key questions for the operator are whether preventative measures that minimise or prevent liability are technically feasible and cost-effective and, if so, how they compare with a policy based on effective waste management, remediation, and monitoring. The operator needs to assess how far preventative measures can go in reducing liability (and if significant liability remains despite a proactive approach, whether the additional costs are justified in taking this approach). Unfortunately, one of the least readily described costs is that of long-term liability associated with different waste types and waste disposal practices.[6]

In broad terms, if investment in pollution prevention is to be justified, then one requires accurate costs for the various short- and long-term alternatives. The true cost of current waste management practices needs to be determined to allow a proper comparison with new preventative technologies, but this is both time-consuming and problematic despite the availability of data from a large number of pollution prevention schemes (including specific techniques and case studies).[7] Estimating closure costs is an important factor in determining which approaches are going to be reasonable as far as taking (and paying for) preventative measures is concerned. The type of costs that need to be considered include long-term protective measures, monitoring, rehabilitation, protection of the local community, control of erosion, and the protection of water quality, all of which are interlinked. Attempting to reduce cleanup costs through the implementation of more "environmentally sound" technologies is sensible from an operator's viewpoint if additional capital and operational costs are offset by reduced remedial costs, financial liabilities, and risk of prosecution.

There is currently a broad trend towards making the mining industry liable for the environmental impacts of resource extraction, via the passage of regulations requiring the cleanup of contaminated and polluted sites. This is reflected in the increasing level of legislation pertaining directly or indirectly to mining operations (e.g., see United Nations documentation[8] for a review of recent legislation in developed and developing countries). Regulation is also adapting to incorporate recent technological advances in pollution prevention, thereby reducing the emphasis on remedial (or "end-of-pipe") measures. For example, changes are afoot in the U.S. that may have significant implications for liability and the balance between preventative and remedial approaches. The U.S. Environmental Protection Agency (USEPA) is currently supporting the inclusion of mineral processing wastes in the Toxics Release Inventory (TRI).[9] The TRI has proved previously to be very successful in reducing the release of toxic materials via "aggressive" voluntary measures, mainly due to the fact that the TRI is a public register, and a powerful tool in the hands of local communities.[9] Regulation of previously exempt wastes may add weight to the argument for preventative measures, particularly if disposal at low cost land-based non-engineered sites becomes a more limited option as is currently planned.[10]

From the regulator's perspective, a clear-cut and enforceable framework is essential to effectively regulate the industry. However, the nature of the relationship between the regulator and the operator within this framework can vary significantly, from confrontational to collaborative. Equally variable are the methods of achieving compliance, which can range from "design standards" to "performance standards."

Design standards can stifle innovation by promoting standard technological solutions that do not deliver the optimum environmental or economic performance in some scenarios. It is also incorrect to assume that once a specific technology is in place, compliance will automatically follow: human resource management, training, and other factors will also significantly affect whether or not compliance is achieved. The current concept of design standards also does little to promote the shift from a "compliance paradigm" to a "beyond compliance paradigm" as design protocols are not sufficiently dynamic to identify and act upon the trends beyond the prevailing legislative regime. Ashford[11] postulates that it should be possible to design legislation more effectively to ensure that the technological output is the technology that is required, rather than that which is available. The use of performance standards is one way of achieving this.

Performance standards are non-prescriptive in terms of what is considered the most appropriate technology: that choice is left to the operator, who is assumed to have sufficient expertise to make sound, well-informed decisions. Performance standards are also seen as locally responsive as they can take into consideration on-site conditions (e.g., the local assimilative capacity of the environment and background concentration of contaminants in soils). However, the use of performance standards means that the additional burden of understanding and modelling the proposed process, in a local or regional context, falls largely to the regulator, as does the need for frequent monitoring to ensure compliance.[12] This approach also assumes that there is an effective regulatory and enforcement system in place and that legal recourse in cases of non-compliance is feasible. This is not necessarily the case in many developing countries (and indeed many industrialised nations), where sufficient resources may not be available.

Irrespective of the variations in regulatory framework and the relationship between regulators and operators in different countries, there is one irrefutable fact: the ultimate fate of contaminants and pollutants following disposal is largely unknown, despite substantial research on the development of computer simulations and predictive models. This is not particularly surprising, considering that biogeochemical reactions are likely to proceed over extended — even geological — timespans. It is therefore quite possible that in future years there will be more "surprises" ahead for industry.[13] For example, Aldrich[6] assumes that any landfill, like all man-made structures, will eventually fail. This assumption can be extended to other mining-related waste disposal sites irrespective of the level of engineering. A policy of minimising environmentally harmful contaminants in otherwise inert wastes could therefore be advocated on the basis of reducing future, unknown, and unquantifiable liability, the economic and environmental effects of which could quite possibly fall to future generations: a concept at odds with most, if not all, definitions of sustainable development.

14.4 CLEAN TECHNOLOGY AND WASTE MINIMISATION: PRINCIPLES AND PRACTICE

The concept of clean technology cannot be explored without reference to economic performance, resource usage, and environmental impact.[14] Consequently, clean technology is not necessarily "state-of-the-art" by virtue of the fact it must be able to compete on economic terms in a commercial arena with existing technologies. There is often a significant gap between what is technically viable and what is economically affordable, and considerable research and development effort is devoted to reducing that gap.

Clean technology is an integral component of (and often considered synonymous with) waste minimisation and pollution prevention and has principally been used (and indeed developed) within the process industries. Waste minimisation is also sometimes referred to as waste reduction, clean or cleaner technology/engineering/processing, pollution prevention or reduction, environmental technology, or low and non-waste technology.

Ideally, the "cleanliness" of a technology should also take into account the environmental performance of upstream component or technology suppliers and downstream users and disposers of products, although this is often difficult to determine, and the boundaries of the investigation are instead restricted more closely to the process in question.

Research groups around the world have rigorously described the concepts underlying clean technology. However, for the purposes of this paper, four propositions recently developed by the Waste Management Task Group of the Canadian Council of Ministers for the Environment represent a sufficient description of the generic issues. These are duplicated below (from Redhead):[15]

- *Proposition 1*
  "Our generation must consider that all resources including the environment will be valued by future generations, and that we are obliged to use them wisely."

- *Proposition 2*
  "Used materials should be regarded as resources, and not labelled as wastes until all efforts to find an economic use for them have been exhausted, i.e., maximise their utility."

- *Proposition 3*
  "Disposal of waste represents some risk to the environment. Reducing the quantities of waste requiring treatment and disposal and reducing the hazardous nature of products should, therefore, receive the greatest effort. Re-use, recycling and recovery are other means for achieving this objective. It is generally preferable to treat and dispose of hazardous material, than to inventory it or release it to the environment."

- *Proposition 4*
  "Waste should be a constant reminder of inefficiencies in materials management systems. The goal must be to improve system efficiency through the continuous reduction in quantities of waste directed to final treatment and disposal."

To meet the aims and objectives of these propositions, a waste minimisation "hierarchy" is used, namely elimination→source reduction→recycling→treatment→disposal, with the priority decreasing to the right.

A number of generic (cross-sectoral) strategies are adopted within an overarching management system to facilitate waste minimisation:[16]

a) Improved plant operations (predictive and preventative maintenance, better handling procedures, separation of waste streams to facilitate in-process recovery and volume reduction).
b) Alterations to process technology (modernisation, modification, better control of process equipment). This approach is more capital-intensive so it is often considered to be more appropriate for a new plant rather than retrofitting of an existing plant.
c) Recycling, recovery, re-use of waste products.

   d) Changing raw materials.

   e) Product reformulation.

   In the context of base and precious metal mining and processing, the first three strategies have the most obvious applications, although the capacity does exist occasionally to change the nature of the process input (as per (d)), for example, by selective and more accurate ("right-in-space") mining practices.[17] Strategy (e) is of little direct relevance to the mining industry and is ignored here.

   Where there are a number of competing clean technologies, they can be ranked according to the reduction in the hazards associated with the waste outputs, treatment/disposal costs, future liability, safety hazards, and input material costs. It is also possible in many instances to measure the extent of prior experience in industry, capital cost, changes in operating and maintenance costs, the effect on product quality, implementation period, and ease of implementation.

   The means of determining the cleanliness of a particular technology relies heavily upon the assessment of resource usage and environmental impact using life cycle inventory (LCI) and life cycle assessment (LCA) methodologies. The former attempts to calculate all the environmental burdens associated with providing a particular service to society. It accounts for all energy and material flows throughout the life cycle and is the only tool that considers the whole life cycle necessary to provide the service, to avoid the shifting of environmental burdens from one part of the system to another. A LCI can also be used to identify those changes in a product's life cycle offering the most significant environmental improvement.

   By comparison, LCA uses the inventory data to assess the potential environmental impact of waste generation, resource consumption, and energy requirements throughout a site's life cycle.[18]

   Following this ranking process, a feasibility analysis of each technology can be conducted to answer a range of technical and economic questions:

- Will the technology work (has it worked before)?
- Are suitable utilities and space available to implement it?
- Can it be integrated with current procedures?
- Will production need to be stopped while it is implemented?
- Will it affect product quality?
- Is special expertise required to manage it?
- Does it create any new environmental problems?

   Additional economic analysis can address the costs and benefits of the technology via a detailed accounting of capital requirements, operating and maintenance costs, waste management costs, raw material costs, insurance savings, changes to utilities costs, and revenue from marketable by-products. The economic consequences of implementing a waste minimisation (clean technology) programme can be measured in a number of ways, including definition of the payback period (typically measured in years, with a period of three to four years being considered acceptable for low-risk investments),[7] internal rate of return and discounted cash flow. These are termed "tier zero" analyses by McHugh.[19] "Tier one" analyses include the benefits associated with hidden costs such as reporting, testing, and monitoring. "Tier two" analyses identify costs associated with future liabilities, including potential fines. "Tier three" analyses are the most difficult to determine with any great confidence as these involve the impact of waste minimisation on corporate goodwill, public acceptance, overall corporate risk, and shareholder value.[19]

   Due to the number of site-specific variables, the benefits of waste minimisation vary considerably, but in general terms they include one or more of the following:[16]

- Increased operating efficiency and process reliability and reduced production costs.
- Improved health and safety for workers.
- Reduced risk of breaching consents and prosecution.

- Reduced liabilities.
- Reduced handling, pre-treatment, transport, and off-site disposal costs.
- Reduced waste storage space (smaller site "footprint," greater space for productive uses).
- Reduced on-site monitoring, control, and treatment costs.
- Reduced administration with regard to waste disposal.
- Improved company image in the eyes of the shareholders, employees, and community.

Although the benefits of waste minimisation are clearly wide-ranging, there are often institutional disincentives that may need to be overcome. These include management uncertainty due to (a) unsure investment returns, (b) the potential for production downtime, (c) problems with product quality, and (d) loss of proprietary information to waste reduction consultants. The best way to circumvent these obstacles is to involve top and middle management, plant management, and plant operators (i.e., the complete corporate structure),[20] and to ensure that sufficient incentives are in place to overcome any personal or institutional bias.

The remainder of this chapter will explore if and how these concepts can be applied to the base and precious metal mining sectors, and the implications of a waste prevention approach for closure and waste management.

14.5 FROM WASTE MANAGEMENT TO WASTE PREVENTION

One of the Berlin Guidelines formulated at a round table conference entitled "Mining and the Environment" in June 1991 (involving representatives from industry, governments, and NGOs) was to "*Adopt environmentally sound technologies in all phases of mining activities and increase the emphasis on the transfer of appropriate technologies which mitigate environmental impacts, including those from small-scale mining.*"[21]

This was added to under the auspices of Agenda 21: "*Environmentally sound technologies are not just individual technologies, but total systems which include know-how, procedures, goods and services, and equipment as well as organisational and managerial procedures.*"[22]

However, in the real world, the choice of technology is a complex issue, and the step from written guidelines to a working plant is a large one. A wide choice of process options is available for minerals. To make an informed choice, there are a number of parameters that must be fully investigated and understood:[23]

a) The types of physical, chemical, and biological processes that are applicable to the minerals of interest.
b) How and why these processes are used.
c) How the processes are controlled (managed).
d) How process routes may be designed and developed.
e) The factors affecting the choice of processing route.

Unlike the manufacturing sector, there is only limited scope to change feed materials in the mining sector, although it is sometimes possible to adapt the mining technique to leave in place a problematic part of the deposit.[5] However, this approach may not be feasible where the whole deposit is environmentally "problematic," and in the long term, it is likely that deposits containing some problem component will be exploited anyway (although it is possible that parallel advances in technology will serve to address the problem). Therefore, innovation in mineral processing, metal recovery, and metal refining is a possible solution to the environmental impacts arising from resource extraction and processing.

At present, the long-term environmental impact and liability of wastes is not normally fully integrated into the choice of the process technology or route. However, there are signs that this is changing. An example of this is seen in the estimates for mining-related environmental science and

TABLE 14.1

Environmental R&D Funding Estimates for the Mining Sector, Canada 1990–2000

Environmental hazard	Estimated R&D funding (Canadian $m)
Cyanide recycling/control	6.0
Thiosalts (chemistry, control, and recovery)	10.0
Water recycling	13.0
Water and air treatment technologies	20.0
Acidic drainage	15.0
Advanced mining, milling, smelting, and refining technologies	50.0
Waste utilisation	5.0

technology R&D funding for the 1990–2000 period in Canada (Table 14.1, derived from data presented by Everell and Oliver).[24]

Based on these levels of funding, it seems that there is an implicit understanding that while existing problems are substantial and will require significant research efforts to overcome, new developments are essential to avoid such problems in the future. Innovative (advanced) clean technologies will play a major role in the sustainable development of the mining industry (if sustainable development is defined in this instance as the capacity of the natural environment to support waste discharges from such activities).

In general, the main drivers behind the uptake of waste minimisation procedures and clean technologies have been (a) maximisation of economic returns through the optimal minimisation of valuable minerals and metals reporting to the waste fraction, and (b) minimisation of the costs of waste management. The direct environmental benefits have often been considered important, but of secondary concern. Where pollution prevention has been a primary driver, it has often been in the context of reducing day-to-day discharge concentrations to allowable levels, rather than to minimise long-term, post-closure liabilities. This arises because the true economic liability of waste disposal has not been incorporated into the cost of production, although the concept of widening processing options to take into account the possible downstream problems associated with solid wastes is now being considered more widely. For example, the inclusion of environmental factors such as the potential for ARD generation leads to consideration of a wider array of processing options.[25] This may lead to the choice of a technology that would not otherwise have been considered appropriate or necessary.

The USEPA is moving towards more closely integrating the environmental costs of mineral processing technologies in its mine permitting process.[12] Although the USEPA is peculiar to the U.S., the author is of the opinion that "trickle-down" and "trickle-transfer" of relevant legislation to the international arena from the U.S. will occur with time. This may be particularly true as U.S. companies operating abroad increasingly seek to meet U.S. regulations irrespective of the existence of less stringent national legislation, and U.S. regulatory staff are seconded to foreign regulatory bodies to aid in the development of regulatory frameworks.

Technology options are assessed on the basis of their environmental impacts and whether technological solutions to those problems exist *and* will be economically viable (i.e., remediation of the environmental impact of an operation will not bankrupt the operator, thereby creating an orphan site). If it is not economically viable to treat the likely environmental impacts of specific process options, those options must be reconsidered and possibly redesigned. This iterative process continues until the financial burden arising from the remediation of environmental impacts is sufficiently low as to be affordable to the operator. The incentive for the operator is the minimisation of bonds, levies, or taxes that are required as insurance against remedial works, although this may be outweighed by additional capital and operational process costs. A lesson that could be learnt from this is that effective partnerships with external stakeholders are likely to play a key role in competitive advantage based on environmental performance in the future. In this way, companies are better able to anticipate the potential conflicts at an early stage.[26]

Warhurst and Bridge[5] recently analysed trends in innovation in the mining industry. Although the industry is well known for its conservatism, particularly with regard to radical changes, the authors propose that there has been a process of incremental improvements in both technology and organisation during the 20th century. These improvements have allowed firms to remain viable as ore grades have tended to decrease. The average grade of mined ore has decreased by a factor of two since the 1950s,[27] and therefore the energy requirements per unit of metal have increased, despite advances in technology. This is a trend that is likely to continue as grades continue to fall.

By reducing production costs and allowing lower grade and more complex deposits to be worked, innovation is proving essential for the continued competitiveness of those areas that may only have low-grade deposits left to exploit (e.g., gold and copper in the U.S.). Incentives to innovate have included intensified competition, cyclic variations in demand (high cost producers are less likely to survive during demand "troughs"), and the "environmental imperative." However, mineral commodities provide little scope for product differentiation between individual producers, and there is little sign (with the possible exception of aluminium) that non-ferrous metal producers are seeking a competitive advantage through the marketing of "green" metal products.[5] That said, the push for "green" metals may come at a more generic level via pressure further along the supply chain (for example, from the manufacturing sector).

14.6 RECENT ADVANCES IN MINERAL AND METAL RECOVERY: IMPLICATIONS FOR PLANNING FOR CLOSURE

There is still room for improvement in all mineral processing unit processes, through the development of both theory and practice. However, certain areas probably offer greater potential for major advances than others. Unit processes such as electrostatic and magnetic separation and gravity concentration are probably nearer the limits of innovation than processes such as crushing and grinding (liberation), hydrometallurgy (leaching, solvent extraction), pyrometallurgy, electrometallurgy, flotation, and *in situ* (solution) mining. From a closure perspective, the key to developing cleaner mineral processing technologies is the minimisation of both the waste volume and the loss of metals or acid generating material to that waste. This in turn would reduce the need for remedial actions. Although many advances have resulted in the economic recovery of by-products that would previously have been considered wastes,[28] these advances would not necessarily be considered "clean" but rather a means (and a result) of improving technical and economic efficiency.

There are major and very real advances being made in all fields of processing, metal recovery, and refining, and a number of reviews have recently detailed these events.[27,29–38] Within these, there are excellent examples of the opportunities that are opening up for planning for closure via the application of clean technologies and new reagents. To fully understand these opportunities requires "lateral" assessment of the literature. There are few direct references to the potential application of these techniques in planning for closure, or for the need to consider modification of plant circuits to meet the requirement of minimising long-term costs and liabilities associated with wastes and improved environmental performance. Rather, the emphasis is strictly on technical and economic efficiency and performance. Indeed, analysis of the Mineral Industry Research Organisation (London, U.K.) Annual Report and Accounts for 1995 shows that there is no explicit link between the technical research underway and planning for closure, despite the fact that both clean technology and closure are major areas of research for this industry-led organisation.

Therefore, the purpose of the brief review below is to highlight recent advances in mining and mineral processing technology and the development of process chemicals, and to make an explicit link to closure issues where none exists in the original literature.

14.6.1 LIBERATION

Liberation of mineral phases is the physical splitting of valuable (target) from non-valuable (non-target) minerals or the valuable minerals from one another. Where downstream processes cannot efficiently handle mixed or contaminated feedstocks, effective liberation and separation play an essential role in optimising economic, technical, and environmental performance. The inability to totally liberate the valuable mineral phase from non-target minerals results in additional contamination of the waste fraction. Incomplete liberation of one valuable mineral from another can reduce subsequent process efficiency and increase the generation of contaminated wastes.

Advances have been made in the quantification and prediction of mineral liberation and in the design of new technologies.[29] These new technologies may eventually result in enhanced liberation of valuable and non-valuable phases, and a reduction in the metal loading of tailings and other waste fractions by improving liberation of mineral particles along, rather than across, mineral boundaries. This would also decrease the size reduction required to liberate mineral phases and, therefore, associated production of slimes (particles less than 10 to 15 microns in size that are difficult to handle or recover).

14.6.2 SURFACE CHEMISTRY

One of the obstacles to controlling the loss of potentially toxic minerals to solid and liquid waste fractions is the selectivity performance of current reagents. A better understanding of surface chemistry will ultimately enable more effective targeting of particular minerals in processes, such as flotation via the development of mineral-specific reagents (see Flotation Chemistry, below).

The surface chemistry of a number of solid-liquid systems has been looked at in close detail. In the context of closure, one system of particular interest is the surface chemistry of pyrite, which is abundant in many ores,[30] and the major source of ARD. Analysis of the surface chemistry of pyrite has particular implications for controlled recovery or rejection during flotation and also for novel methods of surface treatment to prevent oxidation and generation of acidity (see Chapter 7 by Mitchell for further information).

14.6.3 FLOTATION CHEMISTRY

Froth flotation is used to treat about 2 gigatonnes of ore each year.[39] The potential to reduce the environmental impact of mining wastes via the use of innovative reagents is, therefore, enormous (see, for example, Mitchell, Chapter 7). New and exciting advances include hydrophobic frothers for improved recovery of coarse and locked particles; electrochemical sulphide mineral collectors for the recovery of sulphides at greatly reduced pH, and with substantially improved pyrite rejection; chelation-based sulphide mineral collectors for improved recovery from gold, platinum, nickel, zinc, and complex copper-containing ores; and the ongoing development of collectors for the selective flotation of oxide minerals as an alternative to leach-based processing.[38]

However, a detailed analysis of innovative reagents is hindered by their lack of market penetration. For example, despite more than 300 non-patent publications in 1993 relating to flotation reagents,[32] and the wider availability of thousands of flotation reagents, only a relatively small number are actually used in plants around the world.[40]

Cappuccitti[41] noted a number of reasons for the lack of commercial uptake of novel sulphide flotation reagents. These included the conservative nature of the mining industry; the difficulty in proving that the reagent enhances performance when other non-reagent related factors may do likewise; the relatively low price of reagents; and the difficulty of penetrating markets in developing countries due to access, payment, cultural, and geopolitical problems. Further key issues proposed by Nagaraj[40] include the high cost of meeting certain regulatory requirements (e.g., assessment of the ecotoxicological and occupational health impacts of new chemicals); slow returns on investment; and difficulties in scaling up the production process from pilot to commercial scale. Until these and related issues are resolved, the opportunities to assess the true potential of what are currently pre-commercial reagents in aiding the closure process are limited.

14.6.4 FLOTATION PROCESS ANALYSIS AND CONTROL

The performance of flotation is notoriously sensitive to process upsets,[33] particularly when treating ore from underground operations where the feed to the mill can be extremely variable.[42] This can lead to valuable metals reporting to tailings, where they represent both an economic loss and a potential environmental risk. Hence, considerable research effort has been aimed at process optimisation and control through, for example, improved measurement and control of reagent additions. Neale et al.[74] examined the automation of reagent addition in order to increase the capacity to run flotation systems at steady-state, and minimise the losses of valuable (and toxic) metals to the tailings. The USEPA[42] also reviewed the applications of online X-ray analyzers in process control during flotation. The same document[42] described a case study at the Doe Run Fletcher mill where the implementation of online analysis had resulted in a reduction in the metal load disposed of in the tailings impoundment (from a total of 0.2% to 0.15%). Improved process control also reduced reagent costs and the level of gangue minerals in the concentrate, thereby reducing the generation of slag during subsequent smelting.

14.6.5 IN SITU MINING

In situ (or solution) mining is one topic of research where the emphasis is on the need to reduce the volume or toxicity of waste products requiring disposal. It has therefore been proposed as a means of reducing the surface impacts of non-ferrous metalliferous mining. The U.S. Bureau of Mines conducted considerable work on the use of stope leaching as an alternative to the traditional method of surface processing. This work targeted the reduction of surface disturbance and other adverse environmental effects associated with the traditional methods and the exploitation of lower grade deposits. However, there are substantial problems associated with this approach (many of which mirror the problems associated with heap and dump leaching operations, see Bridge, Chapter 9). Not least of these is the difficulty in ensuring effective control and recovery of metal-loaded leach solutions in the sub-surface environment to prevent contamination and pollution of groundwater resources. In this respect, the impact may merely be transferred from the surface to the subsurface environment.

Research aimed at the selective dissolution of valuable components relative to the gangue minerals (e.g., at the now-defunct U.S. Bureau of Mines),[43] has failed so far to offer a practical means of limiting the transfer of unwanted species to the lixiviant phase. These unwanted species must be recovered and disposed of prior to or following recovery of the valuable metal, often resulting in the transfer of contaminants from the liquid (e.g., aqueous) phase to a land-disposed sludge. This suggests that even where research is specifically aimed at the minimisation of solid wastes, the environmental impact can merely shift, at least partially, from one medium to another.

These examples serve to illustrate that the development of new processes and reagents (driven by the desire to improve technical and economic performance) might have a significant impact on closure by reducing the environmental burden of the waste products and thereby shift the emphasis from waste management to prevention. However, while incremental improvements in processes and reagents *enable* waste reduction to occur, they do not necessarily *promote* waste prevention. Therefore, the current practice of accepting the environmental benefits accruing from improved technical performance cannot be considered clean technology development, but it should be viewed as a possible step forward as it reduces the emphasis on post-contamination waste management. Incremental changes that occur within the plant are inherently cost-competitive with existing technologies or reagents, as it is on this basis that they are implemented. This sidesteps the issue of technologies or reagents that may have significant environmental benefits, but for which capital and operational costs may be higher; this is addressed in the following sections.

14.7 CYANIDE: ISSUES, ALTERNATIVE REAGENTS, AND TREATMENT

14.7.1 ISSUES AND ALTERNATIVE REAGENTS

Cyanide is an increasingly political issue in the mining industry. It remains an emotive term, and there is a great deal of public concern regarding its use, particularly in the recovery of gold (its role as a process chemical in flotation systems being less "publicised"). Its public reputation appears to be based more on its theoretical toxicity than the number of actual incidents causing injury or death. Cyanide does not bio-accumulate, and is readily metabolised. There is also no evidence that chronic exposure to free cyanide causes teratogenic, mutagenic, or carcinogenic effects.[44] Neither have there have any documented cases of death by (non-criminal) poisoning in North America in the 20th century, or of any cases of illness or death due to cyanide in drinking water.[45]

In many respects, the concern expressed over the use of cyanide would be more usefully directed towards the presence of metals in wastes, as under properly managed conditions cyanide will degrade to non-toxic end-products (a detoxification route that is not relevant to most metal and metalloid compounds). Public concern has influenced the regulation and decommissioning of operations where cyanide has been used and the financial assurance expected for the closure of cyanide-bearing tailings. For example, in the State of Alaska, assurance must be sufficient for closure of the tailings (including treatment, stabilisation, and civil work) plus 30 years of post-closure monitoring.[46]

The chemistry of cyanide is complex. It can be present as free cyanide (CN^-), weak and moderately strong (stable) complexes (e.g., $[Zn(CN)_4]^{2-}$, $[Cd(CN)_4]^{2-}$) and strong complexes (e.g., $[Fe(CN)_6]^{4-}$, $[Fe(CN)_6]^{3-}$).[44] These complexes dissociate at various rates, depending on their stability and external factors such as light intensity, pH and temperature. Dissociation products include free cyanide, and therefore metal complexes can be viewed as a source of additional free cyanide above and beyond that immediately present in the aqueous phase. This has significant implications for closure as apparently stable wastes can in fact leak free cyanide if environmental conditions change in such a way as to promote metal complex dissociation. Of particular interest are the stable iron cyanide complexes. These are non-toxic at the concentrations normally found in wastewaters, but through photolytically catalysed dissociation they can create significant concentrations of free cyanide.[44] Therefore, in part it is their long-term stability that leads to the possibility of long-term environmental risk.

Although cyanidation is the major route used for gold recovery at formal operations (e.g., 100% of formal gold production in South Africa),[47] cyanide is not the only lixiviant with the potential for use in gold recovery. Development of alternative reagents has been spurred on by the environmental concerns regarding cyanide (despite its excellent safety record) and by the existence of certain ore types that are not readily amenable to cyanidation. Certain sulphide deposits contain sufficient concentrations of pyrite, arsenopyrite, stibnite, copper, and zinc to make the ore refractory with respect to gold recovery by cyanidation.[48] For these types of ore, thiourea is a possible alternative: although its consumption is higher and it costs more than cyanide, it is relatively non-toxic, selective for gold, and it has superior kinetics. Thiosulphate has also been suggested as an alternative to cyanide for the treatment of refractory carbonaceous and sulphidic gold ores.[49] However, thiosulphates may generate acidity when oxidised in receiving waters, its use may also result in unacceptable concentrations of dissolved copper in effluents from copper-rich gold-bearing ores.[49]

Other potential alternatives to cyanide include thiocyanate[49] and halogen-based solvents such as chlorine gas, sodium hypochlorite, iodine, and bromine.[50] None of these could be considered a generic replacement for cyanide, but may be useful in the treatment of specific ore types. Each has its benefits and drawbacks that must be equated with the efficiency and potential environmental implications of cyanide use. To be viable, alternatives should have an efficiency equal to or greater than that of cyanide, be cost-effective, facilitate the recovery of gold downstream in the process chain, and be environmentally acceptable.[50] Indications are that for non-refractory ores, halogen solvents are no more efficient or cost-effective than cyanide, although it appears that a thorough comparison of their environmental impact relative to cyanide has yet to be undertaken. For example, halogen solvents may result in the formation of high dissolved salt concentrations or halogenated hydrocarbons.[49] Instead, halogen solvents may find more applications in the treatment of refractory ores, although this is an area that needs to be substantiated by further research.

It is nearly 100 years since cyanide was first used as a lixiviant for gold recovery. Despite substantial research on alternatives, it remains the most efficient and cost-effective means of recovering gold in most cases. Increasing application of cyanide over the years has been paralleled by reductions in consumption per tonne of gold recovered. At the same time, suppliers have taken steps to improve the safety of cyanide distribution, storage and handling.[39] These steps have included training plant employees in the safe and proper handling of cyanide (which could perhaps be viewed as a limited form of product stewardship). In tandem with these steps are a number of technologies aimed at the degradation, recycling, re-use, and recovery of cyanide. Further details can also be found in the USEPA's Treatment of Cyanide Heap Leaches and Tailings.[51]

14.7.2 Treatment of Cyanide

14.7.2.1 Natural degradation

One of the most common approaches, this method relies mainly on the dissociation of metal-cyanide complexes and volatilisation of cyanide as hydrogen cyanide.[52] This is achieved by ultraviolet (UV) irradiation (sunlight) and changes (decreases) in pH. Due to its low cost, this method of treating tailings decant waters is very common.[51,53] However, this approach is not entirely satisfactory due to the presence of relatively stable metal-cyanide complexes, which degrade at a much slower rate. If the influx of tailings to a holding area exceeds the rate of degradation, cyanide degradation will be incomplete (the prematurely buried tailings being protected from UV). Subsequent disturbance or disruption of such contaminated wastes may result in uncontrolled releases of free cyanide if the metal-cyanide complexes are exposed to UV. Ideally, proactive treatment should be employed where metal complexes are likely to be a problem to ensure complete degradation.

14.7.2.2 Peroxide process

Hydrogen peroxide destroys cyanide without the formation of any intermediate compounds. Free and weak acid dissociable cyanides are oxidised to cyanate, which then hydrolyses to ammonia and carbonate.[54] Copper-catalysed hydrogen peroxide treatment was initially assessed at Homestake's McLaughlin Mine (prior to the development of their famous bio-treatment system), but as it did not remove ammonia, thiocyanate, and residual copper, used excessive amounts of expensive chemicals, and required an elaborate plant design[55] the option was not pursued. Degussa Corporation have developed the use of peroxymonosulphuric acid (a mixture of hydrogen peroxide and sulphuric acid) to partially detoxify tailings slurries ahead of discharge to a pond for further natural degradation.[54] A major factor in the development of Degussa's process was the fact that the company is one of the world's largest producers of both hydrogen peroxide and sodium cyanide, and the development was a synthesis of these two areas of expertise. Following the failure of the Omai tailings dam, the Commission of Inquiry recommended (as part of their report) that a hydrogen peroxide treatment plant should be installed.

14.7.2.3 Inco's SO₂/Air process

Using dissolved copper as a catalyst, SO_2 and air are passed through the cyanide-contaminated effluent. Free cyanide and metal-cyanide complexes are oxidised to cyanate, with the metals released during the oxidation process acting to precipitate complexed iron cyanide as insoluble and stable ferrocyanide salts.[56] Any excess metals in solution are then precipitated using lime. This process has been licensed at over 45 sites around the world to date. The process arose initially from research to reduce SO_2 emissions from the company's Copper Cliff smelter in Ontario. The need to remove pyrrhotite from the smelter feed resulted in the consideration of a flotation process route that would include cyanide as one of the reagents. The cyanide-contaminated tailings then required treatment prior to disposal and it was this that drove the development of the SO_2/Air process. This is a case of a site-specific requirement that has found far wider application within the industry.

14.7.2.4 Cyanisorb®

This patented process was developed by Cyprus Gold New Zealand Ltd. in 1989, and recovers free and weakly associated cyanide by stripping hydrogen cyanide at near neutral pH.[57] The company had originally assessed existing methods for the destruction of cyanide, but eventually took the route of developing its own alternative (in a similar manner to Inco's SO_2/Air process). The hydrogen cyanide is recovered by passing it through a solution of sodium hydroxide or carbonate. Cyanide recovery at this stage is normally in excess of 99%. However, the hydrogen cyanide must be closely managed due to its extreme toxicity. The process reduces cyanide costs and improves the recovery of gold and silver, and Cyprus has sought to commercialise the process to allow it to recoup its R&D costs. The process has had the dual benefit of solving environmental problems and facilitating future permitting and mine development.[5] It may also find applications as a pre-treatment to natural degradation of residual concentrations, thereby improving the efficiency of the natural degradation process.

14.7.2.5 Alkaline chlorination

This is the oldest and most widely recognised process for the destruction of cyanide,[58] although its application in the mining industry is limited due to a move towards other means of oxidation. The application of chlorine in a high pH environment leads to the oxidation of free and weakly acid dissociable cyanide (including cadmium, nickel, copper, and zinc complexes), but not the stable iron and cobalt cyanide complexes. Intermediate products (including chloramines) and excess chlorine need to be treated using sulphite to dechlorinate the effluent prior to disposal. The unaffected stable complexes may also require secondary treatment.

   Laboratory-scale alternatives to the direct use of halogens (such as chlorine) are being developed; for example, halide-mediated electro-oxidation.[59] This treatment relies on the electrolysis of brine (added to the cyanide-contaminated effluent) as a source of chlorine gas which then proceeds, through a cycle of hydrolysis, to form hydrochloric and hypochlorous acids. The latter acid oxidises cyanide and related compounds (e.g., cyanate, thiocyanate, and ammonia) to nitrogen and carbon dioxide and in the process is reduced to chloride, which then acts as a source of chlorine at the anode. This cyclic system therefore reduces the overdosing of chlorine often observed in standard chlorine systems. Further work is required to determine the potential for large-scale applications and to clarify the economics of treatment.

14.7.2.6 Bio-treatment

The most famous application of biological processes in the treatment of cyanide is at Homestake's McLaughlin gold mine. Here, *Pseudomonas* is used to biologically degrade cyanide in conjunction with bacterial nitrification of ammonia and adsorption/precipitation of free metals on bio-films.[55] Construction of the full-scale plant took place in February 1983, and the plant came on-line in August 1984, since which time it has operated continuously, producing high-quality effluents within allowable contaminant concentrations.

14.7.2.7 Engineered wetlands

Engineered wetlands now have an increasing application in the treatment of effluents associated with base and precious metal mining (e.g., ARD at Wheal Jane Mine in Cornwall, U.K.). The potential degradation of cyanide using wetlands has also been considered,[44] although the mechanism(s) by which the cyanide would or could be degraded remain unclear.

14.7.2.8 Ion exchange

Ion exchange resins can be used to extract cyanide from wastewater. The resins are regenerated by back washing with acid to generate hydrogen cyanide, which is then passed through sodium carbonate or sodium hydroxide solution[44] (cf. *Cyanisorb®*, above).

14.7.2.9 Precipitation of free cyanide

Although iron-cyanide complexes may act as long-term sources of free cyanide under the right conditions, their low solubility has also been used as a means of cyanide removal from wastewaters via the addition of iron salts to the cyanide-contaminated effluent.[44]

14.7.3 Discussion

None of these alternative treatment technologies are suitable under all circumstances and a combination of treatments may be required to satisfy the local discharge consents.[44] The choice of technology is further complicated by the need to treat both cyanide-contaminated tailings (solid-water slurries) and "clear" waters (free of solids). Therefore, each case must be considered on a site-specific basis and the implementation of a particular technology or combination of technologies should be based on laboratory and/or pilot plant trials.[44]

Despite the limitations of cyanide treatment and concerns regarding the long-term fate of metal-cyanide complexes, there is insufficient data at present to support the hypothesis that alternatives to cyanide would be a better environmental option. Fundamental life cycle inventory and assessment research is required on both cyanide and its alternatives to clarify the comparative environmental and technical performance of each. The current availability of a range of remedial treatment technologies must also be considered, as the replacement of cyanide with other reagents may also involve substantial effort in developing alternative or refined treatments for the altered process effluents.

From an industrial perspective, the critical issue is not whether cyanide should be used, but rather how to cost-effectively manage its use in an environmentally acceptable manner. This concern applies particularly to the stable metal complexes, which may act as long-term sources of toxic-free cyanide. In this case, there is a clear interaction between planning for closure and the development of suitable treatments that control or prevent the long-term accumulation of cyanide-contaminated wastes. The development of such technologies has already occurred (i.e., these are *existing* technologies) and therefore the case of cyanide as presented here supports the idea that existing technologies in tandem with current waste management procedures could be sufficient to enable efficient and effective closure. The issue of technology diffusion and dissemination within the industry is, however, a particularly pertinent one with regard to cyanide, as the most common treatment by many major gold producers remains natural degradation, despite concerns over the accumulation of metal-cyanide complexes at tailings disposal sites.

14.8 ARSENIC: OPTIONS FOR RECOVERY, TREATMENT, AND SAFE DISPOSAL

14.8.1 Introduction

With regard to the non-economic or non-targeted metals and metalloids that may be present in an orebody, it is common practice to neither specifically recover nor concentrate them, necessitating their disposal with the bulk wastes generated during milling and subsequent processing. One of the arguments in the past against more proactive treatment of these metals and metalloids has been that a "dilute and disperse" approach is acceptable where the carrying capacity of the receiving environment is not exceeded. This approach, in theory at least, avoids the generation of contamination "hotspots" sometimes associated with dedicated disposal sites, which may represent a far greater and immediate environmental risk if mismanaged. However, this attitude is one that has very little support outside the industry, and the days of "dilute and disperse" for all but a limited number of (bio)degradable organic compounds seem to be numbered. Part of the reason that such attitudes are today given short shrift is the rise of the "polluter pays" principle, through which pressure has increased on companies to internalise environmental costs. Another significant factor has been the development and subsequent commercial availability of technologies that are able to either (a) recover metals and metalloids where this was not previously technically or economically feasible, or (b) chemically and physically stabilise recovered contaminants to reduce post-disposal environmental risks.

Arsenic is a particularly topical example of the problems and issues associated with the sub-economic and toxic contaminants commonly associated with valuable mineral assemblages. It is often associated with copper and lead deposits, a frequent major contaminant of gold deposits, and is a classic case of a common contaminant for which there is only a limited market (principally in the production of wood preservatives, agricultural chemicals, and speciality glasses). The world's largest producers are China and Chile, followed by Mexico and the Philippines,[60] with the problem of arsenic in the Chilean mining industry considered by some to be the most serious and complex of a number of environmental issues there[61] (the annual production of arsenic trioxide in Chile in 1989 was estimated at 50,000 tonnes).[62]

The U.S. imported more than 13,000 tonnes of arsenic trioxide and over 500 tonnes of arsenic metal in 1992[60], and in 1993 imported 100% of its arsenic requirements mainly from Chile, China, Mexico, and France.[63] U.S. production of arsenic ceased in 1985 when ASARCO closed its copper smelter and associated arsenic recovery plant in Tacoma, Washington, largely due to the cost of complying with air quality standards.[64] Therefore, the importation of arsenic might be viewed as an externalisation of the environmental cost by transferring the environmental burden and risks from the U.S. to its suppliers.

Although the recovery of arsenic by-products (e.g., arsenic trioxide) is technically feasible, the potential quantity that would be generated if it was recovered from waste products far exceeds the potential market.[65] In keeping with many toxic and/or carcinogenic elements (e.g., mercury, lead, and cadmium) there is increasing downward pressure on the use of arsenic; in the future, the markets are likely to continue to diminish as less harmful alternatives are developed.

In the absence of specific treatment, a common approach is to rely on the natural processes of attenuation and biogeochemical cycling to minimise the environmental impact of arsenic-contaminated wastes generated during physical, hydrometallurgical, and pyrometallurgical processing of the ore.[65] The form in which the arsenic is finally disposed of depends on the initial ore mineralogy and the subsequent processing route. Sulphidic forms may predominate in tailings from gravity separation and flotation, while ferric and other metal arsenates may be more common in hydrometallurgical wastes. Impure arsenic trioxide is a common waste output from pyrometallurgical processing.

Arsenic mobility in the natural environment can vary in a largely unpredictable fashion. This is due to a number of uncontrolled (or uncontrollable) parameters, including waste type, arsenic concentration, the presence or absence of certain other cations and anions, redox and pH conditions, biologically-mediated dissolution and precipitation reactions, temperature, degree of percolating water, and so on. Therefore, reliance on attenuation and biogeochemical processes in all cases is a flawed concept and certainly one that has a significant risk element attached. Changes in legislation controlling the disposal of wastes to landfill are also likely to necessitate more advanced treatments where wastes are being disposed of off-site, principally to meet increasingly stringent regulation on the leachability of contaminants from solid waste inputs.[66]

As a result of its complex anionic chemistry, arsenic is considered a difficult element to treat successfully. Many of the treatment technologies that have been developed for the safe disposal of hazardous wastes are not applicable to arsenic-contaminated wastes. For example, arsenic is not effectively "fixed" by cement-based solidification/stabilisation techniques and can be lost to atmosphere via volatilisation during thermal treatments. Alternatives are, however, being developed. One such technology that has received considerable attention is SMITE (Synthetic Mineral Immobilization Technology).[66] This technology is aimed at the treatment of inorganic wastes, and is an alternative to the standard practice of using cementitious materials (e.g., Portland cement) to stabilise sludges, soils, and other contaminated solids. Other approaches have included the production of stable ferric arsenates and incorporation into pyrometallurgical slags.

14.8.2 SMITE (SYNTHETIC MINERAL IMMOBILIZATION TECHNOLOGY)

SMITE is essentially a two-stage process, consisting of pre-treatment followed by conversion to the final waste form. The pre-treatment is designed to precipitate or convert the metals into non-volatile forms and to add appropriate "tailoring" chemicals so that during the conversion phase the desired synthetic mineral assemblage is formed.[66] SMITE has been used for the stabilisation of arsenic flue dust (containing arsenic trioxide) in the form of a low solubility apatite-type mineral $[Ca_5(AsO_4)_3F]$. This form was determined by the authors as being the least soluble arsenic-bearing mineral, although previously Twidell et al.[67] considered lead chloroarsenate to be the most stable form. The arsenic trioxide is slurried with lime to form a calcium arsenite compound, intimately intermixed with calcium fluoride and fired at 1000 to 1100° C, thereby oxidising the arsenite (As(III)) to arsenate (As(V)) (a less soluble and toxic form) and converting the calcium arsenate to the final desired apatite-type composition. The final high-density product contains approximately 22% arsenic by weight, with an arsenic solubility (using the USEPA's Toxicity Characteristic Leaching Procedure) of less than 0.5 mg L^{-1}, compared to in excess of 4500 mg L^{-1} for the untreated flue dust, or nearly 800 mg L^{-1} for cement encapsulated waste.[66] As the waste is both chemically stable and low volume, it could potentially be stored on surface. While SMITE is not yet commercially available, indicative data shows that the cost of treatment and disposal are 5 to 10 times less than the cost of treatment using cementation techniques.[66]

14.8.3 FERRIC ARSENATES

Research on the high-temperature precipitation of dissolved arsenic as stable, crystalline ferric arsenates has been underway for a number of years (see for example: Swash and Monhemius,[68] and references therein). Research on the stability of arsenical products arising from the bacterial oxidation of gold-bearing arsenical pyrites has also been undertaken by Adam et al.[69] with the eventual aim of optimising the biooxidation operating conditions to produce a stable arsenic-bearing waste suitable for safe disposal. Krause and Ettel[70] detailed the solubility and stability of ferric arsenates in earlier work.

Ferric arsenates are generated by the dissolution of the arsenic species (commonly arsenic trioxide)[71] followed by conversion to crystalline scorodite ($FeAsO_4 \cdot 2H_2O$) using iron-bearing acidic nitrate (or sulphate) solutions at temperatures ranging from 140° C to 160° C. Van Weert and Droppert[71] describe crystalline scorodite as the least soluble "host" mineral for arsenic, although this is contradicted by White and Toor[66] who consider that apatite-type minerals are more stable. A number of parameters are critical in determining the solubility of the product, including the ratio of iron to arsenic, temperature, the presence of other ions in solution, and the rate of crystallisation. Due to these variables, ferric arsenate can, in fact, be considered a generic term for a whole range of compounds of varying crystallinity and solubility. This fact has been used against ferric arsenate as a suitable compound for disposal, although there is now a greater understanding of the control required to produce a suitably insoluble material for safe disposal.

Irrespective of the scientific uncertainties regarding the chemical stability of ferric arsenates, this approach is being used at plant level to deal with arsenic residues (e.g., from the BIOX process)[5] via precipitation with iron present in the ore (e.g., Codelco's copper processing complex at Chuquicamata, Chile), or by the addition of sufficient iron to fully precipitate the dissolved arsenate anions (e.g., El Teniente's Caletones copper smelter, south of Santiago, Chile).[61]

14.8.4 INCORPORATION IN SILICATE SLAGS

Alternatives to the formation of low solubility mineral phases include the incorporation of arsenic in silicate slags. This route may offer a more practical pathway for arsenic treatment at pyrometallurgical operations than alternative hydrometallurgical routes such as ferric arsenate precipitation,[72] as it reduces the need for additional capital investment. Research has indicated that up to 10% arsenic by weight can be incorporated into "glassy" silicate slags with very low release of arsenic via subsequent leaching.

14.8.5 DISCUSSION

Despite these recent advances, the issue of arsenic recovery, treatment, and disposal is one that has still to be resolved. The lack of markets for arsenic by-products removes the financial incentive for recovery (although recovery is often achieved incidentally during the processing operation) and the financial implications of liability are not sufficiently developed to act as an alternative driver. Clearly, innovative approaches to dealing with arsenic are required and these are being developed: future options may include stabilisation of arsenic in silicate slags (an in-process, clean technology option) and more widespread use of end-of-pipe treatments such as SMITE and ferric arsenate precipitation. The data appear to support the concept that clean technologies currently being developed will offer significant reductions in post-closure liability by promoting waste prevention rather than waste management. However, current technological approaches are typically end-of-pipe, and do not prevent waste but instead aid in the generation of wastes with enhanced chemical and physical stabilities. Such approaches are critical in controlling the environmental impact of mining operations: it is the *rate* of contaminant release from the disposal site to the surrounding environment as much as what is present at the site which determines the capacity of that site to impact local ecosystems. In many cases, untreated wastes can be viewed as uncontrolled wastes, since parameters relating to the disposal site itself may be largely beyond the control of the operator once the site is in operation (e.g., hydrology and hydrogeology, changes in pH, redox conditions, and other chemical or physical factors).

Where there is no market for a particular waste product and the presence of the contaminant in the process feed is unavoidable, the development of a truly clean technological solution is impossible. The alternatives are to (a) not exploit ores containing problem metals and metalloids for which there is no market, or (b) accept that the recovery, isolation, and effective treatment of the problematic elements prior to disposal is the optimal option. Realistically, it is the latter case that will predominate, although there have been instances where proposed mining developments have been rejected on the basis of generally unacceptable risk (e.g., New World Mine on the edge of the Yellowstone National Park in the U.S.). These may set a future precedent for restrictions on mining on or near "sensitive" sites not already "protected" from such activity. At the very least, the unpalatable nature of restrictions on mining may help promote the recovery and pre-disposal treatment of non-target and non-economic metals and metalloids if that can be established as the best environmental option based on rigorous scientific studies.

14.9 CONCLUSIONS

In the mining sector, the distinction between clean technologies and end-of-pipe technologies is rather blurred. This arises because the development of clean technology concepts and practice has occurred mainly in the process industries and many of the factors that were central to this development are not relevant or applicable to the base and precious metal mining sectors. It is an unfortunate fact that the accepted conventional definition of pollution prevention does not appear to fit the mining industry particularly well. This point was made by the Chief of the U.S. Bureau of Mine's Division of Environmental Technology as recently as 1996.[73] There is no doubt that the move towards best practice is an essential one (see, for example, the excellent "Best Practice in Environmental Management in Mining" series by the Australian Environment Protection Agency), but it is *defining* that best practice which remains problematic.

At present, historical precedent is a major factor in promoting waste management in the context of closure rather than substantive scientific studies that demonstrate its relative merits compared to waste prevention. In more general terms, the lack of *quantitative* research comparing alternative process and management options in terms of optimising economic and technical performance *and* minimising environmental impact (i.e., a life cycle approach) is one of the fundamental gaps in the literature. The continued domination of remedial action is potentially flawed as it accepts contamination and pollution (even if limited, and controlled within the site boundaries) without considering the possible environmental benefits of preventative measures. This comment does not presume that preventative measures are always environmentally "better" than remedial approaches, merely that the relative merits of each remain unquantified within the industry at present.

Consequently, as might be expected, there is conflicting evidence as to the significance of clean technology in planning for closure. There is clearly a requirement to continue developing innovative solutions to the non-target and non-economic metals and metalloids commonly found in base and precious metal ores (e.g., arsenic). At present, this seems to be one of the most significant problems, as it has no obvious economic solution or driver. Regulation may be the most effective way forward by forcing the full internalisation of environmental costs into the cost of production. In other areas, waste management rather than a move to new and unproven technologies might prove more effective in limiting environmental impact (e.g., cyanide use in gold recovery).

The draconian implementation of preventative approaches by regulators is not necessarily appropriate in a more general sense given the nature of this particular sector (i.e., necessarily a high-volume waste producer) despite the increasing dominance of clean technology in many industrial sectors. The question must be asked as to whether or not policies that were developed in these other sectors to promote lean and clean production or pollution prevention will work in the mining sector, or even if they are required. Irrespective of definitions, if an alternative to simple disposal is not classified as a clean technology this should not imply that it is ineffective at reducing the threat of harm to human health or the environment, or ultimately delivering cost savings to industry.

That said, if the industry wishes to continue using waste management as a tool into the 21st century, then it must *justify* that approach against a backdrop of advances in cleaner mineral processing. It is no longer sufficient to merely claim that waste management is an acceptable approach and reject alternative routes to planning for closure on that basis. Industry would do well to invest in research that will define the relative requirement for waste management and preventative approaches into the 21st century so that it can debate these critical issues with regulators and other stakeholders on a scientifically sound footing. It is quite apparent that regulation is changing and will continue to change, and that industry must take heed of that fact.

It is absolutely essential that there is a reasoned debate between all stakeholders, but particularly operators and regulators, to rigorously define the best environmental options for operational and closure practice within a framework of economic viability and international competitiveness. The current debate between the National Mining Association and the USEPA regarding the future status of certain mineral processing wastes and the use of leaching tests to define the chemical stability of wastes is an excellent example of polarised and polemical confrontation that benefits neither party in the long term.

Despite the uncertainty, the concept that the industry should be based on clean technology rather than waste management appears to be a sound one, but the need for the two approaches needs to be more explicitly analysed. This analysis must incorporate elements of the social context within which the technology will be applied (i.e., social preferences also have a role to play in defining the most appropriate technology or process route). In any case, clean technology needs to be considered over the whole life cycle of a site, not just within the production phase. If this holistic approach is taken, the split between the production phase and the closure phase disappears, and the use of clean technology to meet the day-to-day discharge consents becomes just one part of its wider application in controlling the environmental impact associated with a site throughout its life.

ACKNOWLEDGEMENTS

The author would like to acknowledge the positive input from numerous colleagues during the development of this chapter, in particular Dr. Gavin Bridge (University of Oklahoma), Prof. Joe Barbour (University of Bath), Prof. Barry Crittenden (University of Bath), Mr. Philip Gray (Royal Academy of Engineering), Prof. Jim Petrie (University of Cape Town), Dr. Mark Whitbread-Jordan (Camborne School of Mines, University of Exeter), and Mr. John Barker (University of Bath). Finally, the author would like to thank Ms. Liz Smith and Mrs. Yvette Haines for their secretarial support.

The chapter was developed as part of the author's research on "Maximising Environmental Protection and Economic Performance at Base and Precious Metals Mines through the Optimised Choice and Use of Process Chemicals & Technology." The three-year project at the University of Bath was sponsored by the Engineering and Physical Sciences Research Council and the Royal Academy of Engineering (1996–1998).

REFERENCES

1. Ripley, E. A., Redmann, R. E., and Crowder, A. A., *Environmental Effects of Mining*. St. Lucie Press, Delray Beach, Florida, 1996.
2. Boldt, J. R., *The Winning of Nickel*, Longmans, Toronto, 1967.
3. Godin, E., Ed., *1990 Canadian minerals yearbook - review and outlook*, Energy, Mines and Resources Canada, Ottawa, Ontario, 1991.
4. Gray, P., Clean working in the mining industry: the tonnage problem, *Mining and Environment Research Network Newsletter*, No. 4, 26, 1993
5. Warhurst, A. C. and Bridge, G., Improving environmental performance through innovation: recent trends in the mining industry, *Minerals Engineering*, 9, 970,1996.
6. Aldrich, J. R., Expected value estimates of the long-term liability from landfilling hazardous waste, *J. Air & Waste Manage. Assoc.*, 44, 800, 1994.
7. Smith, J. L., Interactive computer software — a high-tech answer to industries' waste minimization woes, in *Extraction and Processing for the Treatment and Minimization of Wastes*, Hager, J., Hansen, B., Imrie, W., Pusatori, J., and Ramachandran, V., Eds., The Minerals, Metals & Materials Society, Warrendale, Pennsylvania, 1993, 237.
8. United Nations, Effect of Environmental Protection and Conservation Policies on the Mineral Sector, Committee on Natural Resources, E/C.7/1994/1, United Nations, Paris, 1994.
9. Constan, J., Mining wastes matter, *Clementine*, Autumn, 10, 1996.
10. United States Environmental Protection Agency, Addition of facilities in certain industry sectors: toxic chemical release reporting; community right-to-know, 40 CFR Part 372, Federal Register, 61, 33588, 1996.
11. Ashford, N. A., Legislative approaches for encouraging clean technology, *Toxicology and Industrial Health*, 7, 335, 1991.
12. Bridge, G., Research Officer, Mining and Environment Research Network, University of Bath, personal communication, 1996.
13. Berkowitz, J. B., Environmental cost considerations and waste minimisation in new plant design and process optimization, in *Hazardous Waste: Detection, Control, Treatment*, Abbou, R., Ed., Elsevier Science Publishers B.V., Amsterdam, 1988, 1727.

14. Clift, R., Clean technology — an introduction, *J. Chem. Tech. Biotechnol.*, 62, 321, 1995.

15. Redhead, R. J., Waste management: a materials management/product stewardship approach, *CIM Bulletin*, 89, 47, 1996.

16. Crittenden, B. and Kolaczkowski, S., *Waste Minimization. A Practical Guide*, Institution of Chemical Engineers, Rugby, 1995.

17. Almgren, G., Almgren, T., and Kumar, U., Just-in-time and right-in-space, *Minerals Industry International*, No. 1038, 26, 1996.

18. Young, S. B. and Vanderburg, W. H., Applying environmental life cycle analysis to materials, *Journal of Metals*, 46, 22, 1994.

19. McHugh, R. T., The economics of waste minimization, in *Hazardous Waste Minimization*, Freeman, H. M., Ed., McGraw-Hill, New York, 1990, chap.6.

20. Haas, C. N., Waste elimination options, in *Hazardous and Industrial Waste Treatment*, Haas, C. N. and Vamos, R. J., Eds., Prentice Hall, Englewood Cliffs, New Jersey, 1995, 325.

21. United Nations, *Mining and the Environment — the Berlin Guidelines*, Mining Journal Books, London, 1992.

22. United Nations, Earth Summit: Agenda 21, The United Nations Programme of Action from Rio, 1992.

23. Hayes, P., *Process Principles in Minerals & Materials Production*, Hayes Publishing Co, Queensland, 1993.

24. Everell, M. D. and Oliver, A. J., Environmental technologies for the minerals and metals sector — a Canadian perspective, in *Minerals, Metals and the Environment*, Elsevier Applied Science, 1992, 331.

25. Orava, D. A. and Swider, R. C., Inhibiting acid mine drainage throughout the mine life cycle, *CIM Bulletin*, 89, 52, 1996.

26. Anonymous, Strategic alliances for environmental solutions — the Greenpeace recipe for business success, *ENDS Report*, 19, 1996.

27. Themelis, N. J., Pyrometallurgy near the end of the 20th century, *Journal of Metals*, 46, 51, 1994.

28. Tregenza, C., Mineral extraction: refining the process, *Australian Mining*, 86, 53, 1994.

29. Adel, G. T., Comminution, *Mining Engineering*, 46, 440, 1994.

30. Rice, D. A., Seitz, R. A, Arnold, B. A., and Mulukutla, P., Concentration, *Mining Engineering*, 46, 441, 1994.

31. Scheiner, B. J. and Sharma, S. K., Dewatering, flocculation, *Mining Engineering*, 46, 443, 1994.

32. Nagaraj, D. R. and Akser, M., Flotation chemistry, *Mining Engineering*, 46, 444, 1994.

33. Seitz, R. A. and Jordan, C. E., Flotation process analysis, *Mining Engineering*, 46, 447, 1994.

34. Thompson, R. E., Hydrometallurgy, *Mining Engineering*, 46, 449, 1994.

35. Rule, A. R., Process mineralogy, *Mining Engineering*, 46, 452, 1994.

36. Sohn, H. Y. and Cho, W. D., Developments in physical chemistry and basic principles, *Journal of Metals*, 46, 43, 1994.

37. Sohn, H. Y. and Cho, W. D., Developments in physical chemistry and basic principles, *Journal of Metals*, 47, 60, 1995.

38. Klimpel, R. R., Technology trends in froth flotation chemistry, *Mining Engineering*, 47, 933, 1995.

39. Kral, S., IMPC delegates discuss the future of minerals processing into the 21st century, *Mining Engineering*, 48, 55, 1996.

40. Nagaraj, D. R., A critical assessment of flotation reagents, in *Reagents for Better Metallurgy*, Mulukutla, P. S., Ed., Society for Mining, Metallurgy, and Exploration, Inc., Littleton, Colorado, 1994, 81.

41. Cappuccitti, F. R., Current trends in the marketing of sulfide mineral collectors, in *Reagents for Better Metallurgy*, Mulukutla, P. S., Ed., Society for Mining, Metallurgy, and Exploration, Inc., Littleton, Colorado, 1994, 67.

42. United States Environmental Protection Agency, Innovative Methods of Managing Environmental Releases at Mine Sites, U.S. Environmental Protection Agency, Office of Solid Waste, Special Waste Branch, Washington, D. C., 1994.

43. Kral, S., Hydrometallurgy fundamentals technical sessions, *Mining Engineering*, 46, 124, 1994.

44. Smith, A. and Mudder, T., *The Chemistry and Treatment of Cyanidation Wastes*, Mining Journals Books Limited, London, 1991.

45. Smith, A. and Mudder, T., Cyanide, dispelling the myths, *Mining Environmental Management*, 3, 4, 1995.

46. Fristoe, B., Environmental Engineer, Alaska Department of Environmental Conservation, Personal communication, 1996.

47. Adams, M. D., Group Leader: Process & Environmental Chemistry, MINTEK, Randburg, South Africa, personal communication.

48. Swaminathan, C., Comparative cyanide and thiourea column leach studies on a Central Victorian gold ore, in *Recent Trends in Heap Leaching*, The Australasian Institute of Mining and Metallurgy Publication Series No. 7/94, Carlton, Victoria, 1994, 39.

49. Adams, M. D., Environmental impact of cyanide and non-cyanide lixiviants for gold, Proc. New Developments in the Extractive Metallurgical Industry, SAIMM/MMMA Symposium, Johannesburg, 23 February, 1996.

50. Ramadorai, G., Halogen solvents in precious metal ores processing, in *Reagents for Better Metallurgy*, Mulukutla, P. S., Ed., Society for Mining, Metallurgy, and Exploration, Inc., Littleton, Colorado, 1994, 283.

51. United States Environmental Protection Agency, Treatment of Cyanide Heap Leaches and Tailings, U.S. Environmental Protection Agency, Office of Solid Waste, Special Waste Branch, Washington, D. C., 1994.

52. Ou, B. and Zaidi, A., Natural degradation, *Mining Environmental Management*, 3, 5, 1995.

53. United Nations, Environmental Management of Mine Sites, Training Manual, Technical Report No. 30, United Nations Department for Development Support and Management Services, United Nations Environment Programme, Paris, 1994.

54. Norcross, R. and Steiner, N., Degussa's peroxide process, *Mining Environmental Management*, 3, 7, 1995.

55. Waterland, R. A., Homestake's bio-treatment, *Mining Environmental Management*, 3, 12, 1995.

56. Robbins, G. and Devuyst, E., Inco's SO₂/air process, *Mining Environmental Management*, 3, 8, 1995.

57. Stevenson, J., Botz, M., Mudder, T., Wilder, A., and Richins, R., Cyanisorb recovers cyanide, *Mining Environmental Management*, 3, 9, 1995.

58. Smith, A. and Mudder, T., Alkaline chlorination, *Mining Environmental Management*, 3,12, 1995.

59. Haque, K. E. and MacKinnon, D. J., The halide mediated electro-oxidation of ammonia, cyanide, cyanate and thiocyanate in mine/mill wastewaters, *CIM Bulletin*, 89, 104, 1996.

60. United States Environmental Protection Agency, Arsenic, in Identification and Description of Mineral Processing Sectors and Waste Streams, U. S. Environmental Protection Agency, Washington, D. C., 1995.

61. Wiertz, J. and Gutierrez, M., Arsenic - a Chilean approach, *Mining Environmental Management*, 4, 20, 1996.

62. Crozier, P. C. F., Chile's legacy — pollution, *Mining Journal (Environment Supplement)*, 315, 10, 1995.

63. Anonymous, *Mineral Commodity Summaries*, U.S. Bureau of Mines, Washington, D.C., 1994.

64. Office of Solid Waste, *Human Health and Environmental Damages from Mining and Mineral Processing Wastes: Technical Background Document Supporting the Supplemental Proposed Rule Applying Phase IV Land Disposal Restrictions to Newly Identified Mineral Processing Wastes*, U. S. Environmental Protection Agency, Washington, D.C., 1995.

65. Hopkin, W., The problem of arsenic disposal in non-ferrous metals production, *Environmental Geochemistry and Health*, 11, 101, 1989.

66. White, T. and Toor, I., Stabilizing toxic metal concentrates by using SMITE, *Journal of Metals*, 48, 54, 1996.

67. Twidwell, L. G., Plessas, K. O., Comba, P. G., and Dahnke, D. R., Removal of arsenic from wastewaters and stabilisation of arsenic bearing solid wastes: summary of experimental studies, *Journal of Hazardous Materials*, 36, 69, 1994.

68. Swash, P. M. and Monhemius, A. J., Hydrothermal precipitation from aqueous solutions containing iron(III), arsenate and sulphate, in *Hydrometallurgy '94*, Institution of Mining and Metallurgy and Chapman & Hall, Cambridge, England, 1994, 177.

69. Adam, K., Komnitsas, C., Papassiopi, N. L., Kontopoulos, A., Pooley, F. D., and Tidy, N. P., Stability of arsenical bacterial oxidation products, in *Hydrometallurgy '94*, Institution of Mining and Metallurgy and Chapman & Hall, Cambridge, England, 1994, 291.

70. Krause, E. and Ettel, V. A., Solubilities and stabilities of ferric arsenates, in *Crystallization and Precipitation*, Strathdee, G. L., Klein, M. O., and Melis, L. A., Eds., Pergamon Press, Oxford, England, 1987, 195.

71. Van Weert, G. and Droppert, D. J., Aqueous processing of arsenic trioxide to crystalline scorodite, *Journal of Metals*, 46, 36, 1994.

72. Machingawuta, N. C. and Broadbent, C. P., Incorporation of arsenic in silicate slags as a disposal option, *Trans. Instn. Min. Metall. (Sect. C: Mineral Process Extr. Metall.)*, 103, 1, 1994.

73. Schmidt, W. B., Waste management/treatment and its role in pollution prevention, in *Pollution Prevention for Process Engineers*, Richardson, P. E. and Scheiner, B. J., Eds., Engineering Foundation, New York, 1996, 13.

74. Neale, A. J., Franklin, M., and Cooper, M., The automation of reagent addition systems to enhance flotation plant performance, in *Reagents for Better Metallurgy*, Mulukutla, P. S., Ed., Society for Mining, Metallurgy, and Exploration, Inc., Littleton, Colorado, 1994, 25.

15 Using Financial Assurances to Manage the Environmental Risks of Mining Projects

Kathleen Anderson

CONTENTS

ABSTRACT

The first section of this chapter briefly describes and reviews the main categories of environmental protection policies, concluding with an introduction to financial assurances, the focus for the rest of the chapter. The second section briefly reviews methods of calculating the value of financial assurances. This is followed by a discussion on the mechanisms available to implement financial assurance and finally, a set of policy recommendations.

15.1 PUBLIC POLICIES TO MANAGE ENVIRONMENTAL ISSUES IN MINING

Mining, in much of the world, provides foreign exchange, income and employment, and tax revenues, which can be used to build infrastructure and fund governmental programmes, and are a means to fund economic development. However, mining can also bring environmental degradation, threaten aquatic habitats, and adversely affect other sectors of the economy such as farming and fishing. Government decision-makers use a variety of approaches and public policy tools to prevent, control, and remediate the potential environmental consequences of mining. Common approaches include:

- Command and control techniques that rely primarily on quantitative chemical standards and criminal penalties for enforcement.
- Technology transfer, information, and education.
- Economic instruments which allow firms to make a wider range of decisions about how to minimise environmental degradation.

Many governments are now considering developing another approach: "financial assurances," which redistribute the risks of environmental damage from the public sector to the private sector. Financial assurances typically require mining companies to guarantee financial responsibility for reclamation of mine sites. There are a number of specific financial instruments grouped within the broad category of financial assurance, including several types of bonding, trust funds, and insurance programmes. How countries choose to approach the regulation of mining environmental problems will impact the mining sector, the likelihood of achieving public policy objectives over the long run, and political acceptability. In countries that want both a dynamic mining sector and environmental protection, the thoughtful selection of approaches to managing mining environmental issues is important.

15.1.1 COMMAND AND CONTROL POLICIES

Command and control approaches are typically "media" specific, designed to reduce discharges of specific pollutants into air, land, or water. Most often measured by quantitative chemical or physical standards (other types of standards, such as biologically based narrative standards, are also possible), command and control measures are designed to be enforceable in administrative or legal actions. This approach can be effective and successful in reducing many types of pollution, particularly in circumstances where:

- A single polluter can be identified.
- The discharge of the pollutant can be tracked and monitored with a high degree of accuracy.
- Pollution prevention is both technically and economically feasible.
- There is little variation in the site-specific characteristics of the industry being regulated.

There are many industries whose activities fall within these boundaries. For example, command and control can be the most cost-effective and efficient approach to regulating many manufacturing processes, where climatic variation, rainfall, or altitude have little or no effect, the production process is relatively standardised, and the pollutant is released from the factory via discrete pathways. Even under these optimal circumstances, however, command and control approaches are costly to implement and maintain, require a well-educated and well-trained enforcement team, extensive support from laboratories and consultants, and an effective administrative/judicial system.

There are both advantages and disadvantages to the use of command and control approaches for mining projects. Because every orebody is unique, the standardised measures of compliance that are inherent in traditional command and control may be less than fully effective. Effective environmental protection in mining should allow for characteristics which are specific to each mine site, such as site geochemistry, water balance, presence of fragile ecosystems, altitude, distance from receiving bodies of water, slope and stability of surrounding landforms, and sociocultural values in the host community. Command and control approaches can result in significant under-protection at some sites and create an onerous burden of over-protection at others.

Regulatory compliance often becomes the primary goal of industries regulated under command and control systems. There may be little incentive to develop new pollution prevention technologies and innovative production processes. Scarce social and economic resources are often drawn away from activities such as research and development, and devoted instead to litigation and regulatory compliance.

15.1.2 Technology Transfer, Information, and Education Policies

Technology transfer, dissemination of timely and appropriate information, and education are impor-tant to the successful management of mining environmental issues for several reasons. First, asymmetries between regulators, communities, and the mining industry are reduced. The result can be more effective participation of all parties affected by a mining project, participation which has become increasingly important to successful mine development, operation, and closure. Second, information and education can assist small operators and artisanal miners who often have little or no access to state-of-the-art information about current engineering and environmental practices. Technology transfer programmes, the development and dissemination of databases, capacity-build-ing seminars, training in "best environmental management practices," and establishment of industry boards for self-monitored performance standards are all activities which fall within the purview of information and education.

Technology transfer, information, and education are extremely cost-effective long-term approaches. Significant investments in teachers, training centres, materials, and travel are required. However, much of the investment is in "human capital," and the benefits of education extend to the entire community. Innovation in solving site-specific environmental/mining challenges is enhanced, environmental problem-solving skills throughout the community are strengthened, and opportunities to prevent pollution and increase profits in mining are illuminated.

Basic knowledge about the mechanisms and management of mining-related pollution can be applied in a culturally appropriate, site-specific manner. Experience has shown that effectiveness is limited when information and education programmes focus on the dissemination of rigid, externally developed systems of laws and environmental management approaches. These transposed schemes are often informed by a culturally prescribed set of values and beliefs which may or may not coincide or be relevant to the host community undertaking the challenge of raising the level of environmental performance in mining projects.

15.1.3 Economic and Financial Policies

Economic instruments available to address environmental pollution include taxes, credits, rebates, subsidies, penalties, marketable pollution permits, deposits, and effluent charges, as well as many other approaches such as bubbles, offsets, and emission banks. Economic instruments vary widely, but all are intended to internalise "externalities," those spillover effects that are not incorporated into the costs of either production or consumption of goods. *Unmitigated* mining environmental externalities generate costs for society in general (or a subset of society, such as residents in close proximity to the mine) via impacts to human health, political unrest, degraded water quality, lost habitat for flora and fauna, foregone opportunities for future land uses at the mine site, and many other ways. (Positive externalities associated with many mining projects include such benefits as increased income, schools, roads, hospitals, and public revenue).

Economic instruments are used to transfer the costs of externalities back to the private producer or consumer. Where the character of environmental problems in a given industry varies widely, economic instruments, such as marketable pollution permits, can be particularly effective and efficient.

In general, economic instruments, combined with information and education, can be the most appropriate environmental management approach for mining. There are often many small entre-preneurial operators using limited pollution controls, and a few large corporations using state-of-the-art pollution control technologies; there is wide variation in mining processes from one mineral to another, and there is wide variation in the climatic conditions which determines much of the environmental "risk."

Economic instruments are preferable to command and control when applied to mining for a second, perhaps more important, reason. Mining occurs, by definition, in mineralised regions. These are regions where the background levels of environmental contaminants, such as arsenic, lead, and zinc, are often naturally high. There are often many mines in these regions, and discharges from mining operations are seldom discrete. Impacts to waterways from surface runoff and discharges to groundwater may not occur adjacent to the mine, but many miles away. Where there is more than one mine; the background levels of contaminants are naturally high; and pathways for transport are not discrete, it is very complicated and expensive to identify the polluter and enforce environ-mental standards. Economic instruments offer an alternative way to encourage voluntary action when command and control is not effective or feasible.

Economic and financial incentives can be developed to encourage many environmental goals, such as recycling, conservation, reclamation, or pollution prevention. The dissemination of mining methods that incorporate no-cost/low-cost approaches to environmental protection is an important result of the application of economic instruments. Topsoil storage improvements, engineering-hydrologic modifications to control diffuse run-on and runoff from tailings, and recycling of process waters and reagents can be encouraged with economic instruments.

The primary advantage of economic instruments is that they encourage both producers and consumers to evaluate alternative ways to reach environmental goals; cost effectiveness, personal preferences, cultural values, and access to technology can be incorporated into this evaluation. This is information to which regulators often have no access to, or which is extremely costly to acquire.

The effective design and implementation of economic instruments requires a well-trained and educated regulatory community, but with a different mix of skills and abilities than is necessary for command and control approaches. Additionally, economic instruments may need to be backed up by the command and control aspects of regulation to be successful.

15.1.4   Financial Assurances, a Hybrid Policy

Financial assurance is one such "hybrid" economic instrument, which must be complemented by permitting, environmental assessment, inspection and enforcement, and education to be effectively implemented. Standard financial assurance tools and mechanisms are intended to ensure that a normal range of costs associated with reclamation and closure of mines will be paid for by the mine owner or operator, either directly or through some alternative mechanism which assures their financial responsibility. These mechanisms are typically not intended to insure against catastrophic events. Financial assurance assumes that the costs of reclaiming and restoring mined land to subsequent uses, and protecting the public from safety threats such as open adits, shafts and subsidence, are ultimately the responsibility of the owner or operator of the mine.

A comprehensive financial assurance programme has three parts: permitting, inspection and enforcement, and the financial assurance itself.

15.1.4.1   Permitting as a component of financial assurance

The permitting process is important for three reasons. First, the permit can be used to clarify the standards for reclamation and environmental performance at the mine. Second, the mine plan is the foundation for calculating the amount of financial assurance, which is often inversely related to the amount of pollution prevention which is built into the mine plan. A mine designed for optimal reclamation will have planned at the beginning for topsoil storage, regrading and slope stabilisation, revegetation, water management, ongoing reclamation (if possible), and any other engineering and environmental practices deemed by the regulatory authorities to be good practice.

The third reason is that the permit is often the only way that the environmental performance and reclamation standards can be successfully enforced. The mine permit, or an alternative administrative document tied to the right to continue mining, should clearly and explicitly define the terms for release of financial assurance; this protects both the operator and the regulator.

15.1.4.2   Inspection and enforcement components

Inspection and enforcement are the second essential components of a good financial assurance programme. Ongoing inspections require regular communication between the regulator and the miner, with the goal of identifying problems in their formative stages. Many mining environmental problems, including acid generation, can be successfully mitigated and managed if identified early enough (see Mitchell, Chapter 7). Inspection and enforcement, early and often, will help minimise the problem which occurs when all reclamation, remediation, and closure activities are left until mining is completed, the ore is depleted, and the cash flow available to the firm to reclaim and close the mine is at its lowest.

The enforcement capacity of the regulator must be clearly and explicitly defined. All parties, including the provider of financial assurance provisions (banks, insurance companies, trust fund operators, or others), should be aware of the implications for the financial assurance in case of permit violations, bankruptcy, or other possibilities, including the capacity of the regulator to seize assets or close the mine.

15.1.4.3   Mechanisms

There are many mechanisms for providing financial assurance. These include bonding systems, trust funds, insurance plans, and many other institutional structures, which provide assurance of financial responsibility, each developed to meet specific needs. The primary examples addressed in this paper are surety bonds, bond pools, trust funds, and the financial test. Environmental liability insurance plans are available from a very small number of providers, and provide extremely limited coverage. Guarantees from parent corporations are also used to provide financial assurance. These last two mechanisms are not addressed explicitly in this chapter.

15.1.4.3.1   Surety bonding

Sureties are business entities that agree to be responsible for the debts of another party, or the failure of another party to perform an action, typically a reclamation and closure action in mining. Sureties guarantee either that the costs of reclamation and closure will be paid, or that the reclamation will be performed. Surety bonds transfer the risks of environmental non-performance (which results in negative externalities) from the public to the surety; the surety assures that the miner will perform required reclamation and environmental protection activities.

In its most basic form, the owner/operator of the mine purchases a surety bond, based on current cost estimates of closure, and this mechanism assures that there will be sufficient financial resources to close and reclaim the mine. There may be a requirement that the owner/operator of the mine provide collateral, which in many cases is up to 100% of the cost of reclamation.

As with all approaches to financial assurance, the advantages to using this mechanism must be weighed against the disadvantages, in the context of the regulatory regime where it is being considered. One difficulty which arises with this mechanism is that a sureties industry must exist, and it must, in turn, be regulated itself. The failure rate of businesses attempting to enter the sureties industry in the U.S. has historically been quite high. When the surety goes bankrupt, both the regulator and the mining firm lose.

Oversight of the sureties industry requires an investment in building regulatory capacity. There is, however, a major advantage to the use of sureties as a financial assurance mechanism. Because they have a financial interest in the control of environmental degradation at mine sites, the sureties industry can become a partner of the regulatory community, carrying out inspections and setting standards for environmental performance.

15.1.4.3.2   Bond pools

Bond pools are designed to pay for reclamation and closure costs incurred by bond pool members, in case of bankruptcy or other unforeseen events that render them financially unable to fulfil reclamation and closure commitments. Bond pools are often proposed to meet the needs of small operators, many of whom are unable or unwilling to provide the substantial collateral (often 50 to 100% of the bond) required by sureties firms. Membership in bond pools is voluntary. There is typically a "test" for entry, which includes evaluation of the following: compliance record, including number of permit violations; financial standing; years in operation; and reclamation experience.

Following admission, members may be categorised according to the "risk" which they bring into the pool, and payment into the pool takes that risk factor into account. While payments are often based on the tonnage mined, or surface disturbance, there are other approaches to incorporating the wide variation in environmental costs which can occur. For example, site-specific conditions may dictate that operators in fragile ecosystems, or in regions likely to generate acid, be charged higher membership fees.

Bond pools can be difficult to administer, and have not been successful in the U. S. One of the many problems which plague bond pool schemes is that of the "free rider" — the mining company or operator who joins the bond pool, pays in a minimal amount, and then declares bankruptcy or discloses a serious environmental problem. Responsible firms carry the burden of free riders, and this can deplete the assets of the pool that are available for legitimate purposes.

15.1.4.3.3   Non-surety collateral bonds, or trust funds

Collateral bonds, essentially equivalent to trust funds, are yet another variation in the world of financial assurance mechanisms. These are indemnity agreements made by the mine owner/operator, and they involve the mine owner/operator setting aside collateral, cash, or cash equivalent financial devices, equal in value to the estimated costs of reclamation and closure. These funds or assets are then held in trust by the regulator, the government, a bank, or similar financial institution. It is necessary to make clear from the start whether the interest earned on these funds while in trust is to be returned to the owner/operator or held by the government for the following purposes:

- Compensation for administrative costs.
- Offset of reclamation costs at sites where financial assurance was not sufficient.
- As a source of funding to remediate abandoned mines.

A standard list of cash or cash-certain devices acceptable as collateral would include the following:

- Cash.
- Stocks and negotiable bonds.
- Certified checks.
- Irrevocable letters of credit.
- Certificates of deposit.

However, there are instances where regulatory authorities agree to accept the title to land and/or equipment at the mine, in lieu of cash or cash-equivalent collateral. These types of collateral may be difficult to liquidate, be in poor repair if the firm is in financial hardship, and have little market value by the time the regulator gains possession. Additionally, gaining access to these assets may be particularly problematic once it is clear that the firm has taken, or intends to take, bankruptcy. The regulator may have to queue up with other creditors. The effect of these barriers will vary, depending on the legal authority of the host government.

15.1.4.3.4   Self-bonding or financial tests

This approach is based upon evaluating the financial health of the mining firm, and acquiring assurance from the firm itself that sufficient funds will be set aside to carry out reclamation and closure obligations. Many large firms prefer this method of proving financial responsibility. There are significant advantages for the mine owner or operator in this approach, including tighter control of funds and savings in reduced transaction costs.

One disadvantage to this approach is that there are times when even large companies which look good on paper must declare bankruptcy, leaving the public (government) to absorb the full costs of any reclamation, remedial actions, and closure. There is no redistribution of this risk to a third party. Additionally, firms may account for their contributions to the reclamation funds with accrual-based accounting, and not cash contributions to a "sinking fund." While this is not intrinsically risky with most reputable firms, it is an approach which, in theory, could result in cash not being available at the end of the mine life. The primary disadvantage to this approach is, however, that evaluating the financial health of a mining company is extremely sophisticated business, with the number of ratios and financial indicators used to make this decision expanding rapidly. This type of expertise seldom can be found within government, leaving regulators with the need to contract out this financial analysis, or be at a distinct disadvantage in negotiating the proper limits and constraints to self-bonding. Guarantees from the parent corporation are subject to the same comments.

Among the financial indicators most often used but not recommended as adequate are the following:

- Net worth of the firm.
- Ratios.
- Assets.
- Bond ratings.

15.1.4.4   Calculating financial assurance

The amount of financial assurance that is required should be based on cost estimates derived from the mine plan. From the plan, discrete reclamation and closure activities can be identified. The overall objectives of reclamation and closure usually include the following: removal or closure of imminent safety threats; protection of water quality to the desired level in bodies of water impacted by mining activities; and minimisation of surface erosion potential in affected areas. In general, objectives emphasize restoration of mined land in such a way as to assure that mining is a temporary use of the land and that, subsequently, other economic uses are possible. Specific tasks can be identified, such as re-contouring, replacing topsoil, revegetation, and highwall reduction. The following concise approach is based on that provided by the California Mining Association in its Surface Mining and Reclamation Act, Financial Assurance Guidelines.[1]

"… Following the identification of broad reclamation categories, the component parts of these tasks should be identified. For example, revegetation may include seedbed preparation, seeding and fertilising, irrigation and weed control. Each of these subtasks should be estimated individually to simplify the process. Where grading of pit area is part of the reclamation plan, we recommend that cross-sections and maps of pit areas be used to justify grading quantities.

- Identify the equipment necessary to complete the identified task.
- Identify labour requirements.
- Identify materials to be used.

- Define unit costs.
- Multiply the unit cost (e.g., $ hr^{-1}) by production rate (e.g., cubic yards hr^{-1}) to determine the total cost for each cost item (e.g., scrapers).
- Add the costs for all cost items to find the total cost per category (e.g., equipment).
- Add total cost of all categories (i.e., equipment, labour, materials) to determine total cost of reclamation.
- Add charges for supervision, profit and overhead, contingencies and mobilisation."

Although this provides a general outline of the process used to generate a cost estimate, each mine has distinct and unique operations associated with each phase of development, which need to be assessed. It is important to identify the sources of information in cost estimates. Invoices, contractor estimates, and standard cost manuals are valid sources. Additionally, any assumptions made while generating cost estimates should be clearly and explicitly stated. It may prove useful to develop cost estimates under a range of explicit and distinct scenarios, which reflect high, medium (expected), and low costs. An example of a high-cost scenario may be when tailings piles which were not predicted to generate acid begin to do so, and additional remediation for proper closure is needed.

15.1.4.5   Determining the basis for cost calculations

The policy objectives of the host government should be considered when deciding whether to require financial assurance for the entire mine life, or for smaller increments of time. This decision impacts the economic viability of the mine, particularly for small operators. All costs should be summed using a net present value approach, regardless of the time period for evaluation.

15.1.4.5.1   Full life

This method requires that financial assurance be provided which guarantees sufficient funds for all reclamation and closure activities for the entire estimated life of the mine, along with any post-closure activities that are necessary. This approach sums the ongoing reclamation costs over the mine life. The total costs can be high, even when discounted to present value. This can render a marginal project unfeasible.

The primary advantage to this approach is that it provides maximum security to the risk-averse regulator and host government. It is feasible in a strong market where the host government has many operators competing for the right to mine. Additionally, some regulators may feel that there is less need to revisit and review the financial assurance once it is in place for the life of the mine, thereby freeing up scarce regulatory time and talent for other tasks.

However, reclamation and closure costs may change dramatically over the life of the mine, and if they are not periodically reviewed, the financial assurance may prove to be either inadequate or unnecessarily high. Depending on the financial assurance vehicle, this approach can also "tie up" the mining companies' funds until after successful reclamation and closure has occurred. Surety bonding firms may be hesitant to write bonds for these extremely large amounts, partially because of the uncertainties of predicting 30 years (or longer) into the future. If a trust fund mechanism is used, monies may be misappropriated. Whatever the mechanism, it is necessary to ensure that these funds will indeed be available to be returned to the operator, or used for reclamation and closure at the end of the mine life.

15.1.4.5.2   Two to five year increments

Financial assurance can also be calculated by dividing the life of the mine into shorter time periods, based on the mine's plan for ongoing reclamation. Current practice at many sites dictates that reclamation and closure activities (and their associated costs) occur not only at the end of the mine life, when the resource is depleted, but begin even at the design stage.

The advantage to this approach is that it creates less financial burden for the mining company. Additionally, because of the need for frequent reevaluation, regulators and the enforcement team can anticipate potential problems before they happen. A list of changes that may trigger a reevaluation is included below. All firms may not be willing to submit to this level of scrutiny after the mine is permitted, and all regulators may not be able or willing to participate in such a labour-intensive form of regulation, which requires highly qualified and trained regulators. For this system of incremental financial assurance to work, the regulatory agency must have both the will and capacity to enforce agreements and negotiate changes.

Finally, although this approach minimises the delay of costs until the end of mine life, there may be reclamation and closure activities that cannot be completed until production ceases. It may be necessary to add a small "closure premium," (in addition to the reclamation costs calculated each time increment), which is put into a sinking fund to assure that closure can be completed when the mine is not generating revenues.

15.1.4.5.3   Surface disturbance

In this approach, financial assurance values are calculated based on the total surface area that is disturbed over given periods of time. This approach is subject to essentially the same set of advantages and disadvantages as the above.

15.1.4.6   Periodic review

The effective implementation of all financial assurance programmes requires periodic review. This review should be agreed to by the owner/operator when the programme is set in place. There are a number of circumstances that may indicate a need to recalculate the amount of financial assurance required, or the financial mechanism. Such circumstances include:

- Successful exploration which expands the life of the mine.
- Significant changes in product price, which affects profitability.
- Change in the regulatory requirements or new permitting requirements.
- Installation of new technologies or management approaches which minimise pollution.
- Innovations in modelling and prediction which indicate potential environmental problems different than in the initial plan.
- Significant long-term movements in the factors included in cost estimates, including inflation rates.
- Discovery of valuable historic/cultural/ecological resources at the mine site.
- A change in cooperation and compliance of the mining firm.
- A change in the financial status of the firm (e.g., bankruptcy, changes in credit rating, and lawsuits).

15.2 POLICY RECOMMENDATIONS

The solution to one problem can sometimes create another unintended problem. The unintended problem in the case of financial assurance is that small operators can be driven out of business. The additional cost of purchasing bonds or paying into a trust fund is particularly threatening to the economic viability of small mines.

Artisanal and small mining is a critical and irreplaceable source of income for many workers in much of the developing world. Large mines, which may be able to absorb the additional costs, can gain an even greater share of the market when small operators are driven out by high compliance and bonding costs. This can result in the large mining firms holding a great deal of economic and bargaining power in the host country.

Research is required into alternative financial assurance mechanisms that ensure some degree of reclamation, particularly the closing of shafts and adits, without creating onerous financial burdens for small operators.

Regulators often have a difficult decision to make when attempting to enforce reclamation, closure, and environmental provisions. Overzealous enforcement can force an operator into bankruptcy, leaving him/her free to go on to another activity, and leaving the regulator with a "problem" mine. Most regulators prefer not to operate mines. Providers of surety bonds may have an incentive to do the reclamation and closure work themselves, once the mine owner or operator has withdrawn, rather than to pay the regulator. This can be advantageous to all parties. Deciding when to work with a miner who clearly is not meeting permit requirements and when to call for financial assurance to be exercised requires wisdom and perseverance.

The sureties/financial assurance industry can be a partner in enforcement of permitting provisions. It is motivated by financial self-interest to make certain that mines are operated in compliance with requirements, and it can be viewed as an extension of the regulator, carrying out its own inspections and compliance reviews.

For financial assurance mechanisms to be successful there must be an industry, or perhaps a branch of government, willing to provide these services. This has been a problem in the U.S., as many of the firms that provided financial assurances went bankrupt themselves, leaving governments and taxpayers with financial responsibility. Regulators may want to have a plan in place for when those providing financial assurance cannot, or will not, pay.

Some of the most serious problems with respect to financial assurance can occur when there are no clear and specific terms for release of financial assurance. Additionally, the terms of termination or transfer of mechanisms must be explicit to all parties if these tools are to be used successfully. For example, an operator may want to use a trust fund in early years of operation and shift to a sureties bond in later years. There should be a clear process for appeal.

When large sums of money are held for long periods of time, it may be difficult to ensure that the money is available to be returned to the mine operator, or alternatively used for reclamation and closure, at the end of the mine life. Who is the trustee for the funds? Government, banks, or others? How can it be ensured that funds held in trust are not appropriated for other uses? What is the test of qualification of financial institutions to issue letters of credit, or of insurance firms to issue policies? Finally, the ownership of interest accrued while funds are held should be determined when the financial assurances are put in place.

Perhaps the most important policy question in the field of financial assurance is that of how to design mechanisms and cover costs for catastrophic events; that is, events which cannot be predicted or perhaps even managed. These events have such a low probability of occurrence that any attempt to statistically evaluate them and assign an economic value is meaningless. However, if these events do occur, the costs of remedial work can quickly deplete even well-funded regulatory budgets. Ultimately, the answer to this question will be a value judgment, reflecting the amount of risk which decision-makers in our societies feel they can and should absorb, in exchange for the benefits which mining can potentially bring, particularly in resource-rich parts of the developing world.

In summary, it may ultimately be a combination of command and control, information and education, economic instruments, and financial assurances that is most effective in achieving environmental protection. The questions which regulators and other decision-makers may want to ask when designing programmes include:

- What are the environmental objectives?
- What are the economic objectives?
- Is the approach cost effective?
- Is it feasible, politically and otherwise?
- Does the infrastructure to support the programme exist, or will that also have to be created?
- What is the equitable distribution of risk and risk management, which will protect the environment and still allow for economic activity?

REFERENCES

1. California Mining Association, *Surface Mining and Reclamation Act Financial Assurance Guidelines*, California Mining Association, Sacramento, 1993.

16 Environmental Management Systems for Closure and Best Practice in the Indian Mining Industry

Bharat B. Dhar

CONTENTS

1-56670-365-4/00/$0.00+$.50
© 2000 by CRC Press LLC

ABSTRACT

Unscientific planning for mineral extraction affects all components of the environment. Impacts include acid mine drainage, siltation of streams, pollution of air and water bodies, degradation of land, and reduced biodiversity in the mining areas. In India, the regulatory bodies for environmental clearance are the Ministry of Environment and Forest, Ministry of Mines, Indian Bureau of Mines, and State Government Pollution Control Boards. The initiation of any mining activity or the granting of prospecting leases is referred to these bodies. Their mandate includes overseeing different stages of environmental management, auditing, and rehabilitation of mine sites.

The requirements of planning for closure and the adoption of best practices for sustainable development of mining industry are to minimize or prevent environmental impacts during the operation of a mining site, and to ensure reclamation of the degraded site to a beneficial use when working has ceased. This paper highlights a few case studies within the Indian mining industry which have adopted good environmental practices and which illustrate attempts at planning for closure.

16.1   INTRODUCTION

Mining is a core sector industry in India that plays a positive and significant role in the process of the country's economic development, but not without adverse environmental impacts.

Nature has endowed India with rich mineral resources and a vast variety of landforms, spread over diverse geographic and climatic conditions; for example, the peaks of the Himalayas with fragile foot hills, arid desert and semi-arid areas, vast coastlines, plains, and rich forests enjoying a multitude of biodiversity.[1] Figure 16.1 shows the location of important mineral deposits in India. In view of this, the problems of the mining industry are complex and need site-specific solutions. Reckless and unscientific activities in the past have brought an imbalance in some pockets of environmentally sensitive and fragile areas.

16.2   LEGACY OF THE PAST

Rapid industrialization and population growth followed by nationalization of the coal mining industry (1971–72) gave a boost to coal production without taking care of the environment. These factors together resulted in serious environmental damages to the neighbouring ecosystem. Consequently, vast stretches of mining areas are littered with old mines, large waste dumps, abandoned mining areas and deforested lands. Further it is an irony of nature perhaps, that coal is found mostly in those places which have thick forest cover, streams, and fragile and biodiverse ecosystems. That means exploitation of coal will necessarily result in damage to the natural forest wealth and ecosystems.

16.3   LEGAL FRAMEWORK FOR ENVIRONMENTAL MANAGEMENT

Environmental management in mining is regulated by the Environmental (Protection) Act, 1986; the Forest Conservation Act, 1980; and the Mines and Minerals (Regulation and Development) Act, 1957, amended in 1986 and 1994. For exploration and exploitation of coal and minerals in forest land, prior permission of the government is required under the provisions of the Forest Conservation Act, 1980.

The Environment Protection Act (EPA), 1986 and the rules framed under it provide for the prospecting and mining activities only with prior permission from the Ministry of Environment and Forest for all major projects. The Government of India, to exercise its powers under the Mines and Minerals (Regulation and Development) Act, 1957, has provided for environmental protection under the Mineral Conservation and Development Rules, (MCDR), 1958, through its amendments

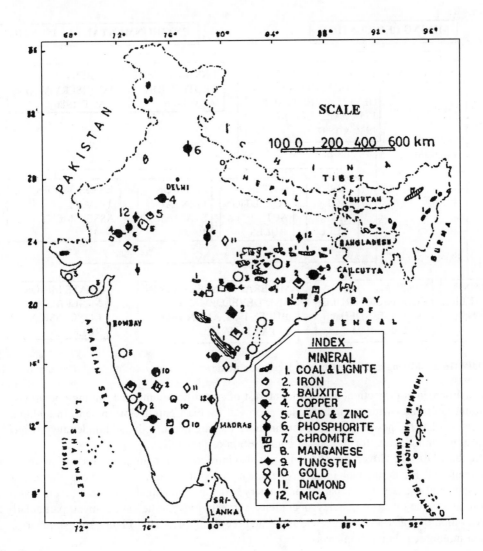

FIGURE 16.1 Important mineral deposits in India.

in 1988 and 1994. Every holder of a prospecting license or a mining lease has to take all possible precautions for the protection of the environment and control of pollution while prospecting, mining, beneficiation, or processing in the area. The leaseholder under the MCDR is duty-bound to take precautions regarding:

- Removal and utilization of topsoil.
- Reclamation and rehabilitation of lands.
- Precautions against ground vibration impacts.
- Precaution against air pollution within permissible limits specified under various environmental laws of the country including the Air (Prevention and Control of Pollution) Act, 1981 and Environmental (Protection) Act, 1986.
- Environmentally safe discharge of toxic liquid.
- Precaution against noise.
- Restoration of flora.

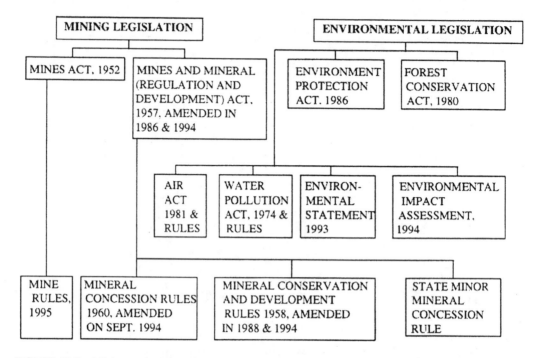

FIGURE 16.2   Mining and environmental legislation framework.

For environmental protection, and in the interest of scientific mining, the mining operations have to be in conformity with the mining plan approved by the Central Government. In addition, the government has the power to reserve certain areas that are not already held under any prospecting license or mining leases for reasons of conservation of mineral(s).

In the 1990s, the Government of India has also introduced a requirement for an Environmental Statement and Environmental Impact Assessment for systematic and periodic evaluation of the environmental status within a particular industry.[2] An Environmental Statement is a management tool to assess the effect of policies, operations, and activities on the environment, particularly the conservation of water and energy, and reuse and recycling of waste.[3] Figure 16.2 summarizes the Indian mining legislation framework.

16.3.1 New National Mineral Policy

The Mines and Minerals (Regulation and Development) Act (MMRD) was promulgated in 1957 and has since been amended from time to time. The major amendments were made in 1986 in order to incorporate provisions relating to environmental aspects in the act. Similarly, the Mineral Conservation and Development Rules (MCDR) were notified in 1958. However, after the major amendment of the MMRD act in 1986, a need was felt to thoroughly revise the MCDR to take into account the new requirements necessary for systematic and scientific mining as well as to incorporate provisions relating to environmental aspects after the government had acquired powers to make rules by virtue of the amendments made in the parent act. Accordingly, the Government of India in 1988 notified the new Mineral Conservation and Development Rules. The Government of India also announced a new National Mineral Policy in 1993 outlining the direction it desired to take to develop the mining industry and the priorities for action in relation to various aspects of mining and environmental protection. Consequently, in order to translate the National Mineral Policy of 1993 into legislation, some changes have been effected in the Mineral Concession Rules (MCR), 1960, the MMRD, and the MCDR in 1994.[4,5]

It is evident from the National Mineral Policy, 1993 that mineral development is central to policy, but this has to be combined with environmental responsibility. This integrated approach is evident in the following aspects of the new policy:

a) The guiding principle in the strategy of development of any mineral or mineral deposit at any location shall ordinarily be the economic cost. The state may, however, undertake the development of any mineral or mineral deposit in public interest to ensure unhindered availability of mineral as a raw material for the realization of national goals.

b) There shall be adequate and effective legal and institutional framework and commitment to prevent sub-optimum and unscientific mining.

c) Conditions of mining lease regarding tenure, size, disposition with reference to geological boundaries, and other mining conditions shall be such as to favourably predispose the leased areas to systematic and complete extraction of minerals.

d) Efforts will be made to promote small-scale mining of a small deposit in a scientific and efficient manner while safeguarding vital environmental and ecological resources.

e) It is necessary to take a comprehensive view to facilitate the choice or order of land use, keeping in view the need for development as well as the need for forest, ecological, and environmental protection. Both aspects have to be properly coordinated to facilitate and ensure sustainable development of mineral resources in harmony with the environment.

f) Prevention and mitigation of adverse environmental effects due to mining and the processing of minerals, and rehabilitating and revegetating affected forest area and land covered by trees in accordance with the prescribed norms and established forestry practices shall form an integral part of mine development strategy in every instance.

g) Mining operations shall not ordinarily be taken up in identified ecologically fragile and biologically rich areas. Strip mining in forest areas should be avoided and it should be permitted only when accompanied by a comprehensive time-bound reclamation programme.

h) No mining lease would be granted to any party, private or public, without a proper mining plan, including the Environmental Management Plan approved and enforced by statutory authorities. As far as possible, reclamation and afforestation will proceed concurrently with mineral extraction.

i) Efforts would be made to convert old used mining sites into forests and other appropriate forms of land use.

j) Whenever mine closure becomes necessary, it should be orderly and systematic and so planned as to help the workers and dependent community "rehabilitate" themselves without undue hardship (see Warhurst et al., Chapter 5).

k) Emphasis should be given to export of minerals in value-added form, and recycling of metallic scrap and mineral waste.

l) The thrust of research and development is to be in the areas of: rock mechanics, ground control mine design engineering, equipment deployment and maintenance, energy conservation, environmental protection, and safety of operations.

Thus, the National Mineral Policy provides a framework for the regulation of the environmental impacts of mining and for promoting measures for ecosystem protection and restoration. Not all the aspects contained under National Mineral Policy have been converted into legislation as yet. However, policy changes brought about in the last seven or eight years do address some of the above issues.

16.4 BEST PRACTICE FOR MINE CLOSURE IN INDIA

The section that follows provides a few interesting stories of best practice in mining in order to prevent ecological degradation. The location of the mining areas is demarcated in Figure 16.3.

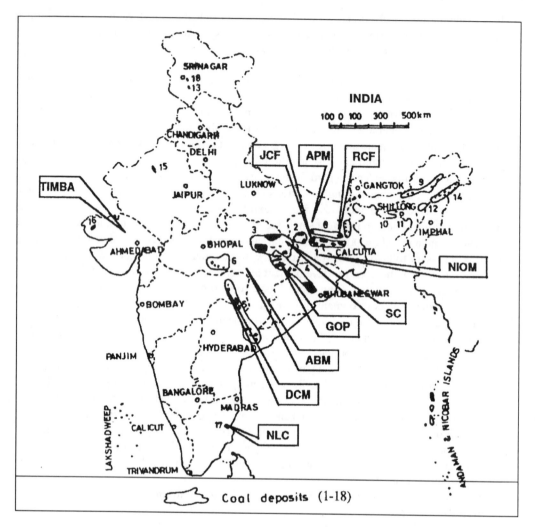

FIGURE 16.3   Location of mining areas.

16.4.1 NEYVELI LIGNITE MINE

The Neyveli Lignite mine is spread over Panruti and Vridhachalam taluks of Tamil Nadu state of India. It is situated between 11°33'–11°35' north latitude and 79°28'–79°32' east longitude and occupies an area of 540 km². The mining activity at Neyveli started in 1957. The concept of reclamation was given due importance even at a time when environmental awareness in the Indian mining scene was in its infancy.

Two external spoil banks were constructed to accommodate waste before internal filling could commence. At the first stage, 78.0 million m³ of overburden was dumped in the northern spoil bank, spread over an area of 195 hectares. With the expansion of the first mine the second external dump, i.e., eastern dumping yard was formed to accommodate 45.0 million m³ waste over an area of 135 hectares. Meanwhile, Mine II started operation in 1981 and a similar external spoil dump was also established there. By 1994, about 1350 hectares in Mine I and 1200 hectares in Mine II had become degraded land.

To control the environmental impacts of mining, Neyveli phased its programme in the following stages:

1. Revegetation of disturbed area.
2. Afforestation of outside spoil banks.
3. Creation of picnic spots.
4. Formation of artificial lakes and ponds.
5. Conversion of disturbed land to agricultural use.
6. Creation of pasture lands.

Neyveli started its afforestation programme in the 1980s by planting saplings of various species of economic and local importance. The corporation has implemented this afforestation programme both in the township and in the industrial area, to cover all the wastelands. Reclamation was carried out successfully both with and without the use of topsoil. Farmyard manure, biodigested cowdung slurry, composted coir pith or vermicompost at 20 tonnes per hectare, together with the recommended nitrogen-phosphorus-potassium (NPK) fertilizer, lignite flyash, gypsum, and humic acid, made substantial change on the physico-chemical properties of the spoil and provided a suitable habitat for growing paddy, pulses, sugarcane, and bananas. Vegetable gardens of local importance were also developed. Neyveli Lignite Mine is the first mine in India to attempt and succeed in converting the backfilled area into agricultural land. Today, this open pit mining area is an example by itself and has even won several national awards in recognition of better mining practices from a restoration point of view.

16.4.2 Singrauli Coalfields

Singrauli Coalfields (SC) is operated by Northern Coalfields Limited, subsidiary of Coal India Limited. It is situated in Rewa district of Madhya Pradesh and Sonebhadra district of Uttar Pradesh states and located between longitude 84°41'–84°52' east and latitude 23°42'–24°12' north. At present, the total land involved in 10 mining blocks is 14,873 hectares, which includes 6615 hectares (44.4%) for mining excavation. 1820 hectares (12.2%) for external overburden dumps and 2528 hectares (17%) for infrastructure.

Environmental management of the mining area involves the following:

a) The area required for external dumping is kept to a minimum.
b) Dragline, shovel-dumper combination and dozers are utilised for filling purposes.
c) The periphery of overburden dump slope is kept at 28° and the height does not exceed 30 m.
d) The initial overburden dumps are to be brought to a specific height and level before biological reclamation is begun.

Through planning, selected plants grown on overburden dumps by Department of Botany, Banaras Hindu University, Varanasi are *Tamarind, Madhuca indica, Amla, Jamun* and several other species.

16.4.3 Durgapur Coal Mine at Chandrapur, Western Coalfields Limited

A reclamation project, the first of its kind in India, has been implemented at Durgapur Coal Mine at Chandrapur. This mine is operated by Western Coalfields Limited, a subsidiary of Coal India Limited. The project covers old barren dumps of stone and gravel extending over an area of about 20 hectares. About 53,600 trees and 113 plant species including teak, sal, mahua, bija sal, semul, and kusum are grown on these spoil dumps. Several grass species are also grown, and these have started rehabilitating the degraded area.

Sludge from the paper industry, and biofertilizers such as Rhizome, Azotobacter, and Endo-mycorrhizal fungi of Glomus Sp. were used to rejuvenate the productivity of the soil and make the system self-sustaining. This played an important role in promoting luxuriant root development to mitigate soil erosion, promote mineralisation of soil nutrients, and the development of a self-sustaining ecosystem.

16.4.4 Gevra Opencast Project

The World Bank has funded the Gevra (expansion) opencast project (GOP) with a capacity of 10 million tonnes of coal and 9 million m³ of overburden per annum. Gevra, located in the Bilaspur district of Madhya Pradesh, is the largest mine in the Korba coalfield of South Eastern Coalfields Limited. It is considered to be the second energy capital of the country. The mine has been developed primarily to meet the coal requirement of the 2100 MW capacity Korba Super thermal power station.

An integrated afforestation scheme is being implemented with the help of Madhya Pradesh Forest Department. A model nursery has been developed at Gevra with a 600,000 sapling capacity and in 1991, 21 different species were being grown there. Afforestation in mining areas is being done in phases. Where quarry edges have attained permanency, tree planting is carried out to screen and reduce noise and dust. Suitable revegetation programmes and drainage arrangements have been provided to arrest soil erosion.

16.4.5 Small-Scale Mining

Small-scale mining was previously considered to be wasteful due to unsafe and unhealthy conditions, brutalising labour practices, and ecological and environmental damage. In developing countries, small-scale mining plays a major role in economic development, because of the small capital investment involved, manageable environment conditions, and more direct employment opportunities offered. The strong support for the small mine sector is a manifestation of the national concern for a rural-based development strategy. Some of the best practices observed in small-scale mining are detailed below.

16.4.5.1 Timba Quarry of Siyaji Iron Works, Gujrat

The quarry area has topsoil of about 15 to 20 feet. Underneath this layer of soil, there is a hard basalt rock. This is mined and crushed for use in road and building construction. The stone is mined down to a depth of 50 to 80 feet, as mining beyond this level becomes uneconomic. While mining, topsoil is removed until the hard rock is struck and thrown on the side systematically. Thus, mining forms deep cuts or gullies or cliffs and hillocks. Because of this operation, the topsoil from the original site goes to the bottom and the bottom infertile soil is placed on top. Usually, this soil is not considered of suitable fertility for the cultivation of crops or for growing trees. The mines leave deep gullies and remain unproductive.

The management (especially the Chairman, Shri Indubhai C. Patel) of the mine took up the challenge of restoring nature as a good industrial ethic and to set up a model which can be emulated by others. The management proposed a concept of establishing afforestation on the hillocks or elevated areas and using mines as water reservoirs. At some places, where open cut mines had a large area, they contemplated development into water lakes.

The mined area looked like a desert 15 years ago. Not even a blade of grass grew in the area. The degraded soil and deep cut mines presented a very pathetic scene where nature had been devastated. The conditions made it difficult for anything to survive: no trace of any flora or fauna, small or big, was visible; no water was available anywhere; and the whole site was radiating heat. There are about 50 quarry works in this area, and all were degraded.

To implement an environmental management scheme, the first and foremost need was water. Suitable arrangements were made to divert and store the rainwater into already quarried areas or areas referred to as cuts or scars. The scattered soil was collected and worked to form small hillocks. Thus, hillocks and lakes were created. Various types of species were planted on these hillocks.

Over a period of a decade, nearly 100,000 trees of various species were planted around the lakes and on the hillocks and were carefully nurtured. Diverse species of trees provided congenial conditions for birds, small animals, insects, and others. By restoring flora and fauna, a natural harmony was established.

Over 240 species of birds, both resident and migratory, frequently visited this developed ecosystem. Since it is an isolated environment, they come in large numbers. It now provides an undisturbed breeding place for a large variety of birds; for example, peacocks, great horned owls, and eagles.

The lakes and water bodies were helpful in nurturing some aquatic plants and attracting water birds. Young fish such as rohu, katla, and other fish were thrown into water to promote aquaculture. Now the lakes have a lot of fish, providing food for birds and other animals. The landscape is totally changed. The quarried area now holds 30 to 40 feet of water. The water is pumped up on top of one or two hillocks and the water falls from a height of more than 60 feet. This has enhanced the area aesthetically and has given a boost to nature tourism. Pedal boats can reach different hillocks. On the hillocks, trails have been established along with bird-watching lookouts.

The area has become unbelievably rich in vegetation, animals, birds, aquatic plants, as well as fish, burrowing animals, insects, and other life. An area once considered as barren wasteland has been turned into a sanctuary. Rehabilitation of this mine site serves as a model for other quarries, highlighting the possibilities of converting a wasteland into a beautiful nature resort.

16.4.5.2 Amarkantak Bauxite Mines

The Amarkantak Bauxite Mines (ABM) of Bharat Aluminum Company Ltd. are located at Amarkantak in the state of Madhya Pradesh. Reclamation work here started in 1979; filling the void areas with lateritic overburden and covering with a topsoil layer have restored more than 150 hectares of land. These restored areas have already been planted with about 3.5 lakh trees. In addition, 18 to 20 ha of mined-out areas are being reclaimed every year with 50,000 to 60,000 trees of 52 different exotic and local species. The technique adopted is based on the "pot holes" plantation method.[6] These pot holes are backfilled with the fresh topsoil to which has been added suitable manure and other additives.

Experiments were conducted with mustard seed cultivation using different techniques and varying topsoil layers. The results show that it is possible to biologically reclaim lateritic fill areas in the mined-out zones, through crop cultivation. The ideal condition can be developed by providing a 15 to 30 cm thick dressing of fresh topsoil intermixed with nitrogen fixing legumes and a small, appropriate dose of NPK chemical fertilizers. These techniques are now practiced on a larger scale around the area and are yielding significant quantities of crops for the local people.

16.4.5.3 Noamundi Iron Ore Mines

Noamundi Iron Ore Mines (NIOM) are situated between 22°9'N latitude and 85°29'E longitude on the eastern edge of Saranda forest of Bihar state. It consists of steep to gentle slopes with reddish-yellow to chocolate colour soil. Lateritic soils are very poor in lime and magnesia and are deficient in nitrogen. The soil in general is acidic.

Mining has produced large quantities of wastes and these have been disposed in large dumps. These dumps are loose sandy erodable materials consisting of boulders, devoid of organic matter and moisture retention capacity.

These degraded lands were planted, with different doses of fertilizers and farmyard manures, with 14 local and economically important species in 1991. Among the 14 species, two species, namely *Ficus* and *Dendrocalmus* sp. were completely destroyed by elephants during 1992. However, all the other species have managed a luxuriant growth. Amelioration with fertilizer in combination with organic manure resulted in more than two times the growth of the unfertilized area. Application of fertilizer has also had a positive response on plant growth performance. A tailings dam was also revegetated by adding active fungi and bacteria in farmyard manure and then planting with locally indigenous grasses and deciduous species. Plant growth up to 3 m in height along with the production of fruit has been observed within a period of 2 years. The success of the trial plantation indicates that such greening of wastelands is possible, provided technological support is available for such sites and is suitably employed. The success story with these species may indicate that the planting of these more economical and valuable species may lead to a managed forest.[7]

16.5 CENTRAL MINING RESEARCH INSTITUTE EFFORTS TOWARDS REJUVENATION OF DEGRADED MINING AREAS

A big environmental challenge is that of abandoned old mines, worked out during a period when environmental concerns were not an issue. These mines are still acting as a focal point for local pollution, and are responsible for the degradation of surrounding ecosystems. The Central Mining Research Institute (CMRI) was given a challenging scientific assignment to restore the old coal mine waste dumps lying in the region for between 15 and 30 years. Some of the success stories achieved by this Institute are briefly discussed below.

16.5.1 RANIGANJ COALFIELD

The Raniganj coalfield (RCF) can be categorised as a failure story from an environmental viewpoint, as there are so many non-completed sites that still have to be reclaimed and revegetated properly. Lakhimata and Bonjemchari overburden dumps, Binodi katha, Khas and West Baraboni sites are a few such sites.

CMRI took up the challenge of reclaiming a rocky abandoned dump causing problems for the local inhabitants, including pollution of a nearby pond and aesthetic impact over the last eight to ten years. There was an urgent demand from the local people to reclaim the abandoned dump. Therefore, a scientific approach was adopted by using laboratory pot experiments, followed by field trial experiments with respect to physical and biological parameters. These have led to the successful rehabilitation of the area, in which more than 15 tree species along with the grass species have been planted, with the blackened hard rocky dump converted into beautiful picnic spot for the local people. This rehabilitated area is now attracting a lot of local tourists and picnickers.

16.5.2 JHARIA COALFIELD

Mining activities in Jharia Coalfield date back 100 years, and the coal deposits since then have been extensively exploited. Unscientific and unplanned mining in the past have given rise to numerous environmental problems, including degradation of land, air, and water, subsidence, noise pollution, and mine fires. A total of 2620 hectares of damaged land were reclaimed by biological means out of a total of 6294 hectares. The damage to land from overburden dumps covers an area of 631 hectares, out of which the reclaimed area is only 181 hectares. Jharia has several fire areas in which 569 hectares out of a total of 1732 hectares have been reclaimed by adopting fire control techniques and subsequent biological reclamation.

CMRI has taken up a national mission project "Carrying Capacity of Damodar River Basin" for preparation of short-term and long-term environmental management plans on the basis of assimilative and supportive capacity of the region, in the context of sustainable and ecofriendly development of the region.

16.5.3 AMJHORE PYRITE MINE

The most serious acid mine drainage problem (AMD) in India is being faced at Amjhore Pyrite Mine (APM), located in the district of Rohtas (Bihar). During the course of mining, top and bottom shales were excavated, resulting in the dilution of grade from 40 to 22% sulphur. Shales produced during the mining are manually picked and dumped at different places and contain around 0.3 million tonnes of the materials. These materials undergo biochemical oxidation and cause fire problems. These burnt-out shales still contain a high percentage of sulphur. During the rainy season, water percolates through the shale dumps and increases the acidity of the soil. Higher acidity adversely affects the natural succession of plants and the revegetation process. CMRI has developed a technique for the utilisation of pyritic waste dumps by using lime, phosphatic waste, and industrial cinders and have successfully grown green plants. This makes the area environmentally friendly and aesthetically beautiful.

16.6 PLANNING FOR CLOSURE

A mining project moves from the planning and design stages through development, operation, and finally to closure. Proper planning can reduce environmental impacts, result in good environmental performance, and enhance the public perception of the industry as able to operate in an ecologically sustainable way with successful closure. Adequate baseline information is necessary before mines can be planned in an environmental friendly way. This includes information about the resource deposit, its quality, and geological constraints. Planners need to understand the surrounding environment through a programme of baseline environmental monitoring and data collection in order to identify particular features, attributes, and constraints that need to be considered in mine planning.

To improve community perceptions of mining projects, mine planners need to understand surrounding land uses, regional impacts and community aspirations, i.e., both the biophysical and socioeconomic issues. Once the mineral deposit is evaluated and a proper appreciation obtained of the environmental and social context of the mine, alternative extraction paths can be developed. These will include the rate and direction of mining, alternative process design, alternative facility layouts, and other infrastructure (see Figure 16.4).

Each option should be examined for the following:

- Economic feasibility.
- Resource utilisation.
- Community acceptability.
- Environmental impact.

An Environmental Impact Assessment (EIA) of a mining project identifies, assesses, and predicts the likely impacts of the planned project upon ambient air, water, and noise environment; land; flora and fauna; and the people in the surrounding areas. This information is then used for the preparation of an Environmental Management Plan (EMP) and a final rehabilitated land use for the site. Implementation of a comprehensive plan requires workforce training and awareness, along with environmental monitoring and compliance audits. The project design should ensure that post-mine land use would be achieved for successful closure of the mine. Figure 16.5 shows the schematic diagram of planning for successful closure of mine.

While each mineral deposit is unique, the application of integrated environmental planning procedures is a fundamental component of best practice for environmental management in the successful closure of a mine.

16.7 PROBLEMS AND CHALLENGES OF DEVELOPING COUNTRIES

Poverty, income inequality, illiteracy, depleted and polluted water resources, lack of information and capabilities for proper planning for mine closure and best practices are some of the prime constraints facing Indian industry.

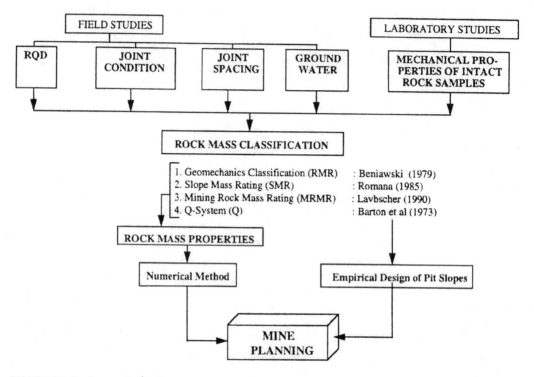

FIGURE 16.4   Approach for alternative mine planning.

When environmental problems in mining and allied industries are seen in the context of society's development objectives, not all resource extraction, deforestation, soil erosion, or land degradation, and air and water pollution can be termed as undesirable. The question is not why all these problems exist, but why do they take a form and magnitude that are inconsistent with society's objectives?

The specific nature of the problems in terms of population explosion, land degradation, misuse of resources, under-use of resources, groundwater depletion, mechanism of pollution, environmental and aesthetic impacts, and physical and chemical constraints on the selection of prevention and control measures are highlighted below.

16.7.1   POPULATION EXPLOSION

Rapid population growth along with increasing economic activity has exerted pressure on mineral resources and created new environmental pressures. Poverty and low incomes have also influenced the way in which the population affects ecology and the environment.

16.7.2   DEPLETION OF WATER RESOURCES

In developing countries, water is a critical factor in both agriculture and industry. Increased agricultural activity, rapid industrialization, and increased income exacerbate demands on water; consequently, water resources are declining rapidly. Southeast Asian countries have an average of 4 km^3 of fresh water per million people, compared to the world average of 7.7 km^3.

Along with the decrease in water supply, there is an increase in water pollution in the form of domestic sewage, industrial effluents, and runoff from land-based activities such as agriculture and mining. Rapid population growth, urbanisation, industrialisation, and different forms of raw material extraction are also threatening marine resources. Forty-two rivers have been declared officially "dead" as a result of industrial waste and sewage in Malaysia.

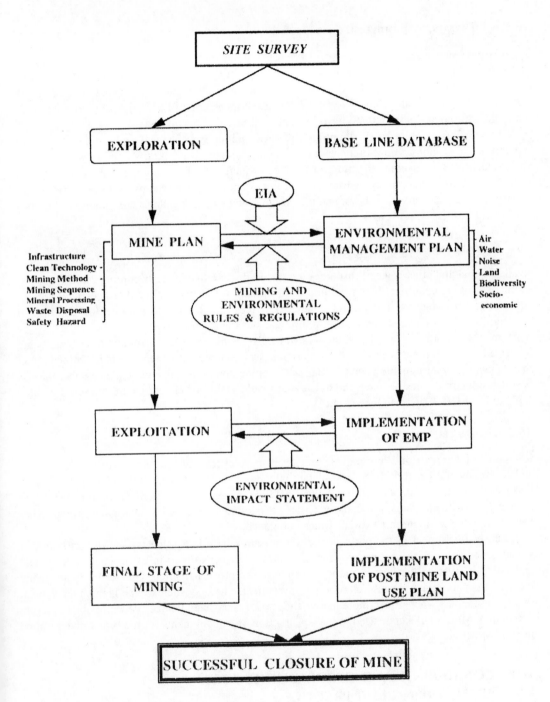

FIGURE 16.5 Planning for successful closure of mine.

16.7.3 Deterioration of Air Quality

Air quality is deteriorating daily not only due to energy consumption but also due to the fuel mix, the exploitation of coal and minerals, and the dumping of rejected material without any biological reclamation. Presently, automobiles, burning of coal, and forest harvesting mostly affect air quality in the urban sector.

16.7.4 THREATS TO BIODIVERSITY

Third World countries are one of the world's richest regions for biological resources, both terrestrial and marine. Many of the native species are being threatened by deforestation, land degradation, draining of wetlands, and air and water pollution, some of which include mining activities. Densely populated Bangladesh and Hong Kong have lost more than 90% of their original wildlife habitat, while the Philippines and Vietnam have lost about 80%.

Governments of Third World countries still need foreign exchange and timber-related jobs and industries for economic development, and corporations still want profits and growth for their shareholders and to ensure their future in the face of increasing fierce competition. This is why it is argued that economic policy instruments are important to mitigate the pressure on forest biodiversity. Calls for conservation, laws, and the designation of protected areas are important, but they are limited in their effectiveness.

16.7.5 MISUSE OF RESOURCES

Available coal and other mineral resources are misused, wasted, and are becoming scarcer in most of the developing countries. In agricultural land, farmers suffer from water shortages while others suffer from waterlogging. Increasingly scarce resources are also put to inferior, low-return, and unsustainable uses, when superior, high-return and sustainable uses exist.

Renewable resources are also being treated as extractable resources. Tropical forests are being harvested at several times their rate of regeneration, which suggests that they are being mined rather than managed for sustainable yields. In most developing countries, mine lease areas where the soil is suitable for fruit trees or other perennials that would yield higher returns are often not planted for years and left in an abandoned state.

16.7.6 UNDERUSE OF RESOURCES

Resources of minerals and by-products are not being recycled, which would generate economic and environmental benefits. In Sumatra, state factories that are allowed to dispose of their waste into rivers free of charge are throwing away palm oil by-products, which can be processed into marketable fertilizers. Thus, a socially profitable economic activity is lost, and palm oil waste has become Sumatra's single most severe water pollutant.

Lack of investment in the protection of the resource base is another missed opportunity that would generate a positive return. For instance, a mine owner fails to invest in land development and soil conservation that would reduce erosion and improve irrigation. Greater costs are being incurred than necessary because of a lack of investment. The opposite extreme is mismanagement in many collieries, where much more labour and capital is engaged than is needed to obtain current production. Mine spoil can be used in the construction of embankments for roads and underground mine stowing, etc.

16.8 CONSTRAINTS AND OPPORTUNITIES FOR TRANSFERRING BEST PRACTICE

Eco-friendly mining approaches for a region or a country will have to be developed and established on the basis of local realities, and not simply by copying the models and strategies of developed countries where the "quality of life" and priorities are different. "Imported approaches" need not always be a solution to fast-developing countries where the need for development is a priority. However, there are a number of environmentally friendly technologies available that may be of use in dealing with the problems faced by developing countries. The question then becomes how best can "appropriate" environmentally friendly technologies be transferred?

Major constraints faced by developing countries to the transfer of best practice are

- Lack of infrastructure.
- Inadequate environmental education and training, and thereby shortage of trained manpower.
- Lack of resources.
- Ineffective information technology.
- Policy failures.
- Political barriers.

Developing countries, together with international agencies, can strengthen the process of clean technology transfer and diffusion along the lines recommended in the "First World Mining Environment Congress" held at New Delhi, India.[8] Another approach to transfer is through greater South-South cooperation. India is well able to play a lead role here.

Today, India has multi-disciplinary human resource and infrastructures facilities for an eco-friendly mining sector and well-recognized research and development facilities. India, with its known skilled manpower and technology in the region, can play a pivotal role in transferring best practices which will be more economic and suitable for the needs of other Third World countries because of the commonalities involved.[2,9-11]

16.9 CONCLUSION

Environmental considerations should be fully integrated into mine planning at every stage of a mining project. Planning needs to address issues to achieve economically viable and locally relevant projects. In addition, mine planners must have an adequate understanding of the resource deposit and rehabilitation philosophy.

To make the revegetation programme a success story, plant species should be screened scientifically so that they are compatible with soil quality, or the soil should be amended to suit the species selected. Manure or minerals should be applied to change the soil fertility. Mine discharge water should be properly treated, so that the pollution of adjacent water bodies is avoided.

16.9.1 FUTURE AREAS OF RESEARCH

(i) Restoration of degraded ecosystems:
  a) Development of techniques for the restoration of mine-affected areas, arid lands (cold and hot deserts).
  b) Rehabilitation of abandoned dumps and pits in major coal mining areas.
  c) Stabilization of dragline dumps slopes in major opencast coalmines.
  d) Quantification of stabilisation activity by biological reclamation on dump slopes and effective utilisation of the technique.
(ii) Development of biological intervention for pollution prevention and control:
  a) Improvements of processes for low or no wastes or clean technology.
  b) Development of microbial and other biological methods for treatment of solid and liquid wastes.
(iii) Conservation of biodiversity:
  a) Development of tissue culture techniques for conservation of threatened/endangered medical plant species.
  b) *Ex situ* and *in situ* conservation of biodiversity.
(iv) Use of pesticides and fertilisers:
  a) Development of eco-friendly approaches and products to reduce hazards of pesticides.
  b) Assessment of ecological effects of use of chemical fertilisers and development of microbial/organic fertilisers.

(v) Development of strategies/technologies for prevention and/or control of pollution:
   a) Development of technologies for preventing mine fires.
   b) Development of effective measures to control haul road dust.
   c) Development of clean technologies.
   d) Appropriate methodologies of EIA for mining areas.
   e) Regional EIA/EMP, employing a Carrying Capacity Approach for the region where a cluster of mines are in operation.

REFERENCES

1. Gautam, N. N., Status of exploration and future outlook, Proceedings of National Seminar on Status of Mineral Exploration in India, Khuntia, G. S., Ed., organised by Mining Engineering Association of India, Delhi Chapter, New Delhi, India, 1995, 72.
2. Dhar, B. B., Environmental impact assessment for mining — the Indian scenario, Proceedings of the Seminar on Capacity Building for Environmental Management in Asia/Pacific Mining, Jakarta, 6–8 September, organised by UNCTAD and Ministry of Mines and Energy, Indonesia, 1994, 36.
3. Dhar, B. B., Environment Auditing. Case Studies Illustrating Environmental Practices in Mining and Metallurgical Process, UNEP/ICME, Paris, France, 1995.
4. Kumar, K., MMRD & MCDR vis-a-vis environmental aspects of mining, Proceedings of Mines, Environment & Mineral Conservation Weeks, organised by IBM, Hyderabad, India, 30 January–5 February, 1994, 41.
5. Federation of Indian Mineral Industries, Background paper, Seminar on Promotion of Investment Opportunities in Mineral Sector, Bhubaneshwar (23 March) and Hyderabad (10 April), 1996.
6. Hassan, A., Reclamation by crop cultivation methods experiments in central Indian bauxite mines, *Indian Mining & Engineering Journal*, Special Issue on Mining and Environment, 31, 1993.
7. Rao, V. S. and Sharma, R. M., Re-establishment of self sustaining vegetation on mine wastes, *Second National Seminar on Minerals and Ecology*, Banerjee, S. P., Ed., Oxford & IBH Publisher, New Delhi, 1994, 179.
8. Dhar, B. B. and Thakur, D. N., Eds., *Mining Environment*, Oxford & IBH Publishers, New Delhi, India, 1995.
9. Dhar, B. B., Impact of globalisation Indian mineral industry, Proceedings of International Workshop on Mining in the Changing World, organised by Indian National Committee of World Mining Congress, Calcutta, 2–3 December, 1994.
10. Dhar, B. B., Business climate and legal provision for mining investment in India, Proceedings of 4th Asia Pacific Mining Conference, Jakarta, organised by the ASEAN Federation of Mining Associations, 26–29 October, Indonesia, 1994, 242.
11. Dhar, B. B., Globalisation and strategies to foreign investments in mining sector — a trend, Mining Geological and Metallurgical Institute, Calcutta, 9 July, 1994, 1.

17 Current Regulatory Approaches to Mine Closure in the United States

Luke Danielson and Marily Nixon

CONTENTS

1-56670-365-4/00/$0.00+$.50
© 2000 by CRC Press LLC

17.1 FRAMEWORK FOR THIS CHAPTER

17.1.1 INTRODUCTION

Nearly two centuries of intensive, widespread mining in the U.S. have produced enormous wealth, but have also left a legacy of externalised environmental costs and damage to the productive capacity of the environment. In response, the U.S. has developed a complex body of laws and regulations directed at control and management of these impacts.

This complex body of laws includes provisions directed at the impacts of prospecting and exploration, construction, operation, and closure of mines. This chapter focuses on the last of these topics, the legal regime for mine closure. Of course, the requirements that apply to other phases of the mining project have a very significant influence on the conditions of closure. The laws directed specifically to this subject are in the U.S. generally termed mine reclamation laws.

Serious attention to the environmental effects of mining has come in waves, resulting from specific incidents or problems which have concentrated public attention: a long series of slope failures and floods in the eastern coal mining regions, culminating in the Buffalo Creek disaster; the Douglas Smelter emissions; the tailings dam failure at Church Rock; discovery of the hazards of low levels of radioactive emissions from uranium mill tailings which had been used widely as construction material; and most recently, the bankruptcy of Colorado's Summitville mine. A recent example is the attention being received by environmental conditions in Butte, Montana (see Dobb).[1] Each of these highly publicised events catalysed the enactment of major changes in the body of regulatory law affecting the industry — but these sporadic intense bursts of public interest in environmental problems have not always translated into the steady vigilance needed for effective long-term management of environmental problems.

There are few better examples of the need for maintaining a consistent long-term approach to environmental aspects of a mining project than effective management of mine closure. By necessity, mine closure planning is not something that can be done at one point in time and put on a shelf. It is an ongoing, dynamic process which may span many decades, during which many variables will certainly change, including the political context, legal systems, and the nature of industry-specific legal controls. If there is one key lesson from this experience, it is that failure to establish a plan at the level of conceptual engineering before project operations begin makes the process more expensive, less effective in managing post-closure environmental impacts, and more likely to generate conflicts. At the same time, new information will always be generated during operations, including all kinds of data about the mine site, the characteristics of mine wastes, and the functioning of the ecosystem in which the mine exists. The system must be flexible enough to incorporate this information into the planning process.

As will be explained below, mine closure for hard rock mines is dealt with at the state level in the American federal system. There is a great deal of variation in the state experience. Some states have had programmes for well over two decades, with very experienced staff, a reasonable level of resources, and significant practical experience with the nuances of mine closure planning; other states have only adopted programmes recently and have yet to fully institutionalise their systems.

17.1.2 CLOSURE PROGRAMME CONCEPTS

The fact that there are so many different systems of mine closure in the U.S. obscures the extent to which these many systems have gravitated toward some basic concepts. All, or virtually all, of the existing systems share these elements in common.

17.1.2.1 Requirement of a plan

New mining operations will not be authorised unless they first submit a plan showing, at the level of conceptual engineering, what the site will look like when mining is completed, predicting the likely impacts of the mine in that configuration, and proposing specific mitigation measures to deal with those impacts. In general, operations existing at the time the law became effective are required to submit plans, but have been given some period of time to comply.

17.1.2.2 Evolving technical standards

Each agency has evolved technical standards for plans based on its experience and the particular conditions of climate, geography, and mining operations in the area. These standards are almost necessarily somewhat general, because of the enormous differences among mining operations, but do serve to give mine planners some general objectives for their efforts; to help sort out approaches that have worked from those which have not; and to ensure some limits on administrative discretion.

17.1.2.3 Plan approval

The plan is not approved until after a technical review to verify its compliance with legal standards and technical requirements. There is almost always public participation, consisting of notice, an opportunity to review the plan and submit comments, and frequently a public hearing. In general, however, this is not an opportunity for the public to challenge basic decisions about land use, or the appropriateness of mining at this site. The issues are limited to the question of whether the closure plan complies with requirements and will be effective in achieving the post-mining condition called for in the plan.

17.1.2.4 Plan modification

Mining is a dynamic process. Most laws recognise that changes will inevitably occur in the mining operation and provide for easy flexible amendment of the closure permit. A company with a good internal environmental management system will, as part of its normal procedures, evaluate any proposed change in operation to understand what its impact will be on the closure plan, and change its closure plan if necessary.

17.1.2.5 Bonding

In order to ensure that funds will be available to implement the plan, and to encourage ongoing reclamation efforts during the operational phase, almost all systems require the operator to submit some form of financial guarantee of the implementation of the closure plan. This is not a guarantee of any and all obligations that the company may incur, but simply an assurance of compliance with the defined plan. Existence of the guarantee system requires that the plan itself be sufficiently detailed to allow for engineering cost calculations to be done. In most cases, the guarantee is adjusted periodically to reflect the amount of work remaining to be accomplished.

17.1.2.6 Monitoring, reporting, inspection, and enforcement

As a mining project evolves, changes may affect the practicability and adequacy of the closure plan. Unanticipated problems may occur; problems that were anticipated may not arise. This may affect the adequacy of the guarantee. All of this requires regulators to remain aware of conditions at the site.

When the company believes it has completed the reclamation in accordance with the plan, the regulatory body must inform itself adequately to ensure that it agrees that the plan has been complied with, so that the guarantee will terminate and the company is freed of further obligations at the site.

The principal means for achieving these goals is reporting by the mine operator. Most systems require a report on the status of the mining operation to be submitted annually, or at some other specified interval. A further report is submitted when the company believes it has complied with the plan and is entitled to have its guarantee terminated. But these reports are supplemented by periodic inspections of the mine site by government personnel, designed to review the state of the operation and any factors that may affect the adequacy of the plan or the guarantee.

These elements are hardly unique to the U.S. They exist also, for example, in South Africa,[2] the major mining provinces of Canada,[3] Australia,[4] and other countries. While most developing countries have not at this point established all of the elements typically found in U.S. mine closure legislation, more complete regulatory schemes are clearly emerging.[5]

17.1.3 RESULTS AND CHALLENGES

In most respects, the U.S. experience with legally required closure planning has been very positive. Often, significant environmental benefits have been achieved at little or no cost, simply because proper advance planning results in pollution prevention. For example, it may cost nothing to place tailings outside a stream bed, not to dump used crankcase oil on the ground, or to achieve an acceptable slope on a waste dump *if these steps are planned from the beginning.*

By contrast, moving the tailings out of the river, cleaning up petroleum-contaminated soil, or recontouring an established waste dump can be enormously expensive if no attention is paid to these issues until the mine closes. Further, mines almost always close when they are losing money and their operators are strapped for funds and facing a variety of other challenges. This is a poor time to be doing closure planning, and almost a guarantee that it will not be done well. And no one has ever been very good at predicting when closures will occur. One vivid example is the Exxon Colony oil shale project, in which the company invested over $2 billion. The mine and processing plant were largely constructed, but closed *before* they every reached commercial production.

There is not generally in the U.S. an explicit requirement that the closure plan result in conditions that will meet general environmental quality standards. There are several reasons for this.

First, limits for discharges of pollutants to water, air, and soil are generally set at the national level, while mine closure planning is a requirement of state law. Further, though the state may well have responsibility for enforcing these environmental quality or emission standards, this is usually not a function of the mine closure agency.

Second, it is often quite difficult to predict precisely the levels of emissions from a given site ten, twenty, or more years into the future. For example, while general sediment loadings may be easier to predict, the future existence and level of acid mine drainage at a site are notoriously hard to predict.

Third, industry, government, and civil society have been slow to confront the extremely difficult policy choices in those situations in which it appears that no realistic plan can result in a "walk-away" closure of the mine site. In a true "walk-away," the potential future existence of environmental problems of significance on the site is so improbable it can essentially be disregarded.

The concept on which much of the existing closure legislation was built was that with proper advance planning, mine sites can, within a relatively short time after operations end, be brought to a condition which meets all closure objectives *and* which provides reasonable assurance of long term, maintenance-free compliance with air, water, and soil quality norms. Where this goal can be reached — and it often can — neither the company nor society at large need continue to expend resources on the site. The company can, in this ideal world, disregard the possibility of future liability for environmental conditions on-site, turn out the lights, and "walk away."

But not all mine sites have been found able to achieve this ideal objective. Those mines that have not been able to meet this objective fall into several categories.

17.1.3.1 Monitoring required

In a relatively large number of cases, there is sufficient uncertainty about conditions on-site such as water discharges, slope stability, or other parameters, that some period of monitoring is necessary to ensure that closure criteria have been met. In cases where noncompliance would present serious hazards, this period of monitoring may need to be quite lengthy, and can involve significant cost.

17.1.3.2 Passive care

In a smaller but still significant number of cases, some measures are needed to maintain acceptable conditions on-site, either because failure to take these measures would result in noncompliance with air, water, or solid waste rules, or because other closure objectives would not be met. A very common example is that valley fills of tailings or waste rock may create pollution or stability hazards if watercourses are allowed to flow directly into them. Generally, drains, tunnels, or ditches have been constructed to divert watercourses around these workings. But these drains, tunnels, or ditches will naturally tend to fill with fallen rock, sediment, or other debris over time. Where the clogging of these diversions will lead to unacceptable problems or hazards, they must be periodically maintained.

What characterises these measures as "passive" is that they are not continuous operations. On a day-to-day basis, the system functions without human intervention. However, someone (either the mining company, or someone else) must *monitor* to ensure that unacceptable conditions do not arise, and *maintain* the system to ensure its continued functioning.

17.1.3.3 Active care

In a yet smaller group of projects, passive care and monitoring cannot produce results that meet closure objectives and ensure compliance with environmental quality norms. "Active" measures, such as continuous operation of a water treatment facility, are needed to meet these goals. Obviously, this can imply very significant levels of expense.

It also raises the questions of how long such measures will need to be applied, and who will apply them. While every operator who finds a need for active measures hopes to be able to progress promptly from active measures to passive care to monitoring only to "walk away," the fact is that there are some cases in which this is not possible, and in which very lengthy periods of active care, and high expense levels, will be necessary.

This raises a number of difficult issues. First, an ongoing obligation for active care has a bad effect on the balance sheet of the mining company. Decades of future operation of a water treatment plant at a non-operating mine property can hardly be looked on as other than a negative from the company's perspective.

Second, since mining is a highly competitive business in which the losers do not prosper, and prices are notoriously unstable, the state is faced with the question of whether or how to ensure that the resources continue to be available for active care at the mine site for as long as may be necessary.

And this leads to the most politically charged question of all: whether, if we can identify a particular project as posing an extremely high risk of a perpetual active care requirement, that whole cost must be internalised: somehow guaranteed "up front," or — in the views of some — the deposit simply not be mined.

These are the questions of the future for mine closure programmes in the U.S. and are dealt with in more detail in Section 17.6 below. While individual states have come up with reasonable solutions at some sites, satisfactory solutions of general application are yet to emerge.

The U.S. experience has been that closures tend to come in waves, in response to economic or political conditions. Thousands of silver mines closed in 1893–94, large numbers of uranium mines closed in the 1960s and then again in the 1980s with little or no warning, and these experiences have been repeated with virtually every mined commodity: gold, molybdenum, and copper, to name a few. This implies that the regulatory body, too, is likely to be stressed at the time closure occurs, suddenly having to cope with large numbers of virtually simultaneous unanticipated closures.

In dealing with the relatively conventional problems such as geotechnical stability, erosion control, revegetation, dust control, and control of hazards to public safety, closure planning has been very successful. But in the case of the relatively small but important number of mines with long-term care problems, most U.S. laws have not yet developed clear or consistent principles.

17.1.4 MINE CLOSURE SYSTEMS IN THE U.S.

17.1.4.1 Coal mine reclamation

While the principal subject this chapter will explore is the system of closure requirements for metal or "hard rock" mines, the U.S. mine reclamation system has its origins in the experience of the Eastern coal fields, and the development and dissemination of ideas about closure necessarily starts with some description of that difficult history.

Coal mine reclamation, for historical reasons, is subject to a national system of regulation, based on national performance standards developed by the Office of Surface Mining, an agency of the U.S. Department of the Interior.[6] While states may apply for, and often receive, authority to administer the programme within their own borders, if they choose to operate the programme they must do so in compliance with these extremely detailed national standards, and under close national supervision. There is in U.S. environmental law no better example of "command and control" regulation than reclamation requirements under the Surface Mining Control and Reclamation Act (SMCRA) of 1977. The national regulations provide hundreds of pages of detailed requirements for construction, maintenance, and reclamation of each specific feature of the mine site.[7] While state programmes can apply more stringent standards than the federal rules, they cannot be less stringent. And whether proposed state standards are or are not equally or more stringent is decided, at least in the first instance, by the national agency, which has had a very low tolerance for differing approaches. After several attempts to promulgate rules that contained a limited number of differences from the federal rules had been rejected by the national Office of Surface Mining (OSM), the State of Colorado suggested rules *identical* to the federal rules. It was advised by OSM that these were unacceptable, because the state's proposed explanatory preamble differed from the federal preamble, and therefore might lead readers to interpret the rules differently, despite the fact that they were identical.

The national agency not only approves all aspects of a state programme, including staffing levels, but also approves any changes in the state programme, and has an ongoing role in monitoring the way the state programme operates. Federal rules prescribe minimum frequency of inspections; federal inspectors appear unannounced to accompany state inspectors on some inspections; and they review state inspection reports to determine whether they meet standards. The national authority can revoke a state's authority if the state fails to meet standards on an ongoing basis.

The public is a watchdog over both state and federal inspectors. Members of the public who have made complaints are allowed to accompany inspectors during inspections. Detailed written records of all inspections must be kept and are available to the public. Complainants and mining companies both may avail themselves of a complex array of administrative and procedural remedies ranging from informal conferences at which all may be present to appeals of most types of decisions, usually to more than one administrative level. Individual discretion is very tightly controlled. There is little room for an inspector to "overlook" a violation of the rules based on an operator's informal promise to remedy the problem promptly; and there are prescribed schedules of monetary penalties for almost every conceivable violation.

It is important to note that, despite the current trend away from "command and control" regulation, this system exists only because the industry failed to take advantage of the many opportunities it had to live under more flexible regulatory regimes. And the system has had some incontrovertible successes. The past environmental legacy of the coal industry, particularly in the eastern coal states, is regarded as unacceptable by contemporary community standards. Current practice, particularly at some of the large western coal surface mines, is generally if not always good. There are many examples of excellent reclamation practices in coal mining: the Trapper and ColoWyo Mining Company operations in northwestern Colorado are two fine examples. Much of the credit for this improvement is due to the people who created and who operate the current regulatory system. Also, the U.S. has remained at or near the top of world coal production and industry productivity while this has been done. The effect of all these rules has been, *inter alia*, to force most operators to hire competent trained personnel who really understand reclamation. Getting these professionals inside the tent at mining companies has had benefits far beyond just regulatory compliance.

17.1.4.2 Non-coal mining

Closure planning outside the coal industry has been influenced by the experience with coal, though more by way of reaction than emulation. The approaches are quite dissimilar. Mine closure or reclamation programmes for non-coal minerals exist at the state level, and each state is free to make its own choices, or even to have no programme at all. The standards imposed by the national land management agencies, the Forest Service, and the Bureau of Land Management, for operations on the nearly one third of the national territory of the U.S. which they control are important, but hardly amount to systems of mine closure. The standards of these land management agencies "today serve as a floor" basically in cases where state requirements are inapplicable or not well enforced.[8] Some states, at the time of the major wave of U.S. environmental legislation in the 1970s, adopted reclamation laws. Others chose not to. This of course did not mean that no environmental standards applied to mining. Air quality, water quality, and solid waste laws profoundly affected the way mining was conducted during the operational phase, which has enormous implications for closure planning. A mine that operates cleanly is much less of an effort to close. What it did mean was that if there was any comprehensive, ongoing effort to plan for mine closure, it was being done at the initiative of the mining firm. Typically, the programmes adopted were based not on the kind of tight prescriptive standards that characterise the coal programme, but on very general performance requirements, implying high levels of discretion for programme administrators. They were enacted not over the bitter opposition of the mining industry, but with the support of large segments of it, who preferred what were seen as industry-friendly programmes at the state level to a less understanding regulatory approach from Washington. Specifically, the industry has seen the type of regulation that exists in the coal industry as undesirable, and has promoted state regulation based on general performance standards as an alternative.

In recent years, there has been a push from the environmental community and others to enact national mine reclamation legislation for hard rock minerals. Passage of something like the national coal reclamation legislation for metals mining has been regarded as an industry nightmare. As national reclamation legislation in some form came closer to passage, proponents pointed to the absence of reclamation laws in some important mining states as an example of the need for a federal law. Just as this issue was heating up in Congress, the unplanned emergency closure of the Summitville Mine, with its serious environmental implications and enormous cost to government, created even more pressure on the states with no closure requirements.

As this process continued, the important mining states that still lacked reclamation laws discovered, often with some encouragement from industry, that they wanted to adopt reclamation laws. As one author noted in 1994, Colorado, New Mexico, Oregon, and Washington "dramatically revised their reclamation statutes and regulations within the last three years. As a result of the Summitville debacle, others are in the process of revising theirs."[8] Thus, while all the important mining states now have such programmes, some have adopted them only recently. For example, Arizona adopted its reclamation statute only in 1994. However, Arizona, the leading state in mineral production, did have an Aquifer Protection Permit programme relating to groundwater protection, obviously a critical issue in that desert state. This was far short of a requirement for comprehensive planning for closure.[9]

At this point, then, the nature, effect, and requirements of mine closure planning, to the extent it is a regulatory requirement, are still largely determined by state law. For a general discussion of the problems associated with closing a mine under U.S. law, see Williams.[10] These state programmes exist in the context of a very complex, highly developed system of national and state environmental legislation in air quality, water quality, and solid waste, which profoundly affect the way they are structured. Closure plans are simply one tool of environmental management. They cannot be expected to address all concerns or solve all problems. Proper development of closure plans is dependent on recognising the point at which closure planning is not the appropriate tool for a particular purpose, and turning to a more appropriate method. Systems of air, water, and soil quality norms and their related monitoring, reporting, and enforcement mechanisms are important management tools in every country with a developed system of environmental law, and form part of the setting in which the closure plan is developed. Part of the challenge is coordinating the appropriate use of these and other tools, such as environmental impact statement requirements. But any overall requirement for planning for mine closure is still a matter of state choice.

17.1.5　Focus of this Chapter

While this chapter will discuss many aspects of this complex and varied set of laws, it is not intended as a comprehensive survey of every aspect of the subject. It will treat many different state systems, but its most immediate focus will be the mine closure programme of the state of Colorado: Colorado passed its mine reclamation law in the mid-1970s and has two decades of experience with its administration. Colorado has one of the more mature state programmes. It is also a substantial programme, with over 1900 mines now holding active permits. It is also the state in which the unhappy events at Summitville, which have had such a profound effect on regulatory systems, unfolded. Colorado was one of the first states to react with new legislation and regulatory approaches to the clear inadequacies exposed by those events, and it reacted with significant reform legislation that passed both houses of the state legislature unanimously, with the support not only of the Sierra Club, Trout Unlimited, and other environmental organisations, but also the Colorado Mining Association, the state's Rock Products Association, and other industry associations, as well as the state regulatory agencies. In contrast with the Colorado experience, the effort to develop better mine closure practice in the U.S. has been marked by much conflict, with a few exceptions.[11]

One of the authors, Mr. Danielson, was for most of the last decade, which included the *dénouement* if not the origins of the Summitville problems, a member of the Colorado Mined Land Reclamation Board, the state commission with responsibility for mine closure programmes. He was also one of the authors of the state's recent legislative and regulatory changes.

In turn, the chapter will treat questions about the basic structure of mine closure programmes (Section 17.2); the mine closure plan, its contents, review, and approval (Section 17.3); financial warranties and systems of guarantee (Section 17.4); reporting, inspection, monitoring, and enforcement (Section 17.5); current issues of concern (Section 17.6); and examples of current best practice (Section 17.7).

17.2 THE STRUCTURE OF MINE CLOSURE PROGRAMMES

17.2.1 WHAT ARE THE PROGRAMME'S GOALS?

There is hardly a question more important than definition of the programme's fundamental goals. Yet there is hardly a question with such a diversity of answers from the several states. Some of these stated goals seem to focus on the mine site itself, while others are directed not so much to the mine site as to its impacts on surrounding lands.

The stated goal of the Colorado general mining reclamation programme, and of its specific construction materials legislation, both discussed below, is to "reclaim land affected by [mining] operations so that [it] may be put to a use beneficial to the people of this state...."[12] Other goals stated in state legislation aim to prevent unnecessary and undue degradation of the environment,[13] to establish plant cover and stabilise the soil,[14] to minimise post-closing visual effects,[15] to protect public health and safety,[16] to achieve a self-sustaining ecosystem following closure,[17] or to provide that mines "no longer pose a threat to water quality."[18]

The national Surface Mining Act, applicable to coal mines, contains a long list of performance standards, which include restoration of mined land to a condition capable of supporting pre-mining uses or acceptable higher or better uses; restoration of the approximate original contour of the land (with limited exceptions); stabilisation of surface areas to control air and water pollution; minimisation of the effects of mining on the hydrologic balance and the quality and quantity of water in surface and subsurface systems; and protecting offsite areas from slides or damage.[19]

For all their variously stated goals, the general principle underlying all these laws is that the choice of the post-mining use of the land affected by mining resides with the private landowner, and is not subject to dictation by the government, except to the extent that any land is subject to local zoning, planning, or other land use regulations. The closure authority concerns itself not with the end, but with the means: whether the plan proposed provides sound technical means to reach the final use which is chosen by the operator, usually in conjunction with the underlying landowner. However, there is a limit on any principle: if the chosen land use does not provide for creation of a "self-sustaining ecosystem" or to "establish plant cover and stabilise the soil," or is not "beneficial to the people of the state," that limit is reached. A proposal to make the site a "museum of poor mining practices" would thus probably be unacceptable. This example is not as fanciful as it sounds: state efforts to remediate environmental problems at abandoned mine sites have a number of times run into stiff opposition from historic preservation advocates who have resisted reclamation programmes because they would affect dangerous historic head frames, unguarded historic mine shafts, or even historic acid drainage.

17.2.2   How Many Different Regulatory Systems are Needed?

Out of necessity, those American states which mine both coal and non-coal resources have two separate sets of requirements. If the state has decided to create its own programme, under federal oversight, to manage coal mine operation and reclamation, the coal programme will be one of innumerable specific rules, procedures, and requirements, closely monitored by national authorities for any sign of lack of diligence. One of the common critiques of the coal reclamation programme is that its highly prescriptive and detailed requirements come largely from the experience in the eastern U.S., and fail adequately to take regional variation in climate, land use, and other circumstances into account. The state's reclamation programme for non-coal resources is not subject to any national oversight, and states are free to construct their own requirements. There are thus in every such state at least two sets of regulations. It is not uncommon for states to divide regulatory responsibility further, and have more than two separate regulatory schemes.

Every state, province, or nation which has decided to develop mine closure legislation has faced the fact that the mining "industry" is in fact *several* different industries — each with its own environmental issues and closure problems. The industry is divided by the products it produces and also by the scale of operations. While the focus is often on the enormous scale of the modern mines of multinational mining companies, there are other producers of significance in the industry, and they often have different needs and capabilities.

The economics of the regulatory process are also important. Regulatory approaches that are cost-effective and appropriate when dealing with the largest segment of the industry are inefficient, expensive, and difficult at smaller scales.

Finally, its environmental issues divide the industry. These are a function of the physical surroundings in which the industry operates, the processes it employs, and the managerial and technical capacity of the mining enterprises. Few manufacturing industries locate plants below sea level, at altitudes of 4000 meters or more, in swamps, or in the centre of great deserts. Yet mines are found in all these locations.

Few types of products are produced by more than one of a handful of alternative manufacturing processes. Yet the range of technologies employed in mining is enormous.

There is no such thing as artisanal production of cars or computers. Yet the mining industry, while it includes some of the largest and most sophisticated firms in the world, also includes thousands of individual "pick and shovel" miners, even in as advanced a country as the U.S. A regulatory system that ignores the limits on the capabilities of the regulated community is headed for conflict.

Some segments of industry are seen to raise specific concerns, which may well lead to continued tightening of standards. Idaho, for example, has a specific law governing mineral recovery operations which use cyanide,[20] and another specific code for placer and dredge operations.[21] California has developed a system where requirements vary substantially depending on whether the mine wastes are classified as Type A, (hazardous), Type B (a variety of different types of wastes which could cause water quality violations), and Type C (not expected to cause adverse water quality effects other than turbidity).[22] Montana has a specific law for open pit operations, the Opencut Mining Act.[23] Many of the controls that focus on hazardous materials, or cyanide in particular, are a result of the situation at the Summitville Mine in Colorado. Environmental contamination at the Summitville Mine has led to a tightening of regulatory requirements throughout much of the U.S. Colorado itself, in the wake of the events at Summitville, adopted major revisions to its mine closure law, imposing specific requirements for a new classification of "Designated Mining Operations," which include those that have significant onsite chemical processing, or high acid drainage potential (see Proceedings of the Summitville Forum '95, Colorado Department of Natural Resources, 1995, for further information on the Summitville incident).

Other portions of the industry have resisted a tightening of their own rules, in the wake of Summitville. The Colorado construction materials industry, for example, successfully argued that it should not be burdened with the tougher post-Summitville requirements, and secured passage of its own separate closure law, the "Colorado Land Reclamation Act for the Extraction of Construction Materials."[24] Previously, all mining other than coal mining had been subject to the same law.

The result of all this diversity in the industry has typically been that states develop more than one type of permit system. The distinctions among the different types of permits are typically based on (a) differences in the technologies employed; (b) differences in the environmental sensitivity of the area where the mine is to be developed; (c) differences in the scale of operations, or some combination of these factors.

Colorado thus has, in addition to a Coal Mine Reclamation Act,[25] the previously referenced Colorado Land Reclamation Act for the Extraction of Construction Materials, which applies to sand, gravel, stone, borrow material and other construction materials, and a Colorado Mined Land Reclamation Act,[26] which applies to all other forms of mining, including metal, or "hard rock" mining. There are three levels of permits for construction materials for large, small, and very small operations, and a special category that applies to mines developed for construction materials for government highway projects. Colorado's metal mining reclamation law recognises, in essence, five types of permits, which are distinguished both by their size and by the environmental risks, e.g., use of acid or other toxic chemicals onsite, or potential for generation of acid or toxic materials.[27] Montana has the above-noted "small miner exclusion" to its Metal Mine Reclamation Act. However, the small miner is not exempt from having to reclaim to general statutory standards, simply from having to obtain a specific permit. The small miner files a statement annually, verifying continued observance of the limits of the small miner limitation. This is simply another example demonstrating that approval of closure plans by a regulatory agency is simply one tool of environmental management. It cannot be expected to address all concerns or solve all problems: promulgation of general performance standards, without requiring submission or approval of a specific plan, is an alternative option seen in many environmental programmes. There is in Montana yet another category, or an exemption to the exemption: if the small miner uses or stores cyanide, a special cyanide permit must be obtained.[9]

Montana also has a separate Opencut Mining Act,[28] and specific permits for cyanide use. Idaho has its ordinary permit system plus special permits for placer/dredge operations and for cyanide operations; in Nevada, "abbreviated permit application requirements apply to small scale, pilot, testing, placer, or other facilities that rely solely on physical separation methods to process ore."[29]

Arizona still has a unitary system, with only one kind of permit. However, this may simply be a reflection of the fact that Arizona was the last of the major mining states to adopt mine closure planning, and the forces which have pushed other states in the direction of diversification may not yet have had time to exert themselves. Arizona, which had resisted such a system for years, almost certainly acted as a result of a threat that national level legislation would be imposed, and the fact that advocates of that legislation were pointing to Arizona's lack of such legislation as an argument for its passage.[30] While the proponents of reform of the national mining law have not yet achieved that goal, they deserve a great deal of credit for motivating states to significant improvements in their programmes.

Arizona's system for implementation of its law is still in development. The law applies to surface disturbances of five acres (about two hectares) or greater, and only to metal mining. Under the state's new Mined Land Reclamation Act, all new metal mining operations, whether in the exploration phase or the production phase and which meet the size requirement, are required to obtain approval of reclamation plans and financial assurances.[31]

A final reason for the variety of permit sizes and types relates to the fees that government charges for the various steps in the process: application fees, permit review fees, annual fees, and the like. States have discovered that the costs of application, review, inspections, and other oversight activities are much higher for certain kinds of permits than for others. Distinguishing among different categories of operations allows government to allocate the costs of its activities among the regulated enterprises in a more economically efficient manner.

17.2.3 WHAT ARE THE PROPER LIMITS OF THE MINE CLOSURE PROGRAMME? WHAT IS "MINING?"

There are important and complicated issues regarding the limits of the activities that should be included in mine closure planning. These issues arise from differences of scale among mining operations (the *de minimis* concern), and from the differences in types of mining activities.

17.2.3.1 Lower limits of regulation

At very low scales of effort, there are serious questions about whether closure planning is the appropriate tool to deal with perceived environmental problems. Regulatory programmes face these questions constantly: are weekend gold panners "operators" who need to have reclamation plans? What would their closure plans consist of? Although, as in the case of mercury contamination of streambeds and aquatic biota by artisanal placer miners, there may be serious environmental issues that need to be addressed, mine closure requirements may not be the appropriate tool. Closure plans are simply one tool of environmental management. They cannot be expected to address all concerns or solve all problems. Proper use of closure plans is dependent on recognising the point at which they are not the appropriate tool for a particular purpose, and turning to a more appropriate tool, such as exempting small operations so long as they are not using dangerous substances (e.g., mercury and cyanide), or requiring attendance at educational programmes as a requirement for engaging in the activity.

Many American mine closure systems have some lower limit on the scale of the activities to which they apply. As indicated above, Arizona requires an operating permit only for surface disturbances of more than five contiguous acres (about 4.5 hectares).[32] Nevada employs a very similar exemption from reclamation requirement for small-acreage operations.[33] In California, a Surface Mining and Reclamation Act permit is not required if the disturbed area is one acre (about 0.4 hectare) or less and if the amount of overburden is less than 1000 cubic yards.[34] Montana allows a mining operation to operate without an operating permit if the operator will not remove material in excess of 36,500 tons in any calendar year, and if the operation is one site that disturbs and leaves unreclaimed less than five acres (or two sites disturbing and leaving unreclaimed less than five acres, approximately 2.25 hectares, so long as certain restrictions apply).[35] Wyoming imposes reduced requirements on operations that mine less than 10,000 cubic yards and affect less than ten acres (approximately 4.5 hectares) per year.[36]

17.2.3.2 What is mining?

There clearly are some circumstances in which, while closure plans may be useful, they are not *mine* closure plans. Closure plans exist for many kinds of facilities other than mines: landfills, electrical power stations, nuclear waste disposal sites, and chemical plants, to name just a few. In our view, key indicators that closure plans are likely to be effective tools of environmental management in a given circumstance are (1) that environmental costs are externalised not just in space but in time; i.e., that a significant portion of the uninternalised social cost will occur after the revenue-generating activities have ended; (2) a potential for very significant environmental effects; (3) reasons to believe that those effects would be very expensive or impossible to reverse once experienced; and (4) actors with the technical capacity to apply the concept effectively. Clearly, mining is not the only human activity that disturbs large areas of the earth. It is not as easy as it might seem to distinguish between "mining" and other activities such as reservoir construction, road projects, agricultural improvements, sanitary landfills, and even urban developments, because these activities often involve selling sand, gravel, or borrow material (or even alluvial gold contained in these materials) and because these are frequently claimed by opponents to be subterfuges to avoid mine closure requirements. Colorado has had some interesting examples of road projects which do not seem to go anywhere except into the centre of aggregate deposits, or reservoirs that are being constructed on a time scale of decades, by people who own no water with which to fill them, right in the centre of valuable mineral deposits — the products of which are being sold in the process.

And the questions only proliferate. Is peat, as an organic material, a "mineral" subject to the Act? Does it really matter in an environmental management sense if "extracting" it is doing long-term environmental damage, and closure planning is a good way of dealing with it? If leveling of land for a building site produces excess material that is sold, is this a "mine"? Is hand collection of decorative stone from fields, without use of tools, a "mining operation"?

The Colorado mine closure agency, the Division of Minerals and Geology of the Colorado Department of Natural Resources, has spent a remarkable amount of time on such issues over the years. In the author's view, the issue should not be legalistic haggling over the definition of "mine" or "mineral," but the more functional questions of (a) identifying the circumstances in which mine closure programmes are effective management tools, the circumstances in which they are not, and (b) the alternative tools most appropriate to employ when there *are environmental impacts*, but closure plans are not the best way to deal with them. The definitions of the terms "mine" and "mineral" need to be shaped based on this judgment, rather than the scope of the programme being shaped by fine legal arguments over the definitions. This comment applies equally to the other common issue of this type: "custom" milling operations. Under most, if not all, U.S. systems, the closure requirement applies to milling and processing operations conducted at a mine site by the mining company. But closure requirements may not apply to an independent milling operation, which may be under different ownership, remote from the mine site, and may accept ore from a number of different mining operations. Here, the logic would seem to favor imposing closure requirements, since the character of the environmental effects is hardly different because of accidents of ownership.

17.2.3.3   When does the mine closure obligation begin?

Mining is a process that begins with a generalised search for minerals, and continues with more specific investigation of a particular occurrence, definition of reserves, bulk sampling, and production. Defining the point at which the legal obligation to submit a mine closure plan attaches is not as easy as it might seem.

Many states have specific reclamation requirements for exploration sites.[37] These requirements generally do not apply to prospecting methods such as aerial magnetic surveys that do not disturb the land surface, or to the odd geologist chipping a sample off a rock outcrop. They apply to "significant" surface disturbances.

There are of course restrictions which landowners, private or governmental, may impose on activities involving entry on their lands. The obligation of the operator who holds a mining lease to the landowner, or of the owner of the mineral right to the owner of the surface right, while important parts of the closure equation, are outside the scope of this chapter.

But recognising that exploration may have impacts of a type which closure plans can helpfully address, most states do require some sort of closure plans for prospecting activities which exceed certain thresholds of disturbance. This may be through either individual permit requirements, or generally applicable statutory prescriptions.

17.2.4   Treatment of Existing or Abandoned Facilities

Some provision needs to be made for mines in operation at the time a mine closure system goes into effect, to take into account past choices; the operator's investment in those choices; and the economic consequences of trying to change them *post facto*. Standard international practice in environmental management now recognises a hierarchy of steps required for good project planning.

1. Environmental impacts must be clearly identified, ranked in importance, and evaluated for positive and negative aspects. This ranking usually recognises criteria of intensity of the impact, probability of occurrence, reversibility, and anticipated duration.
2. All reasonable steps are taken to avoid negative impacts entirely, by, for example, changes in project design.
3. Negative impacts that cannot reasonably be avoided must be minimised, in a systematic and rigorous analysis.
4. If negative impacts of significance still exist after the process of minimisation, they need to be mitigated, preferably by physical measures; and to the extent these are not available, by management or institutional measures.
5. Negative impacts remaining after this process must be compensated for; there is a hierarchy of the types of desirable compensation measures.
6. Positive project impacts should be enhanced, where possible, also through a rigorous system of analysis.

The decisions as to what is "reasonable" to accomplish at each step are heavily influenced by economic considerations. When a mine is in operation, some decisions have already been made. If 100 million tons of tailings exist in a particular location, that has to influence the analysis rather heavily. Thus, closure planning for a mine in operation tends to focus more on minimisation, mitigation, and compensation and less on avoidance.

The alternatives available are obviously much greater when the analysis is applied to a mine not yet underway than to a mine with five, ten, or thirty years of operations behind it. Alternatives are reduced even further in the case of abandoned mine workings, which may have continuing impacts on the environment. Deciding how, whether, and to what extent these belong inside a mine closure programme depends on a series of judgments, based on factors such as:

1. Whether it seems economically inefficient, and environmentally shortsighted, to spend available resources imposing a very high degree of control on impacts from existing operations when the same level of resources would give much greater environmental returns if devoted to improvement of conditions at abandoned sites.
2. The fact that failure to address closure issues at abandoned sites may sometimes mean that enforcing expensive regulations at existing or new sites creates little environmental value. If the river is contaminated by effluent from abandoned mines to a level at which aquatic organisms cannot exist, or at which downstream communities are going to pay excessive water treatment costs regardless of the level of control at new mines, imposing controls on new mines hardly makes sense, *unless there is some commitment to addressing closure issues at abandoned mines.*
3. Recognition of the fact that there are an infinite variety of circumstances in which mines have been abandoned. At one extreme are very old mines which were abandoned a century or more ago, which complied with whatever legal requirements (if any) were effective at the time, and whose owners, if natural persons, have long since gone to their reward. At the other end of the spectrum are mines recently abandoned, which *did not* comply with regulations in effect at the time of their closure, and whose former owners, or stockholders, are walking the streets of Sydney, or Vancouver, or San Francisco with pockets full of money. At the former end of the scale, it seems that implementation of closure systems at abandoned sites is clearly a social cost to be borne by the state. At the latter end, this is not so clear. And where we are to draw the line is not simple. The U.S. experience, for example, is that identifying and imposing liability retroactively on owners can have very high transaction costs which cannot be ignored.

When Nevada created its system, it provided that all facilities in existence on September 1, 1989 had to obtain a permit within three years of that date. No new facilities or modifications could be built after July 1, 1990 without permits. Each "process component" was required to meet whatever regulations were in effect at the time its construction commenced.[38] New Mexico's Rule 5 provided what amounts to a phase in of the law's requirements for "existing mining operations," which were defined as "an extraction operation that produced marketable minerals for a total of at least two years between January 1, 1970 and June 18, 1993."[39] When Colorado began its current programme, it required permits for all new mines and gave operators of existing mines three months to apply for permits.[40] With this advance warning, mines could operate up to the legal deadline without incurring reclamation liabilities. A few in fact closed the day before the deadline in order not to have to deal with the new regulation. But any portion of the mine works that was operated after the deadline required a reclamation plan. The World Bank has suggested a distinction, which could be useful in this regard between "contamination flows," which are an integral part of ongoing production at a site, and "pollution stocks" which are not. The reclamation obligation at an existing site could be seen as applying to everything except pollution stocks in existence on the date of effectiveness of the law.[41]

Obviously, this approach can lead to disputes over whether or not there were operations after the cutoff date and whether reclamation is therefore required. The state's failure to appreciate this soon enough and consequent lack of good baseline data have led to many disputes which could have been prevented.

Whether because state law explicitly says so, or as a result of administrative practice, mines which were in operation at the time the closure permit law became effective are not required to develop closure plans as if they are new mines, but to develop plans based on the existing reality. Each of the various "transitional provisions" has its own set of consequences. There is no significant state law reclamation requirement for operations that ceased activity before the law became effective.

Thus, to summarise, the state reclamation programmes provide (a) for closure planning from the outset of new mining operations or major modifications to existing ones; (b) reduced or "phased-in" requirements for operations which were already in operation when the individual state's law became effective; and (c) little or no reclamation of sites where mining ceased before the law became effective.

There is a fund — the Abandoned Mine Reclamation Fund — established by Title IV of the federal Surface Mining Act originating in a national tax on coal mining, which is distributed to qualifying states for use in reclamation of abandoned mine sites. The coal industry in general feels that it is unreasonable to ask it to pay all the cost for reclamation on all types of mine sites. One obvious reason to base this fund on a coal tax is a pragmatic one: given the relatively low value to weight ratio of coal, transportation costs are a very high fraction of total costs, and the market is more regional than global. Thus, a tax on coal is not as likely to have an adverse effect on the competitive position of the domestic industry as would, for example, a tax on gold production. Reflecting that perceived unfairness, the fund prioritises reclamation of abandoned coal mines over other types of abandoned mines, regardless of the relative environmental or safety problems at the sites. Further, it prioritises anything classified as a "safety" problem, no matter how petty, over "environmental" problems, no matter how acute.

Colorado's share of these funds was adequate to provide for construction of 202 engineered closures of dangerous abandoned mine openings in 1996 out of an estimated total of 23,000 which existed in the state when the programme began.[42] As the programme goes forward, an increasing part of its budget is spent on repair and restoration of its own previous work which has been subject to vandalism or other misfortune; the amount of money provided each year is in any case declining.

Other than a small amount of money originating in grants from EPA's Clean Water Act non-point source programme for demonstration projects, essentially nothing is spent by government on dealing with the enormous environmental legacy of abandoned mines, except under the Superfund programme. And most of the remedial work under the non-point source programme has been stalled, precisely because state authorities are concerned that making any alterations at old mine sites will expose them to enormous liabilities under either CERCLA, the statute which creates Superfund, or the Clean Water Act itself. These problems are discussed in Section 17.6 below.

It seems self-defeating to create regulatory schemes which require industry to spend enormous amounts of money on environmental controls to reach, for example, very stringent water quality discharge standards when the receiving waters are severely contaminated by discharge from abandoned mines over which there is no control at all; and where absent some commitment to remediation at those sites, streams will be devoid of life forever from that cause regardless of the discharge standards applied at new mines. And that commitment, except to a limited extent under the Superfund statute and related solid waste laws, has been lacking.

Superfund was principally designed to deal with orphan chemical dumpsites and the like. It has had limited application to mine problems, and is somewhat cumbersome for that purpose. Nationwide, it has been employed at perhaps several dozen sites. "Approximately 39 western U.S. mining sites have been proposed for inclusion on the NPL ([National Priorities List) since 1981. Thirty-one of the 39 sites were formally listed on the NPL as of May 1994."[43] This represented three percent of the Superfund Priorities List at the time, but no data were found available for the number of eastern mine sites which might be on the list. Some of these comprise more than one mine property. The Clear Creek-Gilpin site in Colorado alone is comprised of hundreds of individual mines, mainly dating from the 1880s and 1890s.

Colorado has the dubious honor of being home to quite a number of these, including Leadville's California Gulch (the Yak Tunnel), the Eagle Mine, Clear Creek-Gilpin (including the Argo drainage tunnel), Idarado, and of course Summitville. While the programme has made some gains at these and other mega-sites, such as Butte in Montana, it is not the solution to remediating the tens of thousands of abandoned mine sites in Colorado, or the hundreds of thousands in the west as a whole. Inclusion on the National Priorities List is based on a Hazard-ranking System developed by the United States Environmental Protection Agency (USEPA).[43] The majority of abandoned mine sites, even those presenting serious environmental degradation, are unlikely ever to get close to making the list. The result, sadly, is that staggering amounts of money will be spent cleaning a relative handful of sites to "background" levels, while nothing at all is spent on the other sites.

Building a constituency for cleanup for even the high priority sites is difficult. Consciousness of the problem is not high in many areas, at least in part because no one has ever seen these mining districts in their natural condition; moreover, it is hard to build enthusiasm to pursue the villains responsible, because many of them, if villains they were, are long dead.

Thus, the focus of all state programmes is overwhelmingly on the problems of new mines, or mines which have been in operation since the respective effective date of the laws.

17.3   PREPARATION, REVIEW, AND APPROVAL OF THE MINE CLOSURE PLAN

This section focuses on the contents of the mine closure plan and the circumstances of its review and approval.

17.3.1   What Baseline Information is Needed for the Closure Plan?

One of the most significant sets of issues in mine closure revolves around the question of what baseline information is required to obtain a permit. The variety of judgments that may have to be made in the future about a mine site is staggering. Are there violations of emission rules? What will be the environmental impacts of modifications in processes or the mine plan? Have there been increases in contaminant levels in environmental media? What is the source of unexpected drops in water levels — are they effects of natural cycles or something new? Are mine operations somehow related to changes in fish or wildlife populations? *Are* there in fact such changes?

Without a good idea of pre-mining conditions, there is no way to define success in closure efforts. This can be critical, because successfully completing closure requirements has major financial consequences for the mining enterprise. And when, as is often the case, the mining company does not own the land on which it mines, the owner has certain expectations about the condition in which the land will be returned to him or her; long, expensive conflicts can ensue regarding whether the closure has successfully met those requirements.

The only hope for rational resolution of these conflicts, or to make the necessary technical judgments on the soundest possible footing, is to have reliable information on conditions prior to disturbance by mining. Careful gathering of good scientific data is also the clearest way to learn and improve our understanding. Most states, and the national coal programme, have detailed descriptions of baseline data requirements.

The problems are, first, that the necessary information often relates to natural cycles: hydrologic cycles, life cycles of living species, climate and weather cycles, and the like, which cannot be gathered in a hurry. A year seems to be the absolute minimum period for which data should be gathered for many of these variables, and a year's data may be inadequate for some of them. It is questionable, for example, if adequate data for prediction of acid drainage potential can be gathered in a year.[44] Some kinds of data, such as those related to storm event impact on water quality, may be hard to get without very frequent or even continuous monitoring, and perhaps a bit of luck. Data can be expensive to gather, too. A number of enterprises have spent a considerable amount of money on studies of acid generation potential, sometimes without developing useful predictions. Second, the question of which data to gather is highly site-specific, as a function of the local environment, the proposed mine plan, processing technology, and other variables.

The concern is clear from the point of view of industry: enormous expenditure on exploration leads to a prospect which is laboriously defined, expensive and time-consuming studies of everything from workforce requirements to tax structure to transportation, electric power, water availability, the market, and a host of other variables are undertaken; a profitable mine is defined; the project is ready to go; and someone identifies a seemingly trivial piece of data — the breeding habits of a rare mouse, iron concentrations in water at time of low flow — which cannot be gathered quickly, and the project is stalled, with extremely expensive consequences.

There is probably no perfect way around this problem. The data needs *are* site-specific. The consequences of not having them *can* be disastrous. They *do* take time to gather. Our experience leads to three less than fully satisfactory observations:

1. The environmental variable as a management issue for the firm is much like other management issues. It is a series of risks and opportunities for the enterprise. It is rarely possible to eliminate all the risks, or to capture fully all the benefits of the opportunities. The enterprises that manage these risks and opportunities best tend to be more competitive in the marketplace. All variables are managed best when there is adequate good, current information available to managers.

2. It is sometimes possible to observe a "negative feedback loop." Mining companies that do not understand the environmental permitting process, or the goals of closure requirements, sometimes avoid contact with regulators to the maximum possible extent, postpone the necessary studies as long as possible, and do not start the permitting process until they have resolved all of the other feasibility issues relating to the mine. This not only makes it very hard for the enterprise to go through the systematic process of identification of impacts and their avoidance, minimisation, mitigation, and compensation, as described above, but much increases the chance that baseline data issues with a high likelihood of disturbing the project schedule will arise late in the day. When they

do, it can reinforce management's negative impression of the regulatory process, and disinclination to deal with environmental issues any sooner than they have to.

3. One of the benefits to an early and open approach to mine development is that baseline data needs and related closure issues tend to get identified earlier, minimising the chance of unpleasant surprises at later project stages. The idea that it is better not to develop too much information early in the process because problems could be uncovered seems to be losing favour as companies experience the consequences of those problems surfacing unexpectedly later, at a much more costly phase. And it is a remarkable concept of management that managers are better off with incomplete or unreliable information when making decisions involving millions of dollars. If it is good not to gather "too much information," it is hard to understand why this concept should be limited to the environmental sphere: if the company is better off with spotty, limited environmental information of poor quality, why is this not also a good idea when dealing with ore reserve information? After all, better studies might indicate that the reserves are less than expected, perhaps not even adequate to support the project, and then the project might not go forward. It would be interesting to apply this principle to transportation studies, energy supply studies, and the like.

There is a final issue. The environmental impact statement in the U.S. is generally a creature of federal, not state law (there are some states, such as Montana and California, which do have state-level environmental impact statement requirements — the majority of the major mining states do not). There is no guarantee that the baseline information needed for the federally required environmental impact statement system will be the same as that needed for the state system. Though much of it may overlap, there may be specific items required at one level which are not required at the other, or which must be gathered according to different methodologies.

17.3.2 WHAT ARE THE CONTENTS OF THE MINE CLOSURE PLAN?

When Colorado first established its programme about 20 years ago, many of the early mine reclamation plans were, literally, sketchy: hand drawn, not-to-scale diagrams of the intended post-mining configuration. As both the agency and the regulated community have gained experience with the system, the requirements, particularly for the category of permits which includes the largest mines, and the mines which use toxic processing agents, have rapidly become more sophisticated and complex.

Colorado's detailed requirements now include submission of an index map, pre-mining maps, a mining plan map, a mining plan, a reclamation plan, a reclamation plan map; information on water resources, wildlife, soils, vegetation, and climate; an estimate of reclamation costs; a list of other permits and licenses which the applicant is seeking; a demonstration that the mining company will have legal right to enter the premises to perform reclamation work; identification of all owners with an interest in the property; identification of all municipalities within two miles; proof that local government officials have been provided with notice of the application; identification of all man-made structures in or adjacent to the area to be mined; a geotechnical stability analysis; and, for "designated" operations which are thought to represent particular hazards because of acid drainage potential or use of hazardous chemicals in processing, an "environmental protection plan." Where hazardous chemicals are present, this last item must include an emergency response plan in case of spills or accidents. Each of these elements must be prepared according to prescribed procedures. These are reasonably typical of requirements in most other states.

Basically, there are two plans which need to be coordinated to work together. The reclamation plan defines the beneficial use for which the site is destined post-closure, with the details of final configuration, measures for geotechnical stability, drainage control, erosion control, topsoiling, vegetative cover, and the like. The operating plan is designed to produce ore from the mine in an efficient and economical way, leading to a configuration close to what is called for in the reclamation plan, so that fulfillment of closure obligations becomes as easy and inexpensive as possible once production ceases. Colorado requires submission of both plans as part of the permit application.

The basic configuration of the mine, location of storage piles, overburden, roads, and waste dumps all have to be planned from the outset with closure in mind, and the mine closure agency must of necessity concern itself with a plan which includes the operational phase.

The part of the state regulatory programme that is of most direct interest to mining companies is the set of technical criteria which the state applies in deciding whether a plan is adequate. Almost always, these technical criteria are subject to exceptions if they are deemed impractical in the circumstances of a particular project, but generally they are a good guide to what the agency expects: that absent compelling reasons to the contrary, for instance, that waste dump slopes should be no steeper that 3h: 1v; that to prevent runoff from reaching excessive velocities and creating gullying problems, there should be terraces at intervals of no less than 50 vertical feet; and that topsoil should be stockpiled separately and protected from wind and water erosion until it is reused in the final cover for the reclaimed site.

Most states have developed many such criteria. They are generally known to the operators in the state, or consultants who aid in the preparation of plans, but in few cases have been published. They tend to be specific to individual states because the problems in each state tend to vary significantly with the type of mining, climate, wildlife resources, vegetative cover, and so on.

17.3.3 WHAT IS THE REVIEW PROCESS FOR THE PLAN?

The principal elements of review common to all states' systems include notice provisions; a procedure for soliciting the opinions of various other organs of government with a potential interest in the project; staff technical review; an opportunity for public participation; and either administrative appeal of the decision by affected parties, or recourse to the court system, or both.

The review is often constrained by time limits. In Colorado, which has different time limits for different kinds of permits, the permit is deemed approved if it is not denied within specified time periods. This is typical of most states.

17.3.4 HOW DOES THE PUBLIC PARTICIPATION PROCESS WORK?

Permit processes in the U.S. typically provide for public participation. This begins with some requirement that the operator of the proposed facility notify potential interested parties, often by publication in newspapers, posting notices in local government offices, posting signs near the proposed mine site, or mailing notices to adjacent landowners or other designated persons. Most systems employ some mix of these types of notice. Some states define relatively narrowly the class of people who are entitled to receive notice: in Colorado, for example, only owners of land within 200 feet of the affected area are entitled to notice; in Wyoming, the figure is one half mile. In general, it appears that states with very broad notification requirements are more flexible in overlooking technical defects in notification, and states with narrow requirements tend to interpret them inflexibly, though generalisation on this subject is difficult.

Typically, following the notice is a period for written comment. Most systems also provide for some form of hearing at which those who object to some aspect of the proposed project may appear and make their views known. In general, the agencies apply a very broad definition of "interested parties," calling any doubts in favor of allowing participation. The level of public participation varies tremendously. Colorado has had a number of permit proceedings where hundreds of people and organisations have commented or spoken at public hearings for or against projects, and one — relating to a proposed gravel quarry in the Denver suburbs — at which over 3000 people requested formal status as parties to the proceeding. Yet many permit proceedings draw little or no attention from the public at large. The key factors that determine the level of public participation seem to include perceived threats to the value of neighboring property, water supplies, the distur-bance and annoyance of having an active mining operation nearby, and the proximity of the mine to sensitive areas, such as national parks or wilderness sites. Frequently, the state agency is faced with what are essentially land use issues relating to perceived incompatibility of mining operations with surrounding land uses. The state agencies that administer mine closure requirements typically do not have the authority to decide issues on compatibility of land uses, but are constrained by law to base their decisions on technical considerations of adequacy of the closure plan, or environmental mitigation measures. Typically, organs of local government or federal land use agencies decide the land use compatibility issues. One of the common criticisms of the U.S. Mining Law is that it does not provide balancing of mining against other potential land uses. Most federally managed land, except for parks and wilderness areas must, under this 1872 law, be available for mining if the claimant has discovered a valuable mineral deposit. The conflict over mining on the federal lands is exacerbated by the relative difficulty of mining hard rock minerals on private lands in the U.S. under the common law system of "fee simple" ownership, and the lack of a right on the part of the discoverer to exploit those deposits without landowner agreement.

17.3.5 BY WHAT STANDARDS IS THE ADEQUACY OF THE PLAN JUDGED?

Most state mine closure laws are structured so that the applicant is entitled to a permit unless the agency in charge makes specific findings, stated in the law. In Montana, the permit issues unless there is a finding that the operation would violate the state's clean air or clean water laws or the regulations thereunder, or the application fails to specify an "acceptable" programme of reclamation.[29]

In Colorado, the reclamation plan is tied to achieving a specified post-mining land use. Appli-cations have been denied for failure to give the appropriate public notice; because the application fails to contain all the required contents; because the applicant had failed to pay outstanding fines for previous violations; because the plan fails reasonably to assure that the post-mining land use will be achieved; for failure to comply with specific provisions of the Act or implementing rules, such as the requirement that the plan ensure that it will "protect the hydrologic balance"; and for other reasons.

In general, the decision to grant or deny a permit is based on technical evaluation of the proposed reclamation plan to determine whether it will achieve the post-mining land use proposed by the applicant in a manner consistent with the requirements of the law and regulations. The decision is also based on whether there has been compliance with procedural requirements. Typically, as noted above, the state mine closure agency has no authority to decide whether mining is an appropriate land use, or consistent with surrounding uses. It is rare for applications to be denied outright: generally, agency staff communicate any reservations to the proponent, who then makes appropriate changes to the proposal in order to make it "approvable." Sometimes the agency's outstanding concerns are also dealt with by attaching conditions to issuance of the permit.

17.3.6 HOW CAN THE PLAN BE CHANGED?

One frequently voiced objection to the concept of closure plans is that mining is a dynamic process, involving constant refinement of the understanding of the orebody, changing definition of reserves, developments in technology, and so on. It is suggested that conditions may change too frequently for a defined plan to hold. This concern has been addressed by creation of flexible and easy amendment processes. Where the mine is sold, typically the new operator simply acquires the permit and assumes the liabilities of the old operator; this is not an occasion for wholesale review and revision of the permit. However, in several states it does trigger a review of the adequacy of the bond amount.

There is here an enormous difference between permit changes triggered by the operator's voluntary choice, and changes triggered unilaterally by the state. The latter are regarded as anathema by much of the industry, which hopes, in submitting to the regulatory system, to gain in return some certainty about what requirements it will have to meet in the long run. Most state programmes provide that the permit is valid for the "life of the mine," or some similar formula (reclamation permits "shall be effective for the life of the particular mining operation if the operator complies with the conditions of such reclamation permits and with the provisions of this article and rules...."),[45] which means that so long as the operator is in compliance with the regulations and the permit terms, the state cannot impose additional closure requirements. This creates a powerful incentive for mine operators to remain in compliance with their permits.

One other phenomenon is worth noting. When a new mining operation is first proposed for a community, it seems local officials, anxious for the revenue and employment, are often among the most vocal proponents of relaxing environmental laws to let the project proceed. Yet in the closure phase, where the local payrolls and revenue are coming from reclamation work, these officials have been known to change direction dramatically and insist that ever more complex and costly reclamation requirements be imposed.

Colorado recognises two types of changes to existing permits: the amendment and the technical revision. The former implies a substantial change in operations, such as adding new land to the permit area, and is accompanied by most of the same information, notice, and public participation procedures used for issuing new permits. The latter is generally handled at a staff level, and is designed to keep the permit and the mining operation in conformance with each other in cases where the proposed change does not fundamentally affect the resources being impacted, the hazards of the operation, or the costs of reclamation.

17.3.7 DURATION OF THE PERMIT

Permits are usually issued for the "life of the mine," and are not subject to expiration or unilateral revision by government agencies barring unusual circumstances so long as the operator is in compliance with the plan. Colorado's law now, in the wake of Summitville, allows retroactive changes to permit conditions, but only after a special proceeding of which all affected parties must have notice, and only when it is determined that failure to do so will pose unreasonable environmental risks.

17.4 PERFORMANCE WARRANTY AND GUARANTEE

One of the most difficult groups of issues surrounding mine closure is the issue of performance warranties and guarantees. These issues are difficult for several reasons.

First, it is hard to run an effective programme without them. Prices of mineral commodities are more volatile than prices in general; mine closure is usually more a function of commodity prices than of physical exhaustion of resources. This means that precisely when the mine is closing, there is often serious financial stress on the enterprise. Further, it means that mine closures tend to come in waves, something with which the U.S. has, like all mining countries, ample experience.

Colorado has experienced, just to name a few, the silver boom of the 1880s, followed by the Silver Panic of 1893; the closure of the gold mines in response to World War I; the boom in oil shale of the early 1920s, followed by the bust of the mid-1920s; the enormous uranium boom of the early 1950s, followed by the crash occasioned by the end of government purchases, followed by recovery in the 1970s, followed by another, seemingly terminal crash in the early 1980s; another oil shale boom in response to the Arab oil embargo and the Iranian revolution and resulting oil shortages, followed by a complete cessation of the industry in the early 1980s; and various booms and busts in other mineral commodities, notably molybdenum, of which the state has long been a leading producer. Each of these events has left its mark on the landscape; Colorado's experience, and the experience in the West as a whole, is that mine closures have come suddenly and in waves.

If closure obligations are to be fulfilled, there also has to be some viable means of assurance. The association of large numbers of mine closures with sharp downturns in mineral prices also means that they are likely to occur at a time when general economic conditions in the mining region are not good, and if the area is one heavily dependent on the mineral economy, it is likely to be a time when public resources are in scarce supply.

Second, maintaining adequate financial guarantees is expensive for industry. Even where the company is allowed to self-insure, an option usually only available to the large, publicly traded company, it requires maintenance of a certain margin of uncommitted assets, limiting the company's options. Freezing cash in a certificate of deposit is expensive, but so are the various forms of insurance company or bank guarantees. These institutions generally insist on collateral. An excellent guide to the economic consequences to companies of various bonding alternatives can be found in Hayes.[8]

Third, the programmes are difficult and cumbersome to administer. Accurate calculation of bond amounts, as with any engineering cost estimate, is fraught with difficulty; the consequences of being wrong can be serious for the enterprise or the state. Performing all the legal steps necessary to perfect and maintain the government entity's legal rights in the guarantee is a difficult and time-consuming task, often beyond the capabilities of agencies typically staffed with engineers, geologists, or others without the necessary training in finance.

17.4.1 WHAT DOES THE ENTERPRISE GUARANTEE?

One common misconception about bonding for mine closure is that the bond insures against any possible accident, problem, emergency or undesirable development — but the financial assurance cannot address all these possibilities. Calculating accurately the cost of completing the closure plan is hard enough; trying to define possible accidents and their consequences in financial terms is an order of magnitude harder. Further, it is hard to understand why the mining industry should be singled out in this respect. The mining industry is hardly the only sector that can have expensive accidents. Until the chemical plants, dams, and electrical stations of the world insure the public against credible accidents (which might be a good idea), there seems to be no compelling need to require mines to do so. And since the "worst case" accident will in many instances as a practical matter have an enormous price tag, bonding for that eventuality — even if somehow desirable — is enormously expensive. Nor should the reclamation bond be confused with another instrument, discussed in Section 17.6: the bond for long-term performance of post-closure monitoring, maintenance, or environmental control on the site. In conditions where there are no practicable technical means to achieve acceptable post-mining conditions (such as some acid drainage conditions), there may be a need for very long-term activity on the site (such as perpetual operation of a water treatment plant). Some jurisdictions are starting to explore various forms of guarantee for performance of these obligations. This is very different from the bond here considered, which is simply a guarantee that the reclamation or closure plan will be implemented.

17.4.2 WHAT ARE THE ACCEPTABLE FORMS OF GUARANTEE?

17.4.2.1 Under current law

Typically, states will accept one of a variety of specific financial instruments to guarantee the closure obligation. Usually, a combination of different forms of guarantee is also acceptable. The types of guarantee instruments most often accepted include surety bonds, certificates of deposit, trust funds, irrevocable letters of credit, insurance policies, deeds of trust or mortgages, security agreements encumbering real or personal property, and cash deposits with the state treasurer. Some states allow companies who are able to meet specific tests of financial soundness to provide certificates of self-insurance, or a self-guarantee.

The issues surrounding the type, nature, and conditions of guarantees are critical, though not always well understood. First, whatever the law says about the guarantee, when major problems arise with a mining company or mine site, what ensues is going to be a negotiation. The company is likely to want to scale back its reclamation commitments, or postpone them, or propose alternatives that may be less desirable, all in the interest of saving money.

The form that negotiation takes depends substantially on the adequacy of the guarantee and whether it is a "real" guarantee — something that can effectively be turned into money in a reasonable period of time in an amount adequate for the government agency to implement the closure plan itself if the firm cannot or will not. If the guarantee is "real," the agency has the opportunity to consider any ideas that may be proposed, knowing that it is not going to face the unhappy situation of a failed mining company and an unreclaimed mine site, without the funds to get the job done. If the guarantee is not "real" the consequence is predictable: a long agonising process of compromise, indecision, retreat, and abandonment of standards, as occurred in the Summitville situation, where much of the state's hesitation to act was a function of its knowledge that the financial assurance it held came nowhere close to what the closure costs would be.

An abandoned, unreclaimed mine site can deteriorate quickly. Particularly where there are harsh climate conditions, or conditions which may impose seasonal limits on construction activities, it is important that a guarantee be convertible to cash efficiently, and within a reasonable amount of time. One of Colorado's post-Summitville reforms was an amendment allowing the agency to reject any proposed form of guarantee that is not convertible to cash within 180 days. If an unplanned closure is the result of a long deterioration of the financial condition of the enterprise, by the time of abandonment the mine site is likely to have a long list of deferred maintenance items, neglected drainage systems, plugged culverts, leaking pipes, safety hazards, and the like. Prompt action is required if the cost of solving the problems is not to escalate out of control. The need for a state emergency response fund to deal with issues of this nature was another of the important lessons of Summitville.

Another point that has been learned the hard way is that the value of the guarantee should not be dependent upon the economic success or viability of the mining operation. If the mining operation can be operated profitably, it probably will be, and there will probably not be a closure. If the mining operation is unprofitable in the hands of the operator, the chances that a state regulatory agency can move in and make money from it are virtually nonexistent. Yet many mining companies often want, for example, to offer a mortgage or other encumbrance of the mine itself as a guarantee — and not at the value it would have as pasture land or in some alternative use, but at the value it has as a mineral property, usually using some optimistic estimate of future mineral prices as a base. This poses obvious problems as well as problems that are not so obvious. Under the federal Superfund legislation, or the Clean Water Act, as explained later in this chapter, it is possible that owners of the site may take on liabilities far in excess of the value of the land, a situation which more than once has deterred state agencies from realising on their guarantees by foreclosing mortgages or otherwise taking possession of land offered as a guarantee. It of course limits the interest of potential purchasers as well. Even property not directly related to the mine site could have a value dependent on the success of the mining operation. The Mid Continent mine in western Colorado gave a deed of trust on a rock dust plant as collateral for its closure obligation. Mid Continent went bankrupt. It turned out that the rock dust plant, which had a substantial value as a going concern, had to be sold for next to nothing, because the one customer for its product was — the Mid Continent Mine. It developed its own environmental problems as well — a real concern for agencies which do not want to inherit sites that have to be cleaned up before they can be sold.

Second, certain forms of guarantee require substantial due diligence and maintenance by the closure agency if they are to maintain value. In the U.S., unpaid real estate taxes become a lien senior to mortgages. If taxes go unpaid for several years, the value of the agency's mortgage is much reduced. If real estate is accepted as a guarantee, there needs to be a way to determine its realistic value. Letters of credit expire. Someone needs to watch the expiration date to make sure a replacement letter is posted before the old one expires. Where some form of self-guarantee is accepted, someone has to maintain vigilance over the financial condition of the enterprise, to make sure that it continues to meet the financial requirements for the self-guarantee. The fact that an enterprise is a large one does not immunise it against failure, or sudden adverse financial events.

These financial issues are often very difficult for mine closure agencies because they often lack the business and financial expertise to understand all the issues and pitfalls. Colorado has had numerous examples of things that can go wrong with bonds: deeds of trust which were never properly recorded before the operator sold the land they were intended to encumber; banks failing and being unable to honor letters of credit; poorly done or inflated appraisals of both real estate and personal property; and environmental contamination reducing the value of the pledged property. The list is a long one. These problems have been reduced over the last several years, but the lesson is that the agency needs someone with commercial experience who knows how to deal with these types of potential pitfalls.

17.4.2.2   The concept of risk pools

There has been a great deal of discussion in recent years about pooling of risk in a way to allow smaller and medium mining enterprises to post financial warranties without incurring prohibitive costs. This concept may have significant merits and should be thoroughly explored. To date, it has been discussed mainly by regulatory departments, engineers, lawyers, and geologists.[47] The right financial and actuarial experts have yet to be brought adequately into this discussion — and there needs to be a certain sense of realism about the concept.

Consider the following issues:

- There is no doubt that the credit standing of the smaller players in the industry, medium national mining companies, and artisanal gold panners alike would be enhanced by having the mining giants of the world stand behind their obligations. This is, however, a role these companies have not rushed to embrace. In part this may be because many of the systems which do require bonding allow self-bonding for the bigger companies. But there are other factors, too. "… [T]he pools insure a high-risk group of operators. Further, operators paying into a bond pool may be tempted to consider their bond pool payments to be an adequate substitute for reclamation itself. It may be because of these difficulties that states with bond pool arrangements have been rather slow in implementing their programmes."[8]
- Without the big international companies, the risk pool at best becomes a large mass of companies — some good, some not so good. In fact, more than half of the companies probably fall into the not-so-good category, because the well-run companies seem disproportionately able to make their own individual bond arrangements, just as they are disproportionately able to make profits, or find minerals.

The result is thus a group of smaller companies — some good, many not — all exposed to the same basic risk: commodity prices. As the companies get smaller, the transaction costs of analysing the risks present at each specific site and monitoring the ongoing performance of the insured — both important functions in any insurance-like arrangement — become more significant, another deterrent to such programmes. While on a worldwide basis this risk can be diversified because of the numerous commodities that are produced and the numerous markets in which they are sold, the mining industry in any one state, country, or province is likely to be less diversified. The risk pool is not worth having if it is not strong enough to survive most foreseeable events, including sharp mineral price downturns. If it is not that strong, maintaining the risk pool is just a way of pretending there is security when there is none, and setting things up for a real crisis.

- If the risk pool is backed by commercial insurance arrangements, it still has to be solvent or the commercial insurance will be as prohibitive as the bond was in the first place. Most insurance companies are not willing to write millions of dollars of insurance against the risk of changes in mineral prices.

These realities have limited state experimentation with this concept. The most substantial experience with such programmes appears to be Nevada's state-run bond pool. The bond pool was established by statute in 1989, and is administered by the state Division of Minerals, within the Department of Business and Industry.[46] The programme is designed to serve operators of small and medium size mines who encounter difficulties obtaining bonding from private bonding companies or for whom the financial burden of obtaining a bond from a private company would be prohibitive. To qualify for the bond pool, the operator must have been rejected three times by bonding agencies at current market rate premiums and must have been required to post more than 15% collateral. Additionally, the bond pool administrators conduct an analysis of each applicant and make on-site inspections before enrolling an operator in the programme. The bond pool is not intended to compete with private bonding companies — it is intended only to provide bonding for operators who are unable to obtain bonding from private companies. The maximum coverage offered by the bond pool is $1 million per operator for mining operations or for both mining and exploration operations, and $250,000 per operator for one or more exploration projects. The operator pays the bond pool a 15% deposit and an annual premium equal to 5% of the bond coverage. Interest earned by funds in the bond pool is credited to the pool. If an operator covered by the bond pool is required to forfeit a bond, the bond pool pays the agency collecting the bond amount directly, then the state can sue the operator to recover the amount paid by the bond pool.

According to the Nevada programme's director, since its inception the programme has enrolled eight mining companies for a total of nine projects. About the same number of companies have applied for the programme, but did not enroll for various reasons. One company enrolled in the programme has since become large and financially strong enough to exit the programme and use a corporate guarantee for warranty purposes. The bond pool programme has experienced no forfeitures of bonds. A few other states have initiated bond pool programmes similar to Nevada's, or have considered doing so.

17.4.3   How Can the Level of Guarantee be Changed?

The amount of the guarantee should be adjustable at any time, so long as (1) it continues to be based on sound engineering cost estimates; (2) there is opportunity for public participation; and (3) companies whose guarantees are adjusted upward have a reasonable time to post any additional guarantee needed. Obviously, one major concern for mining investors is that they do not, in mid-project, suddenly want to face enormous increases in the bond costs, and protection against administrative arbitrariness is critical if industry is to support the programme.

These principles, which have been recognised in the Colorado law since the 1970s, allow for such approaches as phased bonding, in which the bond amount changes as the project proceeds. This encourages concurrent reclamation, in which parts of the mine that have reached their ultimate post-mining configuration can be closed, and the corresponding portion of the guarantee released, while mining continues in other parts of the project area. This extremely beneficial practice should be encouraged, and allowing adjustment of the guarantee encourages the promptest possible reclamation of portions of the mine site, and avoids imposing unnecessary costs on industry.

The experience in all American states is that reclamation gets better results and costs less if done as soon as possible. If nothing else, it is a general requirement that soils be salvaged and stockpiled, and it appears that the useful biological activity in soils deteriorates over time. Plus, concurrent reclamation can reduce dust control costs, and reduce the risks inherent, for example, in seeding the whole site at once at the end of the project, only to find that the seed mix or soil amendments were not appropriate, or that it was the driest year in the last 20. Concurrent reclamation therefore has multiple benefits: it allows for controlled, systematic experimentation with various reclamation techniques to identify and refine the most effective and economical ones; it reduces the environmental impacts during the operational phase; and it reduces the overhanging risk of an enormous "all at once" reclamation programme at the end of the operational phase, to the benefit of the enterprise, the state, and the environment. This is reflected in savings on bond costs: if the bond is posted as a single amount at the outset, as was the case with much past practice, it is calculated based on the point in the operation at which reclamation costs would be highest if the operator defaulted. This point only exists once during the life of the mine, yet fixes the bonding cost for the whole operational phase.

In general, operators are not only free to seek, but encouraged to seek, partial release of the bond whenever they can demonstrate partial achievement of the closure plan. If the reclamation liability can be reduced as the project moves forward, the result is less money being posted for less time (see Morrey, Chapter 13). The caveat — and there always is one in such cases — is that the state must spend more time in inspecting and evaluating such sites, in order to make certain that the lower bond amount is really justified by physical reclamation results on the ground, rather than paper reclamation plans. There have been a number of unhappy surprises in this area.

In a similar sense, many bonds are calculated for a particular phase of the operation. Rather than bond for anticipated expansion at the outset, the company may bond only for the initial phase and agree that it will not start the next phase until an additional bond is posted.

Another aspect of the bonding problem is that costs change, and inflation exists. Mine projects may have a long life, and a guarantee that is perfectly adequate at the outset will, if there is no way to adjust for inflation, certainly be inadequate in time. Colorado has had trouble on this score. In the early days of the programme, the agency lacked experience in cost calculation and many bonds were set at inaccurate figures, most often too low. Now that the agency has developed much more precise techniques for bond calculations, there are still many older bonds which never were adequate to begin with, and which are even less adequate after years of inflation.

The agency has made enormous efforts to catch up with this backlog, while also observing some other principles: (a) the bond should be calculated again any time there is an inspection indicating a deviation from the mine closure plan; (b) the bond is reevaluated whenever there is a transfer of the property to a new owner; and (c) there must be a recalculation any time that there is a change to the permit that may affect costs of completing the approved closure plan. It may still be some time before the agency has completed the task of updating all bond amounts; there is a fair amount of adverse fallout from the effort, as smaller marginal operations learn that they must now increase their bond amounts in some cases by several hundred percent. Further, though it is clearly necessary to base bond amounts on sound engineering cost estimation, this has one disadvantage compared to wild guessing or the "negotiated" bond amount: it takes time to do it right.

The lesson from this experience is that such a situation should not be allowed to develop. Much easier would be a plan where the bond amount, once established, is adjusted each year automatically by some appropriate measure of inflation, perhaps the Construction Cost Index.

17.4.4   WHEN AND HOW DOES THE GUARANTEE TERMINATE?

It is important to mining enterprises to know that when the prescribed closure steps are completed, their financial guarantee can be terminated promptly, and that a clear dispute resolution mechanism exists to resolve any disagreement on this subject. Several states have systems aimed at achieving such certainty and promptness in guarantee termination.

In Colorado, a company may report that it has completed its obligations under the reclamation plan. The state has a limited time to do whatever inspections or investigations it believes necessary and to raise any resulting objections, after which the company's bond is released automatically if the state fails to raise specific objections.[47] Similar mechanisms exist elsewhere. For example, in Arizona, an owner or operator may apply for a release from all or part of the financial assurance mechanism. The state must then release all or part of the financial assurance except for any amount necessary for reclamation within 60 days of receiving such a request.[48] Similarly, in Nevada, the state must respond to an owner or operator's request to release the surety partially or fully within 30 days after receiving a request for release.[49]

One concern of mining firms has been that it is still impossible to assure prompt return of the bond, because many states allow the landowner, or affected citizens, to object to or appeal a bond release decision, and the bond is not released until the appeal is resolved. In practice, however, appeals of bond release decisions have been rare, at least in any of the states with which we are familiar, and they tend to be resolved quickly: in Colorado, generally within one or two months.

17.5 REPORTING, MONITORING, INSPECTION, AND ENFORCEMENT

There are needs for ongoing monitoring and inspection at the mine during the operational phase which are beyond the scope of this paper. As they impact mine closure, the principal functions of reporting, monitoring, inspection, and enforcement are two. First, since the closure plan is closely linked to the methods of operations, and the financial guarantee is dependent upon the plan, departures from the approved operating plan can have drastic impacts both on the feasibility of the closure plan and the cost of its implementation. If a mine is approved for four hectares of surface disturbance, and bonded for reclamation of that amount, there needs to be an "early warning system" to bring to the attention of the agency the fact, e.g., that the mine has now grown to 20 hectares and changed its processing system without amending its plan. Second, mining impacts are difficult to predict; if important impacts, not foreseen earlier, develop during operations, or if mitigation measures are not proving effective, it is important to be sure this is recognised early.

In systems that require financial guarantees, the cost of maintaining the guarantee and the advantages of lowering or terminating it are strong incentives to begin the closure programme and to pursue it diligently. In systems without a guarantee, the incentive is completely reversed, and it is very important to have some outer limit, after which mines that have shown no sign of operating must begin the closure process. This is one reason that states which have not had financial guarantee requirements have tended to adopt them. It is another important function of the inspection and reporting programme.

An additional reason for the system is of course to detect people who should be in the system but who are not.

17.5.1 INSPECTIONS

Mines are in some ways more difficult and expensive to inspect than are many other types of facilities. Largely, this is a function of location. Mines are not, like many industrial facilities, conveniently clustered around major urban centers (the exception to this statement is probably construction material mining: the transportation costs of sand and gravel, for example, give certain advantages to location near urban centers). On the contrary, mines are often in remote, difficult to access terrain. Obviously, mines are inspected for a variety of reasons by a variety of agencies. The focus here is on inspections related to the ultimate objective of mine closure.

Mine inspections undertaken to ensure compliance with mine closure-related requirements are not a quick check of one or two parameters, but can be quite involved and include many aspects of the operation. The goals could be stated perhaps as (1) identifying any areas of noncompliance with the operating plan that will affect closure; (2) identifying any unanticipated environmental problems that are developing on-site; and (3) identifying any factors that could affect the adequacy of the financial warranty. If these problems are identified quickly, while the level of deviation is small, they can often be dealt with informally before any environmental problems become acute; before conditions occur which make the closure plan infeasible; before the amount of the existing financial warranty is greatly exceeded by additional unforeseen closure costs; and before it becomes necessary to invoke legal sanctions.

Every mine inspection programme in the country has its stories of the mine which, when inspected, turns out to be at a different location than specified in its plan; or is considerably larger than it is supposed to be; or which has unexpectedly encountered great quantities of water; or which has made radical changes in its production or processing technology. "Daylighting" a tunnel and suddenly becoming an open pit operation, deciding one morning to try some cyanide to see how it works, and other deviations from the approved plan are hardly unheard of, particularly at the small end of the industry.

Frequent inspection is one solution to the problem. The federal laws covering coal mining and state laws passed to comply with its requirements prescribe minimum frequencies for inspections. State inspectors may be required to visit every coal mine subject to their authority a minimum of once a month. On a certain percentage of these inspections, without advance notice, federal inspectors will appear and accompany state inspectors to ensure the state inspections are adequately conducted.

There is nothing nearly so frequent or so rigorous in state inspections of non-coal mines. Some states do not even seem to keep reliable statistics on frequency of inspections. In Colorado, which we believe is above average among state programmes in the number of inspections conducted, inspections are targeted on certain types of facilities that are believed to present more than average risks. For example, Independence Mining Company's Cresson Project, a recently developed gold cyanide project, had a state inspector on-site virtually full time during the critical phase of liner construction. However, many types of facilities are, on average, inspected less than once a year.

17.5.2 MONITORING AND REPORTING

Of necessity, then, great reliance must be placed on monitoring and reporting by the firm itself. The specific things that must be reported and the frequency of reporting are to a large extent site-specific and dealt with in individual permits. Where information is not provided as required, administrative penalties can be levied in most states. Some of the more stringent penalties levied in recent years in Colorado were against Battle Mountain Gold's San Luis Project for failing properly to report cyanide levels in ponds which exceeded its permit limits. Company management responded forcefully and vigorously to this situation, installing new systems of internal controls to ensure improved future reporting. The most severe sanctions have generally been applied where there has been knowing concealment of information or provision of false information in order to mislead regulators. Among these are penal provisions, which may include incarceration for intentional or willful violations.

17.5.3 PROCEDURES IN CASE OF VIOLATIONS

Many state programme staffs report to boards of appointed citizens with powers to hear disputes and make determinations, including imposition of penalties. Where there is serious noncompliance with the statute, regulations, or permit conditions, the Board that oversees the state mine closure agency staff or the staff itself typically has authority to issue an order requiring compliance, and to file an action in court to enforce the compliance order. Additionally, the Board or mine closure agency may take a variety of further actions, usually after a hearing at which the firm and any affected members of the public may present their positions. Options include suspension or revocation of permits or reclamation plan approval, orders requiring specific corrective actions, forfeiture of the financial warranty, and imposition of significant financial penalties on the operator. In severe cases, the state attorney general's office may institute a lawsuit or seek criminal penalties against the operator.

17.6   CURRENT ISSUES

The mine closure plan, while not a contract in the legal sense, is an important understanding between the government and the company, that if the company fulfills the steps outlined in the closure plan — and no major unexpected problems crop up — that is all the company will have to do. The state has the advantages of implementation of the plan; the mining enterprise has the advantage of knowing its target and being able to assign cost estimates to it with some confidence. This has been the traditional concept of mine closure planning.

There are challenges emerging to this basic concept, from two principal sources. One is the increasing recognition that in some cases implementation of the closure plan will not result in a "walk-away" closure: a maintenance-free, self-sustaining site which will comply with environmental norms in the long run without further intervention. In short, a post-closure phase may be needed which may consist, as noted in the introduction, of monitoring, passive controls, or even of active measures. Some of these, at certain sites, may need to be employed for a very long time at a very great cost.

The second challenge is a set of legal developments outside the framework of mine closure laws — principally from the solid waste laws, RCRA and CERCLA, and under the Clean Water Act.

17.6.1   POST-CLOSURE OBLIGATIONS

While any model oversimplifies, it might be appropriate to conceive of the U.S. Clean Air Act and Clean Water Act in the early 1970s as representative of a first phase of environmental management. These laws heavily emphasised identification of pollutants, and reducing or eliminating them through emission limits or ambient air or water quality standards. It was understood that some enterprises might not be able to meet standards and might have to close, but there was a general concept underlying the structure of the law that *if the operations which caused pollution stopped, we would at least be free of the pollution, because it would stop, too.*

There was a second wave of environmental legislation a few years later based on a more sophisticated appreciation that environmental impacts sometimes can continue even when operations cease. Important mine reclamation legislation began to be enacted in the mid-1970s, such as Colorado's Mined Land Reclamation Act. These laws seem in retrospect to have been based on the concept that even if post-operational impacts might exist, they could always, with proper planning, be reduced to an acceptable level. There was little recognition that any attention was needed to mine closure issues beyond the point at which reclamation was deemed "completed." The underlying idea was that *with proper planning, acceptable environmental conditions could be achieved at an affordable cost.*

Colorado's Mined Land Reclamation Act contemplates that all sites will be reclaimed and bonds returned to operators within five years after operations cease.[50] Other state reclamation laws seem to be based on this same general concept, in that they simply fail to make provision for situations in which all the king's horses and all the king's men (even with most of the king's money) simply cannot reduce the impacts to levels considered acceptable.[51]

Similar in underlying conception is the federal "Superfund" legislation, which imposes liability — often without fault — on broad classes of entities and individuals, with the goal of cleaning up existing contamination, even very old existing contamination. A key difference between state mine reclamation laws and the Superfund statute is that the Superfund legislation applies liability *retroactively*, while the mine closure legislation generally does not. One of the concerns often expressed about the Superfund legislation is that it requires cleanup of contaminated sites essentially to background level[52] and essentially regardless of cost; thus, it is based on the very similar concept that *existing environmental impacts can always be remediated at an affordable cost.*

These concepts are adequate and appropriate in the great majority of cases: in most cases, with proper planning, impacts can be reduced to an acceptable level and a "walk-away" closure achieved.

To the extent that there has been contamination which needs to be remediated before closure is deemed complete, it usually can be remediated at an acceptable cost.

In the discussion that follows, therefore, it is important to keep in mind that we are focusing on problems, which while difficult, apply only in a minority — perhaps a small minority — of cases. Perhaps Colorado's experience with mines that were abandoned in the past without planning can be some guide. If we accept for purposes of order-of-magnitude estimates the figure of 23,000 abandoned mines in the state, it appears that acid drainage, clearly the most serious post-closure impact, occurs in perhaps 2000 to 3000 sites, or about 10%. Of course, many of these other mines do have problems, such as safety hazards or blowing tailings, which should be remediated. The point is that in most cases this remediation can be accomplished with known technology at a reasonable cost, and it is quite possible to imagine that with a little energy this could be done. Further, it appears that the great bulk of this problem occurs at no more than 200 to 300 sites, or roughly 1% of the total. Yet this problem is a severe one which affects an enormous part of the state, leaving rivers and streams virtually devoid of life, increasing water treatment costs, and limiting tourist development, with very serious economic consequences.

In other words, conventional closure planning should be able to result in a "walk-away" closure at the great majority of sites. In this sense it has been and is a very significant success story.

The critical problem is to deal with that limited number of sites which present very serious adverse environmental effects that cannot be remediated to a maintenance-free state with any known technology, at least not at a cost anyone can be found to pay. The focus is how to respond to situations where there are no technical solutions which will reduce pollution to acceptable levels, or where the cost of such cleanup is absolutely prohibitive, and how to determine at the outset which sites those are. One answer, for new projects, is that such situations are to be avoided unless there are very compelling reasons to the contrary. In other words, if there is no way to return land affected by mining "to a use beneficial to the people of the state," or to "establish a self-sustaining ecosystem" on lands affected by mining, or to prevent undue risks to water quality, or whatever the individual state's formula may be, then the permit cannot be granted and mining may not proceed. Pennsylvania has apparently taken this position in the case of coal mining.

An example can illustrate. The San Juan Mineral Belt is historically one of Colorado's most important and productive. It was the centre of the state's enormous silver boom of the 1880s.[53] Yet it, like some other mining districts, is rich in acid-generating sulfides, and an area of heavy precipitation to boot. There is a significant existing water quality problem in the rivers in the area, to which historic mines make a significant contribution.[54] This may not affect every potential mine in the region, but let us suppose that a mining project is presented which has a high potential for creating a long-term acid drainage problem to which the only technological solution capable of meeting water quality standards is ongoing operation of an active treatment plant, for at least many decades, if not centuries, and then a need for ongoing maintenance of passive treatment systems after that.

Who is going to pay to operate that plant indefinitely until that hoped-for time when water quality improves to the point where it meets standards without treatment? Who is going to pay to rebuild or renovate the plant every 20 or 30 years? The enterprise has this burden, not under reclamation laws, but under water quality laws. As has often been pointed out, though, a company that over the years acquires more and more such obligations becomes less attractive to investors and less competitive with companies without such obligations. Also, perpetual operation of water treatment plants has not been regarded as part of the "core business" of mining companies.

It is tempting to say that this obligation should be guaranteed, because no one really believes it is sensible to rely on private corporations to continue to exist and fulfill obligations — particularly such unattractive ones — for centuries. We are here talking about something quite distinct from the traditional reclamation guarantee, which is simply a guarantee of compliance with the closure plan. But there currently is no mechanism in the law to accept a bond beyond the period in which reclamation is complete, because the law simply was not conceived based on an understanding that such situations may exist. And the law simply does not recognise situations in which the reclamation plan cannot be taken to "completion" within a relatively short period of time after operations end.

The cost of such a perpetual bond would be enormous. In the Superfund situations where something like this exists, the cost has been tens of millions of dollars in immediate cash to create a fund to continue to finance the obligation.

But the alternative of saying that important parts of one of the state's most important mining districts, which has produced phenomenal amounts of silver, gold, and other minerals, cannot be mined is startling to say the least — and hard for many people, in and out of the mining industry, to accept. The issue is hardly unique to the San Juan Mineral Belt.

Colorado has appointed a Mine Water Quality Task Force to study these issues and, if possible, recommend legislative, regulatory, or policy changes to deal with them more effectively. They are very similar in some respects to the problems emerging under the national Clean Water Act, dealt with below. Colorado's process is a continuation of dialogue that began with the adoption of major legislative reforms after the Summitville events, and continued through the development of the state's new implementing regulations. Bluntly, even after consensus had been reached on a host of other issues, the acid rock drainage issue remained intractable. While ultimate solutions have not emerged from this process, the fact that the issues are being discussed straightforwardly by industry, government, environmental representatives, and technical experts is in itself a promising development.

17.6.2 LEGAL DEVELOPMENTS UNDER NATIONAL LAW

There has been much written about the regulatory problems of the U.S. mining industry, and the extent to which regulatory issues are responsible for the problems of the industry. One thing is clear: mine closure legislation is not the major impediment to mining development in the U.S. The key regulatory problems that are troubling the industry are *not the result of mine closure laws*, but of other separate legislation, principally water pollution legislation and the Superfund law.

The U.S. has been a pioneer in dealing with many of these issues: everyone involved in the process has realised that the best solutions would be arrived at only by a process of successive approximation; part of the genius of U.S. environmental law has been its willingness to pitch in and try innovative concepts where some others have seemed to wait indefinitely for a train that never comes. The challenge now in U.S. environmental law is to show that same willingness to try new concepts and, building on past experience, to revise the statutory schemes it has created to make them better.

17.6.2.1 Clean Water Act issues

There has been much debate and uncertainty in recent years over whether and how the national Clean Water Act applies to inactive or closed mines. The Act originally was written very generally; it does not differentiate between active and inactive phases of a project or activity, perhaps because, as discussed above, there was an underlying assumption in 1973 that cessation of the economic activity giving rise to the pollution would generally stop the pollution, and an assumption that any residual contamination can be remedied. As a result, the Act leaves much room for interpretation in how it applies to mining in general, how it applies to closed or inactive mines, and how it applies when water quality standards are very difficult or even impossible to achieve. It has become clear in recent years, however, that both the U.S. Congress and the USEPA intend to apply the Clean Water Act to mining operations, both active and inactive, very aggressively. This application of the Clean Water Act can have significant impacts on the mining industry.

17.6.2.1.1   The Dodson letter

The National Pollutant Discharge Elimination System (NPDES) permit requirement is the centerpiece of the U.S. Clean Water Act.[55] In general terms, the Act prohibits any person from discharging pollutants from a "point source" into the "waters of the U.S." without a valid NPDES discharge permit.[56]

One of the issues which affects mining the most is the question of how broadly the term "point source" should be defined in the case of mining operations. The Act defines the term as "any discernible, confined and discrete conveyance," and gives such examples as a pipe, ditch, channel, tunnel, conduit, well, discrete fissure, or container.[57] The courts have interpreted the term "point source" broadly. In an important early case, the Tenth Circuit reversed a lower court's finding that mining is inherently a non-point source.[58] The Court went on to hold that an unplanned overflow from a reserve sump used in gold leaching operations qualifies as a "point source."

In 1993, the State of Montana's Water Quality Bureau requested the USEPA to clarify its position on several issues regarding NPDES permitting of hard rock mines. The USEPA's response, commonly referred to as the Dodson Letter, contains a very broad interpretation of the term "point source," and takes an aggressive stance toward regulation of inactive or abandoned mines.[59]

The USEPA made two major assertions in the Dodson letter. First, the agency asserted that not only mine adits, but also such "less obvious" sources as "seeps and other ground water discharges hydrologically connected to surface water from mines" qualify as "point sources" and require NPDES discharge permits. The USEPA reasoned that "it is more the mine or the facility itself that is subject to NPDES regulations," so that any seeps coming from "identifiable sources of pollution" at a mine would need to be regulated by discharge permits. This was an aggressive position for the agency to take, given the statutory language, but is consistent with the agency's policy of broadly construing the definition.[60]

Second, the USEPA took the position that NPDES permit requirements apply to inactive and abandoned mines as well as active operations. Specifically, the agency stated that mine adits at active, inactive, or abandoned mines fall under the NPDES permit programme, and that its current practices for permit issuance "incorporate historic mine drainage into NPDES permits for active mines if the active mine influences the pollution discharge from the historic area," and that if the firm operating an active mine "owns or has control" over an adjacent historic mining area, the firm must also apply for an NPDES permit for the discharge from the inactive area.

At least some of the consequences of this approach are that permits must be maintained for present mines once they enter the post-closure phase *so long as they are discharging pollutants.* Since the permit conditions must be adequate to prevent water quality limits from being exceeded, the implication is that expensive active treatment systems, adequate to meet all water quality standards, must be kept in operation *until the day when untreated effluent meets all applicable standards without treatment,* or at least meets them with passive measures alone. Suffice it to say that at some sites this could be a very long time, and could imply a very great deal of money: it is hard to call a site at which a million dollars a year are being spent on water treatment "closed."

17.6.2.1.2   Storm water regulations

For many years, the USEPA refrained from actively regulating stormwater runoff under the Clean Water Act. However, in November 1990, the USEPA published new regulations establishing a separate type of permit for pollutants contained in storm water discharges.[61] Storm water is defined as "storm water runoff, snow melt runoff, and surface runoff and drainage."[62] Under the storm water programme, permits must be secured for all stormwater discharges from a point source which are associated with an "industrial activity."[63] The regulations expressly state that either an active or an inactive mine is an "industrial activity."[64]

This affects the mining industry two ways. First, types of water discharges which were previously unregulated now are subject to permit requirements, again probably until the untreated effluent from the mine site meets standards without active treatment, which implies a long-term obligation. Second, so long as there is a stormwater programme, industry would like to include in the stormwater discharge permit various types of runoff, because in practice the standards in stormwater permits have been less stringent than those in traditional NPDES permits. Although, technically, stormwater permits are to include numerical limitations based on effluent standards and water quality standards, just as traditional NPDES permits do, the first round of stormwater permits issued typically have not included such numerical limitations. Instead, the permits have included "best management practices" that the particular permit holder or industry group is required to implement. These best management practices in turn derive from a stormwater management plan that the permit applicant develops. The perception is that stormwater permit requirements are easier to meet than the numerical-based requirements of traditional NPDES permits.

In the Dodson Letter, the USEPA took pains to emphasise limitations on the types of discharges that it believes are subject to the storm water programme. According to the letter, storm water discharges are limited to those "directly associated with a precipitation or snow melt event." As a result, "any dry weather flow from mine adits, seeps, french drains and culverts are mine drainage or wastewater," and require permits under the traditional NPDES programme, which results in more stringent limitations. Under this EPA interpretation, most areas at an active mine require traditional NPDES permits, because the contributions of contaminants from storm water flows were considered in setting the effluent limitations which the NPDES permit standards are intended to help achieve, Thus, according to EPA, only certain ancillary areas of active mines, as well as inactive areas, can be handled with storm water permits. All other areas require the "traditional" NPDES permit.

The issue, as it is with the state reclamation laws, is that if there is going to be a requirement for perpetual or near-perpetual active treatment of effluents, there is no mechanism established in the law for ensuring that this is done. It is simply not reasonable to think that today's mining companies will be, or will want to be, around operating a growing list of water treatment plants at non-revenue generating sites; and there is little concept of what kinds of institutions, other than mining companies, can or should do this work. As agencies and companies develop *ad hoc* solutions to the problem, there are no solid legal criteria for deciding what kinds of arrangements are adequate.

17.6.2.1.3   Solid waste laws

Added to the concerns about requirements to treat water from historic abandoned sites, and the possible need to acknowledge a perpetual treatment obligation under the Clean Water Act, are the concerns (dealt with in part above) about liability under CERCLA, the statute which creates the Superfund. This is not the place for a treatise on the complex features of this statute, but a few comments are in order.

The Superfund concept was developed in response to a realisation that the country was home to a number of "orphan" landfills and chemical dumps, which were in some cases presenting a threat to public health or the environment, or were contamination sources whose effects were unknown. The consequences of applying this law to old mine sites were perhaps not considered carefully.

CERCLA[65] creates a system to identify potentially hazardous sites and list them on a National Priorities List. Sites are subject first to investigation. If the investigation confirms the existence of a serious problem, the process continues culminating in the design and implementation of a remedial programme.

The law also creates a federally controlled fund, the Superfund, which was formed by a tax on the chemical industry.

When a site starts into this process, the USEPA notifies all "potentially responsible parties" (PRPs) it is able to locate. These parties, who are the ones on whom the government will seek to impose liability if the fund itself is forced to pay for the investigation and cleanup, have the option to conduct the investigation and cleanup themselves, subject to government oversight. They often do so, because most PRPs have felt they can control costs — which they will ultimately have to pay — most effectively if they are managing the process themselves.

The entire process: site identification, investigation, listing, evaluation of possible cleanup options, and conduct of the cleanup are subject to numerous opportunities for public participation both for the PRPs and any other interested members of the public. This results in a somewhat slowed process, which has been one of the major criticisms of the act.

At bottom, the law has two distinct features. The first is taxing a specific industry that is thought not to have internalised its environmental costs in the past, in order to create a fund. The second is the attempt to identify and pursue specific parties at each site to reimburse the fund for its costs at that site. The former seems much less problematic and controversial than the latter.

The potentially responsible parties fall into four categories: the generators of any waste disposed of at the site, the operators of the site, transporters who brought any of the waste to the site, and owners of the site (at the time it was contaminated and after). Each of these four categories is subject to very broad definition; suffice it to say that almost any action on the site (drilling boreholes, moving waste material around, digging holes) which even arguably has added to the contamination, helped it spread, or made it less manageable is adequate for membership in the group.

Liability is retroactive, without limit. It is "joint and several," which means that any one or group of the PRPs may be liable for the whole cost of investigation or remediation if the others cannot be found or lack assets. And liability is without fault.

Obviously, no one likes being a PRP at a Superfund site. Just as obviously, *some* sort of mechanism to deal with cleanup at abandoned sites is an absolute necessity. There are some specific concerns which crop up in the mining industry, however, which are very significant at some sites.

One is that the history of the mining industry is replete with examples of new discoveries and rediscoveries in historic mining districts. The heavily mineralised zones generally were identified long ago and have historic mine workings, which often have environmental problems. Hypothetically, a chemical company that wanted to start a new plant today using all the most modern available concepts of pollution prevention and control could find many potential sites, which might be good sites or bad, but which at least would not involve becoming a PRP at a Superfund site. An equivalent mining company, wanting to run the cleanest possible operation, would find that many if not most of the places it could operate have previous environmental problems from historic mining which present some risk of Superfund liability. This is to say nothing of the fact that it is bad policy to give powerful incentives to develop new sites, often in healthy ecosystems, in preference to working in areas where impacts already have occurred.

And the risk is in some sense unknowable and unmanageable. Summitville had historic environmental problems before any of the recent mining activity on the site. However, these were probably not great enough to attract the attention of the Superfund authorities or merit a National Priorities List designation. Several mining companies, interested in the property, apparently conducted exploration programmes on the site, some of which reportedly involved drilling, sampling, or otherwise altering site conditions. Presumably, they were in compliance with all applicable laws. They decided not to develop the property, sold out, and moved on. Later, another company, Galactic Resources, and the related Summitville Consolidated Mining Company decided to develop the site. As they say, the rest is history.

When a few years later the then-bankrupt firm abandoned the site, it did have more than enough environmental problems to qualify for the National Priorities List. Galactic and Summitville appear to have no ability to pay for the cleanup. Their principal promoter is outside the U.S., apparently successful to date in avoiding any legal liability for the events, and having failed to date to reimburse the government for its costs.

The various mining companies that did nothing but explore the property, possibly causing some incidental effects in the process, are now either PRPs with joint and several liability for the whole cleanup bill, or at a minimum are under threat of being assigned PRP status in the future. How do you manage risk *when something someone else might do later* makes you jointly and severally liable for a $150 million cleanup?

The final note is that there is no effective programme in the U.S. for dealing with the environmental legacy of past mining. Superfund may address a few dozen of the worst sites, but it is too blunt a tool for the thousands of smaller cleanup actions required. In the absence of any serious funding for this purpose, one of the principal possibilities is remining: recovery of residual mineral values from tailings or other old mine workings, and relocating and reclaiming these wastes in the process. Conducting this kind of operation in a historic mining district with existing environmental problems is, under the current parameters of Superfund, something only for the bold. And Superfund is not the only worry.

17.6.3 FEAR OF RETROACTIVE LIABILITY AND ITS EFFECT ON THE INDUSTRY

An example of the problems that can arise in connection with activities undertaken on a historic mining site is the Penn Mine case from California, decided under the Clean Water Act. In the 1960s, a local utility district acquired part of the abandoned Penn Mine property, intending to build a reservoir on the property. The utility district and the state water quality control board then built a facility to contain toxic runoff from the old copper and zinc mine on the site. In normal conditions, the facility contained the seepage. However, in very rainy periods, it overflowed. Because the runoff containment facility did not eliminate *all* discharges from the site, an environmental group later sued the two agencies for discharging pollutants without an NPDES permit. The suit was successful, and the agencies, which had never been involved in operating the mine, were required to undertake further and very costly cleanup at the site.[66] The Penn Mine case is viewed by many, particularly in the Western states, as a barrier to state and local government and other third parties who may want to remediate abandoned mine sites, because it presents the potential of heavy liability, possibly perpetual, at any site where remediation fails to eliminate *all* discharge of contaminants, even if it represents an improvement over historic conditions.

The decision to develop a mining property requires analysis of a multitude of financial variables, some very hard to know, and taking significant risks. Any step which helps define or quantify the uncertain variables makes the decision easier. However, if the answer is that the obligation is now quantified but very costly, the decision may be easier to make; it is more likely to be no.

The critical regulatory issues in mine closure today in the U.S. do not in our view come from the body of mine closure legislation itself, which is working reasonably well. The real concerns seem to arise under the federal Clean Water Act and CERCLA, the Superfund legislation, and amount to:

1. Lack of clarity as to the long-term responsibility of the mining enterprise in maintaining the quality of water discharged from the mine site. This is critical because it appears that at some potential mining properties, no technology short of ongoing active treatment is going to meet water quality requirements. Many U.S. water quality parameters for such discharges are difficult and expensive to meet. If they have to be met in perpetuity, enormous costs are added. Whether or not the mining industry *should* have to pay these costs, the fact is that it has not in the past done so.
2. Lack of clarity as to how committed the U.S. is to maintaining the principle that, at least in most cases, requirements will not be imposed retroactively. Retroactive requirements may not be so great a problem in, let us say, light manufacturing, where the costs of the capital infrastructure are recouped in 6 to 8 years, and where technology may be subject to rapid change, obsolescence, and replacement. It is quite a different thing in the mining industry. The examples of litigation to require mining companies to pay to clean up problems created in the 1890s cause people in the industry to wonder how far this retroactive application of liability will be carried in the future.
3. Concern that if there is going to be a requirement for perpetual maintenance of water quality, mineral production in the U.S. can remain competitive with production in countries where industry does not face these costs.

The question may be less what is desirable than what is achievable: the world is rapidly globalising; the mining industry is at the forefront of this trend. The real issue is whether these problems can be solved by unilateral national decisions, or whether this is an example of the limits or the ability of one country, even a rich and powerful one, to force global environmental policy changes unilaterally.

There are various ideas as to how to resolve these issues, which are largely beyond the scope of this chapter: however, the so-called "good Samaritan" proposal, deserves some mention, if only as an illustration of the tensions involved. As part of the pending Clean Water Act reauthorisation process, an effort has been undertaken, chiefly by the western states, to persuade Congress to amend the Act to include a "good Samaritan" provision. Such a provision would protect agencies or other third parties, which acquire abandoned or inactive mine sites for purposes of remediation by insulating them from liabilities of certain types after they complete their cleanup efforts.

The party doing cleanup could be exempt from future responsibility for post-cleanup discharge under the Clean Water Act so long as the cleanup met certain requirements. What those requirements are may be problematic: certainly, they should include the idea that someone who currently has liability to maintain water quality or protect other environmental values under the Clean Water Act or CERCLA, the Superfund law, cannot avoid existing obligations under those laws by some sort of half effort. If the goal is, and perhaps it should be, to identify situations in which the cleanup standards of Superfund or Clean Water Act discharge standards need to be relaxed because society cannot rationally justify meeting them, this should be done explicitly.

The other major provision of the "good Samaritan" proposal would exempt the entity cleaning up the site from Superfund liability, by clarifying that any resulting effluent would be a "federally permitted release" for which CERCLA does not impose additional liability.

Several issues regarding "good Samaritan" proposals remain unresolved. These include whether the exemption from liability should extend to any citizen-initiated lawsuits, and whether the exemption should extend to remaining of abandoned sites for profit. For a detailed discussion of the problems of remining historic or abandoned mines under the Clean Water Act, see Barry.[67]

The real issue remains: at many potential mineral deposits, including many in the historic mining regions of the U.S., there is now no technology that can deliver a "walk-away" site at the end of the closure period, i.e., capable of meeting existing clean water laws without ongoing expenditure into the indefinite future. The alternatives seem to be (a) to accept the idea of such perpetual liabilities and fund them by some enormous "up front" contribution deemed adequate to provide all costs necessary for perpetual treatment; (b) to accept the ongoing pollution as a necessary consequence of mining, and to develop some set of reduced, less stringent standards for these situations; or (c) to say "no" to mining at such sites. If the level of consumption by the U.S. consumer is not reduced, this will result simply in exporting this same set of extremely difficult decisions to other parts of the globe.

Finally, proposed reform of the 1872 Mining Law, which governs mining on public lands in the U.S., poses a whole new set of closure issues. Reform of this law is needed, but the consequences of the proposed legislative changes in the context of closure have been among the least examined and least satisfactory aspects of the proposed reform. The principal reform bill would, for example, further Balkanise the already fragmented administration of these programmes by creating yet another closure authority and yet another set of standards — these applicable only to certain types of mines, and only located on federal lands. The spectre is of competing agencies, none big enough to have the technical skill, the variety of experience, or the political clout to maintain an effective programme. One of the principal lessons of Summitville appears to have been that an agency needs a certain "critical mass" and a certain diversity of expertise in engineering, soil science, botany, law, geochemistry, and other specialties to be effective. "Carving out" a limited number of mines on federal lands and subjecting them to a different system does not seem the best solution.

17.7 CURRENT EXAMPLES OF BEST PRACTICE IN MINE CLOSURE

At its best, the U.S. system of mine closure planning has resulted in some significant accomplishments.

While industry commentators, watchdog groups, and regulators disagree about how to evaluate overall success of mine closure planning efforts in the U.S., there are mines that can claim some consensus of approval for their reclamation programmes. Indeed, these companies' reclamation practices are considered by some to represent the state-of-the-art in the mining industry. Examples of mines that have won praise from numerous quarters, including the industry watchdog group Mineral Policy Center in Washington, D.C., are Coeur d'Alene Mine Corporation's Thunder Mountain Mine in Idaho, and Homestake Mining Company's McLaughlin Mine in California.[68]

Coeur d'Alene has received mining industry awards as well as state and federal government awards for its operations and reclamation practices in Idaho and elsewhere.[69] Its Thunder Mountain gold mine in central Idaho, which was closed in 1990 and reclaimed by 1992, has been praised for its detailed reclamation plan and successful reclamation methods. Because the mine was located adjacent to a pristine wilderness area, there had been increased concerns about impacts on the environment — especially water quality — from mine operations and closure. Coeur d'Alene employed techniques in water management controls including silt fences, infiltration basins, and dispersion terraces[70] that some have credited with preventing adverse impacts to water quality.[69] Coeur d'Alene also did extensive work in cyanide management, including treating used ore with chlorine to oxidise the remaining cyanide, and placing a 20-foot deep clay liner in the mine pit before returning the ore to the pit; recontouring the heap leach pad areas, covering them with soil, and replanting the surface; and wrapping sediment from the solution ponds in plastic blankets, and then backfilling the ponds.[71]

At its "state-of-the-art"[68] McLaughlin Mine in northern California, Homestake has committed to achieving specific post-mining uses at various parts of the mine area, including wildlife habitat, recreation, and grazing.[72] Homestake has conducted reclamation throughout the life of the mine. The company additionally plans to achieve post-closure reclamation through use of such tools as covering a tailing impoundment with topsoil and revegetating the area; extensive backfilling, fencing and installing vegetative screening around the mining pit; and using native grasses and woody plants to increase wildlife habitat value and lessen visual impact.[73] Moreover, Homestake is reclaiming the 10,000-acre project site for use as an environmental studies research station, which it plans to develop in cooperation with local universities.[72] Homestake has stated that it will donate information and materials to the centre, including environmental baseline data, monitoring data, mapping materials, and geologic core.

17.8 CONCLUSION

At this point, all the major U.S. mining states have regulatory programmes requiring mine closure planning at most "hard rock" metal mines. While the state programmes differ in details, and while some of the state programmes are very new, the more established programmes such as Colorado's have accumulated substantial experience and have a major influence on mine closure practice in industry. This article has focused on the situations in which the regulatory programmes have encountered difficulties, which have been quite real: at the same time, the existence of these programmes has brought a new level of awareness to not only the large international operators, but the hundreds or thousands of mines operated by the medium and smaller national segment of the industry as well.

The greatest impact of regulatory mine closure programmes may well have been less to push the newest ideas or lead the dissemination of best practice in the industry, although there are examples where that has happened. The greatest impact may have been to encourage the spread of what is at minimum acceptable practice throughout the industry: some "floor" of performance.

These programmes have had the most impact in dealing with new operations that have come on line since the programmes were created. They have also had positive effects on closure planning at existing operations, though these effects have been limited by the realities of previous mine development, which was not planned with closure as an objective in many cases. The U.S. has yet to develop an effective strategy for dealing with the significant environmental impacts at historic mines that were abandoned without closure work. There has been little progress in dealing with any but the most serious of these sites.

Experience indicates that there are three important aspects to a successful regulatory programme.

First is the closure plan itself. The plan must be based on sound baseline information and developed in conjunction with an operating plan before ground disturbance begins. If there are particular hazards on the site (e.g., acids, cyanide, or other toxic reagents) it is advisable to have an emergency plan for the site as well. These plans should be reviewed by an agency that has a variety of expertise in different disciplines, and should be subject to a public participation process.

Second, there is the matter of the financial guarantee. While many states have tried to find alternatives to the system of financial guarantee to avoid imposing these significant costs on industry, all of the major hard rock mining states have eventually found that some form of guarantee system makes the closure planning programme more effective. If there is to be a system of guarantees, the only sound basis for calculating their amount is good engineering estimates; any attempt to cushion the costs on industry by agreeing to bond amounts which are less than these actual costs loses much of the benefit of the system and opens the door to bureaucratic arbitrariness. Obviously, bonds need to be kept adjusted for changes in the operation or simply for inflation.

Third is the information element: baseline data, ongoing monitoring, reporting requirements, inspections, and enforcement. If the mine closure plan is to become an ongoing iterative process, which benefits from new information as it is developed, attention needs to be devoted to the less than glamorous questions of data acquisition, communication, and processing.

Finally, U.S. programmes are facing the challenge of a fundamental conflict. A major attraction to mine closure programmes from the industry viewpoint is the lure of being able to terminate the firm's responsibility for the site, absent unforeseen conditions, when a specific set of well-defined steps have been completed. Mine closure programmes have been developed with this concept as a fundamental underpinning: industry agrees to undertake certain steps, so long as they are required of everyone in the industry, and in return expects those steps to be the limit of its responsibility.

Major mining enterprises almost anywhere in the modern world will experience closure costs. The disadvantage in some places is that those costs are unknown, and thus very hard to anticipate or plan for. The advantage that U.S. regulatory programmes have sought to offer is that those closure costs can be fixed, with a significant degree of certainty.

Now, it is becoming clear for at least some sites that this kind of "walk-away" goal is not going to be met, at least not on a time scale of immediate relevance to those now living. To try to maintain the goal of a prompt, relatively clean closure at these sites may do violence to the nation's efforts to protect the quality of its waters. To "solve" this problem by requiring perpetual operation of active water treatment plants as the price of mining, but without requiring these costs to be paid "up front" is to invite a serious day of future reckoning. Requiring that those costs be paid or guaranteed in advance would be very expensive indeed, and may not, for reasons of cost competition, be something that can be implemented unilaterally by the U.S. in disregard of the regulatory posture of other major mining nations.

Implementing such a system would at least remove some of the current uncertainty over the retroactive application of new environmental laws: that could be brokered into a legislative compromise if nothing else. But if — and the subject deserves careful study — it imposes a cost level which excessively disadvantages U.S. producers, the consequence will be to export the production, and often the exact same problems, to nations with less technical and regulatory ability to deal with them. But profligate use of mineral resources coupled with unwillingness to experience directly the environmental externalities of their production would hardly be the first example where the consumers in the developed world have appeared hypocritical.

REFERENCES

1. Dobb, E., Pennies from Hell: in Montana, the bill for America's copper comes due, *Harper's Magazine*, October 1996, 39.
2. Department of Minerals and Energy, Aide-Memoire for the Preparation of Environmental Management Programme Reports for Prospecting and Mining, Department of Minerals and Energy, Cape Town, South Africa, 1992.
3. Health, Safety and Reclamation Code for Mines in British Columbia, R.S.B.C., 1996, 293.
4. Department of Minerals and Energy, Guidelines for Mining Project Approval in Western Australia, Department of Minerals and Energy, Western Australia, Revised edition, July 1993.
5. Title VII, De Cierre de las Actividades Mineras, Reglamento Ambiental para Actividades Mineras, D.S. No. 24782, Gaceta Oficial de Bolivia, 31 July 1997, published August 1, 1997.
6. Surface Mining Control and Reclamation Act, 30 U.S.C., §1201 *et seq*, 1990.
7. 30 C.F.R., Part 715 (General Performance Standards), 1996.
8. Hayes, C., Reclamation accounting and practical considerations in funding reclamation obligations, *Rocky Mountain Mineral Law Institutes*, 40, 9, 1994.
9. Barringer, S. G. and Johnson, K. A., *Survey of State Mining Regulatory Programmes in the Western United States*, Mineral Resources Alliance, April, 1994.
10. Williams, G., Closing a hardrock mine — can you ever walk away?, *Rocky Mountain Mineral Law Institutes*, 39, 1993.
11. Danielson, L. and Watson, J., Environmental mediation, *Nat. Resources Law.*, 15, 687, 1983.
12. *Colo. Rev. Stat.*, §34-32-102 (1), Supp. 1996.
13. *Alaska Stat.* §§27.19.020, Michie 1996.
14. *Mont. Code Ann.* §§82-4-302, 1995.
15. *Nev. Rev. Stat.* §§519A.100, Michie 1995.
16. *S.D. Codified Laws* §45-6B-20.1, West, WESTLAW through 1996 Session.
17. *N.M. Stat. Ann.* §69-36-7, Michie Supp. 1996.
18. *Cal. Code Regs.*, tit. 23, §2574a, 1996.
19. 30 U.S.C. §1265(b) (1996).
20. *Idaho Code* §39-118A, tit. 47 ch.15, 1993.
21. Id. tit. 47, Sec. 1317, 1993.
22. *Cal. Code Regs.* tit.23, §§2572 (h) (1) (A), (B), and (C) respectively, 1990.
23. *Mont. Code Ann.* §§82-4-401, 1995.
24. *Colo. Rev. Stat.* §§34-32.5-101 *et seq.*, Supp. 1996.
25. *Colo. Rev. Stat.* §§34-33-101 *et seq.*, Supp.1996.
26. *Colo. Rev. Stat.* §§34-32-101 *et seq.*, Supp. 1996.
27. *Colo. Rev. Stat.* §34-32-103 (3.5), Supp. 1996.
28. *Mont. Code Ann.* §§82-4-401, *et seq.*, 1995.
29. McElfish, et al. *Hard Rock Mining: State Approaches to Environmental Protection*, Environmental Law Institute, 1996.
30. Hocker, P., The environmental side of 1872 Mining Law reform, *Mining and Environment Research Network Research Bulletin and Newsletter No. 6*, June 1994, 25.
31. *Ariz. Rev. Stat. Ann.* §27-921, West Supp. 1994-1995.
32. *Ariz. Rev Stat. Ann.* §27-921, West Supp. 1994-95.
33. *Nev. Rev. Stat.* §519A.080, .110, .120, .200, Michie 1995.
34. *Cal. Pub. Res. Code* §2714, 2770, 2776, West, WESTLAW through 1995-1996 Session.
35. *Mont. Admin. R.* §26.4.101B, 1994
36. *Wyo. Stat. Ann.* §35-11-401(j), Michie Supp. 1996.
37. Minerals Exploration Coalition, *Permitting Directory for Hard Rock Mineral Exploration, Minerals Exploration*, Coalition, Lakewood, Colorado, 1992.
38. *Nev. Admin. Code* §445.2428, 1991.
39. *New Mexico Mining Act Rules*, Rule 1.1, July 12, 1994.
40. *Colo. Rev. Stat.* §34-32-109, Supp. 1996.
41. Bell, A. V., Phinney, K. D., Beggs, C., and Guissani, M., Audits for Bolivian investors, *Mining Environmental Management*, 5, 19, 1997.

42. McArdle, J., Colorado Division of Minerals and Geology, personal communication, 1997.
43. Patton, C. A., McGaffey, K. M., Ehrenzeller, J. L., Moran, R. E., and Eaton, W. S., Superfund listing of mining sites, in *Tailings & Mine Waste '95*, A. A. Balkema, Rotterdam, 1995, 55.
44. Robertson, J. D. and Ferguson, K. D., Predicting acid rock drainage, *Mining Environmental Management*, 3, 4, 1995.
45. *Colo. Rev. Stat.* §34-32-109(5), Supp. 1996.
46. Hough, H., A beginner's guide to reclamation bonding, *The Miner's News,* December 1994–January 1995, 3B, 1995.
47. *Colo. Rev. Stat.* §34-32-117, Supp. 1996.
48. *Ariz. Rev. Stat. Ann.* §27-996, West Supp. 1994-1995.
49. *Nev. Admin. Code* §519A.385, 1991.
50. *Colo. Rev. Stat.* §34-32-116(7)(q)(iv), Supp. 1996.
51. Cal. Code Regs., tit. 23, §2574(a), 1990.
52. 42 U.S.C. §9621, 1996.
53. Smith, P. D., *Mountains of Silver, The Story Of Colorado's Red Mountain Mining District*, 1994.
54. Danielson, L. J., Alms, L., and McNamara, A., The Summitville story, a Superfund site is born, *Environmental Law Reporter*, 24, 10388, 1994.
55. 33 U.S.C. §1342, 1996.
56. Id. §§1311(a), 1362(12), (7), 40 C.F.R. sec 122.1(b)(1), 1996.
57. Id. §1362(14), 1996.
58. United States v. Earth Sciences, Inc., 599 F.2d 368, 10th Cir., 1979.
59. Letter from Dodson, M. H., Director, EPA Water Management Division, to Fraser, D., Chief, Montana Water Quality Bureau, December 22, 1993.
60. 55 Fed. Reg. 47990, 47997/1, Nov. 16, 1990.
61. Raisch, J., The Clean Water Act: new regulatory requirements for the mining and energy industries, *Rocky Mountain Mineral Law Inst.*, 37, 1991.
62. 40 C.F.R. §122.26(b) (13), 1996.
63. 33 U.S.C. §1342(p)(1), (p)(2)(B), 1996.
64. 40 C.F.R. §122.26(b)(14)(iii), 1996.
65. 42 U.S.C. Sec. 9605, 1996.
66. Id. at 309-310.
67. Barry, A., Mining and water quality under the Clean Water Act, *Colo. Law.*, 25, 93, 1996.
68. Schwartz, T., Mining Report Card: Homestake Mining Company, Mineral Policy Center, 1992.
69. Richins, R. T., Responsible mine management rewarded: Coeur d'Alene leads, *Clementine*, winter, 1991.
70. Coeur d'Alene Mines, Fact Sheet on Thunder Mountain Gold Mine.
71. Kenworthy, T., Mining Law: benign and benighted, *The Washington Post*, October 23, 1993.
72. Krauss, R. E., The McLaughlin Mine: a 21st century model, *The Chiles Award Papers*, High Desert Museum, December 1993.
73. Homestake Mining Company, McLaughlin Mine: General Information Summary.

18 Catavi: Lessons for Developing and Closing Mines in Non-Industrialized Countries

Fernando Loayza

CONTENTS

ABSTRACT

The exploitation of an ore deposit is a process that eventually comes to a close. The development of a large world-class mining operation in a low-income developing country gives rise to a temporal economic expansion which can be followed by a severe economic recession in the area surrounding the mine, once the ore is depleted. The slump in economic activities triggers off a technological retrogression, with increasing negative impacts to both worker's health and safety and the environment. Therefore, in low-income developing countries, an effective environmental plan to close large world-class mines should be based on both social and environmental impact assessments.

Via a case study of Catavi, Bolivia, this chapter highlights the environmental problems that arise from the need of mine-dependent populations to find themselves alternative sources of income. In this case, the closure of the mine led to intense unsound mining activity in the Catavi area. Only if the socioeconomic impacts of a large mining operation are considered and properly managed from the outset of mining operations can social and environmental impacts such as those observed in Catavi be avoided.

18.1 THE MINERAL DEPOSIT, ITS DISCOVERY, AND DEVELOPMENT

18.1.1 THE MINERAL DEPOSIT

Intijaljata ("where the sun sets") was the name given by the quechuas to the place formed by two mountains that was to become the famous Llallagua mining district centuries later. This desert and rough land in the middle of South America, well over 12,000 feet above sea level with no roads or other facilities, was first explored for silver in the 16th century.

However, silver was not found in Llallagua. Late in the 19th century, the place began to be worked for tin, its main asset. Prospects, diggings, and small adits excavated by the Spaniards almost three centuries before were centres of activity again.

18.1.2 THE MINING COMPANY, THE MAN

Three persons were outstanding tin producers in the district at the turn of the century: Sainz, Minchin, and Patiño. At Patiño's request, the three men built a 74 km mountain road from Challapata, where the nearest railroad station was located, to Uncia, where Patiño had installed his modern beneficiation plant. By 1905, Patiño's Salvadora mine ranked among the most important mines in Bolivia. That year, Sainz sold his 16 concessions, totalling 1070 acres, to a Chilean syndicate. The purchase price of more than US $1.7 million at the time was to be paid partly in cash, partly in installments, and partly in the form of 80,000 shares of a new company to be formed. The Llallagua Tin Company was born in 1906.

A huge dam was constructed to store water from the Chayanta River, producing electric energy for the mine in the hydroelectric plants of Chaquiri and Lupilupi. Tin production in Llallagua was greater than in La Salvadora.

On December 1, 1906, Patiño opened his own bank in Oruro, The Mercantile Bank, with a capital equivalent to £ 1 million pounds sterling and "whose primordial object was to protect the mining industry."[1] Patiño expanded his property by buying up the small group of concessions of his nearest neighbors, the Bebin brothers. Later, he closed a deal with Juan B. Michin, buying the Unc´a Mining Company. By then, there were two companies working in the zone: the Llallagua Tin Company under Chilean hands and Patiño's La Salvadora Tin Company.

After a nine-year struggle, at a cost of US $5 million, the ninety-kilometer, one-meter-gauge Machacamarca-Uncía Railroad was concluded in 1921. One of Patiño's dreams was fulfilled and his mine was connected to the Pacific Ocean via the Oruro-Antofagasta Railroad.

In 1924, his masterpiece was completed. After years of patient but determined work and infinite precautions, Patiño had accumulated, through foreign agents, more than two thirds of the shares of the Llallagua Tin Company, the Chilean mining firm. Once he had complete control of the fabulous tin deposit, a new company was formed in July 1924 under the name of Patiño Mines Enterprises Consolidated Incorporated (PMECI), which became a model of technical and administrative proficiency in the industry. Important programmes and new technology were introduced to keep pace with the requirements of exploitation. Some of the innovations introduced were a complete underground exploration programme, a block caving method to exploit low grade zones, a sink and float process at the concentration stage, and water recycling.

In 1952, the government nationalised the company along with companies belonging to the so-called Tin Barons. The company was renamed Catavi Mining Company and it was run by the state until its closure in 1986.

18.1.3 TIN PRODUCTION AND THE BOLIVIAN ECONOMY

Figure 18.1 shows the importance of Llallagua in the Bolivian tin economy, in particular, during the first half the 20th century.

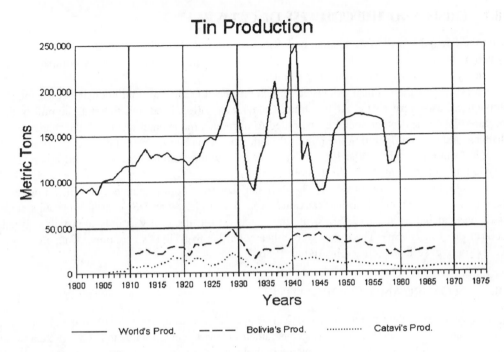

FIGURE 18.1   Tin production: the contribution of Bolivia and Catavi.

18.2   ENVIRONMENT: WATER RESOURCES AND MINERAL WASTE MANAGEMENT

At the time Catavi was developed, environmental concerns were non-existent. There were no agricultural activities or other industries in the vicinity to compete with for land, water, and other natural resources. However, water scarcity, tin contained in the tailings, the need to generate power, and basic health and safety norms were important factors that prompted a remarkable system of water management and tailings disposal. The water shed of the area was kept clear and unpolluted for the downstream Lupilupi power plant expressly built for the mining operations.

Mine water and slime tailings of mineral processing plants were properly impounded, forming three small completely impervious lakes, and the overflow was controlled to reuse water. Mineral waste properly classified in coarse, sand, and fines was piled up separately to ensure physical stability.

18.3   CATAVI'S IMPACT ON ECONOMIC AND SOCIAL CHANGE IN BOLIVIA

Early in the 20th Century, Catavi was the largest foreign exchange earner and contributor to state income from taxes and royalties. On several occasions, the Bolivian government negotiated advance tax payments with Patiño to support budgetary needs. During the war with Paraguay (1933), Catavi lent US $50 million to the National Treasury.

From the above, it is easy to figure out the political power and hold that such a huge mining company had in a poor developing country, as well as the resentment and ideas for social and economic change that it produced. Catavi was the cradle of a very strong leftist (Marxist) labour union that played an important and decisive role in the deep social and economic changes which Bolivia experienced during the 1940s and 1950s. Even today, the presidency of the Central Labor Union (COB), the association of all labour unions in the country, is reserved for a mine leader. This is in recognition of the role played by miners in effecting social change in the past.

18.4   CRISIS AND THE COLLAPSE OF CATAVI

After the Second World War, demand and prices for tin fell. The subsidy paid on tin by the U.S. Government ceased, and the U.S. started selling tin from its stockpiles, thereby competing with the supply coming from the producing countries.

Catavi had by then almost exhausted its rich ore. Feed grade was continuously decreasing and production costs were rising. At this time, Catavi was nationalised and the state administration increased labour and other social costs, failed to introduce the needed technical and management changes, and kept on operating the mine although grade ore reserves were falling continuously. As a result, from the middle of the 1950s, Catavi became an uneconomic operation, and was retained only because of its contribution to foreign earnings.

In 1985, two events occurred: (i) The Bolivian economic policy changed, and (ii) tin prices in the international market fell to half its value (from 5.30 to less than US $2.50/lb). Catavi's cutoff grade was calculated at 2.01% Sn, while the actual feed to the mill was only approximately one tenth of that (0.27% Sn). The losses were a heavy burden to Bolivia's economy.[2] Catavi could no longer survive, so it was closed in August 1986.

18.5   CLOSURE AND COOPERATIVES

By August 1986, Catavi reported insufficient accessible ore reserves to continue operations. The only valuable ore left underground was constrained to small veins, pillars, and some broken ore in old stopes not suitable for the caving method employed in mining. However, these reserves could be economically exploited by small-scale mining.

At the time of closure, over 30,000 people in the Catavi area were dependent — directly or indirectly — upon Catavi's operations for income. To avoid social unrest, the government authorised COMIBOL to lease the Siglo XX mine (Catavi's underground mine) to cooperatives of former workers to alleviate unemployment in the area.

18.5.1   THE COOPERATIVE SYSTEM

A cooperative is formed when people get together for business. This kind of partnership or corporation operates under legal rules and is common all over the country in various activities. Mining cooperatives are among the strongest.

Three cooperatives took over the Siglo XX mine, its mill, and other facilities. Since 1986, Siglo XX Cooperative, 20 de Octubre Cooperative, and Dolores Cooperative, with many workers ranging between 5000 to 6000 (the exact figure is unknown) have produced tin concentrates in the most precarious conditions. Due to a lack of engineering, planning, control, health and safety, the mine — once a model of organization — presents a high risk to workers who go underground every day and mine the ore unscientifically, resulting in a heavy degradation of the environment.

18.6   RESULTS AND PRESENT SITUATION

Some of the environmental impacts currently observed due to small-scale mining by cooperatives are:

1. Water recycling is no longer practiced, since the local streams are sufficient for the water needs of operations.
2. Digging of holes to search for rich ore in tailings dumps have disrupted the functioning of the aerial tramway which was used to transport sand wastes to the tailings piles.
3. Effluents from mineral processing activity are discharged directly to the Ventilla River. This water, once clean and clear with pH near 7, is now heavily polluted and a pH less

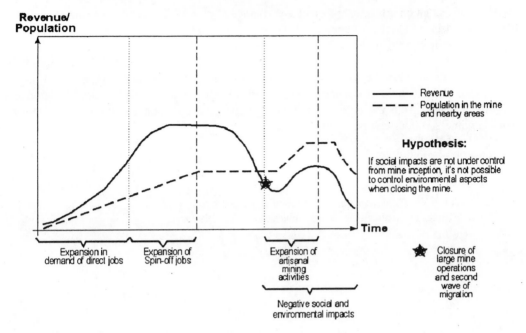

Revenue/Population

Revenue
Population in the mine and nearby areas

Hypothesis:

If social impacts are not under control from mine inception, it's not possible to control environmental aspects when closing the mine.

Time

Expansion in demand of direct jobs

Expansion of Spin-off jobs

Expansion of artisanal mining activities

Closure of large mine operations and second wave of migration

Negative social and environmental impacts

FIGURE 18.2 Changes in revenue and population during the mine life cycle.

than 3.0 was reported. Solids abandoned along the riverbanks have been eroded and deposited downstream.

4. In the underground workings, the cooperatives dig and blast the rich parts of pillars and small veins, picking out the richest parts and hauling them in knapsacks to their homes for final concentration. The low grade ore is left behind in the mine obstructing shafts, adits, chutes, and so on. Therefore, cooperatives are working in a very risky environment. In fact, several accidents, loss of lives, and missing persons are reported every day in more than 800 km of adits.

5. The feverish search for leftover ore is reflected in the moon-like landscape both on the surface and underground because of holes and piles all over the place.

18.7   DISCUSSION, LESSONS, AND CONCLUSIONS

The exploitation of an ore deposit is a process that inevitably ends. Because of this, the development of a large world-class mine operation in a low-income developing country gives rise to temporary economic expansion, followed by a severe economic recession of the area surrounding the mine once the ore is depleted. The slump in economic activities triggers a technological retrogression with increasing negative impacts to worker's health and safety, as well as the environment.

As illustrated in Figure 18.2, during the economic expansion stage, considerable population migrates from low-income rural areas to the mine, which continuously creates direct and spinoff jobs.

Members of the migrant population converted to miners, merchants, or suppliers of services over time lose their roots and links with their original communities, forget agricultural skills, and get used to a new way of life.

When economic recession sets in, the outward migration from the mining area to more promising non-mining regions and/or the retraining of the labour force becomes extremely difficult. Moreover, when the mining operation shuts down, it is almost impossible to keep operations under control since adventurers and jobless people may flood the area searching for opportunities.

Once this situation arises, no adequate environmental plan for closure can be carried out. This is because a lack of alternative sources of income drives jobless people to dig for the remaining high-grade ore left in the middle of low-grade waste by the closing company.

Therefore, in low-income developing countries an effective environmental plan to close large world-class mines should be based on both social impact assessment and environmental impact assessment. Only if the socioeconomic impacts of a large mining operation are considered and properly managed from the outset of mining activities can the great social and environmental damage witnessed in Catavi be avoided.

REFERENCES

1. Geddes, F. C., *Patiño The Tin King*, Robert Hale, London, 1972.
2. COMIBOL, La Realidad De Catavi, unpublished paper, 1986.

19 Lessons from Zimbabwe for Best Practice for Small- and Medium-Scale Mines

John Hollaway

CONTENTS

ABSTRACT

Mining has played a vital role in the economy of Zimbabwe from pre-colonial times, with the emphasis on gold. Although about 80% of Zimbabwe's production comes from major mining groups, the complex geology of the country — in particular the gold geology — coupled with an exceptionally liberal mining law, has led to a large number of small formal underground mining operations. In addition, artisanal gold winning in riverbeds and along the riverbanks has become widespread since independence. The mining regulations require only that on closure there is a degree of cleanup and protection of openings to the satisfaction of the landowner and the mines inspector before the site can be abandoned.

However, these have been recently strengthened with the promulgation of guidelines requiring Environmental Impact Assessments and Environmental Management Systems to be implemented on larger mines. Although the resources for regulation are inadequate, public pressure has caused the formal sector to develop a self-regulatory pattern. However, this approach is inadequate for the burgeoning informal sector and the two main lessons here are

- Government must acknowledge the conversion of artisanal mining into a formal, regulated sector as the only route that will enable control of the environmental effects to be achieved.
- At the same time, disproportionately greater regulatory resources are needed when small-scale and artisanal mining becomes a common rural activity, even though it is operating within existing mining legislation.

19.1 ANALYSIS OF THE PROBLEMS AND CHALLENGES

19.1.1 THE HISTORICAL BACKGROUND

Zimbabwe's mining industry is unusual for a developing country, being composed of about 350 formal mines, almost all of them small by world standards, most of them mining gold, and most of them underground. Artisanal mining, while on the increase for reasons that will be discussed, has never featured to the same extent that it does in countries with a similar geological endowment.

As a result, most of the environmental concerns in Zimbabwe are conventional ones, such as noise, dust from tailings dams, cyanide leakage, and shafts and other openings in the ground. It is this latter aspect which is the most contentious issue when closure is considered, for there have been roughly 4000 mines in the country in the past 100 years; consequently, there are numerous sites where dangers occur. The incidence of other concerns, such as acid mine drainage, unrehabilitated strip mining, and cyanide leakage are (relatively) rare and certainly amongst the larger companies, are matters for ongoing apprehension and control.

Before the colonial period, the country's potential for gold production was known from numerous sources. These included the reports of Portuguese missionaries who were active at the court of the Mhunumutapa in the 17th century, when a gold-based empire, centred on what is now the Zimbabwe ruins, was exporting to the Gulf region, to India, and to the Far East. By the time the immense riches of the Witwatersrand were found in 1886, European prospectors and miners had already penetrated into modern-day Zimbabwe and had confirmed the ancient stories of a country of plentiful reef gold (see for example, Baines).[1]

In 1890, a formal invasion was organised, financed by Cecil Rhodes. He obtained Queen Victoria's signature on the charter for his British South African Company, and emphasised the "Imperial Mission" aspect of the enterprise, but he also made no bones about the attractions of another gold-rich country. It seems probable that there was a subtext here; the realisation that the Witwatersrand was a paleoplacer — a vast, ancient alluvial basin into which gold had been washed — must have suggested to his American mining advisers that a "mother lode" along the lines of the Californian model might still exist. If so, the trickle of primary gold that had come from Zimbabwe up to that time could well be the forerunner of a wealth far greater than the unparalleled riches already found in South Africa.

19.1.2 THE EVOLUTION OF THE MINING LAW

Whatever the unspoken hopes for a bonanza bigger than the Witwatersrand, the mining law for this new Imperial acquisition had to be such as to ensure the rapid discovery and exploitation of the mineral resources. The South African law, at least in the Boer republics of the Orange Free State and the Transvaal, was clumsy — vesting mineral rights (and the rights to the air space) in the landowner.

As a result, a "finders keepers" law along the lines of that in California was adopted for the new country, allowing easy acquisition of mineral title, the obligation to maintain its validity by work or payment, and its easy sale, optioning, or tribute.[2] Surface rights were subordinate to those of prospectors and miners. However, the state (or in this case Rhodes' Chartered Company until 1923) held the ultimate title to all minerals, allowing them to be exploited for a royalty. An important environmental legacy of this facilitative mining law arises from the fact that it is almost as easy to abandon a mine as it is to peg one.

Rhodes' column of pioneers who terminated their trek at Fort Salisbury (now Harare) on the 12th of September, 1890 immediately disbanded and set out to find gold. However, although by 1895 something like 65,000 claims had been pegged, very little actual mine development had occurred. A major reason was that the Chartered Company claimed 50% of the equity of any mining company as a "free ride." This would have been acceptable had the wealth lived up to the standards of the presumptive mother lode, but what was actually found were thousands of shallow "ancient workings" on reef and shear zone structures of archaean and proterozoic age, with generally modest grades. As a result, these requirements severely inhibited mine development. Faced with this situation, the Chartered Company reduced its equity requirement in a series of steps, until it was eventually done away with altogether. Each reduction, it was noted, was followed by a spurt in the number of new mines registered — a lesson that was well taken — and resulted in Zimbabwe being one of the few African countries that does not even levy a royalty on mine output.

19.2 THE MINING INDUSTRY OF ZIMBABWE IN THE ENVIRONMENTAL CONTEXT

19.2.1 THE EVOLUTION OF THE INDUSTRY

By 1908, there were about 250 (gold) mines in operation, of which 175 could be considered "small workings," and output was over 600,000 ounces (19 tons) a year. Thereafter, the picture stabilised, with output peaking at 930,000 ounces (29 tons) in 1916, and then falling slowly over the next 60 years to 387,000 ounces (12 tons) in 1979. Since independence in 1980, gold output has climbed steadily until it is now over 770,000 ounces (24 tons).

These figures suggest a relatively stable gold mining sector. Indeed, the distribution of mine sizes has held remarkably constant over the past 90 years, with a Pareto-type distribution for gold production, 20% of the mines producing typically about 80% of the gold.

In all, mining contributed about 8% of the GDP in 1994, with a total output of about Z$4 billion (US$500 million). It employs about 55,000 persons in the formal sector, with perhaps 300,000 others engaged in seasonal alluvial gold winning ("artisanal" or informal mining). Small-scale miners working gold reefs legally are another significant and growing employer of unregistered labour, although the numbers are uncertain.

At present, about 80% of Zimbabwe's production comes from ten major mining groups, as follows: the Anglo American Corporation (nickel, chromite, and gold), Lonrho Zimbabwe Ltd. (gold), African Associated Mines (asbestos), Wankie Colliery Company (coal), Rio Tinto (Zimbabwe) Ltd. (gold), Ashanti Goldfields Ltd. (gold), the Government-owned Zimbabwe Mining Development Corporation (gold), Falcon Mines (gold), Delta Gold, and Casmyn of Canada.

These companies operate about 40 mines between them, including almost all the major ones. However, not only the emphasis on small mines makes Zimbabwe's environmental concerns in the mining sector uncommon; two other features are also significant. One is the diverse nature of the materials mined. While gold still predominates, in a typical year the list covers about 45 different metals and minerals, or about as many as South Africa. The greater part of the value of the output is as refined products or ferroalloys.

The second is that while the geology is diverse, it is not that bountiful.[3] Zimbabwe has probably the lowest-grade nickel, copper and gold mines in the world. Much of its chromite comes from seams only ten centimetres thick. The Great Dyke platinum grades are generally lower than those in South Africa's mines. Zimbabwe has enormous resources of coal, enough for many thousands of years, but most is of poor quality.

Future hopes are currently focused principally on diamonds. Following on the opening of the River Ranch mine near Beitbridge in 1992, virtually the entire country has been covered by Exclusive Prospecting Orders for this gemstone.

Although the excitement over diamonds has faded somewhat in the countries which provided most of the capital for the exploration boom (Australia and Canada), a more recent and tangible sign of foreign interest has been a reduction of gold exploration in favour of the purchase of gold mines and claims. In 1995, there were about six such investments totalling about Z$150 million (US$38 million).

19.2.2   THE ENVIRONMENTAL CONSEQUENCES

The facilitative mining law favoured by the Chartered Company has survived, so that currently any Zimbabwean individual or company can have a block of ten claims (totalling ten hectares) staked for them by the payment of the equivalent of about US$35 for a prospectors licence and about US$50 for the services of a "Licensed Prospector." The latter has undergone a verbal test of his knowledge of the mining law, and only he is allowed to undertake the actual pegging at the selected site.

Mainly as a result of this, at the present time in Zimbabwe there are about 500 mineral producers in the formal sector, over 90% of them gold operations (although many are tiny dump retreatment operations that do not qualify for the term "mine"). Environmentally, this sounds as though it is a recipe for disaster, particularly if, out of this very skewed pattern for the industry, one was asked to describe that mythical beast, a "typical small Zimbabwean gold mine." The answer would be something like this:

- It will be over 80 years old, having been closed and reopened at least twice, and having been worked under a different name(s) by various "smallworkers." They in turn were the successors to the prospector who located, or, more likely, was guided to the 'ancient workings' at the outcrop.
- The geological environment will be that of greenstone schist within a few kilometres of the granite contact.
- The gold will be in a host of quartz of hydrothermal origin, often as stringers only a few centimetres wide. Values will tend to be concentrated in the contact between the greenstone and the quartz, and will be associated with pyrite and other sulphides.
- The operation will be underground, with steep dipping reefs of a complicated structure. Mining will usually be by underhand stoping, but many mines will have evolved their own variations.
- A hoisting rate of about 50 tonnes per day will feed a mill with a capacity of about 30 tonnes per day. The difference will arise from waste sorting by the wives of employees, whose activities will bring the run-of-mine grade up from about 4 g t^{-1} to about 6 g t^{-1}.
- The mill will produce a gravity concentrate (for amalgamation) and the balance of the gold will undergo cyanidation.
- The mine will employ about 70 people in total, none of whom will have more than a technician-level of formal training.
- It would have been refinanced some years ago by a consortium of local businessmen and farmers who are still waiting to see a decent return on their money.

The "smallworker" gold mining sector as depicted here is something of a fixture in the rural economy. With a mining sector being mainly composed of such small underground operations, only the scattered tailings and waste piles normally make any serious impact on surface. These usually have relatively little sulphide mineralisation, and as most of these are often over 50 years old they are commonly well vegetated by natural colonisation. Groundwater pollution is usually also minimal, as the tight schist or granite country rock limits the dispersion of mine water. Arsenic (from arsenopyrite) is probably more significant than acid mine drainage in terms of ground water pollution potential.

The principal negative environmental effect has been to leave a legacy of innumerable pits and shafts, with the original protection (typically fencing or a wall of boulders, sometimes concrete slabs) missing due to theft, corrosion, or illegal entry.

However, a very important point is that for this type of geological environment, big mines develop out of small ones. All the major gold mines (and a number of the larger ones for other minerals) in Zimbabwe have arisen either directly or indirectly from such small operations, and these larger mines are significant contributors to the economy. There is an argument, therefore, that to inhibit the activities of the small mines sector by imposing environmental restraints appropriate to major potential polluters, the health of future mining is placed at risk.

Efforts at raising awareness on environmental problems have been undertaken with some success. For instance, the Natural Resources Board's Mineral Committee's competition for the best tailings dump rehabilitation has led to the remediation of many such dumps. However, for the many smaller mines, much still depends on the environmental awareness of the owner/manager — and of the landowner on which his mine is situated.

Aside from the numerous small formal gold mines in common with most of the rest of Africa, a ranking of environmental activism by the mining industry has emerged in Zimbabwe, generally as follows:

1. Major multinational mining companies.
2. Junior mining companies from outside Africa.
3. Local (formal) mining groups.
4. State-owned and operated mining companies.
5. Informal (artisanal) mining.

The pressures to preserve a good environmental profile under the scrutiny of green activists causes categories (1) and (2) to be devout adherents to international environmental norms. Environmental concerns have also become increasingly important to category (3), with the emphasis changing from being contingent on the personal interest of the relevant mine manager to becoming the formal responsibility of environmental sections or staff in the group's head office. For new or expanding mine operations, the voluntary use of Environmental Impact Assessments (EIAs) and Environmental Management Plans (EMPs) have become increasingly common, with closure arrangements an important part of the latter.

However, state-owned mines are very seldom allowed to close down; political concerns over the loss of prestige and disgruntled voters usually override financial reality. On the other hand, when financial pressures can no longer be disregarded, the poverty-stricken condition of such companies prevents serious expenditure on closure.

The last category above, that of artisanal miners, was not much represented in Zimbabwe until recently. However a number of major concerns in this area have developed over the past decade. The principal worry is alluvial gold winning, which is associated with stream bank erosion and the consequent siltation of rivers and dams further downstream. This is discussed in more detail below.

19.2.3 The Legislative Structure of Environmental Laws in Zimbabwe

19.2.3.1 An overview of the relevant legislation

For Zimbabwe's mining sector, probably the most important recent development was the introduction in 1994 of environmental guidelines by the Ministry of Mines, discussed below. The ultimate intention is to incorporate them in the mining regulations so that they have the force of law.

However, the Department of Natural Resources (DNR) in the Ministry of Environment and Tourism has identified 18 pieces of legislation with an environmental component, as listed in the table below. The coordination of this diverse body of rules and regulations is the responsibility of the technical division of the DNR, which is tasked with monitoring all environmental problems, including mining issues. However, their capacity to monitor and diagnose such problems has been limited, and is further hampered by a lack of data on the magnitude of the various problems.

In addition, the divided management of the existing regulations complicates the monitoring and control of environmental issues in mining. While water effluent standards exist (as regulations under The Water Act of 1976), their administration is controlled through the Ministry of Lands and Water Development by the Department of Water Resources and Development.

The Hazardous Substances and Articles Control Act of 1972 is administered by the Ministry of Health, who therefore also deals with the use and control of hazardous substances on mines. The Atmospheric Pollution Prevention Act (1971) is again the responsibility of the Ministry of Health and is concerned with the prevention and control of air pollution by gases, dust, fumes, and smoke. The Mining (Health and Sanitation) Regulations, 1977, are also administered by the same Ministry, and make provision for adequate hygiene in and about mines, thus providing a legislative framework for ordering gold rush villages — provided that the workers are mining a registered mining location, which, in the case of artisanal workers, they usually are not. The Ministry of Mines administers various other health and safety regulations for mining locations, which include a number of environmental features, such as cyanide effluent control.

Many mining-related plants that are not on a mining location, such as smelters and refineries, fall under the Factories Act regulations, which also have some environmental elements such as noise and dust, and which are administered by the Ministry of Public Services, Labour and Social Welfare.

The only item of legislation that is controlled by the Ministry of the Environment and Tourism through its Department of Natural Resources is the Natural Resources Act of 1941 and its Amendment (1975). The function of this act is "to make provision for the conservation and improvement of the natural resources of Zimbabwe" and it is administered by the Natural Resources Board.

In 1994, the Ministry of the Environment and Natural Resources adopted a policy requiring all proposals for major new activities in Zimbabwe to consult with the Ministry to determine if the activity will require the preparation of an EIA. The list of "prescribed activities" includes "mineral prospecting, mineral mining, ore processing, and concentrating and quarrying."

19.2.3.2 Current closure requirements

The specific requirements in terms of the law are that the holder of a registered block of claims must fill in or otherwise protect all dangerous workings such that the mine inspector and the landowner (or other applicable entity) is satisfied. The latter then gives his written consent to the issue of a "Quittance Certificate." This is often easier than it sounds, as the landowner inherits all the immovable property of a closed mine once quittance has been agreed. If the landowner is not satisfied, the mining commissioner for the area will arrange for it to be examined by a representative of the Government Mining Engineer's office in the presence of the both parties. The Mining Regulations actually specify in great detail the wiring, straining posts, etc. to be used when fencing off an opening, as well as the positioning, height, and width of dry stone walls often used as a substitute.

The Guidelines go further; there are 42 clauses, amongst them the following:

- All mines should be rehabilitated either progressively or at the end of mining. Each site should be left in a safe, stable, well-drained, and maintenance-free condition, blending in as far as possible with the surrounding landscapes.
- Mine operators should ensure that funds are available for progressive and final site (closure) rehabilitation.
- Unless otherwise approved by an Inspector of Mines, at mine closure all machinery, structures, and buildings should be removed from the site and concrete slabs broken up and buried. The site should be ripped, topsoiled (if available), fertilised, and revegetated. Alternatively, if approved, certain structures can remain for the benefit of the next land user.
- As far as practicable, topsoil should be stripped from all areas to be disturbed by mining/milling operations, and used immediately if possible or preserved for later rehabilitation.
- Areas disturbed by mining should be revegetated as far as practicable, using indigenous grass and tree species. However, on sites such as tailings and waste dumps where it is important to establish a vegetative cover as soon as possible on difficult growing mediums, the use of fast-growing exotic species is acceptable. Care should be taken to prevent the entry and spread of noxious plants.
- Unless otherwise approved, at the cessation of mining, or earlier if practicable, the walls of tailings and slime dams should be battered or terraced to a safe low angle and the surface topsoiled (if available) and revegetated. Measures should be taken to ensure that runoff will not cause erosion and destabilisation to the dam or dump, particularly after the site has been abandoned.
- Unless otherwise approved, at the cessation of mining, or earlier if practicable, waste rock dumps should be stabilised by reducing the slope angle and revegetation. Topsoil should be used, if practicable.
- All shafts not being used should be securely capped or otherwise made safe to prevent the entry of persons or stock.
- The final land-use of an open cast or quarry should be determined prior to the cessation of mining. For example, if the site is to be used as water storage then, at the end of the mine life, drainage could be directed into the pit. If the pit/quarry is to be used for any other purpose, then drainage should be diverted around the site.
- Unless otherwise approved, heap leach pads should be rehabilitated after leaching by detoxification, recontouring, topsoiling, and revegetation so that they will be in a stable maintenance-free condition. Alternatively, the heaps could be used to backfill a nearby pit.
- Mine rehabilitation of open cast operations should be carried out progressively to ensure that a minimum of ground is disturbed at any one time. A maximum of two hectares shall be unrehabilitated at any one time, unless otherwise approved.

These voluntary guidelines were drawn up in conjunction with the industry; as a result, they have wide acceptance, a further example of the tradition of self-regulation that has generally been followed in Zimbabwe.

19.3   RESPONSES AND CASE STUDIES

19.3.1   THE INDUSTRY RESPONSE TO ENVIRONMENTAL CONCERNS

A proliferation of small formal mines, operating under a regime of very limited liability, might be expected to pose an environmental threat of an order somewhere between categories (4) and (5) in the classification given in Section 19.2.2. In fact, in Zimbabwe they are generally fairly good in this respect, usually being no worse than their bigger local counterparts in the (very broad) categorisation above.

A major reason (although it would be difficult for the mining sector to acknowledge this) has been the presence of a relatively sophisticated agricultural sector which is considerably larger and has more political clout that the mining industry. Successive governments, to their credit, have not abrogated the basic principles of ready access to mineral rights (as has happened in much of the rest of Africa), but the history of the sector has been one of a steady tightening up of conditions under which they can be obtained, and of increasing public concern about the environmental consequences. The introduction of the system of Licensed Prospectors was the direct result of pressure from non-mining rural interests.

Other reasons for this above-average performance include:

1. Membership of a Chamber of Mines that endeavours to limit the vulnerability of the sector to valid criticism, and which includes an Environment Committee as well as Safety and Training Committees.
2. An active mine managers' association, which ensures that "best practice" is rapidly disseminated through the industry and who are represented on the Environmental Committee of the Chamber of Mines.
3. Vigorous professional associations, including a local branch of the London-based Institution of Mining and Metallurgy, the Zimbabwe Mine Engineers Association, and the Zimbabwe Mine Surveyors Association.
4. The presence of the National Resources Board (NRB), which advises both Government and industry on practical environmental standards. Its Mineral Resources subcommittee deals specifically with concerns and complaints relating to the mining sector. A representative of the Chamber of Mines is a member of this group.
5. The existence, noted above, of an annual competition run by the NRB to select both the best and the most improved mine site from an environmental viewpoint.

All this said it has to be acknowledged that "planning for closure" was, until recently, something of a tautology. As the case studies show, closure has been a largely *ad hoc* process, often unpremeditated in the sense that no specific person or unit was tasked with this work prior to it becoming imminent. Present-day concerns about remnant cyanide compounds and mercury residues were not addressed until a few years ago.

19.3.2   CASE STUDIES

19.3.2.1   Artisanal riverbed gold winning

There has always been a small amount of alluvial gold panning in Zimbabwe, mainly as a dry season activity by women on the rivers draining the schist belts of Zimbabwe's central plateau. However, the real growth of gold winning seems to have started in 1985, when a company used a rope shovel to recover gold from the bed of the Mazowe River, downstream from Shamva. The project was unsuccessful, but not before it had shown that there was gold in the riverbed, not only in the gravels and sands but also — and indeed principally — in the gravels between the alluvial rubble and boulders close to the bedrock. They reported an average figure of 0.57 g t^{-1}, or about 1300 mg m^{-3} in alluvial gold terms. This is a high-grade alluvium; break-even for a typical dredge operation is usually under 300 mg m^{-3}.

As a result, when the formal operations stopped, the local people kept on working this alluvial rubble horizon. This is difficult material to handle, but it is what the geologists would call a high-energy alluvium, requiring turbulent floods cascading off the gold belts for its creation. Under these conditions, the gold that is deposited with these small boulders will be coarse and there will be a lot of it, relatively speaking. News of this discovery spread, driven by a series of droughts that intensified rural poverty. By now, it is claimed that as many as 300,000 persons are active in this area during the dry season and over 2000 kilometres of rivers are affected. Given Zimbabwe's rapid population growth, there is no doubt that this is going to be a permanent feature of the rural economy in future, even in good agricultural seasons.

The principal environmental problem arises because, in most rivers, water still persists below the sand in the alluvial rubble horizon throughout the year, and a common situation is that the panners have to work up to their waists in water. Frustratingly for them, the water level is often greater than the depth of rubble, and the richest pickings of all, in the grooves of the bedrock, have to be passed up.

To overcome this constraint, the panners work out towards the banks, where the bedrock rises up and, consequently, the water level is less of a problem. Eventually, they are drawn into mining into the banks, and it is here that the problem starts, for their undermining causes the banks to collapse when the floodwaters come, bringing material into the river and silting up the dams downstream. Sometimes these excavations collapse on the panners; in 1995, it was estimated that as many as five persons a month were being killed this way.

The response by government to the situation was crystallised in the Mining (Alluvial Gold) (Public Stream) Regulations in 1991. These sought to improve the policing of the situation by handing over responsibility for its control to the relevant local authorities. The latter are permitted to take out "special mining grants" on rivers where alluvial gold was being won, and in turn to issue panning permits for a small annual charge to "approved persons," to cooperatives and to partnerships. The panners are specifically prohibited from mining the banks, from pitting in river-beds to more than 1.5 metres, and from any form of undercutting. About 25 such special grants have been issued to councils so far, but in the drought of 1994–1995 the numbers of panners increased greatly, swamping attempts to make this new control system operate effectively; an article in the local magazine *The Farmer* gives a vivid picture of what is actually happening on the ground.[4]

However, in the last five years there has been a considerable amount of work done on getting to the facts of the situation — and out of this comes the prospect that it can be brought under control. A number of studies have established the grades, recoveries, and practices of the gold panners, but the most important work has been the project funded by German aid and supervised by Professor Jurgen Voss based at the Department of Mining Engineering of the University of Zimbabwe.[5–8]

This work on the Manyuchi River in the Makaha area and the Insiza River near Filabusi has succeeded in demonstrating that riverbank mining can be undertaken in an environmentally respon-sible and successful way. The economics of the Manyuchi scheme has been hampered by a shortage of water and the lack of a flood plain to give good segregation of gold into the rubble horizon, but the Insiza development is running at close to break-even on operating costs, despite its employees (many of them women) receiving the mining industry minimum wage — at about US$70 a month, a figure significantly above most rural incomes.

The concept is to use the mined-out material from one section of bank to backfill an earlier section. The first stage of this is to remove and stockpile the gold-free topsoil and subsoil, and then to treat the −10 mm fraction of the rubble horizon through sluice boxes. Large rocks are left behind, or used to create a dike between the excavation and the river. The +10 mm material is returned to the base of the old pit, to be followed by the sluice tailings, which fills up much of the space between the dike and the end of the excavation. This gravel material drains to give a stable bed, on which the subsoil and topsoil are replaced. Revegetation occurs normally on the latter with the advent of the rains.

The scheme has completely rebuilt the first stretch of damaged river bank that has been treated in this manner, and has shown how such "rehabilitation mining" can be used to avoid river bank damage, and, with a modest degree of mechanisation as shown in New Zealand on similar material, can turn a profit.[9]

The programme has, however, much greater implications. If the existing pilot schemes were extended into full-scale operations on the potential resource away from the river banks and if the average width of this averaged only 50 metres on each side, then about 20 km² of Zimbabwe would be underlain by alluvial paydirt, requiring the work of about 300,000 people over ten years to recover the gold and rehabilitate the surface.

19.3.2.2   Shamrocke Copper Mine[10]

This Lonrho-owned copper mine was shut down in 1978, at the height of the "bush war." It was an underground mine producing 50,000 tons per month with a flotation concentrator. Poor grades as well as security worries contributed to the closure, as the underground ore ran at only about 1.1% copper.

The mine is situated on a steep side of the valley of the seasonal Nyaschere River, in rugged country about 150 km northwest of Harare. In this environment, the tailings presented a particular problem, as the only place to dump them was over the riverbed. The tailings dam was therefore literally that, with hydrocyclones used to classify the coarser particles to form the walls, which were themselves covered with waste rock to prevent erosion. A spillway was built and two penstocks were created by driving adits into the valley sides and raising before tailings deposition commenced. The dam so created has remained stable, and has a large resident population of crocodiles and hippopotamuses. Saligna and vetiver grass has been used to cover the dam, and numerous indigenous shrubs have taken hold, so that although no topsoiling has been done, a reasonable vegetation cover has developed.

However, nearly all the protection items installed, including fencing and the steel plates secured to the shaft collars to close them, have been stolen.

19.3.2.3   Empress Nickel Mine[11,12]

The government of Zimbabwe has been concerned that mines may be shut down without regard for the broader economic consequences. Empress Nickel Mine started operations in 1969 with the run-of-mine ore grading 0.69% nickel and 0.53% copper. The reserve was judged to contain 44,000 tons of nickel and 38,000 tons of copper. Some modest increases in these reserves were drill-proven at depth but in 1981, when 49,000 tons of nickel and 44,000 tons of copper had been taken out, the mining rate had to be reduced because of poor ground conditions following an earth tremor. At this point in time, the mine was engaged in a sand-fill operation in order to mine the remaining pillar reserves. The earth tremor and resulting ground disturbance resulted in the loss of available pillars and reduced operating capability to 35% of capacity. This made the operation uneconomic, but after an enquiry government refused to allow the mine to close and required that the mine be put on a care and maintenance basis.

As a result, although the township was taken over by the army, the mining claims are still current, continuing to be held by Rio Tinto (Zimbabwe) Ltd (Riozim), the local subsidiary of Rio Tinto (formerly the RTZ-CRA Corporation). Permission was given to sell the equipment there, but the company still has responsibility for the site, and has spent several million Zimbabwe dollars on undertaking extensive rehabilitation.

Formal closure studies and planning commenced in 1992, well in advance of the voluntary guidelines described above. Quittance was granted in 1997, with the rehabilitation work being described by the Ministry of Mines as a model for future closures.

19.3.2.4   Pangani Asbestos Mine*

This mine situated near Filabusi, about 100 kilometres southeast of Bulawayo, was started in 1963 by a consortium of South African and Swiss businessmen. It was initially opencast, with a milling

---

* With acknowledgement to the Office of the Chief Government Mining Engineer, Harare.

rate of about 200,000 tonnes per annum, but the grade and quality of the fibre were lower than expected and a comment by the regional mining engineer in 1968 following an inspection was that "it never has and never will run at a profit."

With limited financial resources, little effort was put into mitigating environmental concerns, and complaints of dust from the mill drifting over the mine housing areas were dealt with by putting a restriction on the mill stack opening to increase the discharge velocity. Fortunately, the long-fibre chrysotile asbestos mined in Zimbabwe appears not to give asbestosis except in a few cases where the individual is a heavy smoker.

The decision to opencast the deposit was, it was acknowledged, an error. Pangani is one of a number of mines in Africa where the attractions of a starting with a cheap and simple mining method were allowed to overshadow the future costs associated with high stripping ratios and steep haul roads.

As a result, underground mining was started belatedly in 1970, with disastrous consequences. It appears that the massive blasts associated with open pit mining had caused extensive weakening of the underground ore, and a series of crown pillar collapses occurred. The consequent air blasts killed four workers in one case and injured fifteen in a second. By 1979, only one stope (out of eight) remained that had not suffered this catastrophe and the owners wished to close. However, concern was expressed by the new (post-independence) government over the loss of jobs; about 300 employees were involved, and the mine staggered on until 1982 when it was closed, the machinery auctioned off, and the buildings taken over by the government as a training school.

The tailings dump is the main worry at the present time. This has been slumping over a main rural feeder road, and although the National Resources Board paid Z$17,000 to have barriers made, the heavy rains of the 1995–1996 season gave rise to further problems. At this stage, there are no funds available for the major cost of recontouring to ensure stability.

19.3.3 MEETING THE COSTS OF CLOSURE

As can be seen above, abandoned mines in Zimbabwe can present long-term problems to the Ministry of Mines, even when the landowner and miner agree that protection of openings has been satisfactorily carried out. Although the theft of protection items presents a major problem, the degradation of poorly designed dumps has also meant that money has to be found to restore mine sites to a satisfactory state.

In South Africa, as in several other countries with a sophisticated mining sector, all new operations have to submit an Environmental Management Programme and place money in a trust fund for rehabilitation after closure. This is workable in a country where the mines are generally very large, and whose reserves are well delineated (i.e., closure can be confidently and fairly precisely predicted).

In practice, economic and technical factors make the prediction of closure almost impossible for the bulk of Zimbabwe's mines. As the state has the ultimate rights to the mineral deposits, so it is argued, it has the ultimate responsibility for rectifying environmental hazards.

The reality is that generally in Africa, the state has no resources to undertake this work. Throughout Africa, the outcome of pressure from the IMF and the World Bank to keep budget deficits down, against pressure from voters to keep subsidies up, has led to the starving of ministries that do not have a visible social function; mining is invariably one of these.

However, in Zimbabwe the opportunity exists for the application of an existing levy into the rehabilitation of abandoned mines. The Zimbabwe Minerals Marketing Corporation (MMCZ) was set up in 1982 to prevent the transfer pricing of mineral exports, and finances itself by a charge of 0.875% of the value of such exports. Because most large companies make their own marketing arrangements, the MMCZ usually turns a surplus of the order of US$1 million annually, which is handed over to government. If this profit went into a mine rehabilitation fund, perhaps administered by a joint committee of the Ministry of Mines and the Chamber of Mines, the problem of rectifying environmental hazards from riverbed panning and abandoned mines might be solved.

19.4 CONSTRAINTS AND OPPORTUNITIES FOR TRANSFERRING BEST PRACTICE

The environmental controls on mining in Zimbabwe, as illustrated by the case histories above, are not without their faults. For example, only now are a significant number of larger mines being operated with their eventual closure being taken into their planning. However, they are important in a broader sense because several other countries, such as Tanzania, Ghana, and Burkina Faso, now see the Zimbabwe situation as a model for the development of an indigenous formal mining sector. These nations are representative of a large class of countries whose economic geology features primary gold and alluvial gemstones, and who therefore have the potential for numerous mines on resources that are too small and/or too low grade to interest the international investor.

This said, it should be noted that, despite favourable geology, so far none of the countries that would like to emulate this situation have succeeded in creating more that a handful of small formal mines, with a correspondingly negligible environmental effect. Artisanal mining concerns about mercury and about "mine rush" villages predominate, and as has been noted above, even in Zimbabwe institutional structures have been unable to cope with the artisanal situation there.

This is perhaps the key point; no matter how admirable the intentions and how comprehensive the legislation, nowhere in Africa apart from South Africa are the administrative resources of the relevant ministries and departments capable of monitoring and enforcing environmental ordinances to the level of the developed world. Despite this, ordinances are normally based on the standards of the "first world," and attempts at diluting them are too often met with knee-jerk threats of sanctions. The answer to this obstacle might be derived from Zimbabwe's experience. This suggests that if it is possible to bring about a transformation of the artisanal sector into a formal one, then the problem of weak regulatory institutions can be at least partially overcome by self-regulation.

This idea is spreading to other parts of Africa. In Tanzania, the formation of a Chamber of Mines on the Zimbabwe model was specifically recommended by United Nations advisers as a starting point in this process. Similarly in Ghana, the formalisation of the small-scale mining sector is planned to enable it to meet its environmental responsibilities more adequately. In both these countries, as well as in others such as Burkina Faso and the Côte d'Ivoire, the concept of trans-formation is being supported by the World Bank as a way out of the legal, financial, and technical traps which hedge artisanal mining.

19.5 CONCLUSIONS

Zimbabwe has a large mining sector for the size and quality of its resources, but the negative environmental effects have been relatively slight. Since the type of geology it possesses is repre-sentative of a number of countries whose mining sectors have not developed to the same extent, it is being widely taken as a possible development model.

The country features numerous small formal mines, organised to confer through, be represented by and obtain information from an elected Chamber of Mines, as well as a number of interest groups. Such structures have been found to provide a good basis for encouraging environmental self-regulation, and this is perhaps the most important single lesson from Zimbabwe's experience of this aspect of the mining sector. However, the artisanal sector, which has swollen dramatically in the past ten years, has so far defied regulatory attempts, principally due to the lack of resources the government has been able to apply in that arena. It is suggested that at least part of an existing government levy on the formal sector be directed at control and rehabilitation in this area.

REFERENCES

1. Baines, T., *The Northern Goldfields Diaries*, Chatto and Windus, London, 1946.
2. Hayes Hammond, J., *The autobiography of John Hayes Hammond*, Farrar and Tinehart Inc., New York, 1935.
3. The World Bank, Zimbabwe; performance, policies and prospects, IBRD, Washington, D. C., 1985.
4. The Farmer, Modern Farming Publications, Harare, December 14/21, 11, 1995.
5. Shamu, C. and Wolff, P. Gold panning in Zimbabwe, Department of Mining Engineering, University of Zimbabwe, unpublished departmental review, 1993.
6. Milne, G. and Marongew, D., Eds., Small-scale gold panning in Zimbabwe. Economic costs and benefits and policy recommendations, Ministry of the Environment and Tourism, Government of Zimbabwe, Harare, March, 1995.
7. Maponga, O., Gold panning along the Mazowe River and its tributaries — a report, Institute of Mining Research, University of Zimbabwe, August 1995.
8. Voss, J., Riverbed mining. The Manyuchi and Insiza River test cases, Department of Mining Engineering, University of Zimbabwe, February 1996.
9. MacArthur, N., Cowie, W., and Utting, B., New Zealand alluvials, *Asia Pacific Mining*, August/September, 30, 1990.
10. Eagling, J., Zimbabwe Chamber of Mines, personal communication, 1996.
11. Viewing, K. A. V., The life of Empress Nickel Mine, Institute of Mining Research, Departmental Report C291 for Rio Tinto Zimbabwe and the Government of Zimbabwe, October 1982.
12. Wiley, I. and Mallon, P., The rehabilitation of the Empress Nickel Mine site in Zimbabwe, in *Minerals, Metals and the Environment II*, The Institution of Mining and Metallurgy, London, 1996.

20 Alluvial Gold Mining Constraints and Opportunities for Sustainable Development in Manica, Mozambique

Isidro R. V. Manuel and Tomás Muacanhia

CONTENTS

1-56670-365-4/00/$0.00+$.50

ABSTRACT

The environmental problems related to alluvial gold mining in Mozambique have become a matter of concern as in other developing countries, e.g., Zimbabwe (see Hollaway, Chapter 19), Colombia, Brazil, and Ethiopia (see Tadesse, Chapter 23).

This paper highlights:

- An outline of the main social, institutional, legal, managerial, and economic constraints and their causes.
- An outline of major "nuisance effects" from alluvial gold mining in the Manica district.
- Measures to be taken by the Mozambican Government to address the alluvial gold mining issues or problems.

The conclusion is that several sectoral laws and policies relating to environmental issues are outdated or inadequate. Thus, several sectoral laws and policies relating to the environment need to be reviewed and harmonised. Moreover, community participation is vital for the sustainable development of Mozambique's mining industry.

20.1 INTRODUCTION

20.1.1 BACKGROUND

Mozambique is known as the poorest country today, although it is potentially rich. The struggles for the independence of Mozambique (1975) and Zimbabwe (1980) and 16 years of civil war have had a negative impact on Mozambique's economy.

War and lack of maintenance have destroyed vast areas of previously established socioeconomic infrastructures. As a consequence, the lack of adequate resources is a chronic and social phenomenon in most strata of Mozambique's population. This penurious situation leads to a very high demand for natural resources.[4]

The wartime environment, the need for a socioeconomic rehabilitation period, weaknesses in management, and a fragile national strategic plan for sustainable development are the main constraints causing this scenario.

The cease-fire is still holding, going well. The country moved from a one-party system and centralised economy to democracy and market economy.[11] The prospects of sustainable development appear to be good as sustainability and democracy are intimately linked, and forgiveness, tolerance, and confidence have been building up since the beginning of the cease-fire.

Environmental impact assessment of the mining industry has never been undertaken before in Mozambique. The present work is a result of changes in mentality as challenges on environmental issues become widespread throughout the country. A new culture is emerging from the ashes of war and dictatorship.

Most of the difficulties to date in natural resources management can mainly be attributed to institutional and economic fragility, including the political situation, although the government and the opposition are reevaluating the real chances, prospects, and barriers for reconstruction and development. However, the material basis from which the start is being made is very poor indeed. Thus, Mozambique is obliged to accept the fact that sustainability will only be possible if the institutional, legal, and technical bases are strengthened and keeping diverse options open appears to be the best approach to our unpredictable and non-linear future.

20.1.2 STUDY AREA

The Manica district is in the western central Province of Manica between 19°08' and 19°20' south, and 32°41' and 33°17' east. The district covers an area of about 4395 km². It is a mountainous area with altitudes ranging from 800 to 2000 m. The Revué upstream basin (approximately 80 km long) is within this province. The district has a moderate tropical climate. It is an afro-mountain climate (i.e., the variation in altitude plays an important role in climate changes). The annual average temperature ranges from 19° to 21° C. The annual average relative humidity is 54.3 to 73% and the annual rainfall ranges from 1000 to 2000 mm.

The study area is located in Manica Greenstone Belt, an eastern extension of the Mutare Greenstone Belt, which occurs in the eastern central part of the Republic of Zimbabwe. Both greenstone belts belong to a very well-known stable nucleus "Cratão do Zimbabwe."

Several minerals occur in this ancient geological formation such as gold, bauxite, and mineral water. There are fairly good to very good soils. These soils are prone to erosion hazard.[1] The district is also rich in agriculture, forestry, and livestock, with endemic flora and fauna.

The population density of Manica district is 35.5 inhabitants per km².

20.1.3 OBJECTIVES

The study was carried out not only to highlight the constraints and opportunities of alluvial gold mining in the Manica district and to justify the continuation of research, but also to ensure that mineral resources are evaluated and extracted from a broad and multidisciplinary perspective, including land use, environmental, socioeconomic and cultural dimensions, and bearing in mind institutional and technical considerations.

20.1.4 METHODS

In this study, all actors were involved. Direct participation of the communities, "garimpeiros," small gold mining firms, and institutions from local to central government level and NGOs was ensured during the study. Geological, agro-ecological and socioeconomic and cultural assessments were also carried out. Water, sediments, soils, and fish were sampled, using a range of relevant and appropriate methods.

20.2 CONSTRAINTS AND CHALLENGES

20.2.1 ALLUVIAL GOLD IN MANICA

Historians and archaeologists have noted that gold has been extracted since the period of the Zimbabwe (1250–1450 A.D.) and Monomotapa (1325–1600 A.D.) states. Therefore, gold mining is a traditional activity, together with agriculture and the rearing of domestic animals such as cows, goats, sheep and poultry.

Studies carried out in the Manica district have demonstrated that the extraction of gold was undertaken in a sustainable manner in ancient times (pre-colonial era) since gold mining took place only when there was a need for products not manufactured locally or available in the area, such as clothing. Agriculture has been the backbone of the local economy since ancient times to present.

According to several workers, the sustainability of gold mining changed by the end of the 19th century. When gold was discovered in Witwatersrand, South Africa in 1888, many Europeans moved into Manicaland looking for gold. Gold prospecting then increased. However, gold mining was restricted by the Portuguese for two main reasons:

- Colonial protectionism.
- Limited availability of Portuguese capital.

"Artisanal" gold mining was also banned until 1989.

About 400,000 ounces of gold were reported extracted between 1900 and 1945. Sixty percent of that amount was from alluvial mines and the remainder from hard rock. It should also be noted that from 1949 to 1975, the extraction of gold declined sharply. Most of the mines are open cast in Manica district, so their impact on the environment is potentially serious.

The gold mining activity increased after the independence of Mozambique in 1975. From 1990 to date, gold mining has increased steadily. There are two sectors involved in the gold mining industry in the Manica district:

- The formal sector represented by small companies.
- The informal or "artisanal" sector represented by prospectors or garimpeiros.

The two sectors have different constraints and challenges, and their interests appear to be conflicting in the Manica area.

20.2.2 FORMAL SECTOR

There are three small mining companies undertaking fieldwork (i.e., prospecting) and another two are already extracting gold in both underground and open cast mines. A sixth company is exploiting bauxite. All companies are joint ventures. The local capital holds small shares since most of the capital invested in this sector is from abroad. The companies employ over 500 people. Over 99% of these workers are Mozambicans. There are very few women working for the gold mining companies. The number of women is so small that one can state that gold mining is a man's job in the Manica district, since there were less than ten women employed when the worker survey was completed.

Most of the workers are not from the local communities. The companies say that the local people prefer to work in their "machambas" rather than in the companies — even if the mining companies employ them. As a consequence, the local workforce is being discriminated against. All watchmen are also from other districts. Therefore, there is a potential for conflict. Researchers have warned about the consequences of discriminating against local communities, since the benefits from the mines might not be shared between the local communities and the incoming workers, and relationships between the two sides might be in jeopardy (see Section 20.4).[3]

20.2.3 INFORMAL SECTOR

Since the pre-colonial era, alluvial gold mining has been taking place in the Manica district. However, "artisanal" gold mining was banned by the colonial powers until 1975, when Mozambique became an independent state (see Section 3.1).

"Artisanal" gold mining restarted, *de facto*, in 1990. It increasingly became an important activity. It reached its peak in 1992 due to four major reasons:

- Socioeconomic impact of drought which devastated Southern Africa (1991–1995).
- Gold mining in upstream basin of Revué River that left many peasants without land.
- Civil war and displaced populations.
- General Peace Agreement (4th October 1992) and returnees from neighbouring countries and elsewhere in Mozambique.

Between March and June 1995, 13 "artisanal" gold mining sites were identified in the upper Revué basin. Most of the garimpeiros are peasants. They extract gold during the dry season, after crop harvest. However, several professionals were found working at the 13 "garimpos" where most of the artisanal miners are engaged in alluvial gold recovery: schoolchildren and teachers, carpenters, stonemasons, agricultural, commercial and health technicians, mechanics, and so on. The penurious situation leads to people living under the survival threshold poverty line. According to the *Maputo Weekly Paper* 79.5%

of men are unemployed in Mozambique. However, it should be pointed out that foreign nationals are also panning in the Manica district and elsewhere in Mozambique. Nationals from Zimbabwe, Malawi, Guinea, Somalia, and Nigeria were found looking for gold in Manica. We believe that the situation is volatile, as potential conflicts between Mozambicans and foreigners might erupt due to a discovery of gold, competition for women, excessive alcohol consumption, and possible consumption of drugs.

The "garimpeiros" lack the basic geological and mineral knowledge, and safety in the mines is ignored or unknown.[2] There have been reports of the deaths of garimpeiros due to the collapse of tunnels or mines (see Sections 20.3.2,4).

20.3 ENVIRONMENTAL IMPACT

20.3.1 FORMAL SECTOR

Since the introduction of a market economy in 1987, several companies (national and foreign) have been investing in Mozambique.[11] However, the investors have been confronted with serious problems, as detailed in the following sections.

20.3.1.1 Excessive centralisation

Government institutions are excessively centralised in terms of their decision-making activities. The decisions are made by the central government. The local authorities do not participate in decision-making even if they are concerned with their own jurisdiction. As a consequence, the implementation of some strategies and programmes is not effective.

20.3.1.2 Excessive bureaucracy

The bureaucracy is enormous. Excessive centralisation and technical weakness of most of the civil servants increases the bureaucratic system and leads to an increase in corrupt state officials. Foreigner investors or entrepreneurs may spend large sums of their capital to obtain a commercial or industrial licence. Over 10% of initial capital might be spent in a slow process that may take 3 to 12 months or even more to receive a commercial or industrial permit. It is obvious that many investors may be discouraged from investing in Mozambique.

20.3.1.3 Overlaps of jurisdiction

Inter-sectoral coordination is fragile. The vertical and horizontal relationships are weak. The same task may be carried out by different sectors with overlapping goals. For instance, land might be licenced by the Ministry of Agriculture and Fisheries for agricultural purposes and the same plot of land might also be given to a mineral company by the Ministry of Mineral Resources and Energy. Thus, different institutions may carry out similar tasks with conflicting consequences.

20.3.1.4 Fragile institutional capacity

Mozambique has serious problems in managing its natural resources. The excessive centralisation, bureaucracy, and overlapping of competencies weaken the institutional capacity. It is therefore difficult for the government to implement the strategies of sustainable natural resource use and control programmes properly because the institutions are not fully interlinked or coordinated.

20.3.1.5 Poor national resources inventory

Knowledge of the ecology of the Manica District is scant, as it is for Mozambique generally. However, some work has been carried out on minerals, soils, forest resources, and water, and on some plant and animal species. Systematic work is still needed to update mineral data and on other natural resources as well. The available inventories usually cover potentially rich areas, such as the Manica District (gold) and Montepuez (marble), where these minerals are abundant.

After independence, several geological surveys were undertaken. Several geological maps and reports are available. However, the data are not being utilised widely. It is difficult to establish strategies for sustainable resource exploitation under this scenario. The authors consider that the institutions should use the data and they should also make them available to interested sectors. The exchange of information vertically and horizontally would help in updating the national technical capacity, and would also improve the management of natural resources since it might be possible to determine their viability.

20.3.1.6   Weak technical and scientific capacity

Over 70% of Mozambicans are illiterate people. About 80% of Mozambique's workforce is employed in the agricultural sector (most of them being real peasants).[7] Thus, the technical and scientific bases are weak compared with those of neighbouring countries. Most of the cadres or technicians working for the state apparatus, private sector or NGOs have some scientific weaknesses. Their environmental knowledge is relatively scant. The reason for that is environmental issues are still new in Mozambican society.[5, 6] However, some efforts are being made and the Ministry for Environmental Co-ordinating Action has been carrying out campaigns of environmental education. Environmental issues are also being included in national curricula.

20.3.1.7   Non-involvement of the community

Socioeconomic, political, and cultural decisions are made most of the time without local communities' involvement. All environmental issues are dealt with from above. Ethno-cultural considerations are not made. Therefore, specific characteristics of the communities such as cultural dimensions (language, food habits, marriage, etc.) and local skills are neglected.

20.3.1.8   Outdated and inadequate legislation

The Mozambican legal framework is still fragile. Recently prepared legislation does not address the new challenges presented by market economy and democracy. There are some concerns relating to monitoring mechanisms that need to be addressed, since some legislation is woefully inadequate. The legislation does not clearly address very important matters such as land access/conflicts, soil degradation, pollution, deforestation, biodiversity, water, and pesticides, along with prevention, inspection and enforcement.

The authors believe that the existing legislation should be updated to provide for monitoring and feedback mechanisms such as:

- Local authorities given power to distribute and manage land and solve conflicts.
- Clear rules set up for determining traditional land use and occupation.
- Recognition and protection of traditional rights of returnees/resettlers so that conflicts between returnees and current occupiers can be prevented/solved.
- Involvement of local communities in developmental programmes (i.e., local people must be participants since their collaboration is vital in sustainable management of local natural resources).
- Environmental impact assessment process application, including mechanisms for fair and public participation.

20.3.1.9   Fragile political atmosphere

A six-year cease-fire has been observed. However, it appears that we are still on a "volcano": "social differentiation has increased, with traders, large farmers, corrupt state and military officials and private entrepreneurs gaining from the changes while women, children and the poor in particular are finding their standards of living dropping sharply with devastating consequences for their health and nutritional status."[4] Since Marshall gave the above picture, the human suffering has increased and there is a "potential chance for social and political breakdown." Thus, several constraints do affect gold mining in the Manica district and elsewhere in Mozambique. There are many opportunities for non-ferrous minerals in Mozambique, though our knowledge of mineral availability is still relatively poor.

Gold mining companies are the major taxpayers in the Manica district through the payment of royalties. The companies encourage the improvement of other companies such as Hydroelectric of Chicamba Dam, since the mines are the biggest power consumers in the district. However, the local communities criticise the companies for not constructing socioeconomic infrastructures such as roads, schools, and health centres. Moreover, the communities say that the royalties do not benefit either the communities or the district. Local community frustration may lead to very bad relationships between the companies and the local communities.[3] The communities may not collaborate with the companies. Furthermore, the local communities compete with the companies in gold deposits, leading to serious conflicts in some gold mining sites (e.g., Andrade).

When gold mining activities restarted in 1989, several peasants lost their agricultural land or "machamba" in Revué valley. The peasants lost arable land and the mining companies destroyed their crops. Food security was broken down. The scenario was aggravated by severe regional drought in the period 1991–1995.

The normal Revué flow declined due to drought that devastated Southern Africa from 1991 to 1995. The companies increased alluvial gold extraction, which aggravated the reduction of water quality and availability. Water flow in Revué River then reduced sharply. The water became brownish-orange or reddish-brown. As a consequence, people and domestic animals were prevented from drinking the Revué water (see Section 20.3.2.3).

Alluvial gold mining has also had a negative environmental impact via stream bank erosion and siltation of the Chicamba Dam, and it is also causing a reduction in the fish population. As the mines are open cast, physical environmental changes were obvious. Erosion and irreversible landscape alterations have taken place. The gold mining companies did not rehabilitate the trenches, boreholes, and galleries contributing to the susceptibility of the sites to erosion.

The surveys carried out on alluvial gold have shown that the gold deposits are usually small and scattered throughout the valleys. The gold mining companies are utilising machinery for spread and small gold deposits. So, we think that there is a need for technological change. The machinery should be small (light equipment) and very mobile so that it can be used in different geological conditions, i.e., the equipment can be easily adapted to different gold sediments or deposits.[2]

As stated above, the companies usually abandon the mines without rehabilitating in accordance with Land Law and Mines Law. When the mines are abandoned without proper rehabilitation, the garimpeiros move in looking for the remaining gold (see Section 20.3.2). The institutional fragility, inadequate legislation, poor technical and professional capacity and lack of community involvement and/or powerless local authorities mean that the companies and the government do not fulfil their legal obligations.

There is an urgent requirement to establish a self-monitoring process for assessing and reporting environmental conditions associated with mining activities (see Section 20.3.1.8). Cooperation between all actors is essential in the Manica district and elsewhere in Mozambique.

20.3.2 INFORMAL SECTOR

The informal sector or "artisanal" mining is an activity that has been underway for centuries. Some researchers have brought to light information that appears to confirm that gold mining used to be a complementary activity until the end of last century. Gold mining was then a sustainable activity as agriculture and domestic animal raising were the basis of local economy (see Section 20.2.3).

However, alluvial gold sustainability is jeopardised by "savage" extraction. Civil war, drought, and the introduction of a market economy have presumably changed the behaviour of local communities.

"Artesian" gold mining or "garimpagem" is one of the main means of subsistence in the Manica District nowadays. There are several thousand garimpeiros involved in this sector (see Section 20.3.3.). The garimpo activity has become one of the factors that alleviate poverty in some districts, e.g., Manica, Lago.

The authors were told that "artisanal" gold mining, informal trade, and good harvest were the deterrent factors that were removing needy urban dwellers from Manica town. Indeed, the authors did not encounter a single urban beggar there.

There are several constraints to "artisanal" gold mining. Negative environmental impacts have been observed.[8] The main constraints and negative environmental impacts are summarised below.

20.3.2.1   Weak technical capacity

There is a lack of basic geological knowledge and mineral techniques. The garimpeiros utilise rudimentary technology.[2] There is also a lack of basic principles in resource management. Thus, the "garimpeiros" do not know how to maximise their productivity and how to save money. The "garimpeiros" are at risk and panning is also a risky business. However, there are some "garimpeiros" who use money more appropriately from selling gold. They build new houses, buy bicycles, agricultural tools, and so on.

20.3.2.2   Degradation of gold deposits

"Artisanal" gold mining is a risky business in the Manica District. If a panner locates or finds a gold deposit, all "garimpeiros" around will make holes as close as possible. The site becomes dangerous as the boreholes may collapse anytime. Thus, several boreholes, trenches and/or galleries degrade gold deposits and make the place unsafe.

The authors found over a thousand garimpeiros extracting alluvial gold in a very small plot of land (approximately 1 hectare) at Andrade, on 9th June 1995. They were digging holes, trenches, and galleries (inside the trenches and boreholes). Trenches and boreholes were scattered all over the small plot. Disorganisation and rudimentary technology were evident, as well as bad management. Site and gold deposit degradation were in fact an environmental menace there.

20.3.2.3   Poor water quality

Boreholes and trenches on Revué and its tributaries' banks, and deposition of sediments had changed the water quality.[8,10] The water became brownish-orange or reddish-brown, depending on the colour of the soils on the river banks. Then, water was no longer suitable for drinking or domestic uses (see Section 20.3.1).

20.3.2.4   Lack of safety in garimpo sites

As boreholes, trenches, and galleries are rudimentary and spread all over the mining sites, garimpo activity becomes a risky job. The boreholes, trenches, and galleries are built without any standard of mining regulations.[10] Everything is done precariously in the Manica District as well as in other parts of the country (e.g., Lago district). Deaths may occur as happened at "Pothole" — (Chua) in 1994 (see Sections 20.3.1.8. and 20.3.2.1).

20.3.2.5   Precarious environmental sanitation

The conditions in panning or "garimpo" camps are environmentally poor. The panners usually live in small groups (4 to 12 persons) in small huts ("cabanas") in humid areas. There is neither a formal kitchen nor latrine. Thus, the panners are prone to endemic diseases such as malaria, bilharzia, and filaria.

20.3.2.6   Child work force

Among the panners, the authors found schoolchildren panning alone, in groups, or with their relatives (see Section 20.4).

20.3.2.7   Informal trade

Informal vendors increase their activities near panning sites such as Minosa. Negative habits were observed in several panning sites. It appears that prostitution, excessive alcohol consumption, and burglary are increasing (see Section 20.4).

20.3.2.8   Lack of promotion and control

"Artisanal" gold mining is legal in Mozambique. However, panning activities are not yet controlled by governmental institutions. Most of the panners do not have mining permits or licences. The panning activity is a spontaneous business (i.e., subsistence activity). The sector is not very well organised. It lacks promotion, and technical and funding assistance. The authors suggest that the government should intervene. Thus, the sector must be promoted, assisted, and controlled because it has devastating effects on environment, society and the Mozambique economy.

20.4   SOCIOECONOMIC IMPACT

The land is the backbone of peasants' subsistence. Gold mining brings little benefit to the local communities as a whole. The peasants do not have any capital. The peasants usually live in poor conditions. The peasants lost their arable land and crops, and they or members of the community are not given any opportunity to work for the mining companies that have destroyed their traditional assets. Thus, the socioeconomic and cultural impacts are presumably very serious in the communities.

The peasants are frustrated, as well as the community at large. Then, the peasants look for alternative activities for subsistence, such as panning and informal trading, which are "boomerangs" for the communities.

Several researchers have stated that poor populations with little means for subsistence have a tendency to degrade more natural resources than "good" poor populations. "Artisanal" gold mining is, to some extent, a result of formal gold mining as peasants lost their arable land in Revué valley. Panning is increasing without any organisation and it is becoming a strong sector without any state apparatus intervention. It appears that food security, mining safety and socioeconomic development are still neglected. We believe that it is not possible to implement any sustainable programme in this kind of scenario.

Gold mining is a risky business for both formal and informal sectors because the deposits are usually small and scattered over a large area. To find the gold is not easy. Production of gold is variable and the profits are usually very low for the "garimpeiros," as they sell the gold on the site. The profits are for the buyers and/or intermediate traders since they sell gold in the big cities (e.g., Beira and Maputo) or abroad (e.g., Zimbabwe) (see Section 20.3.2.3).

The formal sector has generated the informal sector in the Manica District because it occupied the arable land in the valley; after extracting the mineral, the companies abandoned the mine sites without formal rehabilitation. The abandoned boreholes, trenches, and galleries were then occupied by the "garimpeiros" (see Section 20.3.1). Thus, there is a vicious cycle in the socioeconomic and cultural impact of gold mining in the Manica District. Moreover, as the formal sector prefer labour from outside of Manica District communities, it should be pointed out that the formal sector is encouraging conflicts between local communities and incoming workers. This scenario should be avoided or alternative solutions should be considered as soon as possible. The Mozambican Government must act so that national stability and sustainable development are safeguarded in Manica district and elsewhere in Mozambique.

Both formal and informal gold mining sectors generate or encourage informal trade. Alcoholism, prostitution, and robbery appear to be increasing near mining sites (see Section 20.3.2.7).

There is evidence that sectoral and/or inter-sectoral conflicts are aggravating the negative environmental impact of gold mining (see Sections 20.3.1.1 and 20.4.1.5). Decisions are made from above as stated earlier. The local authorities are powerless. The local communities are also not involved in decision making. The sectoral coordination is fragile (see Sections 20.3.1.1.7 and 20.3.1.7). Thus, the horizontal and vertical relationships with the communities as legitimate owners and managers of traditional rights over the land resources should be strengthened so that conflicts can be avoided/solved and sustainable development can also be achieved.

There are conflicting interests between the formal and informal sectors, since both sectors compete for gold sediments. The panners invade the prospecting boreholes of the companies that are used for assessing the availability and viability of gold deposits. In addition, the local communities are fighting back for their traditional rights of land tenure.[9] Serious constraints and conflicts related to land tenure in post-war Manica Province have been reported. Over 200 peasants also informed the authors that the mining companies have been destroying crops without any fair compensation given to the land and crop losers. They were compensated for the crops only. There is no investment in "garimpagem" and the agricultural family sector, as was noted in previous sections. There are also reports that graves and shrines have been destroyed in the Ndirire area.

Water pollution is another source of conflict between the companies and local communities. It is already evident that the water is polluted in Revué river. As a consequence, fish availability has been drastically reduced. It should be pointed out that both formal and informal alluvial gold mining are the polluters. Relationships between the companies and local populations or communities are therefore at stake.

The Chicamba Dam is under threat from serious siltation for some time to come. The threat is caused not only by alluvial gold mining but also by other upstream anthropogenic activities such as agriculture, fire, and forestry since the latter uses exotic plants (eucalyptus and pine) which leave the slopes exposed to erosion. The alternative project of water supply to Chimoio town might be in jeopardy.

Fragile institutional and legal frameworks aggravate all socioeconomic and cultural constraints described here. The Constitution provides that "all land is vested in the State." The Land Law, which is now undergoing substantial debate in parliament, reaffirms this position and provides that "the State is responsible for determining who may have the use and benefit of the land and how it may be used." As was noted earlier, it is imperative that the institutional framework be strengthened and that the existing legislation should be updated and provide for monitoring mitigation and feedback mechanisms that everyone can participate in, since everybody's collaboration is vital in sustainable management of natural resources in Mozambique.

20.5 CONCLUSIONS

The concept of sustainable development is new in Mozambique. Sustainability is now widely accepted as having five dimensions: ecological, social, economic, cultural, and political. "Sustainability should be understood and applied within a changing society and within an open society." Sustainability may be applied within a stable and an open society. Sustainability might work within a challenging and "changing" society. Our society is at stake. We think that sustainability is applicable in societies where political will plays an important role in promoting technical, scientific, socioeconomic, cultural, and political development.

Sustainability would be possible if local communities are participants and collaborators in sustainable management of local natural resources.

The ethno-cultural diversity, biological diversity, and mineral resource diversity makes Mozambique a potentially rich country.

The Mozambican state is still very fragile. The legal and institutional framework is very fragile. The health and nutritional status of large numbers of Mozambique's population (women, children, and the poor at large) is at stake. Corruption of state officials and private entrepreneurs is devastating the socioeconomic, cultural, and political framework.

The authors think that sustainability might be possible if investors negotiate with the communities as holders of land rights in the areas they want to invest in. They will simply have to deal with the "communities" as legitimate owners and managers of rights over the land resources they want to use so that conflicts can be avoided/solved.

Capacity building should also be a national baseline for sustainable development, so that the fragility of legal and institutional framework can be overcome. Then sustainable development can be achieved in Manica district and elsewhere in Mozambique.

Further work and research is needed. The inventory of mineral resources should be carried out to locate, quantify, and qualify minerals so that economic viability can be available for all those interested in exploiting.

Geologically rich areas (gold) should be identified in Manica district, and mineral licences should also be issued so that the garimpeiros can be organised and promoted.

Finally, we think that poverty, vulnerability, and political instability are constraints for sustainable development at local or national levels. All actors involved in socioeconomic development have poor or fragile relationships. Urgent mitigating measures are needed to overcome the long-standing economic crisis in Mozambique.

ACKNOWLEDGEMENTS

We would like to express our gratitude to Hon. Dr. B. P. Ferraz, the minister for Co-ordination of Environmental Action, as he was instrumental in arranging funding for the "assessment of environmental impact of gold alluvial mining in Manica district." We are very grateful to SIDA/ASDI (Swedish International Development Co-operation Agency) for funding the project. Our special thanks to Hon. A. Canana, Governor of Manica Province, for the friendly assistance during the field work. Finally, thanks to all who assisted us in many ways in transferring this delicate work to the public arena.

REFERENCES

1. Folmer, E. C. R., and Francisco, J. R., Assessment of nutrient depletion hazard in Mozambique, presented at the 21st Meeting of the Sarccus Standing Committee for Soil Science, Inhaca Island, Mozambique, 27 November–1 December, 1995.
2. Intermediate Technology Zimbabwe, Going For Gold: The Story Of The Shamva Mining Centre, Intermediate Technology Zimbabwe, Harare, 1996.
3. Macdonald, M. and Sithole, B., Discussion paper on the social and cultural disruption associated with mining activities in SADC, SADC Mining Sector Coordinating Unit Workshop On Mining And Environment, Lusaka, 13th December, 1992, 1.
4. Marshall, J., Structural adjustment and social policy in Mozambique, *Review of African Political Economy*, 47, 29, 1990.
5. Ministério Para A Coordenação Da Acção Ambiental (MICOA), The costs, benefits and unmet needs of sustainable biological diversity in Mozambique, draft report, MICOA, Maputo, 1995.
6. Ministério Para A Coordenação Da Acção Ambiental (MICOA), *Programa Nacional De Gestão Ambiental*, MICOA, Maputo, 1995.

7. Muacanhia, T., Faba bean (*Vicia Faba L.*) resistance to black bean aphid (*Aphis Fabae Scopoli*), unpublished doctoral dissertation, University of Reading, U. K., 1994.

8. Muacanhia, T., Environmental impact of gold mining in Manica, presented at the Urban Environmental Quality in the SADC Region Workshop, Windhoek, Namibia, 27–30 November, 1995.

9. Myers, G. W., Land tenure issues in post-war Mozambique: constraints and conflicts, unpublished report, Land Tenure Center, University of Wisconsin-Madison, 1993.

10. Vicente, E. M., Modificações Fisico-Ambientais Provocadas Pela Explorção Do Ouro Aluvionar Na Bacia Do Revué — Distrito De Manica, Tese De Licenciatura, Universidade Eduardo Mondlane.

11. Wellmer, G., Mozambique: the economy, Centro De Estudos Africanos, unpublished report, Universidade Eduardo Mondlane.

21 Planning for Mine Closure: Case Studies in Ghana

Peter C. Acquah and A. Boateng

CONTENTS

ABSTRACT

Mines are known to generate environmental impacts both during their lifetime and after they have ceased operation. At closure, the mine is obviously not generating any income and therefore it becomes difficult if not impossible to find the necessary financial resources to carry out any meaningful remedial works. It is therefore important to plan the mine closure phase of the operation and make financial provisions for it while the mine is producing, so as to meet all environmental requirements. It has become the acceptable practice in many countries to make it an obligation during the life of the mine to adopt appropriate measures to ensure that after closure of the mine all worked-out areas are left in a condition which will minimise long-term lingering problems.

Good planning is the key to successful control of long-term environmental problems associated with the post-operational phase of mining. Planning for mine closure should be an integral part of the general mining plan.

In Ghana, operating mines are required to operate in accordance with, *inter alia*, the Minerals and Mining Law (PNDCL 153), Ghana's Mining and Environmental Guidelines, Ghana Environmental Protection Act 1994, Act 490, and also follow internationally recognised best environmental practice.

The case studies included in this chapter illustrate the conceptual plan of two gold mines in Ghana.

21.1   INTRODUCTION

Ghana is a moderately sized country of about 239,000 km^2 with an estimated population of about 16.5 million (1995 estimate). It is located along the Guinea Coast of West Africa.

The history of mining in Ghana dates back many centuries. The exploitation of gold in particular dates back to the 5th and 6th century B.C.

Minerals form an important component of Ghana's export earnings. In 1994, minerals accounted for more than 48% of the export earnings of the country. From the early 1970s to the mid-1980s, the minerals industry experienced a steady decline, which has been attributed to a combination of factors including unfavourable and inflexible economic policies in the country at the time. By 1983, the value of mining had dropped to about 60% of its 1971 value. The government then launched the Economic Recovery Programme in 1983 to arrest the decline and restore growth.

In 1986, the Minerals and Mining Law (PNDCL 153) were promulgated with the view to providing the necessary framework for increased investment in the mining sector. The Minerals and Mining Law made provisions for dealing with environmental aspects of minerals processing. Under the Minerals Commission Law (PNDCL 154), the Commission was established to formulate recommendations for national policy on exploration for, and exploitation of, mineral resources in Ghana.

As a result of the government's new policies, there was an upsurge of investment in exploration to establish new mines. Between 1986 and 1996, more than 14 new surface gold mines, including the Ghanaian Australian Goldfields Limited (one of the main case studies included in this chapter), have been established.

In order to avoid irreversible damage and assist the mining industry in adopting sound environmental management practices, the Minerals Commission and the Environmental Protection Agency (EPA) have through their collaborative efforts developed guidelines for the mining industry. In a booklet entitled *Ghana's Mining and Environmental Guidelines*[1] all the relevant areas of mining, including mine decommissioning are covered.

21.2   ANALYSIS OF PROBLEMS AND CHALLENGES

The challenges facing a mining operation after the cessation of production principally relate to restoration or rehabilitation of mined-out areas and the management of long-term environmental problems. The resolution of these issues is essential in developing a mineral resource in accordance with the tenets of sustainable development. In recent times, mines have been urged to include the maintenance of biodiversity in their quest for environmental protection. This is particularly important where the concession area includes ecosystems with rare or endangered species that could be affected by the mining operations. Good environmental planning and management in the mining industry thus includes measures aimed at mitigating the adverse environmental impacts of mining, and preserving biodiversity.

The degree and diversity of environmental impacts associated with mining vary with the type of mining and the characteristics of the mineral deposit being exploited. Surface mining is generally more destructive than underground mining. The former involves removal of vegetative cover and the destruction of soil profiles. Soil degradation in the form of accelerated water erosion and soil compaction results in the loss of soil fertility. The removal of large tonnages of overburden and waste rock and subsequent placement in spoil tips or mined-out pits can alter the topography and physical stability of the landscape.

Dust from unvegetated spoil surfaces, stockpiles, and tailing impoundments can be a problem particularly during periods of low rainfall and high winds. The dust may contain metallic compounds, which can cause widespread contamination of land and surface waters. Erosion of sediments from spoil tips, stockpiles, and so on can result in the transfer of sediment to surface waters and other drainage channels. The chemical quality of surface waters may also be adversely affected if the appropriate control measures are not taken, particularly where acid mine drainage is a potential problem.

In metal mining, residual chemical reagents and heavy metals in the mine wastes and effluents are gradually released to the natural environment for long periods after the mine has closed. The combination of these complex metals could pose a threat to human health and also have deleterious effect on aquatic life. The organic reagents that occur in the mine effluents may not be toxic when they occur at low concentrations. However, at higher concentrations they can be toxic.

At closure, if a mine is located in a groundwater recharge area, the quantity of the groundwater can still be affected. This is because the removal of ore and replacement with spoil may increase or decrease recharge rates.

The major risks associated with underground mining are long-term subsidence, which can cause extensive damage to surface buildings and structures. Other risks are the old, abandoned mine openings such as shafts, adits, incline, and pits which were abandoned without adequate safety measures and whose existence may not be known. Abandoned mine openings which are subsequently camouflaged by natural vegetation are a potential danger to life.

Surface excavations, waste dumps, tailings dams and subsided areas created by mining can sterilise land and exclude it from other land uses. Since mining is a temporal use of the land, it is important that consideration be given to post-mining land use.

21.2.1 Aspects of Guidelines

As already stated, guidelines have been prepared to cover most of the main aspects of mineral exploitation in Ghana. With particular reference to decommissioning the following stipulations, among others, have been made: the company (mining company) shall prepare a conceptual decommissioning plan as part of the EIA (Environmental Impact Assessment) or EAP (Environmental Action Plan). The decommissioning plan shall:

- Nominate the end-use(s) of all lands affected by the mining project.
- Describe the fate of all fixed equipment.
- Describe how public access will be managed after mine closure.
- Describe the type and duration of post-decommissioning monitoring.
- Ensure that the company shall submit to the EPA and the Mines Department, through the Minerals Commission, a detailed decommissioning plan at least two years before the

planned abandonment of the mining project. In the case of a mine located in forest/wild-life reserves, a copy of the plan should be submitted to the Forestry Commission.

- Ensure that the company shall honour all commitments made in the detailed decommissioning plan, except where written permission is given by the EPA in the light of new field evidence.

Since the guidelines were published in 1994, no company has submitted a detailed decommissioning plan for review. In accordance with the provisions of the guidelines, the mining companies have only prepared conceptual decommissioning plans with scant information as part of their Environmental Impact Statements. This situation has not made it possible to include any detailed information on the decommissioning plan of the two companies, Ghanaian Australian Goldfields Ltd. and Siam Goldfields Limited, which are the subject of the case studies in this chapter.

21.3 CASE STUDY 1—GHANAIAN AUSTRALIAN GOLDFIELDS LTD.

21.3.1 DESCRIPTION OF THE MINING PROJECT

The mining concession of Ghanaian Australian Goldfields Ltd. (GAG) is located in the western region of Ghana. The region is characterised by heavy rainfall and high temperatures. Mean monthly rainfall ranges from 30 mm in January to 280 mm in June. Average monthly temperature is about 26° C.

The concession is located some 240 km west of Accra, near Tarkwa. GAG's mining interests cover two concession areas, namely the Iduapriem Mining Lease (IML) area at the southern section of the concession and the adjacent Ajopa Prospecting Licence (APL) area in the northwest. The IML stretches over an area of 46.3 km² including 11.5 km² of the Neung Forest Reserve. The adjacent APL covers an area of 56.6 km². The concession has a number of scattered small settlements, but three villages — Diwobrekrom, Iduapriem, and Adeyie Junction — were considered at the planning stage of the mine to be most affected by the mining activities. Mining commenced in the Iduapriem concession in the early part of 1991 and was extended to the APL area when the company delineated some ore bodies there.

The topography of the concession area consists of a series of ridges and hills that contain the auriferous ores. In the IML area, the ridge runs east-west with a northward bend at its western end to form the northern hills. The highest point is about 164 metres above sea level toward the east. In the APL area, two parallel ridges run southwest to northwest, with the highest point reaching 251 m above sea level.

Drainage of the two concession areas is by a system of rivers namely Ankobra and Bonsa and their numerous tributaries. The principal geological formation underlying the concession area is the intensely folded metamorphic and argillaceous sediments, which are referred to as the Birimian-Tarkwaian system.

21.3.2 MINING AND ORE PROCESSING

The GAG operations use conventional open pit mining methods involving drilling and blasting and the use of excavators, wheel loaders, and heavy trucks. Ore from the Iduapriem concession is mined from two pits and the waste is deposited in a valley north of the pits. It is estimated that about 1 million tonnes per annum of waste is deposited in the valley. At Ajopa the ore is from a pit of approximately 47.5 hectares, with a projected depth of 72 m. The waste dump area is 76 hectares. An ore stockpile of 1000 tonnes is maintained at Ajopa. The ore from Ajopa mines is hauled by 25-tonne trucks to Iduapriem mine site, a distance of about 11.3 km, for processing. Fifty-tonne trucks are used to handle waste and low-grade material out of the mining pit. The waste dump is located to the northwest.

The gold is extracted from the ore using the carbon-in-pulp (CIP) method. The capacity of the plant has been increased significantly since the Ajopa mine came into production. The capacity of the tailings dam has also been increased to accommodate the increased tailings input by raising the dam wall. A new site is being considered for another tailings dam.

The ore processing operations include crushing, milling, and cyanidation. The gold is adsorbed onto graded carbon that is later removed and desorbed to recover the gold by electrowinning. The tailings containing 65% solids with cyanide content of 150 mg L^{-1} is pumped to the tailings dam at a rate of 6000 tonnes per day. About 600 L of hydrochloric acid used in washing the desorbed carbon is discharged with the tailings every 2.5 hours. Decant water from the tailings dam is recycled at the rate of 1000 tonnes per day and stored in the process water dam.

GAG proposes to develop a heap leach facility at its Iduapriem mine to process low-grade gold ore that will otherwise not be economical using the current CIP technique. The company has already commissioned a consulting firm to undertake an EIA study of the proposed facility.

The heap leach facility will have a heap leach dump and process solutions ponds. The planned operating life of the heap is five years, from July 1996 to June 2001. It will be located near the present plant site. About 8.52 million tonnes of low-grade ore (0.75 g t^{-1}) will be treated to provide an estimated 142,000 ounces of gold. The facility will complement the existing CIP processing facility and enable a planned annual output of 180,000 ounces. Feed material will be obtained from ore stockpile and from future material from Iduapriem and Ajopa.

A leach pad of 180,000 m^2 will be completely underlain by a high-density polyethylene (HDPE) liner incorporating a leak detection system. The ore will be placed by truck and sprayed with a sodium cyanide solution to extract the gold.

Eight process ponds will be built to collect and handle the collected heap leach solution. The ponds will have an engineered soil foundation and two layers of HDPE separated by an HDPE geonet. The geonet provides a pathway for solution to flow to a sump between the liners if the primary liner is accidentally punctured. The secondary liner will then contain the solution. Monitoring of leakage is by means of a pipe between the membranes to the sump. The leak detection system can be converted to a solution recovery system. The gold will be recovered using the existing recovery plant.

When required, the solution will eventually be discharged into the raw water storage dam, which has enough holding capacity for the cyanide to degrade to acceptable standards before discharge.

21.3.3 ENVIRONMENTAL IMPACTS

The variety of environmental impacts that arise from the operation of Iduapriem and Ajopa mining projects are categorised as indicated below.

21.3.3.1 Landscape and topography

In addition to loss of land through direct land-take, the mining operation has drastically changed the general landscape and topography of the area where it is sited by excavating the existing ridge system to a level below the valley floor. This has implication for upland habitats and the drainage pattern of the surrounding areas, which in turn leads to increased sediment load in watercourses.

21.3.3.2 Hydrology

Most settlements within the concession depend on streams and rivers that take their sources from the ridges for domestic water supply and other aquatic life. The water supply system of the settlements has been disturbed by mining activities of the four mining companies operating in the area, namely Teberebie Goldfields Ltd., Ghanaian Australian Goldfields Ltd., Ghana Manganese Company, and Goldfields (Ghana) Ltd.

The mining activities have led to changes in the drainage pattern of the surrounding areas and transport of increased sediments into watercourses through runoff and pit dewatering. At the early stages of the GAG mine, flow of ground water into the pit did not pose a problem. The main concern regarding pit dewatering then was surface water inflow. To address this, a raw water storage area was established north of the waste dump to accommodate a runoff volume equivalent to a typical storm event.

Later in the project life when groundwater flow to the pits occurs and a dewatering system is implemented, it is envisioned that the water supply system of surrounding villages and forests could be further affected. The mine proposes to prepare an EAP to address this and other subsequent problems.

21.3.3.3   Land-use

There are land-use conflicts between the mines and nearby settlements, populated mostly by farmers. The main impacts are loss of productive land and displacement of farmers. The most significant land-take has resulted from mine excavations, placement of waste dumps, and location of tailings dams. At Iduapriem, 54.9 hectares of land cultivated with cocoa, oil palm, and cassava had been taken by the mine and 25 farmers were displaced. At Ajopa, eight hamlets with a total about eight hectares of plantations were to be displaced.

21.3.3.4   Forest

About 11.5 km² of the Neung Forest Reserve falls within the Iduapriem Concession. The direct impact of the operations on the forest has been a direct loss of some primary and secondary forest as a result of mining, waste dump placement, construction of access roads, mine camps, raw water storage dam, and other ancillary mine works. The forests are not only important for their ecological value and their role in maintaining sediment and water balances, but also provide sanctuary for at least two rare species of birds in Ghana and act as a corridor for other birds (e.g., great blue turaco) and mammals (e.g., white-nosed monkey). GAG has estimated that at the completion of the Ajopa project, more than 62 hectares of primary forest and about 99 hectares of secondary forest would have been lost.

The loss of the forest resources also has implications for the local community members, who depend on it for essentials like fuel wood, meat, plant medicine, and materials for basketry.

It is noteworthy that the company purchased forested land in a non-reserve forest area and gave that to the government to re-designate as a forest reserve — thereby indirectly "replacing" the forest lost to mining.

21.3.3.5   Dust

The severest dust conditions are observed in the dry season. However, the impact on the local population is minimal since no settlements are located along the haul-roads.

All drill rigs are equipped with dust collection systems, and drill operators are routinely supplied with nose masks. Maintenance of the rigs is carried out regularly to ensure that they continue to operate correctly. At present dust four water bowsers using 73 million litres of water per annum carry out suppression in the mine and on haul roads. The bowsers are being monitored closely and, when necessary, additional bowsers will be provided by the company.

The company undertakes an occupational dust programme weekly. Where dust levels exceed the occupational limit of 10 mg m⁻³, nose masks are provided. GAG has also installed a dust gauge between the plant site and Block 1 and 2 mine pits to collect airborne dust. Other monitoring stations are to be installed at sensitive locations, including population centres. Dust exposure is regarded as a serious issue as the rocks being mined are siliceous. Long-term exposure can give rise to respiratory diseases.

21.3.3.6 Blasting and vibration

One of the major and frequent complaints received from villages and settlements near the mine is the effect of blasting on the communities. The company maintains that it employs good blast design and techniques in all its blasting operations and that blast monitoring results are within internationally acceptable standards. GAG believes that part of the complaint is from the surprise factor. The company insists on giving sufficient audible warning to local residents, informing them of intended blasts. The company has also intensified its environmental education exercise to explain blast vibration and damage to property to the communities, and what it has been doing to bring them under control.

21.3.3.7 Socioeconomic impacts

Some of the major socioeconomic impacts of the project on the area have already been discussed. These include loss of farmlands and displacement of farmers, along with loss of forest and forest products. Other impacts that are being considered are population increase in the area, health issues, cost of living, and changes in community structure.

The expectation of the predominantly farming communities living near the mine, particularly younger people, is to find employment at the mines. However, this has not materialised since the local people are largely unskilled. This is a common problem with all the mines in Ghana.

21.3.4 PLANNING FOR ENVIRONMENTAL CONTROL AFTER CLOSURE OF GHANAIAN AUSTRALIAN GOLDFIELDS LTD.

Planning for the closure of Iduapriem and Ajopa mines has the long-term objectives of ensuring that:

- People can live in the area and continue with farming or other economic activities without health risks and damage to life arising from the previous mining activities.
- Within a reasonable number of years, all pollution sources in the area will be eliminated and the environmental conditions returned to their pre-mine conditions.

The main areas of concern are

- Mine pits.
- Waste dumps.
- Tailings dams.
- Dust from various sources.

Residual chemicals, particularly cyanide present in spent heaps, can be a problem if not properly managed.

No formal request has been submitted to GAG or any of the mining companies operating in the country to post a bond to meet environmental costs after mine closure. It is not known, therefore, how much funding has been set aside by any mine for this purpose. However, GAG has indicated that once mining in the pits has ceased, they will eventually be filled with water and, in some cases, tailings material. The slopes bordering the pits will be terraced to assist revegetation and to facilitate the productive use of the area. Experience from other similar worked-out pits from the 1950s has shown that such areas become well vegetated naturally, including the steep pit slopes. The possibility of using abandoned pits for fishery will also be investigated.

The wastes from Block 1 and 2 pits are deposited in the forested area that does not have high quality forest to minimise ecological damage. However, where rare plant species such as the uncommon small palm are identified, they will be removed and replanted in unaffected areas. If mining is carried to the west of Block 2 pit, the waste will be deposited in non-forested areas.

The waste dumps will consist mainly of large rocks. Restoration or rehabilitation of these areas will commence once the dumps are completed. Runoff is to be diverted from the steep slopes to avoid gully erosion.

Prior to revegetation, the slopes of the dumps will be regraded to facilitate establishment of vegetation. Overburden is used as a cover and also graded.

In 1994, the company initiated a revegetation programme for which it retained a consultant to provide the necessary technical advice. Consequently, a nursery has been constructed and is being developed as an important element in the implementation of the programme. The tailings dam is located in a secondary forest where the ecological impact will be relatively low. Residual cyanide and other pollutants in the tailings dam are monitored. Seepage from the dam may occur through the dam wall and through the underlying strata. Five permanent boreholes have been located downstream and adjacent to the dam walls to monitor this. Piezometers are installed to monitor the levels of water within and beneath the dam wall. In addition, interceptor channels are provided to divert runoff from the surrounding slopes from entering the dam. Filter drains in the dam walls are also provided.

The dam has been designed so that water from peak storm events can safely be stored within the dam walls, and subsequently released through a spillway. The dam walls will be raised to accommodate increases in ore input.

Consideration is to be given to installing a water treatment plant to treat effluents from the dam before discharge. Since acid mine drainage (AMD) conditions are not present at the mine, the tailings dam is to be revegetated to prevent water and wind erosion.

The pollution from dust will be greatly minimised when mining activity ceases and the mine is closed. The potential dust-generating points such as waste piles and tailing impoundment will be revegetated. Long after the mine has closed, certain social changes introduced into nearby communities will continue to operate, and certain amenities such as infrastructure for potable water will begin to breakdown. The company will undertake the training of youth in the communities to establish small-scale enterprises, so that when the mine has closed these communities will not disintegrate.

21.3.4.1 Heap leach facility

No heap leach facility has been decommissioned in Ghana yet. Ghanaian Australian Goldfields Ltd., like other mining companies involved in heap leaching, is now developing comprehensive plans for heap leach closure. Ghana is in the process of developing standards for mining wastewater discharged into the environment. The company will therefore be guided by internationally acceptable standards. For the present, the company's plans for decommissioning its proposed heap leach facility entail the analysis of drainage from all spent dump areas, as it becomes available. It is anticipated that rainfall can be relied upon to remove residual cyanide from the spent dump. Laboratory investigations and field monitoring will also be necessary to determine how long it will take for the leachates from the spent dump heaps to reach permissible discharge limits.

Toxic effluent from mine waste, tailings impoundments, and spent heaps (unreclaimed) are primary sources of potential contaminants. Even solutions retained within the rinsed and reclaimed heaps (which may contain complex metals and cyanide) are potential contaminants.

21.4 CASE STUDY 2 — SIAM GOLDFIELDS LTD.

Siam Goldfields Ltd. is a joint venture company between Ekosco Mining Company of Ghana and Rodam Resources of the U.K. The company owns the Esaase Gold Concession (EGC), which is located about 10 km west of Nkawkaw in the eastern region of Ghana.

21.4.1   Site Description

The concession area of EGC is 69.12 km^2. There are eight major settlements within the concession: Esaase, Besease, Mpeyon, Kwadua, Nkwatuokow, Ahantanana, Odumasua, and Apradan.

The concession has a proven ore reserve of 1,007,360 tonnes and an average grade of 4.71 g t^{-1}. Mining at a rate of 165,000 tonnes per annum, the mine has a life of 6.2 years. Further exploration is being carried out to bring the probable and possible ore reserves into the proven category.

21.4.1.1   Types of ore

There are two types of gold-bearing ore in the concession, namely oxidised and primary ores (with a transition). The zone between the oxidised ore is composed mainly of quartz, kaolinite, and sericite, with a small fraction of other minerals. The primary ore is mainly a gold-pyritic type.

21.4.1.2   Mining methods

The mine will run three shifts, and is projected to employ about 360 people. Both open pit and underground mining methods will be used to mine the ore. The open pit operation will include the use of hydraulic backhoe shovels for stripping and mining, and 20-tonne dump trucks for haulage of ore and waste. Excavators will mine the oxidised ore, while the sulphide ore will require drilling and blasting. The mine will operate as open pit before the mining activities are extended underground. The underground mine will have a shaft and four levels, with the distance between each level being 50 m. The underground mining method adopted will vary depending on the thickness of the orebody.

21.4.1.3   Ore processing

The ore will be processed by the carbon-in-leach (CIL) method, which includes in its circuit crushing, milling, cyanidation, carbon adsorption and desorption, electrowinning, and smelting. Tailings from the ore treatment will be impounded in the tailings dam.

21.4.1.4   Water supply

A dam will be constructed on the Odumasua River to store water to be used on the mine. The total water requirement of 3470 m^3 per day of the mine will be obtained from boreholes, the dam, and recycled water. Potable water for domestic purposes will be extracted from a 200-mm diameter deep well at a rate of between 100 to 150 m^3 per day.

21.4.1.5   Tailings pond

The tailings dam is sited in a valley southwest of Odumasua village. The dam will cover an area of about 846,700 m^2 and will be 6.71 m high and 280 m long. Approximately 750 m^3 of tailings containing 8.2 mg L^{-1} of cyanide will be discharged into the drains. Decant water from the tailings dam will be pumped to the process water pond for recycling.

21.4.1.6   Pit dewatering

Constructing interceptor ditches along the outer boundary of the pit will control inflow of runoff into the mine pits. Water that collects in the pit from rainfall and groundwater will be pumped out from the pits, using submersible pumps.

21.4.1.7   Waste dumps

Waste rock from the open pit mine will be dumped at a low-lying area near the pit. The height of the dump is projected to be about 3 m. Waste rock from underground will be used to fill the surface excavations of the open pit mines.

21.4.2   ENVIRONMENTAL IMPACTS

The major impacts of the Esaase Gold Concession will be:

- Visual intrusion.
- Atmospheric pollution.
- Land degradation.
- Noise and vibration blasting.
- Water pollution (e.g., AMD, cyanide and sediments).
- Socioeconomic impacts.

21.4.3   PLANNING FOR ENVIRONMENTAL CONTROL AFTER CLOSURE OF SIAM GOLDFIELDS LTD.

At the close of the mine, the main residual impacts of concern that will be addressed will be:

- Visual intrusion.
- Degraded lands.
- Aqueous pollution problems.
- Dust.
- Socioeconomic impacts.

In the decommissioning programme for the Esaase concession, the company will dismantle all surface installations, machinery, and plant before leaving the concession. Other mine buildings and structures that cannot be dismantled will be, in consultation with the District Assembly, the EPA and other relevant agencies, left behind to be put to a fitting use. The surface excavations will be filled with waste rock from underground after ascertaining the waste is not acid generating. Some overburden from the waste dump will also be used as fill material for the mined-out pits. Topsoil from previous surface working will be spread on the filled area and revegetated to minimise visual intrusion and dust blow, and also control acid generation in the reclaimed area.

The waste dump will be regraded and potentially acid-generating waste will be covered by oxides and topsoiled before revegetation. The objective of covering the waste is to minimise the transport of oxygen to the waste to generate AMD. Siam Goldfields has not given a detailed revegetation programme of the disturbed mining areas, but it is envisaged that the company will establish a nursery to provide the necessary plant seedlings. An expert will also be required to oversee the implementation of the programme.

To prevent dust from the tailings dam from becoming wind-borne during the dry Harmattan month, the tailings dam will also be revegetated.

All accessible mine entrances will be closed or fenced to prevent possible accidents. Mine shafts will be completely sealed or filled with inert material. The possibility of long-term subsidence and resultant damage to surface structures will be evaluated.

Unused explosives, fuel, lubricants, and other chemicals will be removed from the site. Siam Goldfields Ltd. will undertake environmental investigation to assess the effectiveness of the decommissioning plan. Thereafter, it will monitor the restoration process for four years before returning the reclaimed land to the affected communities.

21.5 ANALYSIS OF BEST PRACTICE AND FUTURE TRENDS

This section is rather sketchy in view of the reasons given above; i.e., mine closures in Ghana were hastily done during the first and second world wars. However, the Minerals Commission and Ghana's EPA are working steadily to put proactive measures in place.

The practice in some leading mining countries is to follow laid-down principles of mine rehabilitation. First of these principles is to prepare rehabilitation plans before mining commences. *Ghana's Mining and Environmental Guidelines* require all mining companies to prepare an initial reclamation plan, which should encompass all land on the concession to be disturbed as part of the EIA and EAP. The EIA should also include the company's conceptual decommissioning plan. Both plans are subject to approval by the EPA before the company can proceed to execute them. The reclamation plan should be designed to achieve the following minimum objective:

- For restorable land, to physically stabilise and "clean" it of chemicals then leave it in a safe condition and return it to a similar land capability as prior to mining if feasible.
- For non-restorable land, to physically stabilise and "clean" it of chemicals, then leave it in a safe condition and encourage revegetation.

Some basic principles to be observed in developing a reclamation plan that the guidelines espouse are

- Identify land which is restorable (e.g., tops of waste dumps and land used for stockpiles) and non-restorable lands (e.g., deep open pits, steep faces of waste dumps).
- Identify quantities of reclamation media required.
- Identify applicable reclamation techniques.
- Develop a planned approach to progressive reclamation as an integral part of the mine plan.
- Commence reclamation as soon as possible after commencement of operations.
- Where practicable, undertake reclamation trials to refine techniques.

The guidelines require companies to develop a formalised reclamation plan after the first two years of operation, and submit copies of the plan to the EPA, Mines Department, and the Minerals Commission. If the mine is located in a forest/wildlife reserve, a copy of the plan should be submitted to the Forestry Commission.

A final reclamation plan is required to be developed by the mines prior to the five-year period before planned closure. It is necessary that the mining companies will honour all commitments made in the plan, except where a written permission otherwise has been given by the EPA.

The government of Ghana reserves the right to request mining companies to post reclamation bonds. The guidelines also require that the conceptual decommissioning plan of mining companies shall:

- Nominate the end-use(s) of all lands affected by the mining project.
- Nominate the end-use(s) of all buildings, housing, and other mine infrastructure components.
- Describe the fate of all fixed equipment.
- Describe the steps required to make the area safe.
- Describe how public access will be managed after mine closure.
- Describe the type and duration of post-decommissioning monitoring.

The guidelines are consistent with best practice in other leading mining countries. The starting point is always a good rehabilitation plan, which should be an integral part of the mine plan from exploration decision to mine closure. The use to which the worked-out area should be put is influenced by many factors including soil type, climate, and topography. Post-mining land-use plans are prepared in consultation with relevant government agencies, local government officials, and the landowner.

Obviously, it is not feasible to restore all areas of the mine site. Restoring or backfilling large open pits is very difficult and often uneconomic. Such sites are sometimes put to other uses, such as water storage purposes. Knowledge of restorable and non-restorable mine sites is essential.

Planning for rehabilitation requires knowledge of the volume, physical, and chemical characteristics of the overburden and topsoil available for that purpose. Where topsoil is available it should be stored for subsequent reuse. Unless found undesirable, the preferred approach is to reuse the topsoil immediately.

If topsoil is not available or suitable, other viable alternatives such as overburden should be tested and reused. Experience in Australia has shown that some waste rock weathers rapidly to produce suitable rehabilitation materials.

Land disturbed by mining, particularly waste dumps, should be reshaped and regraded to render it stable and well drained. Long-term visual intrusion is a feature of mined-out areas that can be minimised by creating a similar pre-mining landform, which blends with the surrounding landscape. Topsoil should be spread over the completed landform to begin the revegetation process.

21.5.1 REVEGETATION

Rehabilitation of mined-out areas requires the establishment and maintenance of vegetation on the disturbed areas. Before the establishment of vegetation, wind and water erosion should be controlled. The appropriate time to commence the revegetation is determined by the local climate conditions. The selection of species for the revegetation is dependent on the land use plan for the area, soil, and local climatic conditions.

Experience has shown that care must be taken in the introduction of exotic species which could become weeds for local agriculture. Carefully applied fertiliser may be needed to accelerate the growth of the species. Revegetation trials can be carried out for species diversity and growth rate.

21.5.2 MONITORING AND MAINTENANCE

The rehabilitated areas need regular monitoring and maintenance until they reach self-sustaining conditions. Areas not developing satisfactorily should be reworked. A typical monitoring programme for the rehabilitated areas could cover the following:

- Stability of the area, including drainage systems.
- Species diversity, canopy covers, and returns of fauna.
- Water quality (surface and groundwater).
- Public safety.

The monitoring programme should also embrace sociocultural issues.

21.6 SOME CONSTRAINTS AND OPPORTUNITIES FOR TRANSFERRING BEST PRACTICE

It has been argued that most developing countries, because of their weak economies, cannot afford and are unwilling to enact comprehensive environmental legislation for fear of losing needed investment capital. This argument can be misleading in that it has been established that for most international mining companies, investment decisions are least influenced by stringent and yet pragmatic environmental regulations. In many developing countries, the appropriate environmental legislation may exist but because of the limited monitoring capability this legislation is not effectively enforced.

Many countries tend to make up for the limited monitoring capabilities of regulatory agencies by asking mining companies to monitor themselves and submit periodic reports to the regulatory agency. However, surprise inspection by the agency is necessary to uncover any irregularities.

A few "fly-by-night" mining companies tend to regard environmental regulations as an unnecessary cost burden. However, experience has shown that for the more dynamic companies willing to undertake technological change and innovations, both environmental and production costs can be lowered.

The main problem in Ghana seems to be with the state-owned mines and the small-scale mining groups. These two categories of mines are faced with production inefficiency and are severely constrained technically and financially.

The measures Ghana has adopted are to solicit loans and grants from donor agencies, including the World Bank, to rehabilitate the environment damage caused by these mining groups and to put all the state-owned mines on divestiture.

It has been recognised that for the mining industry to achieve environmental efficiency, all joint venture and loan agreements involving mining companies should include a clause on transfer of best practice and training.

21.7 CONCLUSION

Mine closures in Ghana during this century have been linked with the outbreak of the two world wars. For that reason, these mine closures were inevitably unplanned and problematic.

The mining industry in Ghana is currently required to plan for an environmentally sound closure through the EIA and Environmental Management Action Plan requirements, and *Ghana's Mining and Environmental Guidelines*. One area that needs further research is how to effectively address socioeconomic and cultural impacts of mine closure. Such impacts include disruptions of the set patterns of communities and the "bust town" effect that is created by the closure.

There is therefore the urgent need to critically reappraise the national requirements for sustainable exploitation of the country's mineral resources *vis-à-vis* planning for mine closure.

ACKNOWLEDGEMENTS

The authors would like to give special thanks to the senior technical personnel at Ghana Australian Goldfields Ltd. for the invaluable assistance given during the preparation of this paper. Our gratitude also goes to Dr. George Manful and other EPA staff for their support.

REFERENCE

1. Environmental Protection Agency and Minerals Commission, *Ghana's Mining and Environmental Guidelines,* FLOENT Ltd., Nyaniba Estate, 1994.

22 Socioeconomic Impact of Mine Closures: A Case Study of Ghana

Edward Nyamekye

CONTENTS

1-56670-365-4/00/$0.00+$.50
© 2000 by CRC Press LLC

ABSTRACT

Ghana has a mining history dating back to 1400. However, during half of this period, the territory was a colony of Britain. Mining activities during the colonial period were organised essentially as enclaves. Consequently, mining operations were never integrated into the local economy. Some of these mines were forced to close down during the first and second world wars and no attention was paid to the socioeconomic impacts that resulted from these closures. Environmental impacts resulting from the operations of these mines were also neglected at those times.

Some mine closures that occurred recently clearly demonstrated the potential negative socioeconomic impacts of permanent unplanned mine closures. Significant impacts that could result from unplanned mine closures include:

- Degraded vegetation and agricultural lands — long exposure to oxides of arsenic and sulphur.
- Haphazard discharge of untreated liquid effluents that affect the aquatic environment.
- Loss of jobs.
- Dilapidated social infrastructure.
- Loss of taxes to government.
- Loss of captive markets for ancillary industries.

Mine closures could be efficiently managed by putting together appropriate legal and fiscal frameworks. The following critical issues would need to be integrated into contemporary mine closure management guidelines:

1. Defining and monitoring the depletion rate of ore reserves.
2. Establishing ancillary industries that could support socioeconomic activities within the mining communities after reserves have been mined out.
3. Integrating mitigating programmes into mainstream mining activities.
4. Establishing a fund to finance the cost of ancillary industries and mitigation schemes.
5. Providing supportive fiscal incentives for addressing closure expenses.

Since decommissioning expenses eventually affect the economics of mining projects, it would be beneficial to take care of such expenses at the mine planning stage rather than later. The appraisal of mining projects should, therefore, take into account the socioeconomic impacts of mine closures. The main focus of this chapter is an understanding of the socioeconomic impacts of mine closures, and learning from these, how to plan for smooth closures in future, taking the case of Ghana. The chapter begins with a background section on the mineral sector's role in socioeconomic development in Ghana. This discusses the share of minerals in export earnings and employment, its future potential, and its environmental impacts. Section two discusses mine closures in Ghana, dating from the colonial period. The third section looks at the essential concerns in planning for closure. The final part makes suggestions on how to transform a mineral-rich developing economy into an industrialised economy through action in the main mining areas.

The chapter concludes with a note that countries like Ghana have little to show for the exploitation of its mineral wealth. A transformation process developed into a sound environmental policy can change this.

22.1   THE MINERALS SECTOR IN SOCIOECONOMIC DEVELOPMENT IN GHANA

22.1.1   Primary Export-Led Growth

The exports of most developing countries comprise primary commodities — food crops, agricultural raw materials, and minerals. In the case of Ghana for instance, cocoa, minerals, timber, and other agricultural commodities accounted for 86% of total foreign exchange receipts in 1995.

TABLE 22.1A
Value of Exports (US$ million) 1989 to 1995

Year	1989	1990	1991	1992	1993	1994	1995
Commodity							
Cocoa — total	407.8	360.6	346.6	302.5	285.9	320.2	432.8
Beans	381.3	323.8	313.5	276.8	250.5	295.0	401.6
Processed	26.5	36.8	33.1	25.7	35.4	25.2	31.2
Non-cocoa — total	414.0	536.1	651.2	683.9	777.8	916.2	1032.0
Coffee	0.9	0.6	0.4	0.9	1.0	1.2	1.2
Sheabutter	0.4	1.3	0.2	0.7	1.0	0.8	0.8
Gold	169.8	201.7	304.4	343.4	434.0	548.6	637.2
Diamonds	5.2	16.6	19.2	19.3	17.3	20.4	15.3
Manganese	11.7	14.2	20.2	16.5	13.9	9.6	6.4
Bauxite	9.2	10.0	8.6	9.5	8.4	9.6	9.7
Timber and timber products	80.2	118.0	124.2	113.9	147.4	165.4	190.6
Residual oil	19.5	28.6	19.7	19.2	14.2	17.7	16.0
Electricity	82.5	88.5	96.0	95.6	69.1	56.4	55.0
Non-traditional exports	34.7	56.7	58.1	64.9	71.7	86.6	100.0
Total exports	821.8	896.8	997.7	986.3	1063.6	1236.4	1464.7

TABLE 22.1B
Percentage Share in Total Export

Year	1989	1990	1991	1992	1993	1994	1995
Commodities							
Cocoa	49.6	40.2	34.7	30.7	26.9	25.9	29.5
Gold	20.7	22.5	30.5	34.8	40.8	44.4	43.5
Timber	9.8	13.2	12.5	11.5	13.9	13.4	13.0
Electricity	10.0	9.9	9.6	9.7	6.5	4.6	3.8
Non-traditional	4.2	6.3	5.8	6.7	6.7	7.0	6.8
Other minerals	5.6	7.9	6.9	6.6	5.2	4.7	3.4
Total	100.0	100.0	100.0	100.0	100.0	100.0	100.0

Table 22.1A provides statistics of the total value of merchandise exports from Ghana for the period 1989 to 1995. A plot of the trend indicates that from 1992 on, gold replaced cocoa as the leading gross foreign exchange earner. Table 22.1B provides details of the percentage contribution of gross value of goods exported. In 1995, value of gold export amounted to US$637.2 million equivalent to 43.5% of gross value of exports. The total value of all minerals exported amounted to US$668.6 million, equivalent to 45.6% of total value of exports.

22.1.2 EMPLOYMENT AND SOCIAL CONTRIBUTION OF MINERALS SECTOR

The minerals sector represented by large-scale mines employed a total of 19,500 people in 1995. The gold subsector alone accounted for 18,599 employees, equivalent to 95% of the total labour force in the minerals sector. On the other hand, the small-scale mining sector — mainly gold and diamond winning — employed over 100,000 people during the same period. Tables 22.2A and 22.2B provide

TABLE 22.2A

Social and Economic Impacts of Operating Mines (1995) — Gold Mining

Category Company	I	II	III	IV	V	VI	Wage Bill US$(,000s)
Ashanti Goldfields Co.	11,480	X	X	O	O	O	47,637
Dunkwa Continental Goldfields	383	X	—	X	O	O	213
Gold Fields (Gh) Ltd.	1459	X	—	X	O	O	4225
Teberebie Goldfields	941	—	—	—	—	O	6240
Prestea Goldfields	1580	X	X	X	X	O	—
Billiton Bogosu	1058	O	X	X	—	O	4769
Obenemase Goldfields	259	O	—	O	—	—	840
Bonte Goldfields	323	O	—	O	O	—	1111
Ghana Australian Goldfields	530	O	—	—	—	O	1514
Cluff Resources	407	X	—	—	—	O	1609
Prestea Sankofa	184	—	—	—	—	—	247
Subtotal	18,604						68,405
Small-scale miners (gold)	60,000						16,245

Key
I—Employment
II—Clinics
III—Schools
IV—Water
V—Electricity
VI—Housing
X = Available to the community
O = Mine staff exclusively
— = Not provided

an overview of the social and economic implications of mines operating in Ghana. All the industry towns developed around the mines and facilities such as clinics, water supplies, schools, and electricity provided by the mining companies were sometimes taken for granted by the community. The mining companies also provide residential accommodation and clubhouses for their employees.

22.1.3 THE FUTURE POTENTIAL OF THE MINERALS SECTOR

The resource base of the large-scale minerals sector in Ghana is given in Table 22.3. The operating mines cover a total concession area of over 1200 km², of which gold mining enterprises cover about 1000 km². The remaining concession areas have been allocated to bauxite (13 km²), diamond (153 km²), and manganese (40 km²). A total concession area of 35.4 km² has so far been assigned to the small-scale mining (SSM) sector. Information on the resource base of the SSM sector is not available, since the sector operates without any information on reserves.

Current estimates put the total value of Ghana's mineral wealth at nearly US$15 billion. Gold alone accounts for over 93% of this stock of capital estimated at almost US$14 billion. The average life of the extractable gold reserves is conservatively estimated at 21 years.

In view of the current high interest in gold exploration and the effectiveness of Ghana's minerals policy, the future potential for the gold mining subsector is very encouraging.

22.1.4 ENVIRONMENTAL IMPACTS OF MINING

Ghana, known previously as the Gold Coast before its independence in 1957, has a long history of mining (especially for gold), dating back to the 5th century BC. Gold mining on an industrial

TABLE 22.2B
Social and Economic Impacts of Operating Mines (1995) — other sectors

Category Company Name	I	II	III	IV	V	VI	Wage Bill (US$,000s)
Ghana Bauxite Company	551	X	X	X	X	X	
Ghana Manganese	732	X	X	X	X	O	
Ghana Continental Diamonds	1,087	X	X	O	O	O	
Subtotal	2,370						
Small-scale miners (diamond)	10,000						7,314
Grand Total	90,974						
(Tables 22.2A and 22.2B)							

Key
I—Employment
II—Clinics
III—Schools
IV—Water
V—Electricity
VI—Housing
X = Available to the community
O = Mine staff exclusively
— = Not provided

TABLE 22.3
Estimated Proven Mineral Reserves

	Gold	Bauxite (Mt)	Diamond (carat)	Manganese (Mt)	Total Value (US$M)
Quantity	34,765,000 oz	31.0 Mt	8.0 M carats	5.05 Mt	
Estimated value ($M)	13,906	612.0	194.0	186.0	14,898
Estimated average life (years)	21	62	15	25	
Annual exploitable value ($M)	662.19	9.87	12.93	7.44	692.43
Percent	93.3	4.1	1.3	1.3	100.0

scale, however, began in the latter half of the 19th century. Alluvial diamond deposits have been mined since 1919, while mining of manganese and bauxite commenced in 1924 and 1941, respectively. The operations at the famous Ashanti Goldfields date back to 1897.

Mining during the colonial period was organised essentially on an enclave basis. As a result of a lack of effective supervision, negative environmental impacts were not addressed, and mine effluents were consequently discharged directly to the environment without prior treatment.

A paper entitled "Study on the Effect of Mining on Ghana's Environment," completed in October 1996, catalogued impacts of mining activities on Ghana's environment. NSR Environmental Consultants Pty. Ltd of Australia conducted the study, which was prepared to assess the level of past environmental damage and define environmental liability as part of the ongoing divestiture process. The Ghanaian government has been saddled with the responsibility of financing the remediation expenses, even though foreign companies owned the mines. The study revealed that considerable damage to the environment had occurred as a result of long-term neglect.

A discussion of some of these impacts follows.

22.1.4.1 Air and atmospheric pollution

Major air and atmospheric pollution has resulted from the roasting of sulphide ores at Obuasi, Prestea, Konongo, and Bogosu. Common air pollutants included sulphur dioxide, arsenic oxide, and particulate matter.

22.1.4.2 Effects on vegetation

Negative environmental effects have resulted from arsenic and sulphur oxide poisoning of the vegetation. As a result of acid rain resulting from fallout of the oxides, these mining areas have lost their natural vegetation. Thus in Obuasi, Prestea, and elsewhere, the vegetation is dominated by a few plant species, especially the palm tree. Timber was also used extensively for underground support. The virgin forest has consequently been replaced in Obuasi and Prestea with plantations of teak, mahogany, and other exotic species.

22.1.4.3 Effect on land

Pressure on agricultural lands has become a major problem in mining communities. Initially mines were located in remote and sparsely populated areas. As settlements developed, non-mining and agricultural activities developed. This has heightened competition for farmland within mining areas.

22.1.4.4 Effect on water bodies

Surface water resources have also been generally polluted by arsenic and sulphur oxide fallout and by other inorganic elements present in the waste rock. In addition, surface mines in Akwatia, Dunkwa and Awaso have rendered some rivers and tributaries unsuitable for household purposes. Additional information is provided in Appendix 22A.

22.2   THE HISTORY OF MINE CLOSURE IN GHANA

Gold mining in Ghana has a checkered past. During the period 1493–1600, Ghana produced 35.5% of recorded world gold. Gold production fell to 22.8%, 8.9% and 9.7% for the periods 1601–1700, 1701–1800 and 1801–1900, respectively. During this period, a total of 22.6 million fine ounces of gold were produced and exported from Ghana. From 1901 to 1934, another 30.43 million fine ounces of gold were produced and exported.[1]

From 1935–1957 when Ghana became a self-governing nation, an additional 13.9 million fine ounces of gold were produced and exported. Thus between 1493 and 1957, about 67 million fine ounces of gold have been exported from Ghana. This excludes gold produced by natives that did not enter international trade. From 1957–1995, another 24.66 million fine ounces have been exported. Thus from 1493–1995, Ghana is recorded to have produced over 90 million fine ounces of gold.

From 1920–1995, a total of 100 million carats of diamonds were produced and exported from Ghana.[2] Ghana enjoyed its first gold rush during the period 1897–1914. During this period, as many as 3613 concessions inquiries were known to have been filed in the then Crown Colony. This phenomenon was galvanised by the Boer War (1899–1902) in South Africa and the eventual annexation of Ashanti Kingdom by the Crown, culminating in the Yaa Asantewa war (1900–1902).

The gold boom was, however, interrupted by the First World War (1914–1918). Most of the mines closed down and only the large ones survived after the war. The second boom occurred between 1925–1935. By 1933, according to Kesse,[1] there were as many as 7000 inquiries for mineral concessions in the colony (Gold Coast), and, as in 1935, 600 of these mineral concessions had been validated with 33 operating mines listed on the London Stock Exchange (LSE). Another 54 companies that were not quoted on the LSE were known to be in operation.

The boom created labour shortages, and fierce competition for labour developed between agriculture (cocoa and rubber) and small-scale indigenous gold miners on one side and expatriate mechanised mines on the other. The colonial government intervened on the side of the expatriates and passed the Mercury Ordinance to prohibit the use of mercury by the natives. This deprived the natives of the technology that was considered to be at the heart of the gold recovery process.

The Second World War (1939–1945) interrupted the second boom. Thus by 1950, only 11 companies remained as a result of political agitation for self-government which began after the war. As the future outlook for economic policy after independence became uncertain, capital flight from the emerging state ensued and most of the mines were deliberately mined into extinction by curtailing investment capital for mine development and exploration. Ghana, then at the forefront of the decolonisation process of Africa, was essentially perceived as an emerging communist state and a strong destabilising factor to foreign investment in Africa.

In 1961, as a result of anticipated negative socioeconomic impacts from unplanned closures, the government created a public company called the State Gold Mining Corporation to protect jobs and safeguard socioeconomic activities within the following major mining towns: Konongo, Prestea, Tarkwa, Aboso, and Bibiani. It was only the Ashanti Goldfields at Obuasi that was kept in operation by Lonrho. Similarly, the government had to intervene to prevent production at the following mines from closing down:

- Diamond mines at Akwatia.
- Manganese mines at Nsuta.

22.2.1   CAUSES OF MINE CLOSURES IN GHANA

In order to plan for smooth closures in future, it is pertinent to draw some lessons from Ghana's experience by examining some of the causes of mine closure. Available records indicate that most mines were closed in Ghana as a result of a combination of these factors:

- Unfavorable commodity prices in the international market.
- The World Wars (1914–1918 and 1939–1945).
- Unfavorable host nation policies.
- Lack of clarity in government policy

22.2.1.1   International commodity prices

Gold, one of Ghana's leading export commodities, was used as basis for international settlement–the gold standard. Thus demand for gold increased as international trade expanded. However, when the gold standard was abolished, the demand for gold was affected and investments were curtailed. On the supply side, the price of gold until 1971 was fixed at $35 per ounce. Thus it became unprofitable to operate certain mines, even when ore deposits had not been exhausted.

22.2.1.2   International conflicts (the World Wars)

The two World Wars, as already mentioned, caused resources — especially labour and capital — to be diverted towards the war effort. Mining activities in the colonies, including Ghana, were therefore starved of resources. Commerce was simultaneously disrupted and this affected the demand for gold, which formed the basis of international settlements.

22.2.1.3   Unfavorable host nation policies

Nationalist movements characterized the period following the Second World War. Most of the colonies that fought on the side of their so-called "colonial masters" agitated for self-governance.

This agitation for self-determination in Ghana and elsewhere was interpreted as creeping nationalisation. This fear resulted in capital flight; consequently, most of the gold mines in Ghana were strategically run down to precipitate premature closure, because investors were apprehensive of the future.

22.2.1.4 Lack of clarity in government policy

No clear policy exists for the management of the process of closure in most developing countries. In the Mining Regulations of Ghana, for instance, only a terse reference is made to abandonment: the regulations merely require the mining company to restore the mining area to its original state.

22.3 PLANNING FOR CLOSURE

The objectives of mine closure policy may differ from country to country. With respect to developing countries, given the crucial role that mineral revenues play in their economies, the three objectives stated below could be considered legitimate and mineral resources should be exploited to:

1. Maximise returns from mineral exploitation to the economy of the host nation, both in terms of foreign exchange earned and as sources of government revenue.
2. Transform an economy based on a stock of depletable natural resources into an industrialised economy.
3. Protect the environment from permanent damages that are associated with mineral resource exploitation.

22.3.1 MAXIMISATION OF RETURNS TO THE HOST NATION

For Ghana, the need to maximise returns from mineral exploitation is deemed crucial for the following reasons:

- It is the leading source of foreign exchange (46%).
- It is a major source of government revenue (13%).
- It is a major source of employment and social services.

Factors that are a prerequisite for ensuring effective management of mineral resources include:

- Ownership of mineral rights.
- Sharing of returns associated with minerals exploitation projects.

22.3.1.1 Ownership and assignment of mining rights

Systems of property rights can and do have significant effects on the way a society uses resources, including nonrenewable ones such as mineral deposits. With the exception of the U.S. and a few other countries, property rights with respect to minerals are normally vested in the state. Ghana is not an exception to this popular international convention. According to Ghana's Mining and Minerals Code, all minerals within the legal boundaries of Ghana are vested in the state and the code creates the Minerals Commission, which is given the responsibility to promote and ensure that the exploitation of mineral resources optimally benefits the state.

Mineral development is a very expensive and risky undertaking. Host governments that are often faced with multiple choices of expenditure responsibilities can therefore ill afford to invest a substantial portion of their limited resources in mineral exploration. Mineral-rich developing countries therefore either have to rely on international mining companies or postpone the development of their mineral resource sectors until they are in a position to exploit them optimally. The strategy that is often adopted to promote the development of mineral resources involves the creation

TABLE 22.4
Rights and Obligations in Mineral Contracts in Ghana

State	Investor
A: Obligations	
i. Guarantee exclusive right to concession	i. Provide risk capital
	ii. Supplies management
	iii. Restores environment to original state after prospecting, during closure, or abandonment
B: Rights and Financial Benefits	
i. Consideration and permit fees	i. 90% equity holding
ii. 10% free equity holding	ii. 65% of net profit
iii. 3% royalty on quarterly assessment of gross value of minerals won	iii. Accelerated depreciation—75% in year one. 50% of residual annually
iv. Corporate tax equivalent to 35% of net profit.	iv. Losses carry over
	v. Retention of 60–80% of value of minerals exported in offshore account
	vi. Import duty relief on machinery, equipment, and items on approved mining list
	vii. Use of property to guarantee loans
C: Management	
i. Board representation	i. Total management and control

of an enabling environment and the appropriate legislative framework to ensure a fair return for both the international mining companies that take the risk and provide financing for exploration and development, and the host nation which owns the minerals.

Ghana uses the concession system for the management of natural resources. It is an improvement over the historical relationship that existed between the investor and the state. Under this system, the investor is given the exclusive right to prospect for a specific mineral within a designated area for an agreed time frame. If a discovery is made, the finder reserves the right to exploit the find. Table 22.4 summarises the rights and obligations of both the state and the investor under the mineral and mining laws of Ghana. The net effect of the legislation is that the host nation assumes a small financial risk, but secures a minimum three percent of the value of minerals won in the form of royalty, in addition to:

- Ten percent of dividends declared.
- Corporate taxes on net profit (35 percent, currently).
- The employees' income tax and social security payments.

The system, as operated in Ghana, has the tendency to encourage:

- The use of capital-intensive technology.
- Rapid depletion of mineral deposits.
- Restriction of the host nation's benefits to the minimum level of royalty payable by the mining company.
- External control of mineral resources.
- Capital flight from the host nation, through the establishment of external accounts.

In view of the government's heavy reliance on income from the minerals sector, manipulative reduction of income accruable to the host nation has serious negative long-term socioeconomic impacts. Since the investor retains total control over the enterprise, it is difficult for the state to predict with any certainty the revenue that will accrue. Receipts from equity participation and corporate tax most often never materialize. This makes budgeting and planning by government cumbersome. In Ghana, for instance, government has been forced to divest its interest in mining ventures as a result of cash flow constraints.

22.3.1.2 Sharing and evaluation of benefits from minerals development projects

One measure used as an index of the host nation's benefit from mineral resource investment is the concept of retained value. Though not a perfect method of assessment, movements in the ratio of retained value to gross value of minerals won — and comparing it to similar situations prevailing in other countries for projects in the same sector — could assist in arriving at effective policy instruments.

Retained value (RV) is primarily defined as the total of all revenues from natural resource projects effectively retained in the host nation as shown in Eq. 1.

$$RV = W_d + C_d + DP(1 - Z) + K_d + T_d + Q_d \qquad \text{Eq. 1}$$

Where   RV $=$ retained value
   W_d $=$ labour income for host country workers
   C_d $=$ proportion of income of expatriate workers spent locally
   DP $=$ domestic procurement of goods and services
   Z $=$ percentage import content of DP
   K_d $=$ capital income for domestic shareholders — including the state
   T_d $=$ taxes, royalties, and other fiscal receipts, received by host governments and their assigns
   Q_d $=$ miscellaneous payments (licences, permits, etc.) received in host country

Often, mining companies base their investment decisions on the return from the project with respect to their total investment exposure. Numerous modes of measurements are in use including payback period, internal rate of return, etc. Host nations, as a matter of strategy, put up minimum risk capital for mineral development projects. Consequently, the use of the conventional investment ranking procedures such as payback period or internal rate of return will yield results that could mask the objectives of host nations.

Though some of these criteria may be useful to the investor, they do not take into account the socioeconomic objectives of the host nation. Host nations are more interested in their share of the revenue from a depletable asset that is available to promote development. The ratio of retained value to gross value of mineral exported therefore provides host nations a more solid basis for assessing the socioeconomic desirability of mineral investment projects.

Retained value targets could serve as benchmarks for monitoring the effectiveness of mineral sector projects. For Ghana at the moment, the retained value ratio for the minerals sector during the 1995 operational year was estimated at about 20%. That is, for every 100 dollars of minerals exported, less than 20 dollars are effectively available for socioeconomic development within Ghana (Table 22.5). If developing countries such as Ghana are to advance their mineral sectors in a sustainable manner, they require assistance to develop an effective methodology and capacity to evaluate the contribution of mineral sector investments to their economy.

22.3.1.3 Internal rate of retention

The analysis based on retained value of mineral exports could be made more sophisticated by taking into account the period during which the value is retained. To achieve this, one would need to discount separately the streams of the annual retained values obtained throughout the assumed

TABLE 22.5
Retained Value Ratios of Minerals Exported (1990–1994)

Year	1990	1991	1992	1993	1994
1) Labour income (W$_d$)—local	3.78	5.3	12.0	19.34	21.10
2) Employers' social security contribution	0.47	0.63	1.37	2.35	2.69
3) Domestic purchases of goods and services (DP)—energy	2.12	2.88	6.55	10.20	17.76
4) Percent import content of DP (Z)	—	—	—	—	—
5) *Capital income of domestic share holders (K$_d$)	—	—	—	—	—
6) Taxes, royalties and other fiscal receipt (T$_d$)	6.65	5.49	6.81	10.53	15.36
7) Miscellaneous payments (licenses, permits, etc.) (Q$_d$)	4.20	4.96	7.49	10.18	29.64
8) Retained value (RV) (cedis)	17.22	19.26	34.22	52.60	86.55
9) Average exchange rate (C/US$)	326.3	326.8	500.2	750.9	956.7
10) Retained value (RV)—US$ million	52.77	53.10	68.41	116.5	90.47
11) Export value of minerals (R) US$	226.40	332.41	368.81	451.39	540.62
12) Ratio of RV/R × 100%	23.3	16.0	18.55	25.82	16.73

* Records are incomplete. Payments are not sufficiently substantial to affect the analysis.

life of the project and the total value of minerals projected to be exported at the project's internal rate of return. It is required to produce a bankable feasibility study before the grant of the mining lease. It is the established process for mining companies to determine the internal rate of return associated with the projects. The mining companies are guided by their internal policies to participate in projects that show a certain minimum internal rate of return.

The sum of the streams of the discounted periodic values of the retained value of minerals produced gives the present value of the stream of retained values (RV$_p$) over the project life. Similarly, the sum of the streams of the discounted periodic values of the total value of minerals exported helps us to arrive at the present value of minerals exported over the life of the project (EV$_p$).

The ratio of the sums of the two streams of present values (RV$_p$/EV$_p$), which may be referred to as the internal rate of retention, could be used by host nations as an additional criteria for project selection or can serve as an instrument for negotiating with international mining companies.

It may be desirable for developing countries to monitor the operations of individual mining companies using the retained value criteria, in order to assess the effectiveness of such projects to the host nation's development process.

For instance, in Ghana, the impression has been created that the minerals sector is the highest foreign exchange earner. This statement may be true in absolute terms. However, the mineral sector is extremely capital intensive and therefore consumes a greater proportion of the foreign exchange generated. Mining companies are, for instance, allowed to retain between 60-80% of the foreign exchange earned in offshore accounts just to service their import requirements. Thus the retained value of most mining ventures may hardly exceed 30%. On the other hand, the retained value within the cocoa industry that depends exclusively on peasant agriculture is over 80%.

22.4 TRANSFORMING A MINERAL-RICH DEVELOPING ECONOMY INTO AN INDUSTRIALIZED ECONOMY

The establishment of a strong mining sector could be used as growth mode for the promotion of rapid industrialisation. This could be achieved through backward integration and forward linkages. The strategy that is appropriate for a particular developing country depends on the stock and level of technology which already exists in that country. Specific sectors of the economy that could benefit from such a transformation process include manpower; materials and chemicals; service and engineering; and manufacturing sectors of the economy.

22.4.1   Domestic Cost Component of Minerals Produced

A desirable objective is to increase the domestic cost component of minerals development projects. This could provide a driving force for the industrial transformation process. The roles that the various sectors of the economy are expected to play in this process will be examined below.

22.4.2   Manpower Planning and the Minerals Sector

The rapid development of the minerals sector leads to the creation of jobs within the economy. Specific skills are required for locals to fit into vacancies that are created. Where locals with the requisite skill and experience cannot be readily located, the mining companies have no option but to recruit expatriates. The employment of expatriates in the long run is expensive to the host nation and constitutes a resource transfer. For instance, a typical mine will have openings for geologists, along with mining, chemical mechanical, electrical, and civil engineers, as well as accountants, lawyers, doctors, middle-level technicians, drivers, caterers, and so on.

Thus a mineral-rich developing country could retain a higher proportion of the total value of minerals exported by having a good mix of skilled, employable, and trainable human resources. For this to be achieved, the host nation has to put in place a good general education system that provides avenues for professional and higher education with middle-level technical and vocational support. This requires effective manpower planning to define the needs of the mining industry in advance, so that resources could be provided ahead of time to stimulate the training of requisite human resources.

22.4.3   Local Manufacture of Mining Inputs

In any typical mining operation, various materials in the form of chemicals, fuels, lubricants, and so on are consumed. In developing countries, due to the weakness of the manufacturing sector, most of these mining consumables are imported. Unfortunately, this situation affects the potential benefits that the host nation could have derived from mineral resource projects. In this process, the ratio of the value of minerals exported that could otherwise have been retained within the economy for socioeconomic development in the host nation is reduced. In order to take advantage of the captive market and thereby increase the retained value of minerals won, it is important to prepare detailed statistical information on all imported inputs, indicating their sources of supply, value, and quantity and demand forecast over a medium-term period of five years. The data could be developed into an inter-industry demand matrix, which could form the nucleus for identifying investment projects. The successful implementation of such captive projects will accelerate the process of transformation, increase the ratio of retained value of minerals exported, generate jobs, and prepare the grounds for the export of manufactured goods.

For instance, as a result of increased production of gold in Ghana, certain raw materials suppliers have established intermediate processing facilities relating to the production of gold. These include:

- Explosives.
- Lime (bulk handling facilities).

Other mining consumables that could be readily produced locally include steel balls, caustic soda, activated charcoal, alum, and sulphuric acid, all of which are imported at present. A shortlist of such materials that could be manufactured in Ghana is provided in Table 22.6.

22.4.4   Domestic Mining and Process Equipment Industry

This is one of the areas where developing countries are most vulnerable. Despite the fact that Ghana has a mining history dating back to the 1400s, almost all machinery has been imported. In view

TABLE 22.6
List of Machinery and Mining Consumables

Equipment

1. Crushers
2. Ball mills
3. Steel balls
4. Tools

Mining Consumables	Suitable Local Raw Materials
1. Hydrated lime	Limestone, clam shells
2. Caustic soda	Salt
3. Alum	Bauxite and sulphuric acid
4. Alumina	Bauxite and caustic soda
5. Sulphuric acid	By-product from pyrite ores
6. Activated charcoal	Coconut shells
7. Leaching pads	Plastic film extrusion plants
8. Explosives	Air
9. Gold refinery	Chlorine and dore
10. Cement	Limestone and gas
11. Rubberised hoses and rubber lined products	Natural and artificial rubber
12. Boots	Rubber

of the high cost of mineral processing equipment, it is worth every bit of effort for the host nation to offer incentives to promote local fabrication of machinery and equipment. In the long run, the development of a vibrant local engineering industry to supply the needs of the mining industry is crucial. This sector, if effectively established, could support activities in other sectors of the economy.

22.4.5 Mobility of Mining Companies

Mining companies have the built-in flexibility to move to countries where mineral resources are available. Unfortunately, host nations have boundaries and a fixed stock of mineral resource deposits within them. Once a nation's mineral deposits are depleted, it would have to rely on imports to satisfy future local demand. Nevertheless, mineral resource-poor but capital-rich countries benefit from mining projects by investing in mineral development projects in mineral resource-rich countries. This option, however, cannot be exercised by financially weak developing countries; this therefore underscores the need for poor but mineral-rich developing countries to exploit such resources efficiently.

22.5 ENVIRONMENTAL MANAGEMENT IN THE MINING INDUSTRY

Environmental management with respect to closure is an area where developing countries remain most vulnerable. Ghana, for example, is at the moment saddled with the problem of cleaning pollution that resulted from the mining activities of foreign companies — dating back to 1493.

Despite the negative environmental effects that the country has sustained in the past, there appears to be no urgency to implement environmental standards to provide a basis for process design. Sometimes, developing countries use the relaxed environmental legal framework as an incentive to attract investment into the minerals sector.

22.5.1 MINERAL DEPLETION POLICY

Mineral depletion policy has some effect on the environment and mine closure. At the moment, Ghana has no official mineral depletion policy in place. There is no mechanism for monitoring the ore reserve policy of the mining companies, an area where developing countries would need assistance.

22.5.2 LAND RECLAMATION

Ghana has good potential for agro-forestry in the areas where surface mining activities are currently taking place. Bauxite, manganese, diamond, and all the gold mines now undertake extensive surface mining operations. The reclamation of mined-out areas could therefore be managed in such a way that the land could be planted to agro-forest that could provide raw materials for future agro-based or forestry related industries.

At the moment, there is no policy to guide mining companies in the area of land reclamation. Consequently, the mining companies undertake land reclamation projects without any guidance. Some mining company representatives who were interviewed by the author indicated that under the terms of leases granted to them, they only hold mineral rights. There is therefore the need to address the issue of the ownership of the agro-forestry resources that are eventually created by the land reclamation efforts of the mining company, as well as how such operations should be financed.

There is also the issue of the eventual utilisation of the resource. The establishment of, for example, a plantation will require maintenance, harvesting, processing, and marketing. By putting in place the appropriate guidelines, reclaimed as well as alienated lands could be utilised with the assistance of mining companies to provide raw materials for pulp and paper, sawmills, fruit juices, sugarcane plantations, and so on. In the case of open pit operations, potential for the creation of vast inland fisheries and water resources could be exploited.

Thus with an appropriate policy framework, land reclamation could be used to establish the basis for an agro-forest-based industry and promote the process of transforming a community based on non-renewable mineral resources into one with a renewable industrial base. Land reclamation could thus become a cost-effective undertaking and yield tangible benefits both to the mining company and the host nation.

22.5.3 MANAGEMENT OF SOCIOECONOMIC IMPACTS AND EXTERNALITIES

Environmental Impact Assessment studies are required to be submitted before large-scale mining projects can be implemented. At the same time, existing large-scale mining companies are required to produce impact and mitigation plans to handle past and projected negative impacts. Environmental impacts have both technical and social contexts. In our experience, the mining companies are able to thoroughly predict and provide solutions for technical impacts, but often gloss over the social aspects. It is important to regard mining projects as an opportunity to solve social problems within the mining areas.

22.6 CONCLUSION

Ghana has very little to show for being a mineral-rich country. The minerals sector operates as an enclave with very few linkages to the national economy. Thus while Ghana is pursuing macroeconomic reforms aimed at liberalising the economy, mining companies have been given exclusive rights to operate offshore foreign exchange accounts. The mining companies are given the right to open an external account, and thus between 60-80% of the value of minerals exported is not available to the economy. The existing mining and minerals laws need to be revisited to ensure that Ghana benefits from mineral projects sufficiently to generate the much-needed resources for socioeconomic development.

It is also vital that measures are put in place that will enable the replacement of mining projects based on depletable resources with agro-industry, industrial, and other projects which are based on renewable resources. Finally, this transformation process should be dovetailed into a sound environmental policy that ensures sustainable development.

REFERENCES

1. Kesse, G. O., *The Mineral and Rock Resources of Ghana*, A. A. Balkema, Rotterdam, 1985.
2. Ghana Chamber of Mines, Annual Report, Chamber of Mines, Accra, 1994.

APPENDIX 22A

IMPACT OF MINING ON GHANA'S ENVIRONMENT

COMPANY — Ashanti Goldfields
MINING OPERATIONS — Underground and surface gold mining
LOCATION — Obuasi
IMPACTS AND ALLEVIATION —

  (a) Lease area — 334 km^2.
  (b) Discharge to the atmosphere comprised oxides of arsenic and sulphur.
  (c) Discharge to land comprised tailings waste rock.
  (d) Open pit mines have affected the vegetation and damaged a substantial amount of land.
  (e) Waste disposal from historic and ongoing processing of underground sulphide and arsenic-rich ores has led to severe water pollution problems.
  (f) Ashanti Goldfields has developed a plantation covering 29 km^2 stocked with teak, cassia, gmelina, eucalyptus, and luecaena.
  (g) Compensation is paid to farmers and landowners.

COMPANY — Dunkwa Continental Goldfields
MINING OPERATIONS — Dredging of the Ofin River
LOCATION — Dunkwa
IMPACTS AND ALLEVIATION —

  (a) Lease area 1 km wide and 75 km stretch of Ofin River and tributaries.
  (b) No major air pollution occurs.
  (c) Flood plains of river have been affected and an estimated 90–120 hectares per annum will be lost.
  (d) Dredging generates sediments and leads to high suspended solids (200 mg L^{-1}).
  (e) Aquatic life has been severely disrupted and river fisheries occur further downstream.
  (f) Gold recovery was by amalgamation. Mercury losses had been put at 40kg per month in 1990.
  (g) The dredging operations have rendered the mainstream of the Ofin River turbid and unsuitable for domestic uses.

COMPANY — Ghana Bauxite Company
MINING OPERATIONS — Open pit mining of bauxite for exports
LOCATION — Awaso
IMPACTS AND ALLEVIATION —

  (a) No air pollution occurs.
  (b) Concession area covers 13 km^2.
  (c) Has the highest aluminum grade (52%) known in the world.

(d) Area within Afao Hills Forest Reserve and hence does not impinge upon cultivated land.
(e) Effects on water consist of surface runoff into rivers. This constitutes no harm since no chemical is employed.
(f) Effluent from washing plant is contained in ponds.
(g) Provides basic amenities including water to the community, which is sparsely populated.

COMPANY — Ghana National Manganese Corporation
MINING OPERATIONS — Open pit mining of manganese
LOCATION — Nsuta (Tarkwa)
IMPACTS AND ALLEVIATION —
(a) No air pollution occurs.
(b) Effect on land by way of conflict is minimal.
(c) The use of settling ponds effectively minimizes discharges into rivers.

COMPANY — Tarkwa Goldfields — now Goldfields (Gh) Limited
MINING OPERATIONS — Underground gold mining
LOCATION — Tarkwa
IMPACTS AND ALLEVIATION —
(a) Lease area of 111 km^2.
(b) No air pollution occurs.
(c) Effect on land is minimal.
(d) Effluent discharges to the natural drainage occur at the tailings dam overflow and from mine dewatering.

COMPANY — Teberebie Goldfields
MINING OPERATIONS — Open pit gold mining
LOCATION — Tarkwa
IMPACTS AND ALLEVIATION —
(a) Lease area of 42 km^2.
(b) Discharge to atmosphere consists mainly of dust.
(c) Compensation is paid for land lost to users.
(d) Rainfall-based erosion and scour of solids in exposed mining and construction areas contribute sediments to local streams.

COMPANY — Prestea Goldfields Limited
MINING OPERATIONS — Underground gold mining - over one century old
LOCATION — Tarkwa
IMPACTS AND ALLEVIATION —
(a) Emissions to the atmosphere comprised oxides of sulphur, carbon, and arsenic.
(b) Tailings and effluent discharged directly into nearby river and streams.
(c) The pollution from oxides of arsenic, sulphur, and carbon has restricted the surrounding vegetation.
(d) The company has a timber concession of 385 km^2 and an extensive teak plantation.

COMPANY — Billiton Bogosu
MINING OPERATIONS — Open pit gold mining old
LOCATION — Bogosu
IMPACTS AND ALLEVIATION —
(a) Concession covers 148km^2.
(b) Mined since 1906; present activity started in 1990.
(c) Atmospheric impacts consist of dust and oxides of sulphur, arsenic, and carbon.

(d) EIA produced to manage impacts.
(e) Compensation paid for farmland crops affected include oil palm, orange, coconut, plantain, cocoa, cassava, and groundnut.
(f) Alternative water supply for villages provided.

COMPANY — Ghana Consolidated Diamonds Limited
MINING OPERATIONS — Alluvial diamond mining since 1923
LOCATION — Akwatia
IMPACTS AND ALLEVIATION —
  (a) Concession covers 153 km^2.
  (b) No discharges into the atmosphere.
  (c) Effluent discharges contain no chemicals.
  (d) Vegetation stripped prior to mining.
  (e) Compensation payments very low.
  (f) Sediments pollute village drinking waters.

COMPANY — Goldenrae Mining Company (1990)
MINING OPERATIONS — Surface gold mining—floating processing plant
LOCATION — Kwabeng
IMPACTS AND ALLEVIATION —
  (a) Concession is two separate leases covering 85 km^2.
  (b) Water polluted by sediments.
  (c) Compensation payment for farmers.

COMPANY — Southern Cross Mining (1989)—now Obenemase
MINING OPERATIONS — Surface gold mining
LOCATION — Konongo
IMPACTS AND ALLEVIATION —
  (a) Concession covers 125.9 km^2.
  (b) Mining has existed since 1907 as underground operation.
  (c) Vegetation affected.
  (d) Compensation payments to farmers.
  (e) Sedimentation of surface waters.

23 The Environmental Impact of the Mineral Mining Industry of Ethiopia: The State of Development and Related Problems

Solomon Tadesse

CONTENTS

ABSTRACT

The developing mineral industry of Ethiopia brings a considerable and irreplaceable input for the country's economy. Mining (due to geological reasons — as far as mineral deposits of metals are concerned) at present is concentrated in the southwest of the country within a relatively small area, where gold and columbium-tantalite ores are being mined by open pit methods and processed. The main targets of the mineral development are: (i) Adola gold field, with its primary gold deposits (Lega Dembi and Sakaro) and numerous rich gold placers; and (ii) the Kenticha province of acidic pegmatite, bearing rare metals.

The country produces 3 to 5 tonnes of gold annually, from ores with an average gold concentration of 5g t^{-1}. The removal, processing, and storage of huge masses of overburden, ore, waste, and tailings alongside the presence of added process chemicals (e.g., cyanide) accompany the gold production. Gold and rare metals extraction is in close proximity to the main surface water arteries, which reach groundwater aquifers in the region. The storage of gold-related tailings is undertaken within river valleys barred by soil dams. The infrastructural patterns of the gold- and rare metals-producing enterprises are superimposed on the local environment, with a resulting high density of impact per unit area. For acidic pegmatites, the presence of irregularly distributed and relatively high (in some parts of the region) natural radioactivity should also be noted.

The impact of the mineral industry on the environment is considerable. Therefore, the most effective, least expensive, and quickest means must be identified and urgently implemented to preserve the environment, or at least to minimise the undesirable damage to it within the region.

23.1   INTRODUCTION

The Lega Dembi primary gold deposit is located in the Adola gold field (Sidamo region, southern Ethiopia), 500 km south of Addis Ababa. It is comprised of gneiss and volcanic-sedimentary rocks which have been deformed, folded, faulted, intruded by basic and ultrabasic intrusive bodies, and regionally metamorphosed to greenschist-amphibolite facies.

The beneficiation of the deposit is the result of general prospecting and exploration activities carried out by the Ethiopian Mineral Resources Development Corporation (EMRDC) in central Adola since 1979. The deposit is now under production, yielding around three tons of gold annually.

The present study has made use of the results of surface, underground, borehole, mine, and processing plant data obtained since 1979 by EMRDC. The aim of the study is to assess the environmental impact from the mine and processing plant, and to recommend environmental protection in the region.

23.1.1 GEOLOGY

The Lega Dembi primary gold deposit belongs to the Upper Proterozoic volcanic-sedimentary sequence, which is located in the central part of the Adola gold field. The rocks cropping out in the deposit area comprise the following litho-stratigraphic units (from older to younger): biotite gneiss (Middle Complex); biotite-feldspathic-quartz and carbonaceous-mica-quartz schists and amphibolites (Upper Complex). The intrusive rocks of the ore field and its vicinity consist of metamorphosed basic and ultrabasic rocks transformed into talc-tremolite schists, talcite, and amphibolized gabbro.

Tectonically, the rock units of the area as a whole, including the mineralized area, are subdivided into two structural stages: lower and upper. These are differentiated from each other by lithology, degree of metamorphism, and types of tectonic deformation.[1] The lower structural stage is made up of gneiss rocks, while the upper one is composed of volcano-sedimentary rocks.

Three systems of faults are widespread throughout the area. One system strikes from north to south and is accompanied by the small elongated body of ultrabasic rocks. The second and third (the younger diagonal faults) strike northeast and northwest, respectively. These later faults cut the package of rocks squeezed between the near north-south-trending faults into several blocks, among which the largest correspond to the sections defined by EMRDC[1] as the Upper, Northern, Central and Southern.

The ores of the deposit are altered (oxidized) to a depth of 20 to 60 m, producing kaolin-rich material, and the development of limonite and secondary sulfide minerals. Gangue minerals are quartz and carbonate. Wall rock alterations accompanying the mineralisation include silicification, carbonatization, sericitization, biotitization, talcization, chloritization, and sulphidization. Field observations revealed that silicification is widespread.

The ore bodies occur as veins, lenses, and stockworks grouped according to an elongated north-south-trending swarm, following important shear zones concordantly with the host rock structure. They are confined to the rock sequences of biotite-quartz-feldspathic and carbonaceous quartz-mica schists, intercalated with hornblende and actinolite schists. The thickness of the ore zone varies from 50 to 80 m and the length from tens of meters to hundreds of meters along strike and down dip. The ore bodies have been tested from the surface to a depth of 500 m in north and upper Lega Dembi.

From a mineralisation point of view, the deposit occurs as a complex paragenesis where gold is associated with copper-zinc-lead-iron sulphides, tellurides, and sulphosalts. The distribution of gold is highly erratic. Gold is not confined only within the quartz body: it (in valuable grades) has been determined also in the host rocks, especially in the septa between adjacent quartz veins or lenses, so that the limits of the ore bodies may be rather variable and not strictly bound to the quartz occurrences. According to EMRDC[2], the gold content ranges from below the limit of detection (0.2 g t^{-1}) to 250 g t^{-1}, the average being 4.5 g t^{-1}.

The Lega Dembi ore bodies represent about 11 million tonnes of ore with a total gold content of about 49,000 kg. The volume of rock to be moved is about 29 million m^3. Present production is around 750,000 tonnes of ore and 3.1 million tonnes of wet waste per year.

23.2 IMPACT OF THE LEGA DEMBI PRIMARY GOLD DEPOSIT ON THE ENVIRONMENT

23.2.1 MINING AND MINERAL PROCESSING

Mining and mineral processing can cause serious conservation problems if not properly controlled. The main environmental problems identified with mining and processing in Ethiopia are water pollution, land degradation, pollution of the air, and socioeconomic difficulties.

23.2.1.1 Impact of mining

The main environmental impacts of surface mining in Ethiopia can be classified as follows:

- Land erosion.
- Noise.
- Emissions to atmosphere.
- Water quality.
- Socioeconomic.

23.2.1.1.1 Visual impact

The area is in a mountainous region on the eastern flank of a mountain that rises to more than 2000 m. Due to the steep relief, the possibility of visual impact is limited to the eastern sector of the area. The visual impact will be mainly one of colour contrast between the bare rock surfaces within the mine site and the surrounding vegetation-covered areas. However, the visual impact will remain very localised, given that the site is contained.

23.2.1.1.2 Effect of mining on the landscape

The mining will entail the complete removal of the vegetative cover over approximately 52 hectares, and modification of the topography. The pits created will lower the topographic level by 100 to 200 m in places.

23.2.1.1.3 Noise impact

The mining operations will entail an increase in the ambient noise levels due to:

- Mine blasting.
- Drilling of shot holes and the general movement of heavy equipment.
- Movement of lorries delivering supplies (e.g., fuel, reagents, and explosives) to the mine, and haulage of the ore to the processing plant and waste to the dumps.

23.2.1.1.4 Dust pollution

The main causes of dust are:

- Shot hole drilling.
- Rock blasting.
- Loading of blasted rock and off-loading it at either the dump (waste) or the stockpile (ore).

However, it can be predicted that the dust caused by these stripping operations will not be dispersed, except at the very beginning of the operation, due to the fact that the mine site will be an enclosed pit.

23.2.1.1.5 Air pollution

Open pit mining produces a certain amount of carbon monoxide and nitrogen and sulphur oxides from blasting, diesel-powered equipment, and so on. These contaminants do not cause external problems, the main reason being that the sources are dispersed.

23.2.1.1.6 Groundwater

The mine does not intersect any significant flowing springs and does not act as a drain to an established water table. Existing springs are located in areas not affected by the workings.

23.2.1.1.7 Surface water

The mine is located in the upper part of the Lega Dembi drainage basin, but will not seriously affect the flow of the runoff waters, especially as much of the rain rapidly infiltrates through the rock fissures.

23.2.1.1.8   Flora
The overall project will involve the disappearance of vegetation over an area of about 52 hectares.
   No regionally rare plant species exist at the site and identical environments are found throughout the region. The likelihood of a species becoming extinct does not arise.

23.2.1.1.9   Fauna — direct impact
Death of individual fauna incapable of fleeing or not being able to escape quickly enough from disruptive or destructive mining activity.

23.2.1.1.10   Fauna — indirect impact
These include the changing of runs used by animals and the cutting of areas frequented for food, which could result in fluctuations in the populations, especially birds. These effects, however, will be relatively small, considering the restricted area of the project.

23.2.1.1.11   Socioeconomic impact — personnel
- Social and economic impacts due to the creation of new jobs and equipment.
- Of special note is the use of women as truck drivers — an important sociological factor.

23.2.1.1.12   Social facilities and public services
Hospital, schools, a club, sport complex, radio communication, post office, and international communication facilities are available.

23.2.1.2   Impact of the mineral processing plant

The impacts of gold processing in Ethiopia include the following:

- Land erosion.
- Noise.
- Air pollution.
- Water pollution.
- Biological.
- Socioeconomic.

23.2.1.2.1   Impact on the landscape
The installation of processing plant has resulted in the disappearance of the vegetative cover and a change in the topography. The tailings are spread behind dams that close off the Lega Dembi valley. The material held back behind these dams also causes a notable change in the topography.

23.2.1.2.2   Noise impact
The various stages of ore processing all constitute sources of noise from a wide range of equipment such as crushers, screens, mills, and workshops.

23.2.1.2.3   Dust pollution
The processing plant produces dust only at the beginning of the cycle: that is, in the crushing circuit. The crushed ore is sent by conveyor belts to several stockpiles, where fugitive dust emissions can occur during the dry seasons.

23.2.1.2.4   Air pollution
Mercury is the most dangerous pollutant emitted by the processing plant; in spite of the steam condensation systems, a small part of the mercury escapes into the air. As a precaution, therefore, the atmosphere in the building containing the amalgamation plant is checked daily.

23.2.1.2.5   Impact on water

The tailings from the ore processing are the main source of potential pollution of surface and groundwater, especially as a sodium cyanide solution is used to dissolve gold and silver. Other potential contaminants in liquid effluents include process reagents such as acids, flocculents, mercury, and sodium hydroxide.

23.2.1.2.6   Biological impact

The construction of the processing plant and its related installations involved deforestation of the area. Similarly, the construction of the tailings dam will result in the disappearance of the fauna and flora from the bottom of the valley.

23.2.1.2.7   Socioeconomic impact

Population growth and infrastructural development, better health care, and economic benefit to the local population are results of the Lega Dembi mine development.

23.3   REMEDIAL MEASURES TO REDUCE THE IMPACT

23.3.1   RISKS OF ROCK FALLS AND LANDSLIDES

- Values of the overall slope angle and bench widths allow the different open pits to be mined safely.
- The use of slope protection devices from runoff is essential (e.g., lined, longitudinal ditches).
- Pumping units should be used to evacuate temporary flooding during the rainy season.

Using such measures, the disturbed land caused by open pit mining will be less susceptible to erosion and silting.

23.3.2   TAILINGS DISPOSAL

- To diminish the quantity of runoff water reaching the tailings pond, diversion ditches should be dug around the dam perimeter.

23.3.3   WASTE ROCK DISPOSAL

Geotechnical considerations applied to waste rock disposal call for:

- Diverting the surface water around the storage area.
- Draining the dumps.
- Proper design of the dump slope as a function of the geotechnical properties of the rocks (which will change with time).
- Planting the dumps to increase their stability, reduce erosion, and diminish the landscape impact.

23.3.4   DUST

23.3.4.1   Surface mining and waste dumps

These usually produce dust as long as they are in active use. Dust emission normally diminishes rapidly after abandonment because the supply of fine particles is exhausted.

23.3.4.2   Tailings dams

While in operation, dust emission problems from the tailings dam are similar to those of the waste dumps. Revegetation is the most effective way to control dust emissions.

23.3.4.3 Haulage roads

Haulage roads — where dust emissions occur through spillage from trucks and through abrasion by their wheels — are a major but intermittent dust source. The usual solution is to spray water on the roads from tanker vehicles.

23.3.5 WATER

Measures are taken to protect surface waters, such as digging ditches above the mine so as to divert the runoff water. The same system should be used at the perimeter of the tailings dam.

23.3.5.1 Monitoring

A laboratory is in operation that provides permanent assessment and control of the quality of the tailings water, especially with respect to cyanide concentration.

23.3.5.2 Water treatment

To ensure a waste product in the natural environment with cyanide content of less than 1 mg L^{-1}, an alkaline chlorine treatment plant has been set up for the oxidation and destruction of free cyanide and various metal complexed cyanides. So as to ensure the safety of inhabitants and cattle in the area, it is planned to fence off the tailings pond and erect panels stating the potential danger of the effluents stored upstream of the detoxification plan. Finally, the area surrounding the mine is considered as a restricted zone, which should enable all new settlement to be controlled.

23.3.6 SOCIOECONOMICS

The Lega Dembi mine guarantees an economic development in the region, both through its own activities and also through the fabric of commercial exchange that it generates.

However, the mine only has a 10-year life, based on the production and capacity of the processing plant. Thus it is necessary, in order to avoid a sudden halt to activities in the area at the end of the project, to feed the mine with ore from other mineralized areas within a 30 to 50 km radius (the economically viable distance at present) of the site. Furthermore, a large exploration programme is necessary to ensure the complement. This exploration programme should be carried out over the next five years, so as to enable a smooth transition between the different supply sources.

23.3.7 FAUNA AND FLORA

Successful site restoration is the most important measure for compensating the damage suffered by the natural environment.

23.3.8 WORKING ENVIRONMENT — SAFETY AND ACCIDENT PREVENTION

23.3.8.1 Health hazards

These include silica (related to silicosis), cyanide (highly toxic), mercury (acutely poisonous as a liquid or vapour), noise (workers in the open pit mine are exposed to high noise levels emitted by drilling equipment, loaders, and trucks; processing plant mills and compressors generate significant noise).

23.3.8.2 Safety measures

Medical control (e.g., chest X-rays, urine analysis).

Safety manual (providing employees with a basic set of rules and procedures which will assist them in working safely).

Training (e.g., safety posters and signs in critical areas of the mine and mill, first-aid training, preparation of materials for safety seminars).

23.3.8.3 Monitoring

Pollution monitoring results and control measures should be subjects of routine examination by the management, and regular discussions should be held with the concerned staff. It is important to bear in mind that prevention is generally far less costly than remedy where pollution is concerned.

23.3.9 SITE RESTORATION

Restoration of the mined area to the desired condition will include landscaping, stabilization, and revegetation. In order to minimize the topographic impact of the mines, partial infill will be carried out (using waste material) so that certain parts of the mine will be restored before the end of the project.

Once the site has been remodeled, the surface will be covered with humic soil preserved from the original stripping and a vegetation programme of the whole area will be undertaken. Revegetation will be accomplished by transplanting and by seeding local species. Fortunately, natural recolonisation will play a major role due to the favourable climate in the region.

Revegetation will be used to accomplish the following:

- Help stabilise easily eroded slopes.
- Control dust.
- Improve the aesthetics of the area.
- Increase evapotranspiration, so that a minimum percentage of precipitation enters to runoff cycle.
- Restore wildlife habitats.

23.4 CONCLUSIONS

It appears from the environmental impact studies of mine projects that three main domains of pollution are to be the subjects of particular care for protecting the environment affected by the mining operation: water pollution, air pollution, along with noise and vibrations. The methods proposed for controlling the different sources of pollution are to be adapted to the site and ore processing conditions. They can require, simultaneously, a water treatment plant; wetting of the ore during mining; reforestation of the waste products (in order to restrict the emissions of dust); the insulation of noisy fixed equipment; and the use of blasting schemes to reduce the noise and vibration levels. These measures are compatible with the mining operation, provided they are programmed simultaneously in order to preserve the environment.

In Lega Dembi, the predictable effects that the mining will have on the various environmental parameters during the course of the project are limited in terms of geographical area and the time (10 years). Most of the pollution will cease as soon as the site is closed.

REFERENCES

1. Ethiopian Mineral Resources Development Corporation (EMRDC), Results of Geological Prospecting and Exploration for Primary Gold in the Bedakessa-Upper Bore And Lega Dembi Areas, unpublished internal report, Ministry of Mines, Addis Ababa, 1985.
2. Ethiopian Mineral Resources Development Corporation (EMRDC), Preliminary Exploration of Northern Lega Dembi, unpublished internal report, Ministry of Mines, Addis Ababa, 1987.

24 Environmental Management and Pollution Control in Mining: A Case Study of the Non-Ferrous Metals Industry of China

Gao Lin, Wang Yishui, Ge Feng, and Zhang Wenmin

CONTENTS

ABSTRACT

This paper consists of four parts. In the first part, the Chinese non-ferrous metals industry is introduced to demonstrate that China now ranks fourth in the world for output of the 10 main non-ferrous metals, with an output in 1995 of 4.25 million tons. This expansion of the Chinese economy and the large-scale exploitation of mineral resources are then linked to a series of environmental problems. In the second part, environmental problems in the Chinese non-ferrous metal industry are described and related to the characteristics of mineral resources, techniques for mining, dressing, smelting and processing, installations, and management capacity. Major pollutants include sulphur dioxide, industrial powder-dust, smoke and soot, solid wastes, and wastes containing heavy metals.

In the third part, environmental management and pollution control is emphasised. A recent shift in industry's concept of pollution prevention is outlined and characterised as a turning away from "end of pipe control" to "whole process control" in order to strengthen environmental management in mines and enterprises. Through technical innovation, process improvements, modernisation of equipment, and implementation of cleaner production, the emission of pollutants not only has been reduced but the recovery of metals from the three principal waste streams (gaseous wastes, liquid wastes, and solid wastes) and the comprehensive utilisation of solid wastes has been undertaken. In addition, the manufacture of acid by using the sulphur dioxide in smog has also been achieved. Through such incremental improvements, the capability and technical level of pollution control has been enhanced.

In the fourth and final part, the policies for reclamation of mined land and the theory, methods, and techniques of rehabilitating wasteland are discussed.

24.1   DEVELOPMENT OF THE NON-FERROUS METAL INDUSTRY OF CHINA

Non-ferrous metal is an important material basis for economic development. It plays a positive role in improving the living standards of people, promoting the development of science and technology, and realising Chinese industrialisation and modernisation. China has abundant non-ferrous metal resources: reserves of tungsten, tin, antimony, zinc, titanium, and tantalum rank first in the world; reserves of lead, nickel, mercury, molybdenum, and niobium rank second; and the reserve of aluminium ranks fifth.

The non-ferrous metal industry has been developing rapidly since the adoption of economic reform policies and the opening up of China to the outside world (Figures 24.1 and 24.2). The establishment of the Chinese National Non-Ferrous Metals Industry Corporation (CNNC) in April

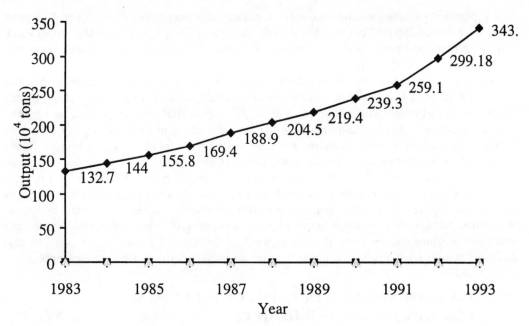

FIGURE 24.1 The total output of ten main non-ferrous metals.

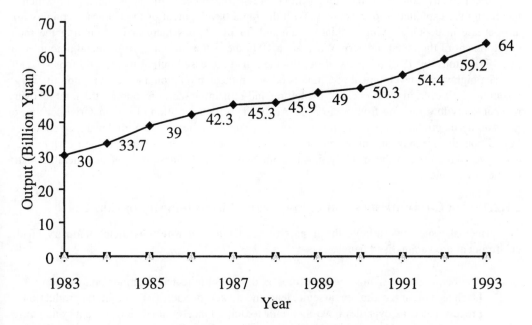

FIGURE 24.2 Gross value of industrial output (fixed price in 1990).

1983 opened a new chapter for the development of the non-ferrous metal industry of China. Productivity of the non-ferrous metal industry has increased by 8.86% annually. In 1995, the output of the 10 major non-ferrous metals reached 4.25 million tons, an increase of over 3 million tons since the establishment of the CNNC, sufficient to rank China fourth in the world in terms of output.

A set of major non-ferrous metal complexes has been set up in China, and the equipment and technology for non-ferrous metal production has greatly improved. It is estimated that 20% of the enterprises had come up to the standards of the advanced, industrialised countries by the end of the 1970s and the beginning of the 1980s. Most old enterprises have made significant progress in their production including the Jianxi Copper Company, Jinchuan Non-Ferrous Metal Company, Baiyin Non-Ferrous Metal Company, Great Wall Aluminium Company, Shanxi Aluminium Plant, Guizhou Aluminium Plant, Shandong Aluminium Company, Huludao Zinc Plant, Zhuzhou Smelter, and Fankou Lead and Zinc Mine.

The non-ferrous metal industry in localities including town and township enterprises has also made much progress in production. In 1994, these localities produced 1.53 million tons of the ten major non-ferrous metals which represented about 41% of the total national production. Up until now, the non-ferrous metal industry has been the main part of the public economy in China, with the coexistence of the collective economy, town and township enterprises, and joint ventures. There are currently 477 state-owned non-ferrous metal mines in China, of which 147 are run by the national government and the remainder by the provinces and counties. There are 510,000 miners in total, with 380,000 employed by the national government. Underground mining predominates at the current time, accounting for 70% of total ore production. A relatively comprehensive industrial system for non-ferrous metal production, which includes mining, separation, smelting, and processing, has been developed in China and promotes the development of the national economy. For example, it is estimated that the input/output ratio of the non-ferrous metal industry is 1:5 in money terms.

24.2   THE ENVIRONMENTAL PROBLEMS OF THE CHINESE NON-FERROUS METALS INDUSTRY

At present, many countries, including China, are in the process of industrialisation, which is increasing the exploitation of resources. With the rapid development of the Chinese economy and an increase in people's living standards, demand for metals has increased, promoting the rapid development of the metal industry. Whereas in 1949 the output of non-ferrous metal amounted to just 13,000 tons, by 1978 30 years of development had increased output to 952,400 tons.

The high-speed development of China is based on the exploitation and consumption of energy resources. However, in the process of exploiting and utilising mineral resources, the total recovery rates are very low, ranging from 30 to 50%. Accordingly, consumption of raw material per unit of gross domestic production (GDP) value is two to four times higher than that of developed countries. In addition, the dressing, smelting, and processing of mineral resources generate their own impacts and have contributed to the severely negative impacts of the Chinese non-ferrous metal industry on the environment.

24.2.1   THE CHARACTERISTICS OF THE NON-FERROUS METALS INDUSTRY OF CHINA

The principal issues determining the impact of the Chinese non-ferrous metals industry on the environment are summarised below:

- The composition of most mineral ores is rather complicated. Components of the ores (such as mercury, cadmium, arsenic, lead and nickel) or additives used in the production process (such as cyanide) have a certain toxicity. Once emitted into the environment, many of these toxic materials may be transferred, diluted or changed but not destroyed.
- Mineral resources in China are dominated by low-grade ores, with average ore grades typically lower than the rest of the world. As a result, there tends to be more waste material produced per unit of metal recovered.
- The consumption of energy, water, electricity, coal, and other additives to the mineral production process is also considerably higher than at equivalent plants in the advanced industrialised countries.
- Because of the diversity of the CNNC's operations — in terms of the complexity of production processes, the range of metal ores, and the large number of different mining, dressing, and smelting techniques — a wide range of wastes are generated.

TABLE 24.1

Data on Discharges of Pollutants in the Chinese Non-Ferrous Metals Industry

	Wastewater (million tons yr^{-1})	Waste Gas (billion m^3 yr^{-1})	Solid Waste (million tons yr^{-1})
Total discharge	560	250	59
Percentage of total pollutant discharges in China	3.5	9.7	10.6

TABLE 24.2

Data on Heavy Metals Contained in Wastewater In The Chinese Non-Ferrous Metals Industry

	Hg (tons yr^{-1})	Cd (tons yr^{-1})	As (tons yr^{-1})	Pb (tons yr^{-1})
Total discharge	5.6	88.0	173.0	226.0
Percentage of total pollutant discharges in China	16.0	48.8	11.3	20.0

- Most of enterprises directly under the CNNC's control were founded in the 1950s and 1960s, and have not received systematic investment or modernisation since that time. As a result, old plants and equipment, inferior quality products, high levels of energy consumption, substandard techniques, and severe environmental pollution are prevalent.

The main pollutants discharged from the non-ferrous metal industry comprise sulphurous oxides, industrial dust, smoke and soot, fluoride, hazardous solid wastes, and wastewater contaminated with heavy metals. Discharges of some typical pollutants are shown in Table 24.1 and Table 24.2.

According to recently released government statistics, the emission of pollutants from the non-ferrous metals industry represents a significant proportion of the total discharge of all kinds of pollutants across China. Based on data from 1990, discharges of industrial wastewater, waste gas, and waste slag from the minerals industry amounted to 0.56 billion tons, 297 billion m^3, and 60 million tons, respectively. This represents 3.5, 9.7 and 10.6% of the annual domestic discharge, respectively. In 1990, the discharge of sulphurous oxides in waste gases from the Chinese non-ferrous metals industry totaled 568,600 tons. Although this represents only 3.8% of the national discharge of pollutants, the ecological damage to some areas has been serious. Releases of heavy metals to the environment in industrial wastewaters are also high. Annual total discharges of mercury, chromium, selenium, and lead from the non-ferrous metals sector are estimated at 56,000 tons, which account for 16, 48.8, 11.3 and 20% of the total releases from all industries, respectively. It is apparent from the available statistical data that the release of liquid, gaseous, and solid wastes from the Chinese non-ferrous metal industry is enormous. The reduction of these wastes should therefore be a key priority in improving environmental protection.

24.2.1.1 The discharge of waste gas and its impacts on the environment

Waste gases are produced in the drilling, blasting, milling, screening, mining, transportation, and smelting phases of metals production. Some of the gas streams are hazardous and, if not handled proficiently, can affect both human health and the local ecology.

24.2.1.1.1 Dust pollution

Air pollution in underground and open pit non-ferrous metal mines can be very serious. This is particularly the case in underground mines due to the cramped working conditions; shortage of fresh air; and relatively high concentrations of carbon dioxide, nitrogen oxides, dust, and radioactive substances. In recent years, diesel equipment has been popular in underground mines, and this has compounded air quality problems.

Hazardous gases and dusts, including carbon monoxide and nitrogen oxides, can be generated in the development of open pits using explosives. The dusts stirred up by trucks can account for approximately 70 to 90% of the total dust produced by a mine, and this often produces serious local pollution. Dust concentrations in the air at a distance of 50 meters away from the side of the road may reach 750 to 800 mg m^{-3}. Dust concentrations in the driver's cab may amount to 6.0 to 15.0 mg m^{-3}.

The formation of hazardous dusts from the combination of toxic gases with dust is one of the main causes of selenium lung disease. Available data on the health impacts to miners engaged in onerous physical activity over a long time in a working environment contaminated with dust and toxic gases indicates that more workers died from respiratory diseases — such as selenium lung disease and cancer — than did those from industrial accidents.

24.2.1.1.2   Sulphur dioxide

Discharges of sulphur dioxide come principally from smelters and, secondly, from the burning of raw materials. Over 90% of the sulphur content of ores is given off as sulphur dioxide during the process of smelting. Where the concentration of sulphur dioxide in the tail gas from smelters is over 3.5 to 4%, it can be efficiently recovered and converted to sulphuric acid. If the concentration is less than 3.5%, it is uneconomical to recover and treat the sulphur dioxide. On average, 1.12 tons of sulphur dioxide will be emitted per ton of copper produced in China, while 0.85 tons of sulphur dioxide will be discharged per ton of zinc. In 1995, however, the production of sulphuric acid from the fixation of sulphur dioxide from non-ferrous metals smelters increased to 2.76 million tons, representing a sulphur recovery rate of 62%. A large amount of sulphur dioxide was still emitted into the atmosphere. Farmland in the vicinity of smelters continues to suffer from damage to crops caused by sulphur dioxide and acid precipitation, to the extent that compensatory payments for agricultural losses have been rising on a yearly basis.

24.2.1.1.3   Waste gases containing fluoride

Fluorine-rich waste gases come mainly from the electrolytic recovery of aluminium. The quantity and chemical form of the fluorine in the waste gas varies, depending on process parameters such as the nature of the electrolytic cells and the temperature of the electrolyte. Over 20 kilogrammes of fluorine can be emitted from the anodes of electrolysis cells per ton of aluminium, and between 16 to 23 kilogrammes can be discharged from pre-roasting electrolysis cells. Hydrogen fluoride accounts for the bulk of fluorine gases, while selenium tetrafluoride and carbon fluoride account for a smaller proportion.

24.2.1.2   Wastewater pollution

Large volumes of wastewater are produced in the processes of mining and mineral dressing. Wastewater streams from the pumping of working pits, process water, and water associated with the discharge of tailings can contain a number of different metals, acid or alkaline-generating material, and bacteria, which together or on their own can cause severe pollution to farmland, rivers, lakes, and drinking water if they are discharged without prior treatment. For example, the use of sodium cyanide in the process of lead and zinc flotation at the Fankou Lead and Zinc Mine Company in Guangdong Province resulted in excess cyanide concentrations in wastewater. Over 60 hectares of cultivated land was polluted and many animals were killed.

24.2.1.3   Solid waste

By its very nature, mining produces large volumes of waste rock in the stripping and processing of metal ores. In open pits, the generation of waste through stripping can significantly exceed the

TABLE 24.3
Soil Pollution by Heavy Metals at Xihuashan Tungsten Mine, Jiangxi Province

	pH Value	Cu (μg g^{-1})	Pb (μg g^{-1})	Zn (μg g^{-1})	Be (μg g^{-1})	As (μg g^{-1})	Fe (%)
Near Mining Area	2.5	1417	160	180	1.801	22	8.6
	5.4	72	80	80	1.801	22	4.1
Xiaban Irrigation Area	5.0	495	100	150	2.703	15	4.75
Along Tongmu River	5.5	30.8	72	60	1.801	4.0	4.05

TABLE 24.4
Analysis of Rice in Polluted Region (μg g^{-1})

Sample	Cu	Pb	Zn
Soil	1025	200	135
Root	312.2	93.75	97.5
Stem	46.75		100
Seed	5.75	1.25	26.5

production of ore: the ratio of stripping to mining at non-ferrous metal mines averages between 5 and 10, and between 13 and 16 in the case of bauxite. Other solid wastes include red mud from aluminium production, various smelting slags, and coal ashes. The reutilization of wastes as resources, through re-mining, re-processing, or alternative uses is very limited in China.

The accumulated total discharge of solid wastes from the Chinese non-ferrous metals industry has already reached a billion tons, with an additional 60 million tons discharged annually. According to statistics, the physical disposal of solid wastes in piles and heaps has recently caused several accidents, and is widely regarded as a particularly acute problem facing environmental protection works. If improperly managed, solid wastes can cause a series of environmental problems. A major issue in China is the degradation of arable land through the generation and dumping of mine wastes. According to statistics, the area of cultivated land has been declining over time. In 1952, the cultivated area was 108 million hectares (0.188 hectares per capita), but by 1987 the total area of arable land had declined to 95.86 million hectares (an average of 0.089 hectares per capita). While there are many contributing factors to this national decline, the role of mining wastes in the loss of arable land can be significant at the local or regional level. At the Shouwangfen Copper Mine in Hebei Province, for example, more than 36 million tons of waste rock were spread out over an area of 60 hectares; 90 hectares of land are occupied by over 2 million tons of waste rock at the Dabaoshan Multi-Metal Mine in Guangdong Province. It is estimated that waste rock and tailings in China take up over 66,000 hectares of land. If one assumes that an average hectare can yield around 6000 kg of grain per year, then the degradation of land through mine wastes has contributed to a decline in the annual grain output by 400 million kg.

A second major issue is the leaching of acid and alkaline wastewaters from the areas where waste rock and slags are stored. These leachates can contain many hazardous heavy metals and radioactive elements and, if untreated, can pollute water sources and soils. In addition to the immediate impacts to health, the pollution of soils with heavy metals can kill soil microbes in large quantities. This results in the soil losing its capacity to decompose and incorporate organic matter, leading to declines in soil fertility, deflocculation, and possible desertification. Tables 24.3 and 24.4 illustrate concentrations of heavy metals in the soil and rice in the vicinity of Xihuashan Tungsten Mine in Jiangxi Province. They show how the concentrations are significantly higher than the norm.

In addition, the collapse of tailing dams and waste slag storage areas caused several serious environmental pollution incidents. For instance, the tailing dam of Jinduicheng Molybdenum Mine in the Shaanxi Province, collapsed in April 1988, leading to the release of about 1.4 million tons of tailings and waste water into the Luo River, Shaanxi province which then flowed into the Yellow River and subsequently dispersed over a total length of 440 km. Heavy metals and cyanide have affected all fishponds and electric power stations, and as much as 14 million RMB yuan in compensation to farmers has been paid.

24.3   ENVIRONMENTAL MANAGEMENT AND ENVIRONMENTAL POLLUTION CONTROL

24.3.1   STRENGTHENING ENVIRONMENTAL MANAGEMENT OF THE MINE AND ENTERPRISE

The non-ferrous metals industry is an important basic industry for the national economy — and also one of the main industrial departments causing environmental pollution. The principal problem that China faces is how to keep the non-ferrous metal industry as a source of sustainable development while causing no threat to the environment. From the time the CNNC was set up over 10 years ago, the state and CNNC have invested a large amount of capital in the control of industrial pollution. Many facilities have been built to control pollution. While this has produced demonstrable economic, social, and environmental benefits, the rate at which industrial pollution has been reduced is much slower than the development of industrial production. This suggests the need for a change in corporate strategies and public policies towards the control and prevention of pollution, to consider the following:

- A switch from controlling pollution through end-of-pipe techniques to integrated pollution control through the consideration of the whole production process. Comprehensive utilisation of resources will be promoted, and the amounts of generation and discharge of pollution will be reduced through energy savings and a reduction in raw materials consumption.
- The control of pollutant discharges should consider not solely the concentration of emissions, but also total quantity over time.
- Policy approaches to industrial pollution need to be extended from scattered point source control to an approach that combines concentration control with scattered treatment.

The following measures can help to promote more comprehensive environmental management at mines and enterprises:

- The adoption of environmental management systems to facilitate pollution prevention, and the integration of environmental management techniques with cleaner technologies for production and remediation.
- The adoption of environmental protection planning and the development of environmental protection systems within the orbit of conventional management, to establish environmental management as an integrated part of project management.
- The implementation of environmental protection laws, the formulation of relevant rules and regulations, and investment in environmental management training.
- Strict implementation of EIA procedures for newly built projections and enterprises. This should seek to integrate pollution prevention into the project from the outset, and should adhere to what in China is known as the stipulation of "three simultaneously." This applies to all new constructions and expansions, and requires facilities for preventing pollution to be designed, constructed, and put into operation simultaneously with the

main body of the projects (Article 26 of the Environmental Protection Law). Since 1979, the stipulation of "three simultaneously" has been basically implemented among large- and medium-scale, plus new and extension projects of the non-ferrous metal industry.

• Key non-ferrous metal enterprises have established long-term environmental monitoring stations, tracking the state of the enterprises' environment and key pollutant sources, and producing environmental reports. Since 1982, these enterprises have accumulated considerable environmental statistical data.

24.3.2 ADJUSTING AND REVISING THE TECHNICAL POLICY ON ENVIRONMENTAL PROTECTION IN THE NON-FERROUS METALS INDUSTRY

Several technical policies on environmental protection in the non-ferrous metal industry have been drawn up since 1983. In 1985, they were completed and formalised as a technical policy: the Blue Cover Book of State Science and Technology Commission (8th). Since 1985, these policies have been actively implemented and have played a positive role in controlling environmental pollution and ecological damage. Nonetheless, the task of environmental protection has grown more formidable and urgent. Some articles of technical policy do not fulfil the needs of a sustainable development strategy for the non-ferrous metals industry. Therefore, it is necessary for some significant adjustments and revisions to the technical policy of environmental protection. Such a revision requires a prior investigation and assessment of the implementation and effect of the technical policy, the problems emerging from its execution, and the opportunities for adapting the policies' contents.

24.3.3 TREATING INDUSTRIAL POLLUTION

In recent years, the following treatment technologies and measures in China have been taken to control the gaseous, liquid, and solid wastes of the non-ferrous metals industry:

24.3.3.1 Waste gas

24.3.3.1.1 Dusts
Many measures have been taken to prevent dust. These, together with changes in working practices, have enabled reductions in the concentration of dust sufficient to reach the industry's health and safety standards in most underground mines. Additionally, much research work on the prevention and treatment of radon and its decay products, as well as the control of waste gas from diesel engines, has been carried out and applied with a favourable effect.

24.3.3.1.2 Sulphur dioxide
From 1981 to present, a successful strategy has been adopted of capturing SO_2 to make sulphuric acid, thereby reducing SO_2 emissions. During this period, a series of acid-making installations has been put into effect, raising the reuse rate of sulphur substantially and partly alleviating the problem of SO_2 pollution. These include the converter for making acid at the Jinchuan Company, Gansu Province; the second series of sulphuric acid plants at the of Yunnan Smelting Factory; the 10,000-ton series of Huludao Zinc Factory; and the complex of sulphuric acid projects introduced at Guixi Smelting Factory, Jiangxi Province. The productive capacity of facilities making acid reached two million tons, with a reuse rate of sulphur of 64% by the end of 1990. Data on the discharge and reuse of SO_2 in the Chinese non-ferrous metals industry are provided in Table 24.5.

24.3.3.1.3 Fluorine-bearing gases
Several aluminium factories have applied a pre-roasting trough and the technique of drying purification. As a result, fluorine discharged per ton of aluminium has been reduced to below 2 kg. For example, after Guizhou Aluminium Smelter introduced intermediate feeding pre-roasting and the

TABLE 24.5
Discharge and Reuse of SO$_2$ in the Chinese Non-Ferrous Metal Industry

Year	SO$_2$ Discharge in Waste Gas (10^4 tons)	Acid Produced from Gaseous Discharge (10^4 tons)	Utilisation Rate of Sulphur (%)
1981	67.8	86.1	45.3
1982	57.6	89.1	50.4
1983	45.7	100.4	58.9
1984	56.0	124.6	59.2
1985	46.4	119.6	53.3
1986	41.3	132.5	54.6
1987	47.5	163.2	64.7
1988	52.7	171.1	63.2
1989	54.5	183.5	64.1
1990	49.0	181.8	63.5
1994	Not Available	223.4	70.9
1995	75.0	276.0	70.0

technique of dry smoke-gas purifying in the 1970s, relevant indicators of environmental performance such as the off-gas gathering rate reached that of the advanced industrialised countries.

24.3.3.1.4   Chlorine-bearing gases

At present, the technology of liquid absorption is used extensively in controlling chlorine emissions worldwide. In China, the technique of alkaline liquid absorption has been adopted to treat chlorine-bearing gases derived from the process of smelting nickel and copper. The by-product sodium hypochlorite is used as a raw material for the chemical industry.

24.3.3.1.5   Mercury- and arsenic-bearing fumes

Copper, lead, and zinc smelters usually remove arsenic and mercury in the purification system of the sulphuric acid plant. When the concentration of mercury is as high as 30 to 40 mg m^{-3}, it may be removed via mixing with iodine, followed by condensation of mercuric iodide.

24.3.3.2   Wastewater

Wastewater from mining and mineral processing complexes can be treated in two ways: it can be reused to reduce the discharge of waste water to the environment, or it can be discharged to the environment once levels of contaminants have been reduced to meet state discharge quality standards.

Increasing the reuse of industrial water can not only save water resources but also improve the level of environmental management. The reuse rate of industrial water has increased year by year: from 51% in 1981, to 67% in 1990, and 70% in 1994. Each year, between 100 and 200 treatment projects of different scales in the non-ferrous metal industry come online. Examples of large-scale facilities include the comprehensive treatment system of the Baiyin Non-Ferrous Metals Company, Yunnan Smelter, Zhuzhou Smelter, Shanghai Smelter, and the wastewater treatment plant of the Daye Non-Ferrous Metals Company, Hubei Province. In addition, a large batch of wastewater treatment projects such as Shenyang Smelter, Huludao Zinc Smelter, Dexing Copper Mine of the Jiangxi Copper Industry Company, Yinshan Lead & Zinc Mine, and the Lingxia Non-Ferrous Metals Smelter have been built or put into operation. Table 24.6 displays data on the reuse of industrial wastewater in the non-ferrous metals industry in China. In addition to recycling, control technologies have been developed to recover valuable metals from wastewater. For example, during the period 1978–1984, Shenyang Smelter recovered 1768 tons of metal from wastewater sludge, the value

TABLE 24.6
Reuse of Industrial Wastewater in Chinese Non-Ferrous Metal Industry

Year	1981	1982	1983	1984	1985	1986	1987	1988	1989	1990	1993	1995
Re-use (10^6 tons)	590	674	668	808	796	909	1021	1168	1288	1449	1532	1650
Rate (%)	51.7	55.1	58.8	58.3	56.7	56.1	59.4	62.4	65.3	66.5	69.3	67.0

amounting to 1.06 million RMB yuan. Dexing Copper Mine has also used the alkaline wastewater from the dressing process to buffer acidic wastewater containing copper from the mining process.

24.3.3.3 Solid waste treatment

According to statistics, the annual discharge of tailings in Chinese non-ferrous metal mines is about 60 million tons, accounting for 11% of total industrial solid wastes. The amount of strip material from opencast mining in China reached 260 to 320 million tons, while the amount of waste from underground mines represented 170 million tons. At present, research on treatment technologies and comprehensive utilisation technologies for waste rock from mining, tailings, and all sorts of smelting slags has been strengthened in China. The main treatment measures under consideration are:

- Using advanced mining technology to reduce waste rock discharge — Zhongtiaoshan Non-Ferrous Metal Industry Company renovated its mining technology from the former mining method to a block caving method. As a result, the loss rate of ore mineral was reduced from 20 to 10%, and the dilution rate of ore mineral was reduced from 20 to 30% to 10 to 15%.

In opencast mines such as the Xiaoyi Aluminium Mine in Shanxi Province and the Pingguo Aluminium Mine, the cost effectiveness and environmental proficiency of the technique of removing and storing soil have been demonstrated. By adopting such a policy in advance of operations, 60 to 70% strip material was returned to the opencast mine, thereby reducing the area of land occupied and protecting the environment.

- Use of waste rock from mining and tailings as backfill material in underground mines—techniques for the use of waste ore from mining as fill material, returned to the underground mine, have been developed. After a grading process to recover refined sand, all kinds of waste ore from stripping at opencast mines and driving in underground mines have been used as both cemented filling material and filling aggregate by the Yunnan Tin Industry Company. These backfilling techniques utilised computers to automatically control production. The system has been in production for six years, and more than 90,000 tons of waste ore have been consumed, representing a filling cost saving of 600,000 RMB yuan. The technique reduced the land area occupied by waste and reduced environmental pollution, while also reducing the cost of mining.

In Fankou Lead Zinc Mine, the process of using tailings as cemented and filling aggregate at underground mines has been operated for 20 years. This has recovered 1.72 million tons of coarse sand-grade tailings; reduced the pressure on the tailings dam; raised the filling production rate 4 to 5 times; and saved both the investment of purchasing land to build a new dam plus management expenses, amounting to 2.26 million RMB yuan. It also saved 20.27 million RMB yuan via avoidance of processing costs relating to rock crushing for fill. Because efficiency is raised by coordination between the mining and filling processes, the profit made by mining reached 0.43 million RMB yuan. Research continues on cemented and filling techniques using tailings at this mine. A target of reusing 90% tailings as underground mine filling has been established.

- Comprehensive utilisation of waste — in recent years, substantial work on the comprehensive utilisation of tailings has been undertaken. A principal area of research has been the use of tailings as a secondary resource by recovering valuable associated elements. This includes recovering tin and copper from tin tailings; recovering lead, zinc, tungsten, and silver from lead and zinc tailings; and fluorite concentrates and sulphurous iron concentrates from copper tailings. The use of tailings to make building material has also been explored, since it may not only prevent land from being lost and the environment polluted, but it may also effectively eliminate tailings. For example, Xihuashan Tungsten Mine, Jiangxi Province, made use of tailings to produce a calcareous brick of high quality, which not only generated new profits and occupied less land, but also helped to reduce tailings pollution. Every year, 1700 m^3 of tailings is utilised and 10 million bricks are produced. At the same time, by-product lime powder may be supplied for building, and the mine also uses tailings to produce cement components such as cement floor brick and cement beams.

Chinese technology for producing cement by the sintering process is already well-respected internationally. The productive capacity of Zhengzhou Aluminium Smelter and Shandong Aluminium Smelter, for example, is near two million tons each and their products have met the state quality standard. This method may not only save raw materials, such as clay, limestone, and so on, but also reduces energy consumption by producing cement through sintering, generally reducing energy consumption by 20% and electricity consumption by 10%.

24.3.4 THE PROMOTION OF CLEANER PRODUCTION

Both the transformation of production and the transformation of environmental management can be achieved through the development and diffusion of cleaner production techniques. By implementing cleaner production, enterprises can initiate a series of critical changes such as product design, raw material selection, process innovation, technical improvement, and increasingly productive management. On one hand, minimising the generation of industrial waste, moving from end-of-pipe control to integrated pollution prevention, and shifting from passive pollution management to active prevention management can reduce industrial pollution. On the other hand, production costs of the enterprise can be slashed, enhancing the opportunities for economic profit.

If cleaner production is to be put into operation throughout the whole system of CNNC, then primary tasks like the innovation of technology, adaptation of new techniques, and the renewal of equipment must be undertaken. Since the 1980s, the level of skills and management capacity in most of the dressing, mining, and smelting enterprises in CNNC have been raised to a large degree. In some enterprises and mines, the level of techniques and installation have already reached an internationally comparable level.

24.3.4.1 Mining

Since average ore grades of non-ferrous metals in China are generally low, China must expand mine production, lift labour productivity, develop mining techniques to capture economies of scale at larger mines, and invest in technical improvements at medium and small mines to meet growing domestic demand for metals during economic reconstruction. Some new methods for open pit mining and underground mining have been applied, and some domestic rock drilling fittings like gear wheel machines and drills have been standardised. Techniques for explosions and explosion control are gradually becoming more developed, and new explosive equipment and instrumentation are being employed extensively.

24.3.4.2 Ore dressing

The development of copper ore dressing across China reflects phases of ore separation common to all non-ferrous metal mines. There are over 120 copper mines in China. These include more than 30 copper ore dressing factories, most of which were established in the 1950s and 1960s, and are characterised by old equipment, poor installation, and small production capacity. During the late 1980s, a large number of newly built, large, modern dressing factories were founded and put into operation, together with modern capacity expansion at existing sites. For example, a dressing factory was built in 1986 which can process 1000 tons of raw materials per day at the Yongping Copper Mine Company; at the Dexing Copper Mine Company, the 3rd Dressing Factory was established with a capacity of 6000 tons of rock processing per day. All of these newly built and expanded large-scale dressing factories have applied more efficient machinery, capable of increasing productivity.

Technical parameters to promote "more breaking with less grinding" were applied to reduce consumption of energy and steel in the process of grinding mineral ore. In addition, dressing plants developed large flotation equipment, improved dressing technology, and increased the degree of ore separation and the recovery rate of metal. In 1983, for example, Dexing Copper Mine Company developed a low-alkali flotation technique, increasing the recovery of copper to 85.9% and the recovery of associated gold from 57.8 to 64.3%. Moreover, research into techniques of asynchronous mixed selection regrinding and redressing has been undertaken so as to further enhance copper and gold recovery to 86.7% and 69.0%, respectively, while reducing the cost of process reagents.

24.3.4.3 Smelting

The considerable progress in the development of copper smelting techniques may be regarded as the epitome of developments in heavy metal smelting in China. Before the 1980s, copper smelting production was dominated by the blast and reverberatory furnaces. These are characterised by high consumption of energy, serious air pollution problems, and a low level of mechanisation and automation. In 1985, a flash smelting furnace at Guixi Smelting Factory, Jiangxi Province, was founded and brought into production, representing a historic breakthrough in the Chinese copper smelting industry. As a large fully automated installation, the techniques and practices of the smelter have reached an international level. Copper recovery has been increased to 97.7%, with sulphur recovery at 95.8%. The equipment is typically operated at 96.3% of capacity, and electric energy generated by surplus heat represents over 50% of electric energy consumption. A series of technical innovation programmes was conducted, including an oxygen-rich flash smelting project and a programme to enhance gold and silver recovery. After completion of these programmes in 1994, copper output has increased by 30,000 tons compared to before the innovation programme, and is currently 100,000 tons per year. The quality of electrolytic copper meets "A-grade" standards.

24.4 LAND RECLAMATION

24.4.1 MINING WASTE

According to statistical data from the National Land Administration Bureau, accumulated wasteland as the result of human activity amounts to about 13 million hectares. Of this, wasteland due to raw material extraction such as mining and brick-and-tile making is around three million hectares, and includes one million hectares attributable to rural and township enterprises and two million hectares to state-owned industrial and mining enterprises. Research shows that in the mining industry, land directly used by mining projects accounts for 59% of the total disturbed; the overburden of open pit mines, 20%; tailings dam, 13%; and wasteland caused by mining-related subsidence, 3%. During the mining of open pits, the area of land destroyed is between 2 and 11 times as large as the open pit itself. This kind of land destruction can transform fertile cultivated land into wasteland or bog, and can lead to serious pollution incidents such as pollution of air and water resources.

According to data released by the National Environmental Protection Bureau, industrial solid waste throughout China amounted to 0.65 billion tons in 1995. The accumulated amount of industrial solid waste has reached 6.6 billion tons and has taken up 55,085 hectares of land. Of this, the dumping of coal rock accounts for 0.15 billion tons, with an accumulated total of 2 billion tons occupying 6000 hectares of land. The annual discharge of coal ash is 50 million tons and its accumulated total takes up 13,000 hectares of land. By contrast, the non-ferrous metals refinery industry discharges 60 million tons of solid waste annually and has accumulated a total amount of one billion tons, occupying 60,000 hectares of land.

In addition to the solid waste issues above, subsidence caused by mining is also a serious problem. According to preliminary statistics, the area affected across China reaches 1700 hectares annually and the accumulated total area exceeds 14,000 hectares.

The degree of land degradation varies from mine to mine. Table 24.7 shows the extent of land degradation at selected major coal mines.

24.4.2 LAND RECLAMATION CONDITIONS

According to sampling statistics conducted by the National Land Administration Bureau, the total area of reclaimed land in China was 133,300 hectares at the end of 1991. In comparison, the accumulated total area of degraded land is estimated to reach two million hectares, giving a reclamation ratio of 6.67%. While small, this has increased three to five times, compared with 1 to 2% before 1989.

Reclamation conditions vary from region to region and mine to mine, however, which is partly shown by Table 24.8. Based on this, the reclamation ratio at some mines can reach 15%, and some as high as 90%. It can be seen from Table 24.9 that the average reclamation ratio of wasteland at aluminium mines in China ranges from 20 to 45.5%. After reclamation, up to 80% of wasteland is transferred to arable land for grain crops.

It is estimated that annual wasteland which is rehabilitated will be over 130,000 hectares in China during the period of "the Ninth Five-year Plan" and the reclamation ratio will reach or exceed 20%, so that the supply of existing arable land area will be kept relatively stable.

24.4.3 RECLAMATION POLICY

24.4.3.1 National reclamation policy

In order to facilitate the process of land reclamation and reduce the degree of land degradation and improve the ecological environment in China, "land reclamation stipulation" was issued in November 1988 and put into effect on January 1, 1989. This policy is designed to guide and direct land reclamation in China and undertakes this role through six principal parameters.

- It defines the polluter pays principle that "whoever destroys land is responsible for reclamation."
- It prescribes the administrative departments in charge of land reclamation and their form.
- It emphasises that land reclamation should be planned from the outset of operations and as part of a comprehensive environmental protection strategy.
- It provides for reclamation expenses for the enterprise and an assessment of the size of the reimbursement for land loss.

TABLE 24.7
Land Degradation at Selected Major Coal Mining Areas

Mine Name	Type of Degradation	Degraded Area (hectares)
Antaibao Open Coal Mine	Land-taking	1092
Shenfu Dongsheng Coal Mine	Land-taking	2184
Kailuan Coal Mine	Land subsidence	8000
Huaibei	Land subsidence	6200
Huainan	Land-taking	100
Shandong Province	Land-taking	665

TABLE 24.8
Reclamation of Derelict Strip-Mined Land in China

Regions	Mining Land Area (hectares)	Reclamation Area (hectares)	% Reclaimed (reclamation ratio)
Fushun	400	61.8	15
Xiaomabei	400	146.7	37
Xialongtan	400	150	37.5
Lingbei	800	200	25
Kebao	133.3	120	90

TABLE 24.9
Reclamation Ratio at Selected Bauxite Mines

Mine Name	Total Area	Total Reclamation Area (hectares)	Reclamation Ratio (%)	Reclamation Depth (m)
Xiaoguan Bauxite Mine	3.2	150	45.5	1.2–1.5
Yangquan Bauxite Mine	0.95	32	34	0.5
Luoyang Bauxite Mine	0.78	13	17	0.5
Fengshui Bauxite Mine	5.9	170	29	0.5

- It enlarges the range of specific best practice techniques for reclamation.
- It rationally stipulates ownership rights and the right to use rehabilitated land.

24.4.3.2 Local preferential policies for reclamation

To meet the site-specific needs of local conditions, the different levels of government and the departments in charge of all aspects of industry have actively developed regulations to be implemented and a series of preferential policies for reclamation. These policies mainly focus on land purchase methods. Specifically, they include policies addressing "using land temporarily (renting the land)" and "exchanging land for land."

24.4.3.2.1 Temporary land use

"Temporary land use" means that the enterprise obtains the productive land for a project by way of renting the land from the local authority. After completion of the project, the enterprise should rehabilitate any degraded land in line with certain agreed standards, so as to return the land to productive use. During renting, the local authority possesses the ownership rights to the land and the enterprise has only use rights. The cost of renting is, however, much lower than the expense of purchasing the land. After production, the enterprise must be responsible for rehabilitating the land to the standards agreed by both sides. These policies can slash the expense of purchasing land, lighten the burden on the enterprise, and provide an incentive for the enterprise to finish reclamation tasks in terms of the quality, quantity, and time presented in the contract so that the high-quality agricultural land can be returned to the local authority again. For example, the Land Administration Bureau of Zhaoyuan city, Shandong Province, asked an enterprise wanting to rent land to sign a reclamation agreement with the Land Administration department and to pledge a reclamation payment as bond (30,000 RMB yuan per hectare) in advance, which will be returned if the content of the contract is implemented, and the farmers can use the restored land.

24.4.3.2.2 Land exchanges

The practice of land exchanges means that, having consulted with the local authority, an enterprise can exchange the land it needs for rehabilitated land from the local authority. In this case, both ownership and user rights to the land are exchanged. In the process, if the rehabilitated land is not sufficiently fertile or the surface is not smooth and the facilities for land cultivation are not fully equipped, the enterprise should reimburse the local authority according to the terms between the enterprise and the local authority. In this way, the expense of purchasing land can also be greatly reduced, and so it is welcomed by the enterprise. In the meantime, enthusiasm of the enterprise for reclamation can be improved and saved expenses can be used for reclamation, thus further ensuring the reclamation funds of the enterprise. For example, in 1994 Xiaoyi Bauxite Mine exchanged about 2.6 hectares of reclaimed land for the same area of arable land from the local village. The ownership and user rights are also exchanged. This is the first case in which the policy of "exchanging land for land" has been implemented in China.

24.4.4 PRINCIPLES AND METHODS FOR RECLAMATION OF WASTELAND

There are several specific meanings of the term rehabilitation, depending on local prevailing conditions. The word "rehabilitation" is used in China in the context of a priority value for agricultural use, while the international use of "reclamation" includes the meanings of rehabilitation and other socioeconomic purposes. In terms of sustainable development, the purpose of management and rehabilitation of mining wasteland should be to establish or restore an artificial ecosystem which is harmonious with regional conditions. The essence of reclamation in China is, therefore, "ecological restoration."

24.4.4.1 Ecological principles

There are a significant number of ecological principles that underpin successful mining wasteland rehabilitation, such as ecological niches and the food chain. The most important of these is ecological succession, which refers to the sequential process of change from one type of ecosystem to another. For instance, when grassland is degraded to bare land, ecological succession follows the process from weeds to sparse herbs and then dense herb grassland. After deforestation, the succession of *Picea* forest is from forested land to grassland, *Betula* forest, then *Populus* forest and finally *Picea* forest. This process takes place over several decades. However, human activities may intervene in and adjust the process, expediting the successional sequence or changing its direction.

In recognition of this ecological principle, planners of the rehabilitation of mineral wasteland should bear in mind the following aspects in order to hasten the ecological succession process.

1. Select drought-resistant, sterility-tolerant, and fast-growing crops or fodder to grow on mining wasteland in order to rapidly return constant vegetation.
2. Once the substrate of wasteland has been restored to a certain extent, plant several kinds of grasses together to enable them to cover wasteland quickly. Alternatively, inter-crop/rotate grass with a leguminous crop to achieve the purpose of combining cultivation with soil fertility restoration.
3. Based on soil element composition and fertility, reconstruct self-sustaining ecosystems with the assistance of proper application of water and fertilisers, especially microbial fertilisers.
4. Plant diversified crops and fruit trees, and integrate agriculture with forestry, animal husbandry, and other sidelines, according to local conditions, to comprehensively utilise mining wasteland.

24.4.4.2 Principles and methods for ecological restoration at mines

The rehabilitation of mining wasteland is a complex process and requires a systematic approach that incorporates the principles of planning for closure. This approach should contain the following six key elements:

24.4.4.2.1 Legislation

Legislation should establish the extent of legal liability incurred by the mineral developer in managing and rehabilitating ecological degradation. Liability laws should follow the "polluter pays principle" and should establish a framework to ensure that standards for environmental quality and the extent of rehabilitation are met.

24.4.4.2.2 Ecological risk assessment

Before the development of a mine or mineral processing complex, ecological risk evaluation must be completed in order to reduce the risk of irreversible environmental degradation and reduce the costs of rehabilitation on closure.

24.4.4.2.3 Minimisation of disturbance

The land area occupied by mining should be minimised to the extent practicable from the onset of the project life cycle. Alterations to the topography, environmental processes, and biodiversity of the site should also be reduced to a minimum.

24.4.4.2.4 Resource transformation

Land transformed by mining activity should be rehabilitated in a way that adds value to the land through, for example, creating conditions for future economic growth.

24.4.4.2.5 Waste treatment and reduction

Mine wastes should be processed and treated to reduce their impact on the environment. Where possible, the principles of waste minimisation, through pollution prevention, should be adopted. Where waste minimisation and reduction are not practical aims, the objective should be to treat wastes so that their impact on the environment is reduced.

24.4.4.2.6 Restoration of ecosystems

The rehabilitation of ecosystems, degraded by mining activity, should aim to restore natural ecosystems rather than generate artificial ones (e.g., agriculture) where possible. Reclamation activity should begin from the outset of the project and be phased across the life of the mine.

24.4.4.3   Reclamation techniques in a typical mining area (Xiaoyi bauxite mine, Shanxi Province)

- Assessment of reclamation needs, determination of post-mining land use, and rehabilitation of wasteland into arable land.
- Establishment of a linked system of overburden stripping and ongoing reclamation.
- Requirement that arable soil is of uniform quality with a thickness of over 60 cm and a volume to weight ratio of 1.5:1.6 g cm^{-3} after reclamation.
- Selection of excellent agricultural breeds with strong resistance to environmental stress, such as corn varieties like Yedan 13 and Zhongdan 104 and the bean varieties Beinong 50098 and Zhongzuo 84.
- Applying trace elements such as zinc, molybdenum boron, and manure containing nodule bacterial have demonstrated very good effects.
- Applying the technology of endomycorrhiza has accelerated the transition from immature soil to "mature soil," which has therefore shortened the period of reclamation. In the period 1991–1995, the total area of reclaimed land exceeded 78 hectares, with a reclamation ratio of 74%. Grain (maize) output has increased 10 to 40% in comparison with local farmland.

24.5   CONCLUSIONS

- The Chinese non-ferrous metals industry has developed rapidly in the period from the early 1980s to the present day. Output of the ten major metals now ranks fourth in the world.
- The non-ferrous metals industry is not only an important basic industry of the national economy, but also one of the most significant economic sectors in terms of environmental pollution. The discharge of pollutants to the atmosphere, ground, and surface water resources and soils has led, in some cases, to the contamination of crops and community health.
- Mine wastes including waste rock, tailings, and smelting slags take up a large area of land. In recent years, the reclamation of land has been undertaken at a rapid pace; to date, 40 million hectares of land have been rehabilitated nationally. Following the promotion of public policies to facilitate the reclamation of land, a series of research exercises to demonstrate the theories, methods, and techniques of ecological restoration at wasteland have been undertaken.
- If the CNNC is to develop along a sustainable trajectory, more attention must be paid to strengthen the capacity for environmental management in order to control industrial pollution and the generation of atmospheric liquid and solid wastes. Most importantly, the paradigm of pollution control, through end-of-pipe treatment, has to be transformed by adopting integrated approaches to environmental management which seek to prevent pollution from the outset, and which plan mine closure from the onset of operations. Moreover, the treatment and prevention of pollution has to be facilitated by adopting public policies and corporate strategies which facilitate technological innovation; improvement, renewal, and modernisation of equipment; and a reduction of pollutant discharge and the diffusion of cleaner production practices. Since the 1980s, technical skills and investment in treating industrial pollution have been enhanced, and have demonstrated that economic profit and environmental and social responsibility can be achieved simultaneously.

25 Planning for Closure: The Case of Australia

Ian Clark

CONTENTS

ABSTRACT

Mining is an activity that has, in the past, been regarded as often leaving land in a condition unfit for further use. As the need for development that is both economic and ecologically sustainable has grown, so has the issue of reuse of land after mining become significant. During operation, it is usual that mine environmental conditions which minimise unnecessary or avoidable environmental impacts are agreed with the regulatory authorities. These conditions are usually set down in such a way that both miners and regulators can interpret them reasonably well. However, the conditions to be met by a mining company before it can be absolved of responsibility of a site are often less clear. The closeout criteria by which the success of rehabilitation programmes will be judged have received little attention in the past. These conditions are now growing in importance as the costs of rehabilitation of a mine site to an agreed post-mining land-use assume greater significance in the economics of mining. This chapter briefly describes the status of closeout criteria presently being applied across Australia, and reviews the legislation and the company-developed codes of practice which are determining the closeout process.

25.1 INTRODUCTION

Decommissioning or closing down a mine is becoming an important consideration in the planning and operation of any mine and mineral processing plant in Australia today. In most states, part of the legal requirements of a company when seeking approval to establish a mine or processing plant involve describing the process for the restoration of the site to its original land-use capability prior to the commencement of mining or processing. In the recent past, the particular requirements for decommissioning were considered as an afterthought when the mining operation ceased. In the more distant past, but not that long ago, it was simply enough to remove equipment and buildings that could be used at new locations and to make the site safe.

This chapter describes the situation as it is developing in Australia today. It briefly reviews the past and the legacy that it has left; it reviews the current and developing environmental legislation as it relates to planning for mine closure; and it describes some examples of site rehabilitation and strategies that are being investigated to attempt to improve the current best practice.

25.2 ANALYSIS OF PROBLEMS AND CHALLENGES

The development of a policy for mine closure practice presents a number of challenging problems. Three groups are involved in the discussions: the mining groups which are responsible to their shareholders; the environmental groups which have made themselves the guardians of the natural environment; and government which is ultimately responsible for the well-being of society.

Three tiers of government are involved in planning associated with the permitting process for major developments in Australia. The Commonwealth (Federal) Government has legislative power over the Australian continental shelf, including the exploitation of submarine minerals. Generally, legislative control of mining activities on the Australian mainland is vested in the individual state and territory governments, and certain planning approvals are in the jurisdiction of local government authorities. There are six state and two territory governments and numerous local governments in each state and territory, each of which impose their own subtle variations on the governing legislation, which results in a degree of unnecessary uncertainty for companies which explore in more than one state.

The mining and mineral processing industry has been a major land-user in Australia since soon after European settlement. The industry recognises that mining activities will cause some kind of impact on the environment, but argues that in the balance of advantages and disadvantages these impacts are significantly offset by the benefits that result from the production of mineral products. In the past, there was no stipulation that required miners to restore the environment, although in many cases simple cleanup operations were carried out on completion of projects. Even today, many mine operators question the need to even consider closure plans, arguing that mining takes up less than 0.02% of the land surface of Australia, and many mines and mineral processing plants are located in remote areas where subsequent development is highly unlikely.[1] This view is not held by the wider community, and standards for adequate rehabilitation of disturbed lands are set by the community and ensured by legislation and "codes of practice."

The challenge for governments, which must legislate for competing land-uses, is to achieve a balance that not only suits the present needs of the community, but also takes into account the future needs. The issue of access to land is one of the major concerns facing the mining industry, conservation groups, and commonwealth and state governments in Australia today.[2] Conservation, aboriginal heritage, recreation, agriculture, forestry, and urban development are some of the land-uses which are competing with mining. In some cases, these may be mutually exclusive and in others they may not. In some cases where there need not be, there is still conflict.

Governments face a fundamental problem with respect to competing land-uses. They must ensure that decisions about the use or combination of uses will provide maximum benefit to society over time. The benefits must come from the full range of values society derives from land resources and their use including, in Australia's case, providing the base for major export industries. Additionally, land resources provide a wide range of environmental benefits that include access to clean air and water, recreational benefits, and preservation of genetic diversity. Placing too high a value on the environment will deter mineral explorers and adversely affect the economy; placing too low a value on the environment will lead to irredeemable degradation. Finding a balance is the essence of ecologically sustainable development.

25.3   PLANNING FOR CLOSURE — A BRIEF HISTORY

The development of closure practices in Australia has taken place over the last two decades and "parallels the rise of environmental awareness."[1] Prior to the early 1970s, there were few requirements placed on mining companies when they wanted to abandon a mining lease. They were simply required to make the site safe by adequately fencing it, and sealing or capping shafts and adits or, in the case of open pit mines, stabilising the slopes. Buildings and equipment were either relocated to another site or sold by auction on the site.

In general, there was little concern for the existing or future environmental impacts which resulted from the closure. Farrell[1] points out that in many cases these early mining operations and their impacts were much smaller than those of today. However, it was not the size of the impact that permitted this neglect, but rather the prevailing attitude of society which placed no value on resources until they were exploited — and no value on the natural environment. There has been little concern expressed about the majority of these past practices, but a couple of notable exceptions have probably led to the development of many of the standards and practices that are applied today.

The increasing public awareness of the environment in the 1980s and 1990s has seen the growth of a strong conservation movement. This has placed considerably more focus on the long-term effects of mining on the environment. The change in attitude saw the introduction of the environmental impact statement (EIS) for new developments, partly as a response by the companies to the sometimes outlandish claims by conservation groups about the potential impact of mining on the environment and partly in recognition of the importance of the environment. The mining industry began to promote the concept that mining is a temporary land-use, thus implying that the land should be returned to its original use after mining ceases. This resulted in a big increase in the research effort into the process and practice of mine site rehabilitation. At this time, the mining sector was the biggest employer of biology graduates in Australia.

Every EIS of this time addressed, at least in general terms, the end-use of the site and the rehabilitation processes that would be used. However, subsequent practice at many of these mines bore little resemblance to what was actually described in the EIS.[3] Some early attempts at environmental awareness demonstrated that the science was relatively young. For example, although the EIS for the Woodlawn mine recognised the problem of acid drainage from spoil heaps containing sulphides, it considered that the pyrite content of the waste rock was too low. This consideration turned out to be a costly mistake, as acid drainage was evident within two years.

At the same time that new developments were being required to provide a detailed EIS, regulators were beginning to take an interest in existing mines and their environmental effects. They did this by reviewing lease conditions to ensure that rehabilitation requirements were included. While many mines already had comprehensive rehabilitation programmes, the aim of these changes appeared to be to make all mines aware of the changing community attitude towards environmental protection and to "ensure that the public statements of industry spokesmen (*sic*) were reflected in actual practice."[1]

At this time, the emphasis was on rehabilitation. The rehabilitation effort was directed towards tailings dams, waste rock dumps, and land disturbed by mining, which was supposed to happen at the completion of the operation. The objective was to return the area to some acceptable state that involved minimising runoff and erosion, and preventing further pollution, especially of surface and groundwater. Although this usually involved revegetation, it did not emphasise the use of local species or the need to restore the land to its original condition. There was also little attention paid to the buildings and processing plants.

At this time, there was also a move by governments to restore abandoned mining sites and considerable sums of money were spent cleaning up some of the environmental disasters of past mining. One notable example was the abandoned uranium mine at Rum Jungle in the Northern Territory, where acid waters containing dissolved heavy metals were contaminating the local river system.

During this time, environmental control and decommissioning requirements of mining operations remained with the government department responsible for regulating the industry, usually the Department of Mineral Resources or its equivalent.

25.4   PLANNING FOR CLOSURE — THE CONTEMPORARY SITUATION

Since the mid- to late-1980s, there has been a move towards expanding the environmental regulatory responsibilities of the Mineral Resources departments, with reviews of the Mining Acts strengthening environmental aspects. At the same time, new departments and statutory authorities charged with environmental protection functions such as the Environmental Protection Authority were established, which also had clear controls over mining and mineral processing operations.

25.4.1   CURRENT LEGISLATION IN AUSTRALIA

This part of the chapter provides an overview of the present situation throughout Australia with respect to what constitutes acceptable criteria for mine closeout in each of the various legislative scenarios and gives an indication of future trends in closeout legislation.

This information is based on a paper by Waggitt and McQuade,[4] in which they summarised the guidelines for closure throughout Australia. They gathered their data by means of a questionnaire sent to the organisations in each state, which have the responsibility for regulating mining.

The Commonwealth Government has no specific legislation in relation to general mine closeout. However, the Department of the Environment, Sport and Territories does have a set of guidelines specific to the closeout of operations involving radioactive materials.[5] These guidelines relate to the disposal of radioactive wastes from the mining and milling of radioactive ores. They contain only a limited set of conditions concerning the general issue of mine closeout.

In each of the states and territories, there are usually a number of agencies involved when closeout criteria are being set. The number varies from six in Victoria to three in New South Wales and one in the Northern Territory. In the Northern Territory, the Department of Mines and Energy operates a "one-stop shop" which includes an Environment Division. The Environment Division has responsibility for the oversight of environmental matters and carries delegations from other agencies for the appropriate legislation.

In most states, the main agency is the mining regulator, in close association with water authorities and environmental protection and conservation agencies. All states and territories have a mining act that has safety as a major element of the closeout process. These acts are primarily concerned with the reduction of the risk of physical danger and containment of chemical risks such as acid drainage and heavy metal contamination. The preservation of water quality for downstream beneficial use is a common theme. The mining regulator is responsible for the coordination of the closeout process, which invariably involves consultation with mining companies, other agencies, and even other levels of government such as local councils on occasion.

A common feature of the legislation is the overall power given to the minister responsible for mining. Ministers often have wide-ranging authority to impose specific rehabilitation conditions on mining titles.[6]

25.4.1.1 Setting criteria

Due to the wide range of final land-uses and the significant variation in soil conditions and climate across Australia, common environmental criteria or standards regarding closeout are not available. However, broad guidelines usually exist, and in Queensland and Western Australia comprehensive documentation is available. The key element in rehabilitation is a requirement that the site must be left in a state suitable for, or appropriate to, the agreed final land-use. In most cases, the final land-use is agreed before approval is given for the operation to begin. This makes it necessary for the final land-use to be determined as an integral part of the initial development proposal or environmental management plan. In Western Australia, however, the Mining Act does not relate to rehabilitation works on privately owned land. The regulators have no involvement in setting rehabilitation targets and only have authority to act in respect of offsite impacts. However, the mining company is responsible for ensuring that the site is physically safe.

In Queensland, the criteria included in the environmental management plan typically include vegetation density and diversity, soil erosion rates, and soil profile development. These types of criteria form the basis for assessing success in rehabilitation, but are not usually prescribed or quantified.

In the situation where there are no legislative requirements or controls for rehabilitation, progressive companies have tended to set their own rehabilitation goals and objectives for closeout based on corporate policy. For example, at the Goldsworthy mine in Western Australia, the area was landscaped to match the surrounding natural landform and revegetated with native species. An enhanced water harvesting system was incorporated into the final design to assist in the establishment and maintenance of the vegetation.[7]

25.4.1.2 Process

Consultation is encouraged; frequently, the process for obtaining approval requires public display of proposals and may even extend to public hearings for contentious projects. The parties involved in consultations include government agencies (often including planning departments), landowners, local councils, and community groups. The overall trend is that lease agreements must contain the rehabilitation goals and define the final land-use for the rehabilitated site. The usual means of reducing the financial risk to the community is to require the operator to put up a bond. "Bonds" as "financial risk" could be substantial in the event that an operator fails to meet its obligations and leaves society with a bill for rehabilitation of the site. The method of calculating the value of the bond varies greatly from state to state. Some are calculated on the basis of a flat cost per unit area, but the trend is to make bonds, or guarantees, project-specific. Queensland offers a discount rate for the bond, based on the past rehabilitation performance and the current financial standing of the company. On the other hand, Western Australia considers that discounting in this way is inappropriate, as such practices increase the government's risk of exposure to financing rehabilitation after failed operations. Progressive rehabilitation is generally encouraged and progressive reduction of performance bond requirements in proportion to the work done and completed to the satisfaction of the regulator is practised by some regulators (see also Anderson, Chapter 15, and Danielson and Nixon, Chapter 17).

25.4.1.3 Objectives

There is a great consistency between all the states as to the primary objectives of rehabilitation and successful closeout for a mining operation:

- Site physical safety.
- Site stability.
- Land condition suitable for agreed end-use.
- Non-polluting, especially with reference to water resources above and below ground.
- Maintenance-free and self-sustaining (where applicable).

There may also be site-specific objectives relating to the containment of particular materials. In addition, old mine workings or buildings may be subject to preservation due to heritage status.

25.4.1.4 Assessment

In all cases, the assessment of achievement of closeout criteria is the responsibility of the mining regulator; because the criteria do not necessarily represent quantifiable standards, the assessment process is based on expert review. The regulators may seek expert advice from other agencies and/or consultants. This is particularly true in Victoria, where expert advice is sought for such items as water quality, weed control, and soil erosion measures. There is a trend amongst regulators to require self-audit by operators. In these cases, the operator is required to produce a final report demonstrating compliance with the agreed closeout conditions. In many instances, this report has to be independently audited and is still subject to final audit by the regulating agency. In Western Australia, if the project has been especially significant, a public review may be initiated. Generally, before the commencement of operations, proponents are required to agree about what has to be achieved at the final close-out, but may reserve the option to revise the plan in the light of changes in technology and the situation, both economic and political. Any such revisions are with the agreement of all the original parties involved. There is also a growing trend for regulators to audit the closeout situation to ensure that all obligations are being met.

25.4.1.5 Timing

Generally, there is neither a minimum nor a maximum time stated for the achievement of closeout. The aim of rehabilitation is that the disturbed site should be returned to an agreed land-use. When the regulator is satisfied that the end point has been reached, then all outstanding performance bonds can be returned. The time taken to achieve the goal will vary with the climate, soils, vegetation, and degree of human intervention. In the case of New South Wales, there is a requirement that the revegetation be self-sustaining, and it is noted that it could be up to 20 years before this could be assessed as having been achieved. However, Bell[8] has suggested that 12 to 14 years should be sufficient time for a judgment to be made regarding self-sustainability. This is based on the outcome of research underway in Queensland. In Western Australia, assessment periods vary between five and ten years. In most other states where bonds are held, the regulators do not expect to retain bonds for less than one year or more than six years after rehabilitation work had ceased. In Queensland, the assessment process may take between five and ten years to be completed. This is because the proponent must demonstrate success in satisfying the agreed criteria which must then be self-maintained by the site for a further two years before release is obtained. It appears that no other state has such a stringent process.

25.4.1.6 Liability

In most cases, the liability for environmental protection during the operational phase of mining lies with the landowner or leaseholder. In Queensland and Victoria, the liability lies with the mining company. In most instances, at closeout the mining or mineral lease is relinquished or extinguished and the landowner, either state or private, resumes liability. The details vary from state to state:

- Northern Territories — once the mining company has quit the site, liability rests with the land titleholder.
- Victoria — liability rests with landholder, but the mining company can be prosecuted if problems related to mining arise and additional cleanup is required.
- New South Wales (coal) — the mining leaseholder remains liable.
- New South Wales (mineral) — the landowner is liable.
- Queensland — liability is with the landowner, but a mining company may be held liable under the Conservation Act if subsequent problems arise.
- Western Australia — land titleholder or the crown.
- South Australia — land owner, unless another arrangement has been agreed at closeout.
- Tasmania — Crown is liable after the discharge of corporate liability with the discharge of lease and closeout.

It is possible that this is an area where some form of national regulation and consistency would be appropriate. For example, for each project there could be an agreed period of time post-mining after which liability passes to the Crown. The concern with returning liability to a private landowner is that some environmental problems may only become apparent some time after the closeout. Acid rock drainage and flushing of pollutants into groundwater are examples of such long-term and "slow to develop" problems. However, it is unrealistic for the landowner (if not the same as the operator) to be expected to assume responsibility for the costs of any required reclamation, especially as they could be substantial.

25.4.1.7 Surveillance and monitoring

The mining company usually undertakes an agreed environmental monitoring and surveillance programme during the operational and rehabilitation/assessment phases. There may also be check monitoring undertaken by the regulatory authorities. These programmes usually concentrate on assessment of water quality. However, prescribed monitoring programmes after closeout generally do not exist. In some states, there is no monitoring programme at all, or only infrequent *ad hoc* programmes after closeout. In other states, notably Queensland and New South Wales, there is a requirement for the operator to continue monitoring after closeout. Such monitoring continues until it can be demonstrated that there is no unacceptable environmental impact arising from the site or as a consequence of the mining operation. In New South Wales, there may be a progressive reduction in the size of the bond as the monitoring programme proceeds and results demonstrate that risks of environmental impact are reducing.

25.4.1.8 New legislation

In all states, regulations and laws relating to the environment are either under review or have been recently reviewed. In Queensland, for example, new legislation is being drafted which will place the liability for the rehabilitation of the mine site environment on the mining company in perpetuity.

The implementation of such provisions will be difficult in the case of smaller companies that are unlikely to exist in perpetuity. An alternative is to make the owner of the land responsible and to create a register of site histories to be maintained by the supervising authorities. This register would include specific environmental concerns, thus enabling any prospective purchaser of a site to become aware of the liability that would be acquired with the land purchase.

25.4.1.9   The future

The idea of national closeout standards is a total anathema to the mining regulators in all states. Even within states, they feel that there is no place for generic standards. Most agree that site-specific standards must be set. On the other hand, several of the state regulators see merit in national standardised processes for both the setting of closeout criteria project by project, as well as for assessing the success of rehabilitation.

Many people were keen to see rehabilitation bonds made transferable. By this, they mean a system whereby the bond held should be transferable from one site to another operated by the same company, at the discretion of the regulator. Also, the notion that bonds should be reviewable is widespread. This would enable bond conditions to be adjusted in either direction during the lifespan of the project on the basis of improved environmental performance, progressive rehabilitation, unsatisfactory performance, or identification of a new environmental concern.

25.5   ANALYSIS OF BEST PRACTICE AND FUTURE TRENDS

Farrell wrote about the contemporary situation in Australia in a paper presented at the Australian Mining Industry Council (now the Minerals Council of Australia) 1993 Environmental Workshop, and published in the Proceedings of the Workshop. The following section has been reproduced, with only minor amendments, from Farrell.[1]

25.5.1   PLANNING FOR DECOMMISSIONING OR SHUTDOWN

The industry, too, has not been idle in developing its own guidelines and codes of practice for good environmental management, including decommissioning and rehabilitation. Chambers of Mines around Australia have been active in developing and promoting environmental policy statements, as well as various specific codes and guidelines.

While these documents may not be overly prescriptive, they are not necessarily widely known among smaller operators in the industry. The mining industry must be able to demonstrate a considerable degree of self-regulation before its policies and documents are fully accepted as potential alternatives to regulatory proscription.

To be successful, mine closure requires considerable planning. The best time for planning is moot. Many say that it is best done very close to shutdown, and others say within the last few years of operational life. A compromise view is that the timing of planning is dependent on the life of the project — for a short project (i.e., less than 5 to 10 years), planning should begin at or near the commencement of the mine, while for a long project it should begin about five years before the end.

In contrast to these views, Farrell[1] believes that planning for decommissioning of a mine or processing facility should begin at the project planning stage, a view with which the author of this chapter concurs.

Many long-term aspects of a mine, and to a lesser extent of a processing plant, are decided on during the project-planning phase. These may be the location of key parts of the operation such as waste dumps and tailings dams, or construction techniques such as whether to include a toe drain or not in a dump. *Proper planning of a site, with a view to considering possible end-uses and future rehabilitation needs, may save considerable expenditure in the future.*

Under most circumstances, mine planning generally aims to minimise haul distances for waste rock and place tailings dams in a suitable valley. This may mean that these two facilities are some distance apart (possibly measure in kilometres). If, however, the only suitable cover material for the tailings after mining has ceased is the waste rock, it will have to be remined and trucked or otherwise moved a considerable distance, which will be an expensive exercise. Similarly, planning a single tailings dam which will not be completed until the end of mine processing means that there can be no progressive rehabilitation, usually a far cheaper alternative. At Pasminco's Elura lead-zinc mine near Cobar, a rock quarry for underground fill was located adjacent to the existing tailings dam, so that it was used for tailings when the dam was filled. While logical for operational purposes, this has also meant that the area for final rehabilitation has been reduced.

Failure to characterise waste material may result in future environmental problems at a mine. Similarly, even the handling techniques for waste material should be considered early in the planning phase. For example, the special advantages of separately storing topsoil were not recognised at many mines, with the result that the topsoil has ended up at the bottom of stockpiles or mixed with other waste, making it no longer suitable for spreading. Careful identification of the various overburden and interburden materials at Queensland coal mines in the early 1980s resulted in selective stripping and placement such that the revegetation success was enhanced.[9]

Environmental planning can often reduce future problems at processing plants. While initially expensive, complete paving of a plant area can reduce the potential for spills to cause land contamination that might lead to a large cleanup bill at the closure of the plant. Strategic location of facilities can reduce duplication and solve future problems. For example, the quite fortuitous siting of a process water supply dam downstream from the processing plant, main waste rock dump, and tailings dam at Pine Creek Goldfields in the Northern Territory has resulted in the water dam becoming an integral part of the site's water management system; it will continue to act as a collector for any contaminated seepage from the waste rock and tailings after mine closure.

25.5.1.1 Planning considerations

Many of the factors that should be considered in mine and mineral processing plant closures have already been mentioned. These include:

- The siting of major facilities.
- The interrelationship of these facilities and environmental protection measures.
- The proposed end-use of the land.
- Characterisation of waste materials to identify potential environmental problems.
- Identification and characterisation of disused waste dumps, particularly those for chemicals and other potentially polluting substances.
- The location of any waste materials removed from the site.

The aim of a closure plan should be to minimise operational and post-operational environmental problems. As well as developing a logical plan for demolition of facilities and infrastructure, and rehabilitation of the site to the agreed final land-use, it should also identify all potential sources of ongoing environmental contamination and how they are to be controlled or eliminated.

25.5.1.2 Strategic planning

The first step should always be to clearly set out the objectives and then to prepare a detailed prescription for the way that the decommissioning is to take place (i.e., a plan). The preparation of a strategic plan allows an operator to focus on what the issues might be in decommissioning, and then to obtain the approval of the relevant regulatory authorities for the plan. The establishment of a plan some time before decommissioning allows problems that might develop to be avoided or circumvented, as well permitting the operator to schedule decommissioning activity into the normal production process of the site.

The plan must be dynamic to ensure that changing circumstances can be easily incorporated when necessary. One way of doing this is to conduct a review of the strategy and plan on a regular basis, perhaps in conjunction with the preparation of an annual environmental management plan.

25.5.1.3 Special considerations

As well as these general requirements for good practice in decommissioning, there are some special considerations that should be included when developing strategies and plans.

25.5.1.4 Heritage and mining history

Apart from aboriginal and European heritage, consideration should be given to preserving our industrial engineering heritage for future generations. In the past, many fine examples of this heritage have been destroyed in the name of progress. The same could be said for mining operations that are totally rehabilitated.

In many parts of Australia, mining heritage has become a tourist attraction. Sovereign Hill, Burra, and Ravensworth are examples. People today are interested in their heritage and want to understand what happened in the past, but most of these attractions are the results of mining carried out in the nineteenth and early twentieth centuries.

It must be ensured that, while decommissioning mine sites to contemporary standards, the future heritage value of the site is also recognised. For example, the Guidelines for Rehabilitation and Environmental Management Plans for Broken Hill Mines[10] specifically state that the planning of rehabilitation and of any subsequent land-use at the site will need to recognise and take into account the tourist, educational, scientific, and historical significance of Broken Hill's geological features and mining structures.

Complete removal of all infrastructures in an attempt to return a mine to a greenfield site could be regarded as an attempt to eliminate a valuable part of our heritage. One compromise in this could be to ensure that the site and the operations carried out on it are fully documented. This will at least allow the heritage value to be documented at a future date, and also provide good records should any problems arise at the site in the future.

Interestingly enough, heritage requirements have, in some instances, impeded mining companies in their desire to rehabilitate a former operational site. At the Lambton Colliery south of Newcastle,[11] FAI Mining, the last of many owners of the colliery, ceased mining operations at the end of 1991. Standard conditions on the mining lease required "rehabilitation of the mining area and removal of buildings, machinery, plant equipment, constructions and works as may be directed by the Minister." Approval to close the Lambton Colliery was granted, conditional upon the submission of a programme for rehabilitation and restoration of the site and completion of the programme within six months. However, when the company applied to Lake Macquarie Council for approval to demolish the buildings and structures, the Hunter Regional Environmental Plan 1989 (Heritage), in which the Lambton Colliery was listed as an item of local heritage significance, was invoked and the matter became the subject of an extended case in the Land and Environment Court. While, ultimately, the company was given permission to demolish the site, the case is a warning to mining companies that the heritage value of a site may be a consideration at both the commencement and completion of operations.

25.5.1.5 Land contamination

Many States of Australia have recently enacted, or are about to enact, legislation requiring the identification and possible cleanup of contaminated land. Contaminated land is identified by either direct investigation or by reference to a standard list of operations that may lead to contamination. Mining and mineral processing operations are usually included on these lists. This means that, as well as acknowledging the requirement to rehabilitate a mine site, a company now must consider the potential for the site to be contaminated and recognise that there may be some associated liability. With the need for environmental audits, the potential for contamination arising from the operations of mine sites and mineral processing operations is becoming well recognised. However, less well recognised is the possibility for waste rock dumps and tailings dams to also be regarded as contaminated land.

Yet to be resolved with any certainty is the issue of who carries the liability for such land once a mining lease is relinquished. A mining company could receive a nasty shock if it is found to be liable for contamination some time after it has relinquished title to the land.

25.5.1.6 Unresolved problem areas

Despite the great advances in environmental and rehabilitation science over the last 20 years in Australia, a number of problem areas still remain. These potentially have great impact on the issue of when decommissioning and rehabilitation of a site are completed, and therefore when security deposits can be recovered and leases relinquished. The major issues facing the mining industry are discussed below.

25.5.1.7 Rehabilitation completion criteria

Perhaps the greatest issue in rehabilitation yet to be resolved to the satisfaction of all parties is that of completion criteria. For the purposes of this discussion, completion criteria can be defined as "those measures, either qualitative or quantitative, against which the success or otherwise of rehabilitation can be measured, so that the objectives of the decommissioning plan can be achieved and the lease relinquished."[1]

The Queensland Department of Minerals and Energy has recently prepared draft technical guidelines for environmental management for mining in Queensland. One of these guidelines covers site decommissioning. This guideline states that the objective of decommissioning a mine site should be to ensure that areas that have been mined and/or affected by mining operations are left:

- Tidy and safe, with public safety risks reduced to acceptable levels.
- Stable and resistant to erosion.
- Suitable for the agreed post-mining land-use.
- At or better than the agreed levels of contamination.
- Within agreed waste discharge levels.

Other states already have, or are moving towards, similar objectives for decommissioning.

Completion criteria are most frequently invoked when the effectiveness of revegetation needs to be assessed. For areas that are being rehabilitated to agricultural or pastoral purposes, they are typically couched in terms of productivity. For areas being returned to natural vegetation, the criteria frequently include measurements of species richness and diversity, or vegetation structural analysis.

There are two problems associated with the use of completion criteria. The first is the question of whether the particular criteria used are appropriate for assessing the rehabilitation. The second is whether the criteria are relevant to the desired end-use and to the community. For example, if the desired end product is a mature eucalyptus forest that may take over 100 years to develop, how can we assess 10-year-old rehabilitation to satisfy all interested parties that the end product will be achieved without waiting the 100 years?

Until these questions are resolved, there will always be difficulties in applying completion criteria to rehabilitation.

25.5.1.8 The ultimate void

Open cut mining commonly leaves an open pit (final or ultimate void) remaining at the end of operations. While the volume of the void can be minimised by careful mine planning, this is frequently not done, and thus often quite large voids can result. Such voids represent a potential danger to people and livestock, as well as a potential source of environmental pollution resulting from water accumulation. The environmentally preferred rehabilitation method is backfilling to create at least a self-draining surface, if reformation of original contours is not feasible. This option is frequently not economically feasible and a satisfactory alternative must be sought.

Alternative uses sometimes considered for voids include:

- Water storage.
- Development as a wetland/wildlife habitat.
- Waste disposal.

There are excellent examples of innovative rehabilitation of mine voids, such as the RGC Wetland Project at Capel in Western Australia. However, in many cases the void is simply made as safe as possible and abandoned.

The potential for water contamination, especially of groundwater, still exists at some abandoned mine voids.

25.5.1.9 Waste dumps

Because the primary aim of any mining operation is to maximise the amount of ore mined, the characterisation, mining, and placement of waste rock can sometimes receive less attention than is required. This has resulted in unanticipated environmental problems arising at mines during and after mining. As with all other aspects of the mining operation, proper planning of waste dumps can minimise or even eliminate such problems in the future.

The disposal of waste materials from smelters has also often led to environmental issues. Smelter slags have been used for landfill, and as a material for reclamation of swampy areas. Because these slags contain residual levels of heavy metals, they are often targeted in land contamination issues. The problems that have arisen because of such use of slags from the Pasminco smelter at Port Pirie and the BHP smelter Whyalla are typical examples of this. Other smelter waste materials such as baghouse dusts, which again can contain relatively high levels of metals, can also lead to problems.

Particular issues to be considered when planning for decommissioning include the siting of waste dumps; the possible future use of the material either as backfill or cover for tailings dams; construction of dumps to minimise infiltration of water; the potential for hazardous leachates to develop; and the long-term stability of the dump. Many of these issues are still to be resolved.

25.5.1.10 Tailings dams

Tailings dams or dumps are effectively a special subset of the waste dump situation. Because tailings are usually very finely ground and contain residual amounts of chemicals used in the processing operation, they are potentially a source of environmental contamination. Tailings can also be a potential source of acid drainage. Because of their physical and chemical nature, some tailings may be particularly difficult to revegetate.

While revegetation of more benign tailings has sometimes been successful at least in the short term, the long-term stability of this vegetation has not yet been demonstrated. More hazardous tailings materials have been rehabilitated by isolating the vegetation from the tailings by means of impermeable clay cappings and capillary breaking layers. The chemical processes taking place within these tailings are not clearly known and may lead to environmental problems in the future.

25.5.1.11   Acid rock drainage

Many environmental scientists in the mining industry regard the production of acid rock drainage as the greatest environmental issue yet to be solved. Techniques now exist to enable the characterisation of various materials with regard to their acid producing potential, and many innovative techniques are now being introduced at mine sites to control acid mine drainage (AMD) and acid rock drainage (ARD). However the problem is not fully understood, nor are the exact processes which can take place within a dump or dam, and neither is there an understanding of how to minimise or eliminate the generation of ARD/AMD in the long term.

25.6   CONCLUSIONS

Decommissioning of mines and mineral processing facilities should now be an integral part of environmental management of these operations. In most cases, it should be a consideration even from the early stages of the project planning. The development and regular review of dynamic decommissioning plans and strategies will enable potential issues to be recognised, and action plans to be developed to address them. By doing this, many potential environmental issues can be avoided or eliminated. There can also be large cost advantages to a mining company in pre-planning waste disposal.

Throughout Australia, the issue of closeout criteria for mine operations is being treated in a variety of ways. There are trends towards more flexible arrangements in determining the size of performance bonds and increased use of environmental audits coupled with self-regulation by the mine operators. However, the criteria which determine that closeout has been successfully achieved are frequently site-specific. For this reason, the criteria must be clearly stated and defined at an early stage of project development. This will enable operators to be certain of their objectives and to plan their operation and rehabilitation strategies accordingly.

The long-term liability for environmental issues at closed-out mine sites is a complex issue which seems to fall to the landowner in most instances. The tendency is for site-specific programmes to be agreed between the parties rather than the prescriptive style of the past. Consideration needs to be given to a national guideline to limit environmental liability before the Crown or a landowner assumes responsibility. After all, it is only the state that is likely to be in a position to take any required action in 10 or 20 years' time. However, the notion that the state will eventually pay the bill should not be taken as a sign that mine planning and rehabilitation can be in the "near enough is good enough" genre. As responsible corporate citizens, all parties — miners, regulators and society — should be working together to ensure that mining becomes accepted as an integral element of sustainable development.

Recent events have shown that two areas must now receive special consideration — those of heritage value and land contamination. Although many advances in knowledge and practice for decommissioning have been made over the last 20 years or so, not all the answers have yet been obtained. In particular, completion criteria, ultimate voids, waste dumps, tailing dams, and acid rock drainage are identified as areas requiring considerably more investigation.

ACKNOWLEDGEMENT

This paper is based substantially on a paper presented by T. Farrell at the Australian Mining Industry Council's 1993 Environmental Workshop.[1]

REFERENCES

1. Farrell, T., Some considerations in planning for mine decommissioning, in *Environmental Workshop 1993*, Australian Mining Industry Council, Dickson, Australian Capital Territory, 1993, 235.

2. Cox, A., Land access for mineral development in Australia, in *Mining and the Environment: International Perspectives on Public Policy*, Eggert, R., Ed., Resources for the Future, Washington, D.C., 1994, 172.

3. Buckley, R., Criteria for a sustainable mining industry, in *Environmental Workshop 1991*, Australian Mining Industry Council, Dickson, Australian Capital Territory, 1991, 222.

4. Waggitt, P. and Mcquade, C., Mine close-out criteria — present guidelines and future trends in Australia, in *1994 AusIMM Annual Conference: 'Australian mining looks north, the challenges and choices'*, Australasian Institute of Mining and Metallurgy, Carlton, Victoria, 1994, 407.

5. Commonwealth of Australia, Department of the Arts, Sport, the Environment, Tourism and Territories, *Code of Practice on the Management of Radioactive Waste from the Mining and Milling of Radioactive Ores 1982, Guidelines*, AGPS, Canberra, 1987.

6. Hollands, K., Lease relinquishment in New South Wales completion criteria, in *Environmental Workshop 1993*, Australian Mining Industry Council, Dickson, Australian Capital Territory, 1993, 183.

7. Smith, P., BHP iron ore protocol for mine closure — Goldsworthy, Western Australia, in *Environmental Workshop 1993*, Australian Mining Industry Council, Dickson, Australian Capital Territory, 1993, 285.

8. Bell, L., Biological aspects of the rehabilitation of waste rock dumps, in Proceedings of the Symposium on the Management and Rehabilitation of Waste Rock Dumps, Riley, S. J., Waggitt, P. W., and Mcquade, C. V., Eds., AGPS, Canberra, 1993, 201.

9. Kelly, R., Prediction of some minespoil characteristics from exploration drill hole logs, in *Environmental Workshop 1987*, Australian Mining Industry Council, Dickson, Australian Capital Territory, 1987, 63.

10. Poseidon Mining Investments, Rehabilitation and Environmental Management Plan for Consolidated Mining Lease No.7, Broken Hill, New South Wales, unpublished report, 1992.

11. Davies, D. H., Lambton Colliery closure — heritage pitfalls, presented at Joint Environmental Workshop — Coal and Minerals, Department of Mineral Resources, New South Wales, 1993.

26 The Smelter in the Park: Changing the Way People Think

Ian Clark

CONTENTS

ABSTRACT

The smelter operated by Portland Aluminium provides an interesting contrast to the mills of the 1950s. The smelter at Portland illustrates how environmental care can be used to change the culture and operation of a large industrial complex to increase international competitiveness and increase benefits to staff and the community.

In 1991, academics from the University of South Australia worked cooperatively with the management of Portland Aluminium to produce a master plan for the "Smelter in the Park." The master plan proposed that the function of the Smelter in the Park be

"To integrate the operation of Portland Aluminium into a biologically diverse, aesthetically pleasing setting which makes environmental care an integral part of the work environment."

The benefits of the Smelter in the Park thus extend far beyond the common industry approach of tidying up around the front gate. People who work at the Portland Aluminium smelter enjoy a greatly improved work environment, access to knowledge, and a chance to participate in a worthwhile and exciting project; the company owners get a much more motivated and knowledgeable workforce leading to high performance, fewer accidents, and higher profits; and the community at large gets real conservation and recreational opportunities, and an environmental laboratory for research and education.

26.1 INTRODUCTION

In the 1940s and 1950s, many small towns in the U.S. and Australia were supported by metal smelters and farming. The local mills were built in the 1940s, using World War II technology. They dominated the community and had a big impact on its life. On windy days, the acrid odour of the coke ovens filled the air and clouds of dust from the slagheaps which surrounded the mills settled on everything. Processing water from the mill caused intermittent fish kills in the local rivers and lakes, and moderated water temperatures.

These mills supported families, farms, and to a major extent the towns. As the leading employer in the region, they made a significant contribution to the regional economy and to the social fabric of the community as well. They certainly had a very large impact on the environment — yet they were not really part of the community. They were the place where the men of the community worked. They were part of everyday life, but the community was not part of the mill. The mill was a smelly, dirty icon of heavy industry. The community accepted the fact that heavy industry meant putting up with the negative to share in the positive.

By the mid-1970s, many of these mills were closed because their outdated technology made them uncompetitive and unable to meet Environmental Protection Agency (EPA) standards.

During the life of the mills, society's expectation of industry changed from one of resource exploitation to an attitude of environmental concern and regulation. However, political regulation of industrial practice has not provided the outcomes from natural resources which society requires.[1] Consequently, during the past decade, industry and other groups in society have begun to adopt an attitude of consultation and cooperation concerning the use of natural resources.

26.2 THE EXPLOITATION OF NATURAL RESOURCES

For many years following the industrial revolution, success in industry was measured in terms of the economic value of commodities produced, costs measured in terms of labour, and technology required to produce commodities. Society's only expectation of industry was that it should create wealth and provide employment through the exploitation of resources.

During this time, little account was taken of ecological and subsequent social impacts of resource exploitation. Resources gained value as they were "developed" to create wealth. Thus, although questionable industrial processes and unsound rural practices caused the resulting degradation of the environment, it was the expectations of society that were ultimately responsible.

The adverse environmental effects which inevitably resulted from early resource development were considered only when they affected the economic viability of particular industries or impinged on the economic value of surrounding land, and then only when they could be easily identified. As long as land and other resources were seen to be underdeveloped, the role of industry was solely to increase the wealth of society.

26.3 REGULATION TO PROTECT THE ENVIRONMENT

Regulation was chosen as the "political" solution to safeguard against overexploitation and environmental degradation. Government legislation was followed by a proliferation of bureaucracies to administer and enforce the regulations.[2] This regulatory process was supported by taxing wealth-generating activities, particularly resource development.

The regulatory approach adopted by governments reinforced the traditional, and in some places still prevailing view that environmental protection and industrial profitability are natural opposites. Modifying industrial operations to comply with environmental regulations was thought to result in decreased profits and increased costs for consumers. Improving profitability was thought to require environmental exploitation and degradation. In other words, society could either have a healthy environment or a healthy business sector.[3]

26.4 BETTER PRACTICE THROUGH CONSULTATION AND
COOPERATION

The institutional regulation of industry has not produced the resource outcomes which society requires. Regulation through persuasion is very costly and alone will not bring about the economic or social benefits that are required. Society and industry have come to accept the need for regulation, but also recognise the necessity of utilising natural resources to produce the goods and services on which society depends. The finite nature of the world's resources and the magnitude of global change that is occurring due to the unwise use of resources are now widely accepted. People are becoming aware of a multiplicity of values that must accrue from the judicious use of valuable natural resources. Society is looking for better industrial practice, and many companies are responding.

An increasing number of corporations are showing that it is good business to integrate environmental care into their operations. More companies are realising that the pollution they produce is a sign of inefficiency. Wastes are raw materials not sold in final products. Companies are beginning to adopt the logic that pollution prevention makes economic sense.[4] Environmental concern leads to more efficient industrial processes, improvements in productivity, lower compliance costs, and perhaps new market opportunities.

26.5   THE PORTLAND ALUMINIUM SMELTER

The smelter operated by Portland Aluminium provides an interesting contrast to the mills of the 1950s. The smelter at Portland illustrates how environmental care can be used to change the culture and operation of a large industrial complex to increase international competitiveness and enhance benefits to staff and the community.

In the early 1980s, negotiations began on an initiative to build and operate an aluminium smelter near the small coastal town of Portland in the Australian State of Victoria. The smelter was negotiated between a consortium of companies and the state government. At the time of these negotiations, Portland was a quiet rural community supported by farming and small fishing and logging industries.

The industrial consortium and the Victorian Government launched an extensive media campaign extolling the benefits that would accrue regionally and to the state from such a large smelting operation; however, many local residents and conservation groups vigorously opposed the development. Residents feared the impact on the social fabric of their town from such a large industrial operation. Conservation groups feared the impacts of construction and operation of the smelter on the natural systems in the coastal environment. The economic imperative finally won out, and the decision was made to proceed with construction of the smelter.

26.5.1   A TECHNOLOGICALLY ADVANCED SMELTER

The largest, most technically advanced aluminium smelter in the Southern Hemisphere was constructed on a 600-hectare site located on a peninsula a few kilometers south of Portland. The management of the new Portland Aluminium Smelter was well aware of the competitiveness of the international aluminium industry, an industry that demands a very high level of productivity — especially in Australia, where labour and infrastructure costs are relatively high. Throughout the construction phase, the Portland Aluminium management was reminded of the divisions within the community about the real costs and benefits of the new smelter. It quickly became obvious that it would take more than size and technology to make the smelter profitable and viable.

26.5.2   CHANGING THE WAY PEOPLE THINK

The management of the company knew that making the most of the smelter's technological potential would require the commitment and skill of their employees. The company proceeded with the philosophy that success would depend on developing in their workers the attitude and the belief that pursuing company goals would lead to positive outcomes for themselves, their families, and the Portland community.

They felt that if staff believe the company has an important role in helping them to achieve important personal and community outcomes, they will be committed to the company goals, and will work together to attain them. This acceptance of company goals is especially important in a small community like Portland, where separating corporate culture from the community culture within which it exists is very difficult.

People do not naturally fit into teams and work together in a safe and productive manner to attain common goals. Long-standing, ingrained attitudes have to be challenged and often the culture that has developed in a community needs to be changed. Portland Aluminium management used the concepts of holistic management, "caring for each other, the community, and the environment" as the vehicle for change.

David Judd, Operations Manager from 1986 until his untimely death in 1993, recognised the importance of holistic management to the viability of the core business (making aluminium). He knew that environmental care must be a key component of operating an internationally competitive aluminium smelting operation in Portland. His vision was for Portland Aluminium to function as an integral part of workers' lives, an extension of Portland community. He said:

"You can talk all you like, but people believe what they see. If they see a smelter in a Park, they see a company which is truly concerned about caring for the environment."

26.6   THE SMELTER IN THE PARK

In 1991, academics from the University of South Australia worked cooperatively with the management of Portland Aluminium to produce a master plan for the Smelter in the Park. The master plan proposed that the function of the Smelter in the Park be

"To integrate the operation of Portland Aluminium into a biologically diverse, aesthetically pleasing setting which makes environmental care an integral part of the work environment."

The master plan points out four major considerations that must be followed in order to achieve this function:

- Extension of the workplace to the park boundary.
- Removal of psychological and physical boundaries between the smelter, the park, and the community.
- Rehabilitation of the park and areas within the smelter complex.
- An experimental approach that develops environmental best practice, including new technologies for clean, efficient operation and land rehabilitation.

Portland Aluminium has made a long-term commitment to rehabilitate the wildlife habitats around the smelter to create an environment for conservation, research, education, and enjoyment for the workers and the wider community.

Most of the land in the park has been extensively modified from its original condition initially for agriculture and more recently by the construction of the smelter complex. Large areas of natural vegetation have been cleared for agriculture, and wetlands have been drained and mined for sand. Large spoil heaps were left behind after construction, and invasion by exotic plants and animals are just some of the changes affecting the land.

Despite all of these changes, the area still retained an obvious potential for transformation into a multipurpose park with considerable conservation value. The restoration of diverse habitats and the creation of the park on this highly modified land will take many years and provides a real challenge for researchers and Portland workers. Since 1991, Portland Aluminium and University of South Australia staff have worked cooperatively to meet this challenge.

26.6.1   THE DEVELOPMENT OF THE PARK

Initially, the smelter complex was surrounded by a large buffer zone, a landfill site for rubbish, large areas of sealed pavement, graveled storage areas, and extensive water drainage systems. These features are of dubious value because they can engender undesirable attitudes and culture:

1. A buffer zone encourages the attitude that it is a place to pollute, where environmental values need not be considered too closely, and where the prying public eye can be discouraged. Ultimately, this results in lax work practices, suspicious neighbours, and loss of pride and commitment.
2. A landfill site or dump leads to dumping; it is much easier to tip something out than to look for an opportunity to recycle it. Once things are in the dump, there are cultural restraints about getting them out and reusing them.
3. Hardstand storage areas result in the unnecessary storing of obsolete material and machinery; this leads to spillage, contamination, and general untidy clutter. Machinery and vehicles are driven everywhere, even off the pavements, with the result that nature is pushed further back and the site looks more and more untidy.
4. Extensive drains and drainage engender the attitude that processing and storm water are wastes which can be used to transport and dilute pollution, suggesting that "waste" water should be drained into the sea and away from the site as quickly as possible.

The Smelter in the Park project has been the focus of a concerted effort to stop these attitudes developing, and for the smelter workers at Portland Aluminium to adopt instead the vision of environmental care and world-class performance. Under this programme:

- The buffer zone was replaced with a conservation/recreation park, open at all times to the public.
- The land set aside for the landfill site (dump) was reduced in size and replaced by a transfer station where material is sorted for reuse and recycling. A programme of reducing waste in which containers are recycled supports this and unnecessary packaging is reduced.
- Gardens, wetlands, and wildlife areas replaced hardstand storage areas. Access for machinery and storage areas is being reduced to an operational minimum.
- Extensive common drains are being replaced by a system to harvest storm water and recycle processing water. Wetlands are being created for water storage and wildlife habitat.

Smelter workers are being encouraged to learn about the park and the associated wildlife, and to become actively involved in development and rehabilitation. Work groups undertake projects in the park during their normal working hours as part of personal development and team-building exercises. This increased use of the park and knowledge about nature has had some very positive effects on the operation of the plant. The most important of these are:

1. *A steady reduction in waste generation.* Increased awareness about where dumped material goes and what it does to land values at that site is a very positive motivation for recycling. The previous attitude that "out of sight is out of mind" no longer prevails. The landfill site is now included in public tours of the plant. Areas once designated as future landfill sites have been rehabilitated and enhanced for wildlife, which really motivates workers to embrace waste minimisation programs. Volunteer groups have been involved in planting for wildlife habitat, nature trail construction, and bird monitoring on these sites, and it has been found that these activities reinforce the efforts to achieve waste reduction. Dumped waste was reduced from a monthly average of over 100 tonnes in 1990 to under 10 tonnes in 1993. These figures are widely promulgated around the plant, as are stories and pictures of wildlife returning to the rehabilitated swamps.
2. *Maintenance of a low level of fluoride emission.* Loss of fluoride from the system is a normal hazard of aluminium smelting. This occurs either as evolution of gas from high temperature areas, or as dust from work areas. It is minimised by adherence to standard

work practices, good maintenance of the fume retaining system, and good hygiene. The operating license (issued by the Victorian State EPA) permitted the lowest level of fluoride emission per tonne of aluminium produced for any smelter in the world at the time that the license was issued. Despite this stringent restriction, the smelter normally operates at about half the licensed emission rate. This could not be achieved without the dedication of the workers. This dedication is maintained in part by knowledge of the effects of high fluoride emission on adjacent gardens. A rise in emissions is readily seen in the health of the plants in the gardens around the smelter, which quickly raises awareness and concern.

3. *Water is considered a valuable natural resource.* A water audit has been completed, which identifies water quantities and changes in quality in all aspects of the smelting operation. As a result, research is underway to investigate methods to use water more efficiently in processing. The goal is to replace potable water that is currently being purchased with stormwater which is harvested, stored, and reused in a single loop recycling system. This will save money and eliminate the dumping of contaminated processing water into the sea. In the park, drains are being blocked and swamps rehabilitated for wildlife habitats.

4. *An improvement in people's perception of the value of the smelter to the community.* Initially, there was doubt about whether the wealth generated by the smelter would be worth the environmental cost. As the Smelter in the Park programme has developed, there has been an ongoing demonstration that the onsite environmental cost of producing aluminium is so low that, even very close to the smelter, some of the effects of past environmental degradation can be rectified. This has certainly improved the relationship with the Portland community, and also improves the feeling of self-worth among the smelter workers. This is considered very important in maintaining commitment and involvement.

5. *A more holistic approach to operational decision-making.* It is not unusual in any organisation for decisions to be made without taking into account the full effects of those choices. The management of the Portland Aluminium smelter adopted the belief that increased environmental awareness achieved through the Smelter in the Park will make people better informed in their decision-making, and more likely to look for downstream effects. In doing so, management actions have tried to banish the concept that the process is foremost, and that other considerations are secondary. Obviously, the company is not better off if an increase in operational efficiency results in a large legacy of environmental damage requiring costly rectification in the future.

6. *Improvements in the effectiveness of work teams.* Choosing and performing a voluntary environmental project tends to develop group awareness and decision-making. An increased awareness of the need to care and take personal responsibility for the environment leads to an awareness of the need and responsibility to care for one another and ourselves.

The benefits of the Smelter in the Park thus extend far beyond the common industry approach of tidying up around the front gate. People who work at the Portland Aluminium smelter enjoy a greatly improved work environment, access to knowledge, and a chance to participate in a worthwhile and exciting project; the company owners get a much more motivated and knowledgeable workforce, leading to high performance, fewer accidents, and higher profits; and the community at large gets real conservation and recreational opportunities, and an environmental laboratory for research and education.

26.7 WHAT ABOUT THE FUTURE?

Portland Aluminium's plan to recreate plant communities and manage diverse wildlife habitats around the smelter is a timely innovation, one example of the changing attitude of industry to conservation and environmental care. The important outcome of this development is that it is not only good for the environment — it is also good for business. The development and operation of the park demonstrates that social, environmental, and economic benefits will accrue as industries adopt practices which include clean, efficient processing and environmental care.

Portland Aluminium and the Research Centre for Environmental and Recreation Management (CERM) at the University of South Australia began to work cooperatively to develop the Smelter in the Park in 1990. Since that time, the park has received national and international awards; all of the staff and the company owners are aware of its benefits; and the environment and the operation continue to improve. However, a number of physical and psychological barriers still exist, and much remains to be done before the objectives of the master plan are met. The master plan points out that:

" . . . . *such a model Park will not be created easily or quickly. This development presents a remarkable challenge!"*

Environmental care is not a fixed point that a company reaches or a promotional concept for which a company receives awards. Environmental care is a key principle in the ongoing operation of a business. It requires continuous commitment from management and other staff to gradually improve the corporate culture and the operation of the core business through actively caring for the environment.

The pressures for international competitiveness and increased productivity can easily hide the less obvious and immediate cultural and economic benefits that accrue from long-term investments in environmental care. Without a long-term commitment, the gains that have been made thus far at Portland Aluminium will be lost. A change in management priorities could easily make the smelter operate just like so many other big industrial complexes located outside small towns; industrial operations more like the mills of old.

REFERENCES

1. Nance, C. and Speight, D., *A Land Transformed: Environmental Change in South Australia*, Longman Cheshire, Melbourne, 1986, Chapter 2.
2. Considine, M., *Public Policy: A Critical Approach*, Macmillan Education, South Melbourne, 1994, 73.
3. Papadakis, C., *The Limits and Possibilities of Social Science*, University of New England Press, Armidale, 1993, 14.
4. Schmidheiny, S., *Changing Course*, MIT Press, Cambridge, 1992.

Richard Isnor

CONTENTS

ABSTRACT

In recent years, conventional wisdom on how governments should approach environmental problems has undergone a distinct shift. The current debate is characterised by an increased emphasis on policies that are anticipatory and preventive, reinforcing those that are reactive and curative. Evolution of the principle of environmental and economic policy integration has suggested that environmental objectives should pervade policy areas not only concerned with the environment but also with issues such as economic growth, industrial development, education, and technical change.

Technological development, change, and adaptation are recognised as crucial factors affecting environmental management in mining operations, including issues associated with closure. Traditionally, however, examinations of government policies affecting closure practices have focused quite narrowly on regulatory approaches. In this chapter, a broader range of explicit and implicit government technology policies affecting mining operations in Australia, Canada, and the United States are briefly explored and compared. In each country, the form and function of sectoral policy communities are examined. Moreover, the chapter briefly investigates whether government policies have fostered the development of sector-specific research and technology development networks that collaboratively address environmental issues such as closure and rehabilitation. Specific examples of how explicit and implicit policies have affected the technological selection environment of mining firms in these countries are also briefly examined.

1-56670-365-4/00/$0.00+$.50
© 2000 by CRC Press LLC

27.1 INTRODUCTION

The environmental and economic pressures on mining firms in Australia, Canada, and the U.S. has made investment in research and development (R&D), technological accumulation, and innovation essential.[1] Ericsson[2] noted that the two main driving forces behind technological development in the non-ferrous metals sector are escalation of costs and environmental concerns. As a result, many mining firms have established specialised research and technology development groups that approach environmental problems strategically.

While primary responsibility for developing and advancing technical knowledge with respect to environmental "best practice" on closure issues lies with the private sector, governments also have an important role to play. Governments can influence private firms through a large variety of policy tools, and therefore should be able to provide support to the development of effective mine closure practices through explicit and implicit forms of technology policy. Schot[3] suggests three general government strategies needed to support environmentally sound technological development. These include:

1. Development of alternative variations. Governments can try to stimulate technologies that are not developed in the marketplace.
2. Modification of the selection environment through regulation and other instruments. Stringent regulation is seen as being important for the creation of new expectations about viable technological futures.
3. Creation or utilisation of institutional links called the technological nexus, between places that produce variations and their selection environments. Such a nexus helps translate selection pressure into criteria and specifications used in the design process. Networks need to develop around new technological options and effectively serve nexus functions.

Government, industry, and social actors and institutions are all responsible for helping to shift knowledge, production, and technical change in a direction that protects the environment. However, the addition of an environmental dimension into the technology dimension of sectoral and macro-level policymaking complicates decision-making for government. Foremost is the problem that responsibility for the environmental and technology policies that affect sectors such as mineral production are usually divided between government departments. Government agencies involved in these traditionally separated activities are often only imperfectly aware of the implications of their departmentalised policies on policies for innovation elsewhere in the political system (whether at an intra- or intergovernmental level). Even when they are aware of these indirect connections, they often do not attach such significance to them, as they are preoccupied with their primary mission (see Rothwell).[4] Some government departments and policy networks have a responsibility for stimulating, encouraging and supporting industrial sectors, while others are primarily responsibile for regulating and controlling them.[4] The policy literature offers little insight into whether these two types of networks can be brought together, so that coordinated approaches encompassing both objectives can occur.

It has been argued that governments and the mining sector need to develop an increased level of cooperation and collaboration in research, technical change, technological accumulation, and innovation.[5-7] Such cooperative efforts are deemed increasingly valuable by governments needing to share expenses and legitimise the value of public sector work. Crimes[8] has claimed that the shift towards collaborative research and development, innovation, and technology commercialisation provide an opportunity for a "new wave of innovation" in the mining sector. Scholars working in the mining and environment policy field have suggested that explicit technology policy approaches which foster transdisciplinary knowledge production aimed at solving problems associated with closure are essential to the long-term sustainability of mining.[1] Research and technical change addressing environmental problems in the mining sector can be extremely complex, requiring technical expertise that is often not consolidated in mining firms. The rehabilitation of mined lands, for instance, can require the integration of landform reconstruction, treatment of toxic wastes, and revegetation. The scientific input necessary for the development and implementation of these technologies spans a wide range of disciplines, from engineering and physical sciences to biological sciences. Collaborative initiatives are often necessary to draw such talent together.

In order to determine whether collaborative governance in support of best practice in ecological management in the mining sector is occurring, it is necessary to examine institutional arrangements and relationships between government and societal actors involved with sectoral policy-making. The concepts of policy community and policy network have evolved from efforts to examine institutional arrangements, as well as relationships between government and societal actors involved in policymaking at the sectoral level. Wilks and Wright[9] have used the term policy community to refer to "all actors or potential actors with a direct or indirect interest in a policy area or function who share a common policy focus and who, with varying degrees of influence, shape policy outcomes over the long run." This definition of policy community will be drawn upon in the latter part of the chapter in an attempt to discover whether institutional arrangements and governing relationships support collaborative and consensual approaches to environmental and technology policymaking in the mining sector. In addition, a range of explicit government technology policies affecting mining operations in Australia, Canada, and the U.S. are briefly explored and compared. The purpose of this analysis will be to determine whether instrument choices and implementation styles affecting closure are representative of the three strategies outlined (above) by Schot.[3]

27.2 SECTORAL POLICY COMMUNITIES IN AUSTRALIA, CANADA, AND THE UNITED STATES

The broad membership of policy communities associated with promoting environmental and technological advancement in the mining sector of Canada, Australia, and the U.S. show diversity in the types of institutions represented, as well as in their complexity (see Figures 27.1, 27.2 and 27.3). For instance, each country possesses a lead technical agency responsible for research and technology development/transfer: the U.S. Bureau of Mines (USBM); the Canada Centre for Mineral and Energy Technology (CANMET); and the Commonwealth Scientific and Industrial Research Organisation Institute of Minerals, Energy and Construction (CSIRO-IMEC), respectively.

Despite the presence of "common" institutions in the mining policy communities of Canada, Australia, and the U.S., there are features that distinguish each (Table 27.1). Australia, Canada, and the U.S. have each exhibited a considerable amount of recent policy activity directed towards environmental and economic sustainability in the mining sector. Policy community-wide discussions have addressed technological advancement on closure-related issues such as land rehabilitation, pollution prevention, and the improvement of metal recovery and recycling efficiency. Indeed, advancing the technological performance of the sector to meet these environmental objectives has become one of the clearest points of common ground between institutional actors in the mining policy communities of these countries.

27.3 EXPLICIT AND IMPLICIT TECHNOLOGY POLICY INSTRUMENTS AFFECTING CLOSURE

Policy instruments affecting closure in the mining sector vary in their specificity (i.e., some affect a broad range of industrial sectors, while others are sector-specific). They also vary in the degree to which they focus on environmental protection or on technological advancement. For the purpose of this analysis, therefore, technology policy instruments directed towards closure in the mining sector can be divided into two broad categories — explicit instruments and implicit instruments.

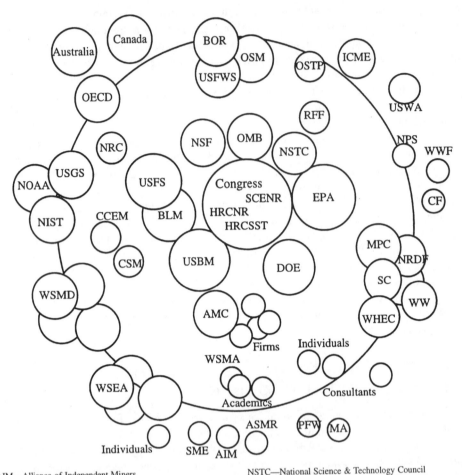

AIM—Alliance of Independent Miners
AMC—American Mining Congress
ASMR—American Society for Surface Mining and Reclamation
BLM—Bureau of Land Management
BOR—Bureau of Reclamation
CCEM—Colorado Centre for Environmental Management
CF—Conservation Foundation
CSM—Colorado School of Mines
DOE—Department of Energy
EPA—Environmental Protection Agency
ICME—International Council on Metals & Environment
HRCNR—House Committee on Natural Resources
HRCSST—House Committee on Science, Space, and Technology
Management Mineral Alliance
MPC—Mineral Policy Centre
NIST—National Institute of Standards and Technology
NOAA—National Oceanic & Atmospheric Administration
NPS—National Parks Service
NRC—National Research Council
NRDF—Natural Resources Defence Fund
NSF—National Science Foundation

NSTC—National Science & Technology Council
OMB—Office of Management and Budget
OSM—Office of Surface Mining & Reclamation
OSTP—Office of Science & Technology Policy
PFW—People for the West!
RFF—Resources for the Future
SC—Sierra Club
SCENR—Senate Committee on Energy and Natural Resources
SME—Society of Mining, Metallurgy and Exploration
USBM—U.S. Bureau of Mines
USGS—U.S. Geological Survey
USFS—U.S. Forest Service
USFWS—U.S. Fish and Wildlife Service
USWA—United Steelworkers of America
WHEC—Wildlife Habitat Enhancement Council
WSEA—Western State Environmental Agencies
WSMA—Western State Mining Associations
WSMD—Western State Mine Departments
WW—Worldwatch Institute
WWF—World Wildlife Fund (U.S.A.)

FIGURE 27.1   The U.S. non-ferrous metals policy community.

27.3.1   EXPLICIT TECHNOLOGY POLICY INSTRUMENTS

A variety of explicit technology policy instruments affect knowledge production in support of closure in the mining sector of Canada, Australia, and the U.S. These instruments are primarily associated with two government strategies — the development of alternative technical variations, and the creation and utilisation of institutional links. A clear trend evident in Australia, Canada, and the U.S. concerns the increased use of explicit technology policy instruments to prevent and remediate environmental pollution associated with mining.

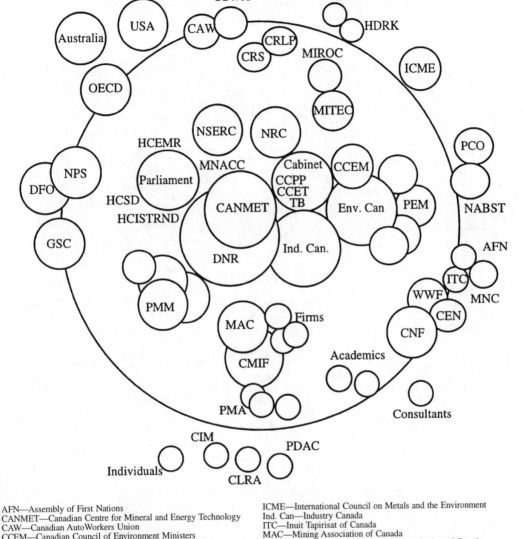

FIGURE 27.2 The Canadian non-ferrous metals policy community.

AFN—Assembly of First Nations
CANMET—Canadian Centre for Mineral and Energy Technology
CAW—Canadian AutoWorkers Union
CCEM—Canadian Council of Environment Ministers
CCET—Cabinet Committee on Economy and Trade
CCPP—Cabinet Committee on Priorities and Planning
CEN—Canadian Environmental Network
CIM—Canadian Institute of Mining, Metallurgy and Exploration
CLRA—Canadian Land Reclamationists Association
CMIF—Canadian Mineral Industry Federation
CNF—Canadian Nature Federation
CRLP—Centre for Resources Law and Policy
CRS—Centre for Resource Studies
DFO—Department of Fisheries and Oceans
DNR—Department of Natural Resources
Env. Can—Environment Canada
GSC—Geological Survey of Canada
HCEMR—House Committee on Energy, Mines & Resources
HCISTRND—House Committee. on Ind., S&T, Reg. & North Dev.
HCSD—House Committee on Sustainable Development
HDRK—HDRK Research Inc.

ICME—International Council on Metals and the Environment
Ind. Can—Industry Canada
ITC—Inuit Tapirisat of Canada
MAC—Mining Association of Canada
MIROC—Mining Industry Research Organisation of Canada
MITEC—Mining Industry Tech Association of Canada
MNACC—Minister's National Advisory Council on CANMET
MRD—Mining Research Directorate
NABST—National Advisory Board on Science and Technology
NPS—National Parks Service
NRC—National Research Council of Canada
NSERC—Natural Sciences and Engineering Research Council of Canada
OECD—Organisation for Economic Co-operation and Development
PCO—Privy Council Office
PDAC—Prospectors and Developers Association of Canada
PEM—Provincial Environment Ministries
PMA—Provincial Mining Associations
PMM—Provincial Mines Ministries
TB—Treasury Board
USWA—United Steelworkers of America

The federal governments of Australia, Canada, and the U.S. each fund university-conducted mining and mineral processing research, much of which is related to closure issues. Government-funded, university-based mining and mineral processing research centres are also common in all three countries. In the U.S., the USBM has funded State Mining and Mineral Resources Research

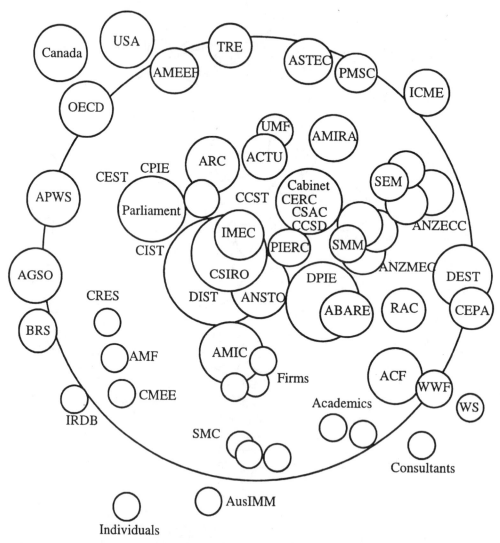

FIGURE 27.3   The Australian non-ferrous metals policy community.

ABARE—Australian Bureau of Agriculture & Resource Economics
ACTU—Australian Council of Trade Unions
AGSO—Australian Geological Survey Organisation
AMEEF—Australian Minerals, Energy & Environment Foundation
ACF—Australian Conservation Foundation
AMF—Australian Minerals Foundation
AMIC—Australian Minerals Industry Council
AMIRA—Australian Minerals Industry Research Association
ANSTO—Australian Nuclear Science and Tech. Organisation
ANZECC—Australian and NZ Environment & Conservation Council
ANZMEC—Australian & NZ Minerals & Energy Council
APWS—Australian Parks and Wildlife Service
ARC—Australian Research Council
ASTEC—Australian Science and Technology Council
Aus IMM—Australian Institute of Mining and Metallurgy
BRS—Bureau for Resource Sciences
CCSD—Cabinet Committee on Sustainable Development
CCST—Coordinating Committee on Science and Technology
CEPA—Commonwealth Environmental Protection Agency
CERC—Cabinet Expenditure Review Committee
CEST—Committee on Environment, Sport and Territories
CIST—Committee on Industry, Science and Technology

CMEE—Centre for Mineral and Energy Economics
CPIE—Committee on Primary Industries and Energy
CRES—Centre for Resource and Environmental Studies
CSAC—Cabinet Structural Adjustment Committee
CSIRO—Commonwealth Scientific and Industrial Research Organisation
DEST—Department of Environment, Sport and Territories
DIST—Department of Industry, Science and Technology
DPIE—Department of Primary Industries and Energy
ICME—International Council on Metals and the Environment
IMEC—Institute for Minerals, Energy and Construction
IRDB—Industrial Research and Development Board
OECD—Organisation for Economic Cooperation and Development
PIERC—Primary Industrial and Energy Resource Committee
PMSC—Prime Minister's Science Council
SEM—State Environment Ministries
SMC—State Mining Councils
SMM—State Mine Ministries
RAC—Resource Assessment Commission
TRE—Treasury
UMF—United Mineworkers Federation
WS—Wilderness Society
WWF—World Wildlife Fund Australia

Explicit and Implicit Technology Policies Affecting Closure   **469**

TABLE 27.1
Distinguishing Features of the Policy Communities Examined

Significance[a]	Australia	Canada	U.S.
Degree of fragmentation in federal technical agencies	*	*	***
Specialised federal public land management agencies	*	*	***
Importance of State/Provincial/Territorial agencies	***	***	*
Intergovernmental coordinating institutions	***	***	*
Relative importance of federal regulatory agency	**	**	***
Relative power of executive branch actors	***	***	**
Relative power of legislative branch actors	*	*	***
Relative influence of trade unions in sub-government	***	**	*
Relative influence of aboriginal groups in sub-government	**	***	*

[a] Absolute measure of significance
* Little significance
** Some significance
*** Very significant

Source: Isnor[23]

Institutes (Mineral Institutes). The USBM Generic Mineral Technology Centers (GMTCs) Programme has included a Mineral Industry Waste Treatment and Recovery Effort at the University of Nevada, Reno. The USBM-funded Mined Lands Reclamation Center (MLRC), based at the University of West Virginia, has also focused on research related to closure. However, in late 1995, the U.S. Congress effectively terminated the U.S. Bureau of Mines. By January 8, 1996, the Bureau's functions were either abandoned or transferred to other agencies within the Departments of the Interior and Energy. The health and safety and materials research functions were transferred to the Department of Energy; mineral information was moved to the U.S. Geological Survey; and Alaska operations went to the Bureau of Land Management. Environmental Research at Pittsburgh has been reassigned to the Department of Energy. Health and Safety Research in Spokane and Pittsburgh were reassigned to the Department of Energy for fiscal year 1996. USBM Research Centers in Denver, Minneapolis-St. Paul, Reno, Rolla, Salt Lake City, and Tuscaloosa were closed, as were Field Operations Centers in Denver and Spokane and headquarters office functions in Washington, D.C.

The effects of the loss of this sectoral technical agency are uncertain at present. The closing of USBM Research Centers will likely have significant impacts upon U.S.-conducted research in the mining and minerals sector. At the very least, it seems that the sectoral innovation system in the U.S. will be weakened. The Bureau has long been the focal point for collaborative industry-government research and technology transfer. It is unclear whether minerals-related research activity will receive much attention in the Department of Energy. Secretary of the Interior Bruce Babbitt voiced his distress over the closing of the USBM, but was pleased that the important health and safety functions would be continued by other federal agencies. From a closure perspective, moreover, it is fortunate that the Pittsburgh Research Centre survived. University-conducted mining research and technology transfer will likely suffer under the new arrangement — a major source of funding was the USBM-sponsored Mineral Research Centers and GMTC's.

In Canada, engineering and mining departments possessing Industrial Research Chairs primarily conduct government-supported university research addressing issues related to closure. In Australia, university research directed towards closure is conducted at specialized university research units like the Julius Kruttschnitt Mineral Research Centre and the Australia Research Council-funded Special Research Centres and Key Centres for Teaching and Research.

Another form of explicit technology policy in support of closure is the in-house research and technology development that governments conduct to prevent pollution or remediate existing pollution. Since 1988, the USBM dedicated an entire administrative unit, the Environmental Technology Division, specifically to pollution prevention and environmental remediation research. In Fiscal Year 1991, the Bureau had a total of 77 environment-related research projects underway.[10] Environmental research was to be elevated in the planned USBM Centres of Excellence reorganisation, but the Bureau was abolished by Congress and its functions divided among other agencies — most notably the U.S. Geological Survey and the Department of Energy. U.S. Government environmental research programmes of relevance to the mining sector are also spread throughout several other departments and agencies — including those which have recently absorbed much of the former Bureau of Mines.

Government research programmes of relevance to closure in the Canadian mining sector are largely the responsibility of CANMET. However, CANMET's programme of environmental research is organised quite differently from that of the USBM. CANMET's environmental research projects tend to be carried out as part of larger government programmes, or as part of projects directed towards improving the economic efficiency of the sector. As a result, there has been an effort to balance productivity-related and environmental research.

Like CANMET, there is no specific environmental division within the Australian Government's IMEC. However, IMEC does focus both on pollution prevention and remediation research. A recent change in direction saw the CSIRO Division of Coal and Energy Technology abandon its work on oil shale and synthetic fuels, and redirect its chemical engineering research effort into environmental process technologies of use to the mineral and energy sectors.[11] In addition, the Australia Nuclear Science and Technology Organisation (ANSTO) has undertaken a considerable amount of research having relevance to mining sector. ANSTO is internationally recognised for its provision of practical technical solutions to mine-related environmental problems, including closure-related issues such as bioremediation and physiochemical rehabilitation of mines, and the design of structures to minimise environmental impacts.

In all three countries, sector-specific collaborative research programmes have become an important explicit technology policy instrument directed towards environmental protection and remediation in the mining sector. For instance, the U.S. Bureau of Mines initiated several collaborative research projects with the private sector through the federal government's Cooperative Research and Development Agreements (CRADAs) programme. The CRADA programme allows federal agencies to provide technical expertise, laboratory facilities, and equipment access to private sector-driven research projects. CANMET's activity in collaborative closure research is focused on the Mine Environment Neutral Drainage (MEND) programme. MEND is financed and managed by three partner groups: the federal government (led by CANMET, but also involving four other departments); a consortium of over 20 firms and associations from industry; and eight departments from five Provincial governments. MEND also relies upon research support from a host of universities and has a number of international associate members, (which includes ANSTO).

While industry was the primary initiator of the MEND collaborative research programme, the participation of CANMET scientists and engineers was actively solicited. CANMET, moreover, was recruited to assume a central role in MEND network administration. The Canadian federal government has been very successful at bringing researchers of fundamental and practical measures (including those in government and universities) into contact with industry representatives, to find out what types of closure problems industry is faced with; what types of solutions industry needs; and what types of new ideas are emerging. In Australia, IMEC is involved in similar collaborative research programmes having relevance to closure. The Minesite Rehabilitation Research Programme (MRRP) cuts across CSIRO Institutes and focuses attention on the characterisation of mine wastes; long-term behaviour and evolution of landforms constructed from mine wastes; and ecosystem establishment. The programme has forged close research linkages with several mining companies and (like MEND) brings teams of research talent from different disciplines together to address complex mine-related environmental problems. IMEC Divisions are also either the leaders of, or major participants in, five Cooperative Research Centres (CRCs) having special relevance to the mining sector. These Centres, established under the 1990 Commonwealth CRC Programme, aim to support long-term strategic research; strengthen industry involvement in Commonwealth and university research; build centres of research concentration in particular areas; and stimulate education and training in industrially relevant research.

The Australian Centre for Minesite Rehabilitation Research (ACMRR) was formed in July 1993 as a joint venture between three major groups carrying out minesite rehabilitation research in Australia: the CSIRO Minesite Rehabilitation Research Program; the University of Queensland Centre for Mined Land Rehabilitation; and the Curtin University of Technology Mulga Research Centre. ANSTO recently became another full-partner in the ACMRR. The Australian minerals industry supports the ACMRR through the Australian Minerals Industry Research Association and through financial support of projects required by specific firms. The ACMRR has developed strong links with Commonwealth-funded Cooperative Research Centres. For instance, a three-year research programme carried out at ANSTO under the umbrella of the Cooperative Research Centre for Waste Management and Pollution Control has resulted in the discovery of a process (patent pending) for photo-assisted oxidation and immobilisation of arsenic. The ACMRR has also successfully marketed itself internationally; the U.S. Environmental Protection Agency (USEPA) has funded an AUS$1 million demonstration project that uses the ANSTO photo-assisted oxidation process to treat wastes from a smelter in Montana.

The governments of Australia, Canada, and the U.S. have all taken steps to provide private sector firms with incentives and support for the development of innovative environmental remediation technologies. Many of these programmes are relevant to closure because they generate technologies to deal with metals-based pollution. The U.S. Government's Superfund Innovative Technologies Evaluation (SITE) programme, for instance, has become a centrepiece in a nationwide effort to clean up hazardous sites. The key objective of the SITE programme is to reduce risks associated with the use of new technologies by providing reliable, realistic data on performance. In Canada, the Technology for Environmental Solutions Initiative (TESI) has similarly supported the development of environmental remediation technologies. Australian environmental remediation technologies have been supported by the Commonwealth Grants for Industrial Research and Development (GIRD) scheme.

Technology demonstration programmes have also become a common policy instrument in all three countries examined, both to foster the use of environmental remediation technologies and to encourage process or design changes that can prevent sector-specific pollution. The joint USEPA-Department of Energy (DOE) Mine Waste Pilot Programme (MWPP) was established in 1992 as a sector-specific initiative to examine technical solutions for mine wastes, tailings, dump/heap leaching wastes, and mine water. Another U.S. technology demonstration programme of relevance to closure in the mining sector is the Department of Energy's five-year Resource Recovery Project (RRP). The RRP was established in 1992 to evaluate and demonstrate multiple technologies for recovering water, metals, and other industrial resources from contaminated surface and ground waters. In Canada, the Environmental Technology Commercialisation Programme (ETCP) has provided financial resources on a cost-shared basis for environmental technology demonstration partnerships and joint ventures. A Technology Transfer Programme (TTP) has also provided environmental technology transfer services to Canadian firms.

In Australia, the Best Practice Development Programme has directed $31 million towards the demonstration of managerial and technical factors contributing to the international success of particular firms to others in industry. The Australian EcoRedesign Programme, established in June 1993, has also helped demonstrate (to industry) the economic and environmental value in redesigning products from an environmental as well as a functional perspective. Government funding has been provided for the demonstration of economic and environmental benefits obtained through the implementation of cleaner processes and the use of advanced technologies in production.

Finally, tax credits have become an important policy instrument, particularly in Canada and Australia, supporting research which can lead to improved closure practices. In Australia, tax incentives for R&D have become perhaps the key policy instrument towards encouraging techno-logical innovation and improved economic performance in industry. The 150% R&D tax incentive scheme in Australia is the most generous in the industrialised world: although there is a threshold of $50,000 annual spending on Australian R&D before the full 150% concession can be claimed, there is a sliding scale of deductibility of between 100 and 150% for annual R&D spending that falls between $20,000 and $50,000. Statistics have shown, moreover, that the level of industry R&D in Australia has risen dramatically as a result of its use.[12]

27.3.2 IMPLICIT TECHNOLOGY POLICY INSTRUMENTS

Implicit technology policy instruments affecting closure in the mining sector are considered to be those directed primarily towards other objectives, such as environmental protection, but which also affect technological choices made by firms. The Governments of Australia, Canada, and the U.S. each employ a wide variety of policy instruments that are directed specifically towards the objective of reducing environmental impacts in the mining sector. The use of these instruments primarily represents a government strategy of modifying the technological selection environment of firms.

Information and education instruments are similar to what Schneider and Ingram[13] call "capac-ity tools." Generally involving a low level of government coercion, these instruments provide skills, information, and education that can help firms recognise the links between technological advancement, pollution prevention, and productive (cost) efficiency. Government agencies in all three countries also assist mining companies to improve closure practices through technical extension services and guidelines relating to water management, minesite rehabilitation, and environmental planning. Initiatives aimed at publicising best practice environmental management include development of educational brochures, videos, and tours of successful operations. The USEPA has started compiling lists of waste minimisation technology relevant to the sector and documenting cases of sector-specific pollution prevention success. The DOE-USEPA Mine Waste Pilot Programme has also provided $300,000 towards the establishment of a graduate degree in mine waste engineering at Montana Tech. Training and education programmes will be directed towards faculty and students interested in mine waste treatment, characterisation, and planning for closure.

Government agencies also often provide funding to assist in the holding of national or inter-national conferences on specific closure issues. A good example was the 1994 International Land Reclamation and Mine Drainage Conference held concurrently with the Third International Con-ference on the Abatement of Acidic Drainage in Pittsburgh. The week-long set of meetings incor-porated the 19th annual meeting of the Canadian Land Reclamationists Association (CLRA), the 11th annual meeting of the American Society for Surface Mining and Reclamation (ASSMR), and represented the first meeting of the International Affiliation of Land Reclamationists (IALR). Both Conferences were jointly hosted and financially supported by ASSMR, CLRA, IALR, the Canadian MEND Programme, the Pennsylvania Department of Environmental Resources, the Tennessee Valley Authority, the U.S. Bureau of Mines (USBM), the U.S. Office of Surface Mining and Reclamation, and the USEPA. Co-sponsors providing financial support for conference functions included no fewer than eleven major mining firms, three U.S. industry associations, and four other U.S. Government agencies (the Forest Service, the Bureau of Land Management, the Geological Survey, and the Soil Conservation Service). Over 300 technical papers (all peer reviewed) were presented and attendance was estimated at 1200 persons from 21 countries. Government support of similar international conferences focusing on environmental and economic policy issues facing the mining sector has become common. The Canadian MEND programme, for instance, has been hosting an annual international conference on acid mine drainage research, technology development, and policy approaches since 1989. USBM-sponsored conferences on closure-related topics became one of the most effective policy tools for technology transfer and encouraging "best practice" in environmental management.

Schneider and Ingram[13] have called voluntary instruments "symbolic tools" because they can motivate target audiences to take action on the basis of their beliefs and values. Moreover, these instruments can provide a powerful learning focus for the target of policy. Increasingly, voluntary instruments are seen by governments as an appealing alternative to regulatory instruments in the mining sector — and are thought to involve significantly lower administrative costs. The basic underlying philosophy of these instruments is that if firms (and other social actors) understand that genuine economic benefits can be realised from implementing best practice in planning for closure, then action will occur without requiring much interference from government. The U.S., Canada, and Australia are all making increased use of voluntary programmes to improve closure practices in the mining sector.

Voluntary toxic substance reduction programmes, such as the 33/50 Programme in the U.S. and the Accelerated Reduction and Elimination of Toxics (ARET) Programme in Canada have been favourably received and led by (in the case of ARET), the mining industry. To date, one of the most unique programmes designed to promote best practice in environmental management and closure is the U.S. Mineral Showcasing Program. This effort, initiated by the U.S. Forest Service in 1989, has begun to alleviate negative public perceptions regarding environmental impacts of mining and mineral processing on federal land. In designating a Showcase operation, the Forest Service enters into cooperative planning with operators, and requires a demonstrated commitment to environmental excellence. Local Forest Service officials help firms achieve environmental goals and concentrate less on regulatory enforcement. Showcase operators, in turn, tend to focus on technical improvements and enhancing the positive image portrayed by the Showcase designation. Mineral Showcasing also involves communication of specific messages to particular audiences, usually members of a local community or area.

A number of implicit technology policy instruments having explicit relevance to closure are related to the planning, permitting, and regulatory requirements associated with mineral development. In all three countries examined, environmental assessment, design approval, remediation planning, and performance bonding requirements are common and have a significant impact upon planning for closure. The use of these instruments provides incentives for mining firms to improve environmental management practices through the development and/or use of new technology. Environmental impact assessment (EIA) processes require industry-sponsored environmental research — increasingly required to focus on closure issues. EIA research can provide great insight into the local ecology of planned development areas, as well as how projects can be designed and managed to minimise adverse impacts on an area. In some cases, the research can suggest specific criteria for rehabilitation. The permits required for mining activity also address issues related to closure, but the fragmented nature of these systems has often prevented them from becoming a central force in comprehensive planning for closure.

The posting of performance (security) bonds is also a common aspect of mine development planning in the U.S., Canada, and Australia. The use of this instrument provides a strong financial incentive for firms to undertake rehabilitation as mining occurs, as well as during closure. In Queensland, Australia, performance bonding has evolved to provide unique incentives for promoting environmental best practice. The amount of the performance bond is determined by the estimated cost of rehabilitation, and a category system has also been adopted to determine the risk relating to leaseholder environmental performance. The categories range from Category 1 leaseholders (who demonstrate that they are fully able to meet their responsibilities in terms of environmental management) to Category 6 leaseholders (who are unable to demonstrate such responsibility). Leaseholders who demonstrate performance criteria below Category 6 are permitted to lodge a percentage of the full security bond. The actual percentage to be paid is based on a sliding scale from Category 1 operators (who are permitted to post only 25% of the bond) to Category 6 operators (who are required to post 100% of the bond).

Liability legislation has become another common and effective incentive for promoting best-practice closure planning in the mining sector. The effect of the liability instrument has extended beyond the firm to financial lending institutions and insurance companies, who now require assurance that their investments in mining operations will not be threatened by negligent environmental management practices. Governments in Canada and Australia have also begun to use the tax system to provide incentives for closure-related environmental expenditures. In Canada, rehabilitation bonds have been included as a tax-deductible expense for mining firms. In Australia, mine rehabilitation expenditures (including plant demolition costs) have become fully deductible company expenses. Tax deductions in the 1992–1993 Australian Federal Budget included costs incurred for environmental impact studies, environmental audits, preparation of environmental management plans, and environmental monitoring. The Australian Commonwealth Government has made use of special taxes on coal production to fund a greater amount of industrially and socially relevant R&D — much of which deals with closure-related issues. Recently, the Commonwealth Government has turned responsibility for the use of the coal levy R&D funds over to industry to increase the relevance of this research.

Finally, environmental awards have become an important policy instrument for promoting best practice in planning for closure. These awards provide recognition, not only for special achievements, but also for building positive corporate identity. Moreover, they foster a sense of competitive spirit among mining firms and employees to receive special recognition for their environmental management and innovation efforts. Environmental awards are not limited to national bodies; they have also been presented by international bodies such as the United Nations (UN). Alcoa Aluminium of Australia was the first company to win a Global 500 award from the UN Environment Programme. Government agencies in Canada, the U.S., and Australia have recognised the potential incentives in environmental awards. The USEPA has presented Environmental Leadership Awards to mining firms for closure achievements, and the Office of Surface Mining (OSM) has initiated awards for outstanding Abandoned Mine Land Reclamation. State and Territory governments in Australia, and Provincial governments in Canada, have also established special sector-specific environmental awards similar to those in the U.S.

27.4   DISCUSSION

The brief overview of instruments presented above suggests that public policy relevant to closure in Australia, Canada, and the United States has evolved beyond simple regulatory approaches. In fact, environmental policy affecting closure practices in the mining sector has become integrated with policy instruments from other "policy areas." A wide variety of explicit technology policy instruments, for instance, are being used to foster the development of technology that will prevent and/or remediate sector-specific pollution. Implicit technology policy choices, moreover, represent a variety of government strategies designed to modify the technological selection environment of mining firms.

27.4.1   THE EFFECT OF POLICY COMMUNITIES ON CLOSURE PRACTICES

Australia, Canada, and the U.S. have each exhibited a considerable amount of recent policy activity directed towards environmental and economic sustainability in the mining sector. These policy community-wide discussions have addressed the issue of technological advancement in areas like land rehabilitation, pollution prevention, and the improvement of metal recovery and recycling efficiency — all of which are important to closure. The interaction of members of policy communities, therefore, has been influential in the development and implementation of explicit and implicit technology policies affecting closure practices.

In the U.S., the Department of the Interior convened a National Minerals Policy Forum in 1992. This forum was intended to be a community-wide policy development exercise involving major industry, government, and interest group actors. However, the forum did not generate lasting policy discussion, and has not been held since. The 1990s have also seen considerable efforts within the U.S. Congress to reform the federal Mining Law. The opportunities provided by this process to develop consensus within the U.S. policy community have for the most part failed. In fact, deep divisions between industry and special interest groups (and even between federal agencies) have been at their most obvious during Mining Law reform. In an effort to heal wounds created by Mining Law reform, DuPont, Inc. has sponsored a Roundtable to bring traditional adversaries associated with mineral and environmental policy together. A news report indicated that the Mining Roundtable's "informal off-the-record discussions" between a small group of miners, environmentalists and regulators, may improve the future of the U.S. mining sector.[14] However, members insist that the Roundtable has no official standing. It does not, moreover, attempt to influence public policy issues; it merely attempts to generate ideas to be acted upon by others.

Canada has had a relatively greater degree of success than the U.S. in its attempts to build national coalitions which support environmentally sustainable economic development.[15] In response to a visit to Canada by the UN World Commission on Environment and Development in 1986, the federal government created a National Roundtable on Environment and Economy to focus on public education and provision of policy advice to government. Similar roundtables have since been established in each province. Canada's Green Plan was another federal initiative involving a nationwide multi-stakeholder consultation process in 1990. When announced, the Plan contained over 100 new policy initiatives, and represented a 5-year, $3 billion federal government commitment to putting Canada on the road to sustainability. Several important programmes, such as the National Contaminated Site Remediation Programme, the Arctic Environmental Strategy, and the Environmental Technology Commercialisation Programme received Green Plan funding. In 1992, the annual Mines Ministers meeting led to a proposal by the Mining Association of Canada for a comprehensive and coordinated plan of action to deal with challenges facing the industry. The Whitehorse Mining Initiative (WMI), named after the location of the 1992 meeting, involved representatives of the mining industry, governments, labour unions, aboriginal peoples, and the environmental community. This assembling of participants was impressive — a policy community-wide effort not previously witnessed for the Canadian sector. The Leadership Council, for instance, consisted of all mine ministers in Canada (federal and provincial); other relevant federal ministers; presidents and chief executives of successful firms; presidents of key industry associations and labour unions; presidents and directors of influential and relevant environmental and aboriginal groups; and prominent academics. In November 1994, the Issue Groups produced over 150 specific recommendations — many of which reflected the overlap between technological advancement and environmental protection in the sector. The WMI Leadership Council Accord contained a vision statement, 16 principles, 70 goals and a statement of commitment to follow-up action. It represented a major step towards a constructive relationship across the whole of the mining policy community. The strategic vision which evolved for the sector from the WMI was not only explicitly directed towards achieving environmental and economic sustainability, but also to the continued evolution of policy-community consensus for the sector's future.

Policy community-wide initiatives to encourage environmental and economic sustainability in the Australian mining sector have almost co-occurred with those in Canada. A 1989 Industry Commission inquiry into the mining and minerals processing industries focused on a wide set of issues and represented a policy community-wide attempt to map the sector's future. The Commission received formal submissions from interested parties, and held discussions with relevant stakeholders from across the sector's policy community. In February 1991, the Industry Commission released a four-volume report which provided an attempt to reconcile these positions in the form of recommendations to the Commonwealth and state/territorial governments. Several recommendations were directed towards encouraging technological advancement in response to environmental problems (like special tax treatment for mine rehabilitation expenditures). In July 1989, the Australian Prime Minister initiated a national Environmentally Sustainable Development

(ESD) strategy. One of the nine working groups subsequently established to develop ESD principles focused on the mining industry. In its dedication of a Working Group specifically to mining, the ESD process provided a policy community-wide forum for consensus-building similar to that which later developed in the form of the Canadian Whitehorse Mining Initiative. The final report of the ESD Mining Working Group contained 88 specific recommendations, as well as detailed suggestions for their implementation.

27.4.2 ENVIRONMENTAL TECHNOLOGY POLICY SYSTEMS AND CLOSURE

In each of the three countries examined, policy instrument choices are being directed towards the three government strategies outlined by Schot[3] in his quasi-evolutionary model of environmentally-relevant technical change. In Schot's model of technical changes, three general government strategies are distinguished as being particularly important for moving technological systems towards an environmentally sustainable future. These include:

1. *Development of alternative variations.* Governments could try to stimulate technologies that are not developed in the marketplace.
2. *Modification of the selection environment* through regulation and other instruments. Stringent regulation is important for the creation of new expectations about viable technological futures.
3. *Creation or utilisation of institutional links.* These are called the technological nexus, between places that produce variations and their selection environments. Such a nexus helps translate selection pressure into criteria and specifications used in the design process. Networks need to develop around new technological options and effectively serve nexus functions.

Explicit technology policies tend to focus on the development of alternative technical variations and the creation or utilisation of institutional links. Governments in all three countries are attempting, through a variety of means, to create or stimulate the development of technologies that do not currently exist in this sector. New *in situ* technologies, applications of advanced control systems, pollution-free smelting technologies, non-explosive blasting techniques, non-chemical flotation processes, and acid mine drainage remediation/prevention technologies are a common focus of government-sponsored research, development, and commercialisation. Reducing the environmental impact of production technologies and creating technologies that can remediate pollution from past production activities are key objectives of this strategy. In many cases, the governments of these countries are also attempting to influence the technological nexus through the development of collaborative networks that can define new technological options for the sector.

Implicit technology policies emphasise modification of the selection environment. Performance bonding is now commonly required for operations; stringency of regulations is roughly equivalent; and all three countries have developed liability mechanisms for environmental damage, thereby providing firms with a financial incentive to prevent pollution. Some of these instruments also affect the technological nexus by helping translate selection pressure into criteria and specifications used in the design process. For instance, in all three countries it is now important for non-ferrous metal-producing firms to communicate honestly about their environmental liabilities and efforts to shareholders. Some companies, such as Noranda Minerals, publish an annual environmental report to inspire public confidence in the company. Banks and insurance companies investing in development projects or guaranteeing performance bonds have established review boards and policies to ensure that the evolving environmental management in their client firms represents best-technological practice. Industrial associations are developing "self-policing" policies that aim to discourage operating practices that can bring disrepute to the entire sector. Since it helps improve project design and reduce public opposition to mining, open communication with the public and avoidance of confrontations with environmentalists are increasingly being emphasised. Many implicit sectoral technology policies, therefore, have created a situation in which non-government actors (including those within firms) are providing a greater degree of environmental influence in the technological selection environment. Acknowledging the environmental impacts of mining, committing to a policy of care and restoration for local ecosystems, and soliciting comments and suggestions which can be visibly responded to are all positive steps encouraged by governments to help firms improve their image. The U.S. Mineral Showcasing Programme, for instance, has focused on recognising company efforts to educate and address public concerns with mineral production.

Authority-based instruments have imposed significant environmental costs on the mining sector. For instance, the environmental costs at Echo Bay's Alaska Juneau project are anticipated to be in the range of $135 million — approximately 50% of the total project capital costs.[16] However, it has been found that, as a percentage of total costs, projects that commit significant resources to environmental protection at the outset have much lower reclamation and decommissioning costs than projects which do not make this investment. For instance, in the Juneau, Alaska project mentioned above, the costs of reclamation are expected to reach just $1.6 million — 0.4% of a total life-of-mine capital cost of $375 million.[16] In contrast, ZMDC's Elvington mine in Zimbabwe spent $16,000 on up-front environmental costs (out of a total initial capital cost of $5.4 million) and reclamation is expected to cost $2.1 million — or 22.3% of total life-of-mine costs. As a result, it can be argued that government policies which provide an incentive for companies to invest in environmental protection (especially those related to closure) at a project's outset are more likely to limit long-term environmental costs than those that do not. In this respect, implicit technology policy instrument convergence in Canada, Australia, and the U.S. may actually save companies money in the long run.

Schneider and Ingram's[13] behavioural analysis of policy instrument choice has considerable relevance to the linking of instrument choice to government strategies for transforming technological systems in support of environmental protection. Schneider and Ingram argue that public policy almost always attempts to get people (or institutions) to do things that they might not have done otherwise, and identify five broad categories of instruments — authority, incentives, capacity-building, symbolic and hortatory, and learning — each of which makes different assumptions about how policy relevant behaviour can be fostered. Table 27.2 shows how common sector-specific instruments in the mining sector currently span these organisational categories.

The fact that policy instruments affecting the mining sector in Australia, Canada, and the U.S. span these behavioural categories supports the notion that governments have adopted multi-strategic approaches towards encouraging technological and environmental advancement in the sector. There is considerable overlap between Schot's[3] vision of multiple sources of government influence over evolutionary technical change, and Schneider and Ingram's[13] notion that a variety of policy instrument choices can address desirable behavioural changes.

While governments in Australia, Canada and the U.S. have used past experience and adapted existing industrial and technology policy instruments to environmental objectives in this sector, it seems that the integration of environmental with industrial/technology policy has created scope for policy innovation. The Government of Queensland's linking of performance bond payment to environmental performance criteria is a good example of this type of policy innovation. Voluntary programmes, such as the U.S. Mineral Showcase Programme and the 33/50 Programme are other examples. Collaborative research and technology development arrangements related to closure issues have developed rapidly, particularly in Canada and Australia. The dynamic role of government in these arrangements is a significant departure from the linear thinking of past government involvement in sectoral research and technology development. Therefore, the integration of environmental, industrial, and technology policy objectives associated with the mining sector has allowed governments to "retool" old policy instruments to serve environmental objectives, as well as invent new types of instruments for this purpose. However, while the trend towards policy instrument retooling and innovation in the mining policy community is partly the result of trial and

TABLE 27.2
Convergence in Behavioural Attributes of Sectoral Policy Instruments

Authority	Incentives	Capacity	Symbolic	Learning
Emission regulatory standards	Research grants	Support for university centres	Pollution prevention division	Use of committee hearings
Environmental assessment requirements	Cost-sharing of research and development	Information provision, surveys, policy research	Increased research expenditures on environment	Science and technology networks
Permit charges	R&D tax credits	Conference support	Awards	Educational centres and programmes
Performance bond requirements	Repayability of performance bonds	Educational centres and programmes	Voluntary agreements	Policy issue networks
Rehabilitation plan requirements	Technology demonstrations	Support of policy forums	Inter-governmental agreements	Support of policy forums
Threat of liability determination	Threat of liability action	Technology transfer, technical assistance	International agreements	International policy activity
Taxes for cleanup	Tax incentives for rehabilitation	School programmes	Establishing conservation areas	
Voluntary agreements	Threat of criminal actions		Cleanup funds	

Source: Isnor[23]

error, and learning from mistakes and past experiences, it is also clearly the result of the evolution of environmental policy thinking in industrialised nations. Moreover, the behavioural assumptions of policy instrument choices affecting closure practices seem to be changing — moving away from an overreliance on coercion to a more balanced approach. The U.S. Pollution Prevention Act of 1990, for instance, is considerably non-regulatory compared to other environmental statutes — due to its emphasis on capacity building and symbolism. This legislation is beginning to transform government thinking about policy approaches to environmental protection in the U.S. and abroad. Voluntary programmes emphasising public-private partnership, technological innovation, education provision, streamlined multimedia pollution control, and decentralised ecosystem management are starting to also characterise U.S., Canadian, and Australian mining policy. However, as Gunningham[17] claims, "it is significant that command and control regulation is still the dominant mechanism invoked in the U.S.A., which has paradoxically also been the most imaginative in its use of alternatives."

All three countries examined, therefore, can be said to exhibit a continuum of policy instrument choices directed towards closure in the mining sector. Moreover, this trend towards use of a wider variety of instruments has occurred because regulatory-based strategies have failed to achieve desired environmental objectives on their own. A common and intense dissatisfaction with regulatory-heavy approaches to environmental problems has emerged among business interests. They, in particular, have pressed governments to devise alternative instruments. Overwhelming reliance on regulatory approaches, in the absence of mechanisms to treat the causes of pollution, has been found to actually create new environmental problems in this sector.[1] In such instances, there is a strong incentive for polluting or high-production cost mining firms to simply declare bankruptcy and abandon operations. This abandonment often causes greater environmental damage than would occur under continued operation since the costs of cleanup are usually transferred to the state (which has neither the resources or cumulative technical expertise to undertake cleanup).

The fact that policy instruments affecting closure in Australia, Canada, and the U.S. have evolved to span Schneider and Ingram's behavioural categories supports the notion that governments have adopted multi-strategic approaches towards encouraging technological and environmental advancement. This finding is a unique one within the environmental policy field. While application of policy instrument choice typologies to environmental policy have been rare, those that do exist have not noted the widening variety of applicable policy instruments from a systemic perspective. Part of the problem is due to the fact that political scientists examining environmental policy continue to maintain a mechanistic view of both technical change and environmental policy. These analysts tend to see technology policy as simply being an information provision-type instrument and, while useful, is not worthy of the attention that regulation and market incentives should receive. The other interesting aspect of examinations of environmental policy instrument choice (indeed, of instrument choice in general) is that they tend to take a macro perspective which ignores the variety and context of policy instruments affecting particular policy sectors. Linder and Peters[18] suggest that closer examination of the sectoral context for policy instrument choice is needed. A systemic perspective can give a better understanding of the different ways that instruments impact on a policy sector as well as their interlinkages. It is concluded here that the three governments examined in this study exhibit convergence in their evolution of more balanced approaches to instrument choices (affecting this sector) than was the case prior to the mid-1980s.

27.5   CONCLUSIONS

The shortcomings of neoclassical, reactive policy approaches to environmental problems have increasingly been recognised in the literature.[19-21] Since the mid-1980s, analysts in the field have pressed governments to develop more systematic and integrated approaches for dealing with environmental problems. Schot et al.[22] provided one of the first empirical studies outlining how governments are beginning to take such an approach. Policy integration, however, still competes with environmental policy approaches based on neoclassical economics.

In this chapter, an attempt has been made to examine policy instrument choices from a dynamic and systemic perspective. The existence and interaction of sectoral policy communities in the countries examined helps to explain emerging similarities in public policy affecting closure in the mining sector. A situation has emerged in which policy instrument choices affecting closure have "diversified" considerably from traditional regulatory approaches. Currently, there is considerable scope for policy "learning" and "lesson-drawing" across countries as governments constantly search for innovative mechanisms to deal with complex environmental issues, while at the same time sustaining, diversifying, or enhancing economic activities.

Therefore, significant developments have occurred in policy approaches towards closure in the mining sector. Governments currently see closure as a much more integral aspect of the total mining process. As a result, considerable government research and technology development attention has begun to focus on closure — an area that was previously governed largely by regulatory policy. Government agencies such as CANMET and IMEC have developed a dynamic presence in research and technology development networks focusing on closure issues. These approaches complement other government policy strategies that have influence on firm-level technology selection choices. Most importantly, they represent a positive step in a direction whereby environmental policies become anticipatory and preventive, while reinforcing those that are reactive and curative.

REFERENCES

1. Warhurst, A., The limitations of environmental regulations in mining, in *Mining and the Environment: International Perspectives on Public Policy*, Eggert, R., Ed., Resources for the Future, Washington, D.C., 1994, .

2. Ericsson, M., Minerals and metals production technology — a survey of recent developments, *Resources Policy*, 16, 284, 1991.

3. Schot, J. W., Constructive technology assessment and technology dynamics: the case of clean technologies, *Science, Technology and Human Values*, 17, 36, 1992.

4. Rothwell, R., Industrial innovation and government — environmental regulation: some lessons from the past, *Technovation*, 12, 447, 1992.

5. National Research Council, *Competitiveness of the U.S. Minerals and Metals Industry*, National Academy Press, Washington, D. C., 1990.

6. Commonwealth of Australia, *Ecologically Sustainable Development Working Groups—Final Report—Mining*, Australia Government Publishing Service, Canberra, 1991.

7. Whitehorse Mining Initiative, Leadership Council Accord, Final Report, November, Department of Natural Resources, Ottawa, 1994.

8. Crimes, P. B., Optimizing scarce resources for mining technology advance, *AMC Journal*, July 5, 1993.

9. Wilks, S. and Wright, M., Eds., *Comparative Government-Industry Relations*, Clarendon Press, Oxford, 1987.

10. United States Department of the Interior, Bureau of Mines, *Compendium of Environmental Activities*, Bureau of Mines, Washington, D. C., 31 October 1991.

11. Anonymous, New directions for CSIRO research, *The Miner*, February 17, 1994.

12. Australian Bureau of Industry Economics, Research Report No. 50, R&D Innovation and Competitiveness: an Evaluation of the Research and Development Tax Concession, AGPS, Canberra, 1993.

13. Schneider, A. and Ingram, H., Behavioural assumptions of policy tools, *Journal of Politics*, 52, 510, 1990.

14. Kosich, D. Y., Roundtable fosters hope for mining, *Mining World News*, 5, 12, 1993.

15. Schmidheiny, S., *Changing Course: A Global Business Perspective on Development and the Environment*, MIT Press, London, 1992.

16. Metals Economics Group, *Meeting the Environmental Challenge in the Gold Industry—Strategic Implications of Environmental Considerations for the Gold Industry*, Metals Economics Group, Halifax, Nova Scotia, 1993.

17. Gunningham, N., Developing an optimal regulatory strategy, *Search*, 25, 98, 1994.

18. Linder, S. and Peters, B. G., Instruments of government: perceptions and contexts, *Journal of Public Policy*, 9, 35, 1989.

19. Irwin, A. and Vergragt, P., Rethinking the relationship between environmental regulation and industrial innovation: the social negotiation of technical change, *Technology Analysis and Strategic Management*, 1, 57, 1989.

20. Jacobs, M., *The Green Economy—Environment, Sustainable Development and the Politics of the Future*, Pluto Press, London, 1991.

21. Pasquero, J., Supraorganisational collaboration: the Canadian environmental experiment, *Journal of Applied Behavioural Science*, 27, 38, 1991.

22. Schot, J., Hoogma, R., and Elzen, B., Strategies for shifting technological systems — the case of the automobile system, *Futures*, 26, 1060, 1994.

23. Isnor, R., Sectoral Governance and Sustainability in Non-Ferrous Metals Production: A Study of Policy Convergence, unpublished doctoral dissertation, Science Policy Research Unit, University of Sussex, U.K., 1995.

28 Environmental Policy and Firm Level Environmental Behaviour: An Indian Case Study

Ligia Noronha

CONTENTS

ABSTRACT

This case study examines ecological management in mining in India through a focus on the iron ore mining industry in Goa and lead-zinc mining in Rajasthan (Hindustan Zinc Ltd - HZL). The main features of the mining industry in Goa are that all operations are in the private sector; there are many producers operating over a small region; the leases are on an average 100 hectares in size; the industry is totally export-oriented; and there is no local market for the low-grade (<60% Fe) ore produced. In contrast, HZL is a public sector company producing lead and zinc for domestic consumption.

More specifically, this paper examines how effective the environmental regulatory regime in India has been by analysing the environmental behaviour of these mining firms. Ecological management in mining can be viewed as a set of practices or measures that avoid or mitigate the reduction in environmental quality resulting from mining, and can include a variety of practices. The central thrust of this study was to identify the factors that prompt different levels of ecological management in mining companies and assess the effectiveness of different practices within the policy context. The case study suggests that within the iron ore mining industry there are two quite clearly identifiable perceptions of environmental management, the "minimalist" and the "proactive": for the "minimalist" firm; the crucial factor prompting improved environmental behaviour is regulation, while for the "proactive" firm it is a combination of several factors. An economic and policy assessment of some environmental management practices reveals that these factors are crucial in explaining the observed choices and practices. The case study generates a number of suggestions for future policy on the environmental management of mines.

28.1   ENVIRONMENTAL POLICY

Policy can affect environmental behaviour in three ways: (i) it can alter the options open to an agent; (ii) it can alter the cost and/or benefits relevant to an agent; and (iii) it can alter the perceptions and significance agents attach to environmental change.[1] Environmental policy in India, as in many other developed and developing countries, has concentrated on instruments that affect behaviour in the first way. The approach has been one of regulating behaviour directly through applying water and air quality standards, bans, permits, land zoning, restrictions, and so on. The policy regime has thus had a strong emphasis on approaches of the "command and control" variety rather than that of the economic or persuasive variety (Table 28.1).

There are four main statutory acts that regulate mining activity in India, which apply to the federating states:

28.1.1   THE WATER POLLUTION ACT, 1974

This provides for the prevention and control of water pollution and the maintenance or restoration of water quality. It vests the authority in State Pollution Control Boards to establish and enforce effluent standards in mines and processing plants. Prior to amendment in 1988, enforcement under the act was achieved through criminal prosecution. After 1988, the authorities were able to close down a defaulting unit or withdraw infrastructural support services if a site was found to be transgressing standards.

28.1.2   THE AIR POLLUTION ACT, 1981

This provides for the prevention, control, and abatement of air pollution. It lays down air pollution standards and is administered by the State Pollution Control Boards. It also empowers the authorities to close down a unit or to withdraw support services if the unit is found to be violating the law.

28.1.3   THE FOREST ACT, 1980

This provides for the conservation of forests and related matters. It is a central act applying to the whole of India, other than the states of Jammu and Kashmir. This act is of particular significance to mining activity in India, as most mines occur in forested areas. The Act requires mining firms to seek Central Government approval for a mining lease to be granted in respect of any forest area;

TABLE 28.1
Regulatory Regime for Environmental Protection in India

Main Elements of the Regime	As Applicable to the Mineral Sector
A: Laws	Water (Prevention and Control) Act, 1974
	Forest Conservation Act, 1980
	Air (Prevention and Control) Act, 1984
	Environmental Protection Act, 1986
B: Policies	National Forest Policy, 1988
	National Conservation Strategy and Policy
	Statement on Environment and Development, 1982
	National Policy Statement for the Abatement of Pollution
C: Administrative Machinery:	Ministry of Environment and Forests at the Federal Level
(i) Main Governmental Body	Department of Environment at the State Levels
(ii) Other Government Bodies	CPCB, NWDB, FRI, CSIR, NEERI, NIO, FSI, BSI, ICFR
	& E*, National Afforestation and Eco-Development Board
D: Fiscal Incentives	Exemptions from income tax
	Depreciation allowance
	Investments allowance
	Exemptions from capital gains tax

* CPCB — Central Pollution Control Board
NWDB — National Wasteland Development Board
FRI — Forest Research Institute
CSIR — Council of Scientific and Industrial Research
NEERI — National Environmental and Engineering Research Institute
NIO — National Institute of Oceanography
FSI — Forest Survey of India
BSI — Botanical Survey of India
ICFR & E — Indian Council of Forest Research and Education

for the resumption of mining operations on the expiry of a mining lease; for information relating to rehabilitation of mine sites, damage to trees, distance of the site from important ecosystems and highways, reclamation procedures, and extent of subsidence expected in underground mines when proposals are made to divert forest land to non-forest use. In approving proposals for the diversion of forestland to non-forest use, the Government of India requires the company to undertake compensatory afforestation on non-forest land. When such land is not available, the government undertakes afforestation on degraded scrub forest land on behalf of the company on an area twice the size of the area diverted, and following payment ranging from Rs. 10,000 to 25,000 per hectare depending on the state in which mining is being undertaken

28.1.4   THE ENVIRONMENT PROTECTION ACT, 1986

This is applicable to the whole of India and supersedes other legislation, including local laws. Under Section 5, it grants the government the power to close down any industrial site if it is found to be violating the law. Consequent to the Act, the Environment Ministry has, among other requirements, laid down the following:

- The standards of emissions for the discharge of environmental pollutants.
- Procedures and safeguards for the prevention of accidents.
- Requirement for a mining plan inclusive of an environmental management plan from mining companies.

Thus the firm is told what it can and cannot do under each act. Regulation is in the form of negative constraints to action, and as such does not have sufficient incentive to allow spontaneous environmental improvements.

Many state-level acts are also powerful instruments for environmental management. For example, legislation related to land seeks to control land use. Town and country planning acts are effective tools for regulating urban and regional development. Municipal acts also have numerous clauses relevant to environmental concerns.

Environmental clearance is a prerequisite for all projects seeking financial support from the government of India. For the private sector, such clearance is required if the mine falls in reserved forest areas, which mines often do. To obtain an environmental clearance from the Ministry of Mines and the Ministry of Environment and Forests (MOEF), a company has to submit: (i) a detailed questionnaire or checklist of impacts; and (ii) an environmental management plan and/or an environmental impact statement.

It is evident from this description that India is well armed with a barrage of regulations. Most of these tend to alter the options available to the agent, but do not in any way change the costs and benefits attached to choices, nor the significance attached to environmental performance. What this study tried to do was examine how effective these regulations have been on the ground by analysing the environmental behaviour of the mining firms in the iron ore mining industry in Goa and the lead zinc industry in Rajasthan.

28.2   ANALYSIS OF THE PROBLEMS AND CHALLENGES

Although iron ore mining and processing does not involve any toxic products, environmental management in this sector in Goa was chosen for investigation because it is believed that the case study illustrates, with a particular clarity, relationships between the mining industry and the environment which are perhaps more generic to the Indian mining industry as a whole. Goa is a good window on these issues because of the relatively high level of environmental awareness, political organisation around environmental issues, and the susceptibility of the Goan environment to degradation. More specifically, the reasons for this investigation are

1. Open pit mining, which is dominant in ferrous mining and which characterises 95% of all mining activity in India, has a particular potential to degrade land and water. In Goa, all mining is by open pit methods. Because of the high overburden-to-ore ratio in Goa, there is a high demand for land that is blocked by overburden disposal. There exist some innovative attempts at addressing the problems of land and water degradation. These attempts are worth studying and assessing, and the lessons learnt worth disseminating.
2. Goa is a state with fragile ecosystems, the mineral lease areas occurring in the forest areas of the Western Ghats with mining operations having impacts on forests, crop land, the khazan lands, and the two main rivers that drain through the state. Khazan lands are a unique estuarine agro-ecosystem. They lie low in the basins of the main rivers and form the "food bowl" of the state. During the monsoon period, paddy is grown on these lands as the rain dilutes the salinity; river water reaching the fields through various waterways has also made these the spawning ground for fish and shrimp. Goa can be considered to have four main ecozones: the Sahaydrian watershed located in the Western Ghat region; the plateau area which is in the Midland region, and the alluvial flats and coastal zone in the coastal region (see Alvares).[2] The mining areas occur in the forest regions of the Western Ghats, and its impacts are felt along all four ecozones in different measures.
3. Leases are on an average less than 100 hectares in size and are clustered together. The concentration of a large number of producers in a small land area creates the problem of how to identify polluters and allocate responsibility.

TABLE 28.2
Production and Export of Iron Ore — 1994

Company	Production (million tonnes)	Export (million tonnes)
Sesa Goa	3.94	3.71
Dempo	1.89	2.21
V.M. Salgaonkar	1.26	2.2
Chowgule & Co.	2.2	1.95
Fomento	0.7	0.99
Others	3.18	3.69
Total	13.18	14.75

4. Active and articulate environmental groups and, in general, a population with a low willingness to accept environmental degradation makes Goa an interesting context to study the influence, effectiveness, and limits of the "command and control" approach dominant in India. Notwithstanding a strong preference for environmental quality over employment often revealed in protest movements, the local population has shown a marked intolerance to degradation from mining activity. The firm stand that Goan NGOs and other groups have taken on environmental implications of some development projects is catalogued and recorded in Alvares.[2] More recently, local protest was sufficient to reject the Nylon 6,6 project of Thapar-DuPont from the state on grounds that it would have serious environmental implications for the region where the plant was to be located.

The iron ore industry in Goa comprises 15 private sector mining companies operating over a region of 500 km². Annual production of iron ore in Goa has averaged around 12 to 13 million tonnes per year, which constitutes over 25% of the country's iron ore production and 55% of its exports. Iron ore export constitutes around 5% of the country's total exports, and yields around Rs. 5,500 million a year. It has been identified by policy as one of the key thrust areas for export promotion. With liberalisation, this mineral has attracted considerable attention from international mining companies, for example, the interest of Rio Tinto and BHP in the iron mines in the state of Orissa. Plans for expansion of domestic steel production also lead to increased interest in iron ore mining. The industry has two types of mining companies: (1) those that only extract ore and sell them to the exporting companies; and (2) those that extract and export the ore. The five largest companies are of the second type, and between them control 75% of the production and 80% of the export of mineral ore. The largest firm controls 30% of the production and 25% of the export of ore as of 1994/95 (Table 28.2).

Ecological degradation in Goa due to iron ore mining activity in general has been of two kinds detailed below.

28.2.1 LAND DEGRADATION

This is evident in the huge "holes in the ground" in the mining areas as land is excavated to extract the ore; in the "artificial mountains" of dumped overburden; and in the crop and non-crop land downstream of mining dumps affected by surface runoff from these dumps. More specifically, mining activity in Goa has resulted in the loss of forest cover, loss of agricultural land, and the deterioration of agricultural soil conditions. Further downstream, ore transportation through the inland waterways has resulted in the deterioration of the khazan land in Goa, considered the most fertile of agricultural lands in the area.

28.2.2   DEGRADATION OF THE WATER COURSES

Surface runoff from waste dumps, the dewatering of mining pits, and the spent water from dust extraction have degraded both surface and underground water. For specific studies on the impact of mining on water courses and the estuarine ecosystem, see Nayak,[3] Parulekar,[4] and Modassir.[5]

Part of the degradation is almost inevitable due to the physical constraints of being mainly in forested areas, involving large amounts of overburden, and being confined to small areas. There is some evidence, however, to indicate that ecological degradation is compounded by mining practices that, if changed, could avoid or limit the degradation.

By contrast to the Goan case study, the mining operations of HZL occur in sub-humid and semi-arid areas, with dry deciduous and thorn forests in the state of Rajasthan. Its environmental groups have become increasingly vocal, but because of relatively few alternative employment opportunities have been more tolerant of degradation resulting from mining. The significant environmental impact from the processing of lead-zinc ores is due to the generation of tailings rich in potentially toxic metals such as lead, zinc, and cadmium. The small particle size of such tailings results in their becoming airborne very easily and dispersed to nearby agricultural areas (see Ahmad et al.).[6] The major source of water pollution is the mill effluent at Zawar, most of which is recirculated for further use in the mill. However, a part of this effluent does find its way to the nearby Tidi River, which is evident in the higher concentrations of metals in the downstream water.

HZL is a public sector company that controls 99% of the lead-zinc mining capacity in the country, all of the lead smelting capacity, and 88% of the zinc smelting capacity. HZL is a multi-unit company, which in a period of less than 30 years (1966–93) increased its turnover from around Rs. 20 million to about Rs. 7,780 million. Its mining and milling capacity rose from 500 tonnes per day to about 11,000 tonnes per day, zinc smelting capacity to 149,000 tonnes per annum, and lead smelting capacity from 3600 tonnes per annum to 30,000 tonnes per annum. This study looked specifically at three of HZL's major lead-zinc mining operations at Zawar, Rajpura-Dariba, and Rampura Agucha, all in Rajasthan. Lead-zinc ore resources in the country are estimated at 416 million tonnes with an average of 4.19% zinc and 1.97% lead content. Of these resources, 95% are located in the state of Rajasthan.

28.3   ANALYSIS OF ENVIRONMENTAL PRACTICES

28.3.1   CASE STUDY — IRON ORE MINING IN GOA

Environmental behaviour of the mining companies in Goa was studied over the period 1981–94, a period that coincides with the framing, enactment, and the enforcing of environmental regulation in India. This case study was based both on primary and secondary information. The study gained considerably from interaction with the mining association, Industry's Environmental Council, local NGOs, and local forest and mining department officials. This period was also one of environmental activism in the state.

From the secondary information available, it is apparent that in Goa the five largest firms, with an annual production between 0.7 to 4 million tonnes are much more significantly involved in environmental management than the smaller firms. This concern with environmental management is also evident in other large iron ore mines such as Noamundi, Donimalai, and Kudremukh (see accounts of environmental management at these mines given in Banerjee and Rangachari,[7] Fasihuddin et al.,[8] and Jain and Reddy.[9] The ten small firms who together produced around three million tonnes in 1994 observe environmental management essentially as compliance with forest regulations. On the other hand, the larger firms try to incorporate some good practices in their operations, over and above those required by legislation, as evident in their dust suppression measures, rainwater management, dump vegetation, and tailings management. However, even among the five there are significant differences, with two firms clearly falling in the category of "proactive" firms, and the other three being between a "compliance mentality" and moving on to a "more proactive mentality." The proactive firms have the following "better practices" relative to the others:

- Their own policies, procedures, and guidelines pertaining to environmental control.
- A proactive role in environmental awareness projects for industry.
- Lead roles in adopting new methods/techniques for treating wastes.
- Investment in R&D.
- Investments in plant nurseries.
- Staff clearly identified and given responsibilities for the development, implementation, and monitoring of various environmental systems.
- Investments in the training and capacity building of its staff in various aspects of environmental management.

The main management measures in the more "environmentally active firms" are:

- Waste dump management.
- Wastewater management.
- Tailings management.
- Dust management.
- Reclamation of inactive ("dead") dumps and tailings ponds.

28.3.1.1   Active waste dump management

Waste dumps can affect ecosystems in three ways:

1. Stormwater runoff can impair the quality of other resources (e.g., land and water bodies downstream of the dump).
2. Land is required to dump the waste, thereby reducing its availability for other uses or reducing the use options available to that land and the area nearby. Over the period 1980–90, 146.75 million tonnes of iron ore were mined in Goa. With an average stripping ratio of 2.5 m^3 t^{-1}, this implies that 366.88 million m^3 of waste were generated during this period, or approximately 37 million m^3 per year. The amount of land required for dumping depends on the height of the dump; thus for an average dump height of 35 metres, approximately 3 hectares of land is required to dump a million m^3 of over-burden waste. Thus if annual waste generation is 37 million m^3, this implies that the mining industry has an annual requirement of 111 hectares *outside* the mining conces-sions in which to dump the overburden. If, for environmental considerations, dump heights have to be less, land requirements for dumping will go up. On a long-term basis, this will become an unsustainable demand for land for a state whose total area is only 3,702 km^2.
3. Their existence reduces the aesthetic value of the region and hence affects human satisfaction, especially that of local residents.

While type (1) impact has to be addressed during the active operational phase of mining operations, type (2) and (3) impacts need to be addressed at the planning stage and/or when the dump becomes inactive. These three impacts constitute the potential environmental damage that dumps can create.

Waste dump management can be of two forms: the first adopts practices that prevent damage from occurring, e.g., ensuring that dumps remain low with gentle slopes, so as to reduce surface runoff. In Goa, this tends to be difficult because land to dump overburden is in short supply. The

TABLE 28.3
Closed Water Circuit Systems: Costs and Benefits

	Rs. (millions)
Benefits (water purchased avoided)	21.25
Treatment costs	
Mine water	1.95–1.97
Tailings water	2.9
Net gain per annum	16.38–16.4

second adopts practices to control runoff, for example, the practice of building laterite walls, check dams, and filtering systems to control the flow of silt.

The *costs* of waste dump management are those incurred by the company: a) to reduce runoff through check dams and protective walls; and b) to afforest inactive or "dead" dumps. The *benefits* are of two kinds:

- Immediate benefits of offsite damage avoided to agricultural fields, water streams close by, and on-site damage avoided to the operational pits worked at lower contour levels.
- Longer-term benefits of ensuring that mined land is rehabilitated and given the capacity to be put to other economic uses at the end of the life of the mine.

28.3.1.2 Wastewater treatment

Treating wastewater immediately addresses two pressing water-related concerns of the mining industry: a) its "pollution" potential and b) its "substitution" potential, i.e., potential to substitute for fresh water. While the costs of wastewater treatment may remain the same, the benefits will vary according to whether both concerns or only one is being addressed.

Some mining companies do not allow any water out of the system, but instead maintain a closed water circuit system. In such a system, mine water is diverted into a settling pond; this water is treated and sent to a reservoir from where the water is pumped to the beneficiation plant. The water used in the beneficiation process is diverted into an open channel and then into a settlement pond where lime and/or other chemicals are used to aid settlement. Tailings are thickened to about 60% solid with the aid of thickeners and this underflow is then sent to a tailings pond; the clear water overflow is sent to the reservoir to be reused in the beneficiation plant. A closed water circuit, if properly implemented, allows no discharge of liquid effluent and allows the treated water to be reused in the system. This is of particular importance in regions of water scarcity, as is the case in India.

The main benefits of the closed water circuit are that the company saves on the purchase of water for the beneficiation process. For example, for the company studied, Rs. 115,000 per day or about Rs. 21.25 million per annum is saved as a result of the closed water circuit system, a savings that the company had not taken into account when making its decision to reuse water. The net gain to the company from reusing its water is given in Table 28.3.

Over and above this direct gain to the mining company, society gains because of the damage avoided through the reduced turbidity and total suspended solids of the treated water, which in the absence of treatment and reuse would have been discharged out of the mine area. This would have polluted local water streams and paddy fields, as is happening in the case of mining companies that are not treating the mine- and wastewater.

28.3.1.3 Tailings disposal

Embankment/dams for the management of tailings are specially constructed to avoid seepage, sliding and erosion. The main benefit of thickening and the impoundment of tailings is that land degradation is reduced. There is no leaching of toxic metals in these mines as the rejects are non-toxic and there is no chemical process involved in ore beneficiation.

Experiments have also been carried out at the laboratory level for the treatment of tailings by microorganisms. Two bacterial cultures, *Arthobacter* and *Bacillus megaterium*, are found to be effective in settling the tailings. Physical methods of treating mine water with horizontal flow roughing filters have also been found applicable to the removable of suspended solids.

28.3.1.4   Dust management

Dust management measures comprise the largest share of the environmental budgets of the mining companies studied. For one company, this share was as high as 48% of the environmental budget. The main measures observed for dust suppression and management are:

- Haul roads and other roads and the active overburden dump surfaces are watered at regular intervals. A fleet of water tankers is hired to water these roads from sunrise to sunset for eight months of the year.
- Dust extractors are placed in the crushing and screening plants.
- Vegetation of active dump slopes and both sides of haul roads as plantlife act as a sink for pollutants.

28.3.1.5   Reclamation of inactive waste dumps and tailings ponds

Degraded land can be reclaimed for a large number of final uses, such as agriculture, housing, recreation, coconut plantations, energy plantations, and so on. It is observed that in iron ore mines in Goa, the favoured reclamation practice of land laid waste by tailings or overburden (OB) dumps has been afforestation, but with no particular end use in mind. This has come about because afforestation allows a company to comply with environmental regulations and be seen to be "green," while covering up a lack of information about the alternative reclamation options available and the technologies that these involve. A general lack of policy and legal incentives that would support a greater investment in rehabilitation of the mine site inhibits a greater commitment to reclamation. Mining leases in India, until very recently, did not bind the leaseholder to undertake reclamation work; neither was there any law that enables the miner to sell the reclaimed land, as the land reverts back to the state on the expiry of the lease. In such a situation, there is little incentive to invest in end-uses since the mining company has little to gain from such investment. So, the tendency is to overuse the leased area while the mineral is being extracted and to underinvest in its rehabilitation.

Afforestation has been considered as the most convenient means of regulatory compliance for the following reasons:

- The level of know-how considered necessary to undertake such programmes is low. In India, there are no service companies providing contractual services or expertise in ecological management. What expertise exists is held in research institutions.
- The level of environmental consciousness of the public in Goa in general is high.
- The experience of the company that being seen to be "green" helps the company obtain quick clearances, other product lines, and governmental assistance.

The trees commonly grown here are acacia auriculiformis, cashew (*anacardium occidentale*), casuarina, soobabul (*Leucaena leucocephala*), neem (*Azadirachta indica*) and khair (*Acacia catechu*). An analysis of the species grown on the dumps and tailing ponds reveals that while cashew dominated the choice of species planted in the mines in the period 1983–87, followed by casuarina, acacia came to dominate over the period since then, with casuarina remaining the favoured second choice.

TABLE 28.4

Comparative Economics of Two Species

Net Present Values	Rs. per hectare	
	Cashew	*Acacia*
5%	304,554	96,454
10%	135,692	42,162
12%	100,154	28,983

The decision on whether to use a single species or a mix of species depends on a number of factors which vary from soil tolerance and species availability, to the selection being based on aesthetic grounds or local demand. The case study indicates, however, that not much thought was given to choice of species other than their ability to survive on degraded soils and grow quickly. Thus acacia came to dominate the choice of plant species over the last five years. There are a number of reasons for this: it is easy to grow on mined land because it has good pH tolerance, good nitrogen fixing ability, and a shallow root system that grows well in soil containing boulders and holes, as found typically at dump sites. It is a fast growing species, which implied that a "green cover" could be obtained quickly to counter the pressure arising from environmental regulations. Economically, however, it is not a very attractive species to grow, as its use as timber is restricted and it does not have much use as food or fodder. It does, however, possess good environmental management qualities in terms of erosion control, shelter, soil stability, improving soil fertility, and so on. The cashew tree also possesses good environmental management properties. Moreover, it can provide valuable economic returns, because cashew nut is a much-valued export item from India. Apart from the nuts, it has value for its fruit and for its fuelwood. Table 28.4 provides a comparative economic assessment of the two species. Net present values are estimated using a 5, 10 and 12% discount rate.

While the analysis reveals that planting cashew trees would be an economically attractive option for the mining company, choice of species cannot be based purely on economic criteria. Economic reasons may make a case for the dominance of a single economically valuable species, but the practice of monoculture may be detrimental to the long-term ecological survival of the region. Various tree species, because of their differing abilities to withstand environmental extremes, provide an insurance against risk. Thus an afforestation programme in degraded land will have to consider both the economic and environmental objectives to arrive at an optimal tree mix.

28.3.1.6 Observations

The case study revealed that a number of factors motivated improved environmental behaviour: regulations, especially the Forest Act, were an important incentive to all firms in stimulating environmental management. The Forest Act was the most strictly implemented, which perhaps explains its effectiveness. However, at the level of the state, implementation of the Air and Water Acts have not been effective. One possible explanation that is put forward for weak implementation is the close nexus that exists between the state and the dominant mining interests, which enables the latter to have access to nodes of power. This is an aspect worthy of further investigation. It is believed that this is a constraint to effective regulation by the state. In addition to these, the need for a "green image," the likelihood of quicker clearances and permits, and increased information about environmental protection possibilities influenced behaviour of the five larger firms. A corporate culture of responsibility for the environment, while not of the same degree of importance. is evident in three of the large firms, while pressure from staff for improved behaviour is observed in only one of the firms.

The "differences" in the environmental behaviour of the firms with just a compliance mentality, as compared to those who are more proactive, enables me to suggest that the factors influencing environmental behaviour are of two types: (1) those external to the firm such as regulation and public pressures; and (2) those internal to the firm. Those factors internal to the firm can be categorised as: (i) those that orient the preferences of the firm toward greater environmental management (e.g., corporate culture, desire for a green image, and staff pressure); and (ii) those that influence its capability to do something about it (e.g., technical and managerial capability, R&D capability, available resources and information). Such a distinction is of importance because it helps policy to concentrate on specific aspects that need to be focussed upon, as discussed below.

28.3.2   CASE STUDY — ECOLOGICAL MANAGEMENT PRACTICES OF HZL

This study examined three of HZL's major mining operations located at Zawar, Rajpura Dariba in the Udaipur district of Rajasthan, and at Rampura Agucha in the Bhilwara district. These three mines together represent 85% of ore production capacity of HZL. Except for the Rampura Agucha mines of HZL, which is an open-cast mine, all other mines are underground operations.

HZL has attempted to mitigate, and in some cases prevent, the degradation associated with lead-zinc mining activities. This is noticed especially in the company's handling of solid waste, effluent treatment, and dust control at the transportation stage for ore at Zawar and at Rajpura Dariba. In these two mines, HZL avoided the problem of having to cope with overburden dumps by utilising the overburden for levelling the hilly terrain around the mine site, for construction purposes, and for backfilling voids. The Rampura Agucha mine is an open-cast mine and is the richest lead-zinc mine in India. For this mine, HZL adopted an environmental management plan (EMP) at the outset of mining operations. The advance preparation of an environmental impact assessment (EIA) and an EMP has enabled HZL to integrate ameliorative measures for impacts. The feedback from these studies has been incorporated in the planning for this mine. Specifically, measures have been put in place for solid waste disposal, noise attenuation, liquid effluent, landscaping, dust control, and the rehousing of displaced people. However, there is no plan for the area upon closure of the mine.

The pollution control measures observed include:

- Dust management.
- Water management.
- Energy savings.
- Recovery of metal values from solid wastes at the smelting units of HZL.
- Tailings management.

28.3.2.1   Dust management

The Zawar mines have implemented dust extraction systems with blowers, and wet scrubbers to reduce air pollution, while drilling water is injected to suppress dust and the contaminated water collected in a sump where it is allowed to settle and then recycled for milling operations. Sumps are provided with traps to control solids. At the Agucha mine, the company has a pressurised water sprinkling system on hoppers and conveyor belts to keep the crushed ore moist, thereby reducing the generation of dust. HZL avoids dust generation during transportation of ore by covering the trucks with tarpaulin sheets. Concentrates going to the smelters at Bihar and Andhra Pradesh have 8 to 10% moisture in them, which reduces the fugitive dusts. Moreover, they are packed in HDPE bags to avoid loss during transport.

28.3.2.2   Wastewater management

In HZL operations, water consumption per ton of ore treated is 4 m^3 in the concentrators at Zawar and Rajpura Dariba. In Rampura Agucha, water consumption per metric tonne of ore is 2.5 m^3. To overcome the water constraint, which became very limiting after 1987–88, due to drought conditions

TABLE 28.5
Benefits of Recycling Water for HZL

	Zawar	Rampura Agucha
Water consumption (million m³ per annum)	2	1.6
Water Cost (Rs. per m³)	5	12
Source of water	Tidi	Banas
Recycled water as % of total water intake	62.5	75
Mine water as % of total water intake	0.5	Used for dust suppression
Decanted tailings water as % of total water intake	62	75
Purchased fresh water (million m³)	0.75	0.4
Savings effected through recycling (Rs. millions)	6.25	14.4

in the state, HZL adopted innovative water management practices. These consisted of: (i) wastewater recycling, and (ii) reduction of water evaporation from open storage facilities through a dispersion of vegetable oil-based speciality chemicals over the water surface.

The water from the underground mines at Zawar is not great in volume, and is used directly for dust suppression. At Rajpura-Dariba mines, the 2500 m³ per day of mine water generated is neutralised with lime. This causes the solids to settle and the heavy metals are precipitated. Clean water is then used for the beneficiation plant. When discharged on land, it is used to support vegetation. Water generated in the Rampura Agucha mine is very clear and is used entirely for dust suppression. The contribution of this water to total dust suppression requirements is, however, only 1%.

The slurry from the beneficiation plants goes to the tailings dam, where the overflow is allowed to drain out through decantation towers in the tailing pond, and the decanted water is then recycled for use in the beneficiation plants. There is no discharge of effluent from the Rajpura Dariba mines. A small quantity of the effluent from Zawar is discharged into the Tidi River. As a result of the closed water circuit in place, the company needs to buy only one third of its water requirements for the Zawar plants and about 25% in the Rampura Agucha mines. The beneficiation plant at Zawar uses 4800 m³ per day of which 3000 m³ is recycled water. Water consumption at Rampura Agucha for beneficiation is 7500 m³ per day. For beneficiation at Zawar, water required is 4 m³ per tonne of ore processed; of this 1.5 m³ (37.5%) is fresh water and 2.5 m³ (62.5%) is recycled water. The company is thus able to save about Rs. 6.25 million per annum at Zawar. The savings at Rampura Agucha are Rs. 14.4 million. Table 28.5 compares the savings effected as a result of the closed water circuit in place.

28.3.2.3 Energy conservation

The major constraints in the production of lead and zinc in India are the shortage of power and water. These shortages have severely affected the production efficiency of the smelters. After 1987, the company sought to cut down its dependence on the grid by: i) increasing the captive power generation through the installation of diesel generator sets; ii) adopting simple energy conservation measures at various stages of the beneficiation and smelting process (e.g., improved practices and also simple improvements in equipment); and iii) adopting a less energy-intensive technology in its new smelting unit at Chanderiya in the Chittorgarh district of Rajasthan. This is the Imperial Smelting Process (ISP) which involves considerable energy savings to the company. Energy savings result in environmental benefits, as less power needs to be generated for that particular end-use, resulting in decreased emissions of sulphur dioxide, nitrogen oxides, and carbon dioxide. As a result of the choice of the ISP technology, there is a primary energy saving of 6.05 gigajoules per tonne of SHG zinc produced and 17 gigajoules per tonne of GOB zinc over the electrolytic route adopted in other zinc smelters in the country. This implies a per annum energy saving of

806.75 terajoules or 72.68 million kWh in the production of zinc. A saving in generation of 72.68 gigawatt hours implies a per annum avoidance of 23,766 tonnes of carbon, 4488 tonnes of sulphur dioxide, and 15,117 tonnes of total suspended particulate emissions.

28.3.2.4   Recovery of metal values from solid wastes at the smelting units of HZL

One of the important practices adopted to reduce the burden on the environment from solid waste is to increase the recovery of metal values from such wastes generated at both the mining and the smelting units. HZL has effected small changes in the flotation process at the Mochia plant at Zawar, which has resulted in higher metal recovery and lower pyrite content in the final product.[10] Cadmium, silver, and sulphuric acid are recovered as by-products at Debari and Vizag smelters, while silver metal is recovered as a by-product at the lead smelter at Tundoo.

At Debari, in the earlier years, loss of zinc occurred through solid wastes generated after leaching of calcine. Such residues ("Moore cake") had high contents of zinc and silver. Subsequently, a residue treatment plant with a silver recovery plant was commissioned during 1984–85, leading to an increase in zinc production of 4000 tonnes per annum and a recovery of 6.8 tonnes of silver per annum.

The sludge precipitated from the treatment of effluent at Debari is being further treated for the recovery of zinc in the form of zinc sulphide, which is used directly with the zinc concentrate for roasting to get calcine.

28.3.2.5   Afforestation

Large-scale afforestation of dumps and vegetative stabilisation of tailing dams are the other environmental measures in place. To counter one of the main problems at the Zawar mines, that of airborne dust, the company has sought to stabilise dry tailings by vegetation over an area of 38.5 hectares. Afforestation programmes at Zawar commenced in 1981 and, until 1984, a total area of 61.8 hectares had been afforested. The main species planted are prosopis juliflora, eucalyptus, pogamia pinnata, acacia auriculiformis, cassia and cynodon dactylon. In Rampura Agucha, afforestation commenced in 1988 with the start of mining operations. By 1994, a total area of 214 hectares at the mining site had been afforested. A wide variety of species have been grown from the outset.

28.4   CONSTRAINTS AND OPPORTUNITIES FOR THE TRANSFER OF GOOD PRACTICE

The main observations that can be made from these two case studies that may be of relevance to other countries are

1. The effectiveness of regulation depends very much on the ability to implement such laws and rules. For example, in Goa, the mining industry gets away with the minimum because of the closeness that exists between company management/ownership and state leadership, both bureaucratic and political, suggesting an interesting insight into the sociology of power.
2. Command and control policies are most effective in cases where there is a single operator, as is the case of HZL. By contrast in the Goan case study, there are many mines and many mining companies, and so emissions and discharges are not discrete. It is therefore difficult to identify the polluter and to enforce regulations.
3. The understanding and information about the relative toxicity and degradation potential of the mineral in question in the basic training imparted to mining staff may explain

differences in environmental behaviour of firms. For example, a worker joining the lead-zinc industry already has a heightened awareness of the environmental implications of lead-zinc mining; this could quite early on create in the worker a sharper eye for problems that may emerge and the need to find site-specific solutions. This could also aid the development of a capacity to innovate and encourage problem-solving skills in the staff. By contrast, a miner joining an iron-ore mine believes that the mineral is comparatively less polluting, and so from the start has a relaxed attitude about the implications of iron ore development. Some mining staff and even company managers find it difficult to believe that the sediment that runs off into local water bodies or agricultural fields can be considered polluting, as there is no chemical toxicity involved.

4. Local pressures for improved environmental behaviour have to be backed by organisational resources to protest against the degradation, and with the capacity to access support at the wider level of the state. Those mining companies that see the attitude of the local people changing and becoming more aggressive with regard to environmental neglect by the mining industry use this perception as a positive influence for more environmental control.

5. Mining companies are still not in the habit of keeping separate records of environmental expenditures, a fact that severely constrained the initial outline of this study. However, it is important that companies do so, as this will enable them to present their environmental record to the public more cogently than the current practice of presenting pictures. It will also enable better monitoring of the effectiveness of governmental regulations and it will help companies keep a track on their own environmental control programmes.

6. Mining companies have not as yet registered the fact that that there does not always have to be a trade-off between profitability and environmental management. It is necessary to highlight this important point to industry to create a self-sustaining incentive, and this study has sought to do so.

7. Both company and government officials pointed to a number of constraints that they faced in improving and promoting improved environmental practices. According to company officials, the constraints were:
   - Lack of information relating to prevention and control options.
   - Lack of environmental consultancy services.
   - Lack of data for advance planning.
   - Other business priorities.
   - No incentive.
   - Absence of ownership rights over mined land.
   - Absence of environmental baseline information.

On the other hand, government officials were constrained by the following in pushing for improved environmental practices:

   - Pressure from the mining lobby.
   - Lack of information about better practices.
   - A view that if regulations were complied with, it was sufficient.
   - Job and income-generation priorities.
   - The foreign exchange implications of mineral development.

28.5 CONCLUSIONS

1. It is evident from the India case study that regulation has been an important factor in inducing improved environmental behaviour. It provided the initial impetus to adopting

improved practices, although without much understanding of its objectives. Over time and through a process of learning and appreciation of the benefits of adopting such practices, there has been a tendency to go beyond mere regulatory compliance. The degree of proactiveness has, however, been a function of the "preferences" and "capabilities" of firms.

2. The Forest Act was very effective in reducing the indiscriminate felling of trees that occurred prior to its implementation, and so one could immediately observe a reduction in the excessive use of forest resources in response to this legislation.

3. Investments in "environmental education" of mining staff are among the most effective means of achieving better mining practices, as the staff begins to appreciate the considerable benefits of simple management practices and the costs of existing ways of doing things. In one mining company, this has lead to pressure and suggestions for improvement from the staff themselves, and attempts at innovative site-specific solutions to environmental problems.

4. There has been very little participation of the local population at the stage where environmental impacts are being assessed and mitigatory measures being put in place. This has a number of disadvantages: i) the valuable input about impact detail and possible remedies or avoidance measures that the local population may be able to suggest is lost; ii) local participation in mitigating and monitoring of environmental impacts can reduce costs to the companies; and iii) local protest after work on the project has begun is avoided or minimised. Such protests, it is argued, lead to time and cost overruns and cannot be afforded by poor countries, such as India. The counterargument is that while such cost overruns are indeed a fact, these are insignificant compared to the costs in the long run if environmental impacts of the project are not considered and the necessary safeguards are not in place. Moreover, the costs of environmental compliance, if any, are private costs and it is in not incurring them that social costs may arise.

Environmental policy in India has so far been very much of the "command and control" variety. What the policy has emphasised is an alteration of the options open to agents, this alteration almost always having been in the nature of restraining certain practices. As a result, environmental policy has come to be seen as a regime of constraints rather than one of creating positive incentives for environmental change. A policy that sought to create change through altering the costs and/or benefits relevant to agents, or an alteration of the preferences and significance agents attach to environmental change, will tend to have a more long-lasting impact on the sustainable development of mineral resources. If policy had sought to change behaviour through an impact on the latter two, it would have placed greater emphasis on the use of market stimuli and/or moral persuasion to influence improved environmental behaviour.

It is evident that the approach followed in India is limited both in its scope and application. "Best practice environmental management" as evident in this case study goes beyond just following regulations; the differences between those with a mere compliance mentality and those that are proactive on the environmental front are explained with reference to factors that are more specific to the firm, such as corporate culture of social responsibility, staff pressures, managerial structure and leadership, competitiveness, and technical and management capacity.

Policy has to target, therefore, those factors that influence preferences through policy measures that include environmental awareness raising, community participation, and a system of rewards and penalties if preference levels are to be raised. If capability is to be enhanced, policy will have to address those deficiencies or constraints currently faced by firms, such as information, old technologies, lack of supportive environmental service firms, and so on. Lack of supportive mining legislation and policy incentives are creating conditions for superficial responses by companies and inhibiting a greater commitment to improved environmental behaviour.

REFERENCES

1. Turner, K. and Opschoor, H., Environmental economics and environmental policy, in *Economic Incentives and Environmental Policies: Principles and Practice*, Opschoor, J. B. and Turner, R. K., Eds., 1994, Chapter 1.

2. Alvares, C., Ed., *Fish, Curry and Rice: A Citizen's Report on the State of the Goan Environment*, Ecoforum, Goa, 1993.

3. Nayak, G. N., Studies on sediment flux of rivers, estuaries and adjoining coastal waters of Goa, west coast of India, unpublished project submitted to the MOEF, Goa University, April, 1993.

4. Parulekar, A. H., Anzari, Z. A., and Ingole, B. S., Effect of mining activities on the clam fisheries and bottom fauna of Goa estuaries, *Proceedings of Indian Academy of Sciences (Animal Sciences)*, 95, 325, 1986.

5. Modassir, M., Impact of current iron ore mining and activities on the environment of Goa and proposed measures to minimise long term environmental and economic damage, unpublished Master of Philosophy dissertation, University of Hull, October, 1994.

6. Ahmad, M., Ratan, S., and Dhar, B. B., The impact of processing of lead-zinc ores on the surrounding environment, in *The Impact of Mining on the Environment: Problems and Solutions*, A.A.Balkema, Ed., Rotterdam and OBH and IBH, New Delhi, 1994, 229.

7. Banerjee, P. K. and Rangachari, G., A status report on environmental management in the iron ore mining sector, in *Proceedings of the First World Mining Environment Congress*, Dhar, B. B. and Thakur, D. N., Eds., Oxford and IBH Publishing Company, New Delhi, 1995, 295.

8. Fasihuddin, M., Sahu, N. K., Sharma, R. M., and Narayanan, B., Towards a total environmental care at Noamundi iron mine, in *Proceedings of the First World Mining Environment Congress*, Dhar, B. B. and Thakur, D. N., Eds., Oxford and IBH Publishing Company, New Delhi, 1995, 353.

9. Jain, V. K and Reddy, K. K., Environmental improvement through industrial waste management — a case study of Donimalai iron ore mine, in *Proceedings of the First World Mining Environment Congress*, Dhar, B. B. and Thakur, D. N., Eds., Oxford and IBH Publishing Company, New Delhi, 1995, 373.

10. Chatterjee, B. N. and Agarwal, B. L., How Zawar beneficiates and exercises quality control for the processing of low grade lead-zinc sulphide ore in lead, zinc and cadmium: retrospect and prospect, technical papers of the International Seminar, Indian Lead Zinc Information Centre and the Indian Institute of Metals, New Delhi, B.72, 1981.

29 Planning for Closure from the Outset: Towards Best Practice in Public Policy and Corporate Strategy for Managing the Environmental and Social Effects of Mining

Alyson Warhurst

CONTENTS

29.1 CONCLUSIONS AND RECOMMENDATIONS

This study sought to understand the factors that explain differences in environmental degradation resulting from mining and mineral processing activities, both within and between regulatory regimes. It sought to understand the constraints governing the diffusion of technological advances in environmental management, and the changes required in both corporate strategy and public policy to resolve those constraints — the project's purpose, emanating from the identification of these development problems was two-fold. First, it set out to examine different practices towards environmental management within a number of case studies across different geographical and socioeconomic contexts from the perspective of planning for closure. A second explicit purpose was, through the above research activity, to develop mechanisms to promote the building research and dissemination capacity of our fast-growing international organisation, the Mining and Environment Research Network (MERN).

Each chapter in this book provides a unique and original empirically based analysis of either a key environmental issue or a case-study of the firm-specific explanatory factors underlying the variation in environmental performance occurring not simply between countries but within the same

1-56670-365-4/00/$0.00+$.50
© 2000 by CRC Press LLC

regulatory regimes. This comparative analysis defines best-practice — and therefore the character-istics of the policies that can be developed to promote it — by the extent to which an operation planned for closure from the outset. The earlier an operation began putting in place a closure plan that incorporated an integrated environmental management approach to the mitigation of negative ecological impacts affecting the combined environmental media of land, water, and air, as well as local communities, the more cost-effective and efficient were its results.

MERN research teams in Australia, Bolivia, Brazil, Canada, Chile, Colombia, Ecuador, Ethi-opia, Ghana, India, Mozambique, Pakistan, Peru, South Africa, Tanzania, U.S., and Zimbabwe undertook case studies.

The research findings demonstrated that appropriate planning for closure incorporates the concept of "pollution prevention pays" through technological and organisational innovation, rather than pol-lution cleanup through costly "add-on" incremental technical change. The research indicated that most regulatory frameworks were still "command and control" in mode and fostered such incremental responses by firms, which, if safeguards are not in place, can lead to substantial post-closure pollution.

The research demonstrated that, with some exceptions, it required proactive initiatives by firms, independent of real or perceived regulatory pressures — although sometimes in response to financial drivers to adopt a planning for closure approach, even where the evidence suggests that there are potential economic benefits to be gained from doing so. Hence, the importance of diffusing lessons from best-practice as summarised within the project's recommendations (outlined below).

The findings from our analysis of regulatory change worldwide were three-fold:

- There is a growing emphasis on market incentives, with a pollution prevention focus, and on environmental management system standards.
- There are new requirements for closure plans to be included in the permitting process, building on more forward-looking environmental impact assessment and including social impact analysis.
- There is an increased use of bonding mechanisms to protect society and oblige firms to set aside finances from the outset to pay for pollution cleanup at the end. However, only a few of these are flexible enough to allow firms to draw on them for ongoing reclamation.

A second set of outputs relates to MERN's collaborative research mechanisms and dissemina-tion vehicles, which were built up through the project and now provide a framework for future research efforts in broader areas of mining and environment, particularly in the field of environ-mental and social performance indicators. These include an annual international research workshop which provides feedback on research in progress and opportunities to plan new research initiatives, and a twice-annual Bulletin to report research in progress and issues of relevance to the global research community in the areas of mining and environment.

The research directly addresses DfID's goal of contributing analysis to enhance productive capac-ity and conserve the environment, in particular through providing tangible evidence of the economic benefits resulting from firm strategies that combine production efficiency with environmental profi-ciency. The research also contributes significantly to DfID's goal regarding inputs to international policy that enhance the effectiveness of the activities of multilateral organisations, which all contribute to the overarching DfID goal of encouraging sound development policies and good governance. The recommendations from the 6th International MERN Workshop, held in Harare (August 1996) serve both as conclusions to this book, and as pointers to the requirements for further research.

29.2   TOWARDS INTEGRATED ENVIRONMENTAL MANAGEMENT THROUGH PLANNING FOR CLOSURE FROM THE OUTSET

The participants of this project and several additional network research members and sponsors met in Harare during August 1996, to debate our research findings, draw our conclusions and recommendations, and define a future research strategy.

These recommendations highlight the generic elements of company strategy and public policy that are needed to underline a forward-looking approach to environmental management. These elements should ensure that integrated pollution control and socially desirable measures are built into mining projects from the outset, and remain robust until closure and, where relevant, beyond.

The workshop brought together senior policy makers, company executives with environmental responsibilities, researchers from different academic disciplines, practitioners, and members of international organisations and NGO representatives. In addition to Zimbabwe, participants represented 20 other countries within North and South America, Europe, Africa, Asia, and Australasia. The Workshop was sponsored by a number of firms and institutions including: Anglo American Corporation; Ashanti Goldfields (Freda Rebecca Mine); Bindura Nickel Corporation; BHP; Consortium of Economic Research; Fundicion Alvarez; IDRC (Canada); Industry Club of the Mining and Environment Research Network, University of Bath; Intermediate Technology Group; Lonrho (Zimbabwe) Ltd.; Mineros de Antioquia; Ministry of Energy & Minerals (Tanzania); Ministry of Economic Development, Bolivia; DfID (formerly the Overseas Development Administration (ODA)), U.K.; Rio Tinto (Zimbabwe) Ltd.; Swedish International Development Co-operation Society (SIDA); UNCTAD; UNEP; University of Cape Town; University of Chile; University of South Australia; University of Zimbabwe; World Bank; and ZIMASCO.

The workshop examined the complex issues surrounding artisanal (informal) mining. Participants particularly emphasised the urgent need for these miners to become organised and for governments to address title issues as a first step towards achieving improved environmental management. The recommendations below cover the formal, organised large and small-scale mining sectors.

The workshop also recommended that different policy mechanisms and company strategies be used to integrate environmental management within the very different contexts of old and new operations. The constraints and opportunities for change are quite different depending on vintage of technology and process route; accumulated waste from past operations; and work practices and company cultures already in place.

The following recommendations are presented in summary form only. Their reproduction here serves to highlight the complexity of the issues at stake, rather than providing a comprehensive set of guidelines. The four areas categorised are interrelated. However, for the purpose of these recommendations, we believe that they warrant separate emphasis on account of the crucial role each plays in efforts to achieve environmentally sensitive minerals development in line with the recommendations of the Rio Earth Summit and Agenda 21, which stress the need for environmentally sensitive business practices; clean technology development and diffusion; and regulatory incentives to prevent pollution from the outset, rather than initiate cleanup at a project's end.

29.3   RECOMMENDATIONS

29.3.1   POLICY MECHANISMS AND INSTITUTIONAL SUPPORT

- The promotion of collaboration between different government organisations to achieve the integration of environmental regulation with policy to promote mining.
- The promotion of regulatory flexibility: e.g., the setting of performance standards, not defining of specified techniques; innovative financial arrangements such as phased bonds.
- The avoidance of uncertainty in policy development.
- The removal of disincentives for cleanup.

- The expansion of financial incentives to promote:
  - environmentally proficient practices.
  - pollution prevention through technological innovation.
  - socially responsible mining practices.
- The development of Environmental Management System (EMS) standards that incorporate social issues as a cost-effective and consistent industrywide generic approach; and that are also linked to the ISO 14000 series and the environmental reporting procedures of financial institutions.
- The use of strategic and regional environmental assessment and planning techniques as an ongoing process.
- The development of institutional capacity to evaluate environmental impact assessments (EIAs) and to monitor and assess subsequent practices and EMS.
- Assessment of the environmental implications of the transfer of title and development of a framework for the sharing of responsibility for environmental liabilities over time.
- The definition of markers that indicate the extent to which progress towards sustainable development objectives have been met.

29.3.2 Environmental/Social Impact Assessment (EIA/SIA)

- The use of more proactive and extended EIAs that incorporate social impacts both temporally and spatially.
- The use of social impact indicators to monitor societal effects over time.
- The promotion of stakeholder participation in the EIA process and within the EMS throughout the life of the mine.
- The improvement of social and environmental baseline information.
- The use of risk assessment techniques in environmental management.

29.3.3 Technology

- The promotion of research and development (R&D), and an appropriate R&D framework to stimulate the development of pollution prevention techniques.
- The development and diffusion of technology solutions between firms across both the small- and large-scale sectors.
- Where possible, the incorporation of rehabilitation measures during the mining process.
- The promotion of remediation based on innovative and economically beneficial end-uses, e.g., the development of agro-industrial plantations and water resource storage schemes for irrigation and aquaculture.
- The introduction of life cycle assessment approaches as a generic tool to assist technology choices and support EMSs.
- The need for greater assessment of the social context of technology development, application, and diffusion.

29.3.4 Company Strategy

- The promotion of the industrywide use of generic EMSs as guarantors of environmental performance over time.
- The continuous improvement of EMSs and use of environmental performance indicators.
- The monitoring of predicted impacts and unexpected events.
- The development of participatory approaches and the involvement of local communities in the EMS, and particularly the closure process.
- The integration of all mine employees within the closure process from the outset.
- The designing of incentives for employees to evaluate and improve environmental practices.

- The introduction of efficiency improvements and good housekeeping measures to support EMS in older operations.
- Collaboration within national and international mining associations to disseminate and promote improved environmental practices.
- The support of, and participation in, regional initiatives towards education and experience sharing.
- The promotion of research initiatives to inform and support public policies and company strategies towards integrated environmental management.
- The definition from the outset of the objectives of the closure plan and evaluation of their degree of achievement. Specifically, this means:
  - reduce the generation of waste in both terms of toxicity and volume, and optimise recycling.
  - stabilise and/or isolate residues from the outset to reduce the potential of acid mine drainage and water contamination.
  - dispose of plant, equipment, and materials safely and contain waste to achieve zero discharge and reduce visual impact over time.
  - establish a rolling programme of remediation and revegetation from the outset.
  - collaborate with governments and local communities to define viable post-mining land use for the region, and to ensure that local communities are not being impoverished or are suffering from ill health as a result of either the mining activity or the closure process.

29.3.5 EDUCATION, RESEARCH, AND TRAINING

- Enhance and create environmental awareness throughout the education system and work-force.
- Train for innovative community relationships, and include training and knowledge acquisition initiatives for staff as well as members of local communities.
- Promote training in EMS within firms across all personnel.
- Encourage demonstration and pilot projects to stimulate the development and diffusion of clean technology.
- Ensure continuing education opportunities in the areas of mining and environment by promoting innovative courses within educational institutions for those already working in industry, as well as for new students.
- Support collaborative policy research and dissemination activities, and the diffusion of best practice.

29.4 ADDITIONAL SOCIOECONOMIC CHALLENGES AND CONSTRAINTS

Finally, the "Planning for Closure," Harare Workshop focused principally on environmental rather than socioeconomic issues; therefore, the authors considered it worth reiterating the conclusion and policy implications from Chapter 5 by Warhurst, Macfarlane and Woods, entitled "Issues in the Management of the Socioeconomic Impacts of Mine Closure: A Review of Challenges and Constraints."

- Social Impact Assessment (SIA) needs to be ongoing throughout the life phases of the mine, and planning for decommissioning, downsizing, and closure needs to begin at the outset based on criteria developed through the SIA.
- Closure planning should address effects and solutions for the remote community involved in supplying the industry, as well as the formal and informal workforce of the mine.

- Profiles of recruits, recruitment strategies, and human resource development through the mine life needs to be included in closure planning to facilitate transition for redundant workers and their families, and to broaden the possibilities of future work options.
- An environmental management plan, post-closure, needs to be in place where there is a threat of ongoing contamination/environmental damage such as acid rock drainage or tailings leaks/slippage, so as to improve possibilities of alternative land uses, particularly farming.
- Closure planning could include alternate uses for housing, facilities, and equipment, along with policies to protect the continuation of social networks and community activities.
- Consultation throughout is paramount, assisted by participative approaches to forward planning so as to involve the community from the outset in addressing eventual closure and future options.
- Financial mechanisms need to be in place to ensure sufficient resources exist at the end of the mine's life to implement closure plans, and fund appropriate compensation and redundancy schemes. Bonding regulatory systems could cover social as well as environmental issues.
- More research is required on the socioeconomic effects of mine closure and their mitigation, and case study analysis needs to inform the drawing of lessons to design best-practice corporate strategy and improved public policy.
- What capabilities do companies need to develop, and how might different areas of expertise be integrated to ensure improved planning for closure from the outset and its subsequent implementation?
- How might research contribute to the development of indicators that might define the quality of closure plans with regard to the predictions made and the effectiveness of mitigation efforts and responses?

In closing, the editors and authors of this book and the research members of the Mining and Environment Research Network (MERN) would like to note that feedback and suggestions would be welcomed to inform the development and implementation of future research and dissemination activities that promote integrated environmental management from the outset through planning for closure.

Index